Diverse atoms

Profiles of the chemical elements

Oxford Chemistry Guides

1. P. W. Atkins **Concepts in Physical Chemistry**
2. John Emsley **The Elements**
3. Hazel Rosotti **Diverse Atoms: Profiles of the Chemical Elements**

The elements in alphabetical order of symbols with names and atomic numbers

Symbol	Element	Atomic number
Ac	Actinium	89
Ag	Silver	47
Al	Aluminium	13
Am	Americium	95
Ar	Argon	18
As	Arsenic	33
At	Astatine	85
Au	Gold	79
B	Boron	5
Ba	Barium	56
Be	Beryllium	4
Bh	Bohrium	107
Bi	Bismuth	83
Bk	Berkelium	97
Br	Bromine	35
C	Carbon	6
Ca	Calcium	20
Cd	Cadmium	48
Ce	Cerium	58
Cf	Californium	98
Cl	Chlorine	17
Cm	Curium	96
Co	Cobalt	27
Cr	Chromium	24
Cs	Caesium	55
Cu	Copper	29
Db	Dubnium	105
Dy	Dysprosium	66
Er	Erbium	68
Es	Einsteinium	99
Eu	Europium	63
F	Fluorine	9
Fe	Iron	26
Fm	Fermium	100
Fr	Francium	87
Ga	Gallium	31
Gd	Gadolinium	64
Ge	Germanium	32
H	Hydrogen	1
He	Helium	2
Hf	Hafnium	72
Hg	Mercury	80
Ho	Holmium	67
Hs	Hassium	108
I	Iodine	53
In	Indium	49
Ir	Iridium	77
K	Potassium	19
Kr	Krypton	36
La	Lanthanum	57
Li	Lithium	3
Lr	Lawrencium	103
Lu	Lutetium	71
Md	Mendelevium	101
Mg	Magnesium	12
Mn	Manganese	25

Symbol	Element	Atomic number
Mo	Molybdenum	42
Mt	Meitnerium	109
N	Nitrogen	7
Na	Sodium	11
Nb	Niobium	41
Nd	Neodymium	60
Ne	Neon	10
Ni	Nickel	28
No	Nobelium	102
Np	Neptunium	93
O	Oxygen	8
Os	Osmium	76
P	Phosphorus	15
Pa	Protactinium	91
Pb	Lead	82
Pd	Palladium	46
Pm	Promethium	61
Po	Polonium	84
Pr	Praseodymium	59
Pt	Platinum	78
Pu	Plutonium	94
Ra	Radium	88
Rb	Rubidium	37
Re	Rhenium	75
Rf	Rutherfordium	104
Rh	Rhodium	45
Rn	Radon	86
Ru	Ruthenium	44
S	Sulfur	16
Sb	Antimony	51
Sc	Scandium	21
Se	Selenium	34
Sg	Seaborgium	106
Si	Silicon	14
Sm	Samarium	62
Sn	Tin	50
Sr	Strontium	38
Ta	Tantalum	73
Tb	Terbium	65
Tc	Technetium	43
Te	Tellurium	52
Th	Thorium	90
Ti	Titanium	22
Tl	Thallium	81
Tm	Thulium	69
U	Uranium	92
Uub	Ununbium	112
Uun	Ununnilium	110
Uuu	Unununium	111
V	Vanadium	23
W	Tungsten	74
Xe	Xenon	54
Y	Yttrium	39
Yb	Ytterbium	70
Zn	Zinc	30
Zr	Zirconium	40

Diverse atoms

Profiles of the chemical elements

Hazel Rossotti
Fellow and Tutor in Chemistry,
St. Anne's College, Oxford

Oxford New York Tokyo
OXFORD UNIVERSITY PRESS
1998

Oxford University Press, Great Clarendon Street, Oxford OX2 6DP

Oxford New York
Athens Auckland Bangkok Bogota Bombay
Buenos Aires Calcutta Cape Town Dar es Salaám
Delhi Florence Hong Kong Istanbul Karachi
Kuala Lumpur Madras Madrid Melbourne
Mexico City Nairobi Paris Singapore
Taipei Tokyo Toronto Warsaw

and associated companies in
Berlin Ibadan

Oxford is a trade mark of Oxford University Press

Published in the United States
by Oxford University Press Inc., New York

A catalogue record for this book is available from the British Library

Library of Congress Cataloging in Publication Data
Data available

ISBN 0 19 855815 5

Typeset by EXPO Holdings, Malaysia

Printed in Great Britain by
The Bath Press, Bath

About this book

Chemistry is immensely varied, mainly because it involves interactions between atoms of many different and very diverse chemical elements: one hundred and twelve of them to date. In order to impose some order on our ever expanding chemical knowledge, we classify the elements according to their atomic structure, since it is mainly this which determines their behaviour. Elements of analogous structure are arranged in a vertical column or group in the periodic table, and it is traditional to discuss the chemistry of an element in relationship to other members of the same group. While this approach is very valuable for emphasizing similarities and minor differences within one group, it is of less help when we want to get a feel for the overall chemical behaviour of one particular element. If we know the atomic structure of an element we can usually predict the main features of its chemistry, but not its diverting peculiarities. We cannot, after all, discern people's characters from the information on a census return, however detailed this may be; their charm, or lack of it, is all their own.

This book aims to use traditional concepts as a flexible framework in which we can appreciate the individuality of the chemical elements as well as their family traits. The elements are arranged 'horizontally', in the order of increasing atomic number, in the hope of avoiding undue emphasis on 'vertical' trends, although middle- and heavy-weight elements are of course compared with lighter members of the same group. I hope that new undergraduate students will find the earlier part of the book an agreeable companion during their elevation to first year university work. Although many of the heavier elements will not cross their path until much later (if at all), these have been treated with the same brevity and informality as their more familiar predecessors; and not only to give the book coherence. These sketches of chemical behaviour should give second and third year students a framework on which to hang their new found, more detailed, knowledge; and there may even be a few (but surely very few) exam candidates who will find the outlines a quick route to 'revision' of knowledge which they unfortunately never had time to acquire.

It is always a pleasure to acknowledge help for a task that has been, on balance, very enjoyable and there are many institutions and individuals to whom I am extremely grateful. My college, St Anne's, has given me generous sabbatical leave during which colleagues have uncomplainingly shouldered extra work on my behalf. The OUP editorial and design teams treat authors with an air of cheerful tolerance which is conducive to creativity; and the book would not have been possible in its present form without the goodwill of all those who generously allowed us to reproduce copyright material.

I am deeply indebted to everyone whose good ideas I have, knowingly or unknowingly, adopted, whether these were acquired from personal contact or from published works. My husband Francis has been a most valuable source of chemical knowledge and of suggestions for improvement; my warmest thanks, particularly for his generous help at the proof stage. Courtenay Phillips has read the entire text, which has benefited greatly from his meticulous but always constructive comments: any errors now to be found must surely be from my more recent additions. Bob (R. J. P.) Williams has helped me with expert advice on bio-inorganic matters, as with much else over the decades. Jenny Harrington converted ball-point scribble to elegant copy with her usual good-humoured perfectionism, which goes far beyond her enviable skills at the keyboard. I am extremely grateful to them all. I should also like to thank Peter Atkins, Victor Christou, Margaret Forrest, Patrick Irwin, Rachel Thompson, and Christopher Viney for their valuable help with specific points.

This book also owes much to the many people who have made kind enquiries about it. Progress would certainly have been much slower and drearier without the cheerful support of husband and family, colleagues, undergraduates (not all of them chemists), and other friends. I hope that they, and the many others to whom I am indebted, will feel that they are part of any success the book may have in fulfilling its twofold aim: of helping those readers who are doing a first degree involving at least some inorganic chemistry, and, more importantly, of enhancing their enjoyment of this ancient but rapidly advancing subject.

St. Anne's College, Oxford, UK H.S.R.
Inorganic Chemistry Laboratory, Oxford University, UK March 1997

Contents

This book is gratefully dedicated to St Anne's chemists, and in particular to:
Adam, Adrian, Ali, Amanda, Andy, Angelos, Ann, Anne, Anton, Antonia, Barbara, Carl, Carolyn, Cathy, Celia, Charlie, Chris, Claire, Dave, Debbie, Deryk, Diana, Dot, Edwina, Ele, Emma, Fran, Frances, Gareth, Garreth, Gill, Glynis, Graham, Gregg, Hazel, Helen, Ian, Irene, J.-P., Jacqie, Janice, Jane, Jennifer, Jenny, Jerry, Jill, Jim, Jo, Joe, John, Jon, Jonathan, Jonny, Julia, Julian, Katie, Kevin, Kirstine, Linda, Liz, Louise, Lucy, Lynne, Mark, Mary, Mick, Natalie, Neil, Nikki, Nina, Pam, Pat, Paul, Pete, Phil, Rachel, Razmic, Rich, Richard, Rob, Robert, Ros, Royston, Russell, Ruth, Sallie, Sarah, Shelley, Steve, Struan, Stuart, Sue, Sunita, Susan, Svend, Teresa, Tueng, Valerie, Warren, Wendy, Yvonne, and Zareen.

I hope it may also bring good fortune to their successors, led by: Jo, Rebecca, Sarah, Tom, and Zaw.

User's guide

The following notes are intended to help the reader to use the book as easily and as fully as possible.

Overall plan

The elements are discussed individually in order of increasing atomic number. Each section of the book deals with one horizontal row of the Periodic Table and is numbered according to the value of the primary quantum number of the s valence electrons: for example Part III covers the so-called Second Short Period from sodium ($3s^1$) to argon ($3s^2 3p^6$). The verbal discussion appears mainly on the right-hand pages, with associated non-verbal matter readily accessible to the left of the text or following it. In both areas, the presentation is not immutably rigid but has been tailored to the chemical behaviour of the particular element(s) being described.

The main text

There is no introductory section on *concepts* or *general principles.* This book assumes that all readers have done some years of pre-18 chemistry, and that their physical chemistry is keeping pace with their inorganic. Ideas that may be unfamiliar to new degree course chemists are introduced into the text where they are first needed and they are often amplified in the accompanying tables and diagrams. Many of the concepts that we inorganic chemists need are classified by separationists as 'physical' and are succinctly clarified in Atkins' *Concepts of physical chemistry*; here such terms are identified by a small superscript A (^{A}). Ideas that are more specific to inorganic chemistry may often be tracked down in Mingos's, *Essentials of inorganic chemistry* or one of the other books from the 'wide-ranging' list on p. xi–xiii.

For smooth reading and easy memorizing *electronic configurations* are written (in the text) with orbitals in order of increasing primary quantum number; so that of nickel, for example, is $[Ar]3d^8 4s^2$.

In the first three Parts, which deal with only *n*s and *n*p elements, the verbal profile of each element is followed by a *summary*; and heavier elements of the s and p blocks are treated in the same way in the rest of the book. Parts IV–VII, however, include the more closely related (n-1)d elements in the three transition series and the even more similar (n-2)f series of lanthanides and actinides, and it is worth looking at how chemical behaviour varies within and between these series. So for the three sets of d transition metals there is a brief discussion of trends and this is placed after the last element in each series to have an incomplete valence shell (copper for 3d, and silver and gold for 4d and 5d, respectively). Each summary of trends is followed by a tabulation of the main features of the chemistry of the individual transition metals in that series.

Amongst the lanthanides and, to a lesser extent, the actinides, neighbouring elements have even more in common than do the d transition metals. So the similarities within the 4f and 5f series are outlined in the discussion of their first member (lanthanum or actinium), while the short profiles that follow highlight individual behaviour. The text for the two (n – 2)f series, as for the (n – 1)d series, ends with a brief discussion of trends and a table summarizing more specific properties; these follow the two f^{14} elements, lutetium and lawrencium.

Illustrations and data

For each element the non-verbal material starts with a diagram showing its position in the periodic table and its electronic configuration, usually followed by a table of numerical data. Similar material is presented for the neighbouring elements (if such exist) in the positions to the left of it, and above it, in the periodic table: these are, of course, its predecessor in the same Period and its lighter analogue. There is also a bar diagram showing the distribution of natural isotopes. This introductory matter is followed by tables, diagrams, and equations that complement the text of each particular element.

The *electronic configurations* are given in box form, so that we may see at a glance which orbitals are filled, half-filled, or empty but available; and the boxes are ranked in approximate order of orbital filling, so that for the first transition series the 4s box precedes that for 3d, contrasting with the non-visual representation used in the main text.

Some of the *data* in the table are more reliable than others. Covalent and ionic *radii* are notoriously sensitive to environment and vary with the number, geometry, and chemical nature of neighbouring groups. Similarly, a value of *bond energy* is strictly valid only for the substance for which is it was measured. But although the numerical values of these quantities should be interpreted with caution, they are, none the less, useful for indicating trends.

The content of the data tables depends on selection as well as availability of data. *Density*, for example, is given only for those elements that are solid at room temperature. *Ionization energies* are usually quoted for removal of electrons only from the valence shell, although for ns^1 ('alkali') metals, the second ionization energy is also given in order to show the energy needed to rupture the outermost filled shell. *Standard electrode potentials* (which refer to pH = 0 unless otherwise stated) appear in the tables only for those elements with rather simple redox chemistry; those for more complicated systems are given in Latimer or oxidation state diagrams in the later individual illustrations.

The source of the illustrations for individual elements is denoted using the system described on p. xiii. The isotopic distributions and almost all the values in the data tables were obtained from Emsley (ref. E), supplemented by data from refs D and SAL. Without such sources this book would have been very different, and the large debt to them is acknowledged extremely gratefully.

Books: used and recommended

The books listed below include both those that have been consulted during the writing of this book and those that can confidently be recommended to readers of it, whether for reference, browsing, or systematic study. The two categories are broadly overlapping, but not coincident; some of the more specialized (or more elementary) books used do not fit neatly into most first degree courses in chemistry, while others now out of print.

Essential reference books, which are fun to dip into but too massive to attempt to assimilate.

CW F. A. Cotton and G. Wilkinson, *Advanced inorganic chemistry*, 5th edn. John Wiley & Sons, New York (1988).

GE N. N. Greenwood and A. Earnshaw, *Chemistry of the elements*. Butterworth–Heinemann Ltd., Oxford (1984).

Books intended primarily for undergraduate students of chemistry: of varying length and depth, they all have something of value to offer.

Physical chemistry

A P. W. Atkins, *Concepts in physical chemistry*. Oxford University Press, Oxford. (1995).

AP P. W. Atkins, *Physical Chemistry*, 5th edn. Oxford University Press, Oxford. (1994).

Wide-ranging inorganic chemistry

HKK J. E. Huheey, E. A. Keiter, and R. L. Keiter, *Inorganic chemistry*, 4th edn. Harper Collins, New York (1993).

MM K. M. Mackay and R. A. Mackay, *Modern inorganic chemistry*, 4th edn. Blackie, Glasgow (1989).

ME D. M. P. Mingos, *Essentials of inorganic chemistry I*. Oxford University Press, Oxford (1995).

OB S. M. Owen and A. T. Brooker, *A guide to modern inorganic chemistry*. Longman, Harlow (1991).

PIC W. W. Porterfield, *Inorganic chemistry*, 2nd edn. Academic Press (1993).

S A. G. Sharpe, *Inorganic chemistry*, 3rd edn. Longman, Harlow (1992).

SAL D. F. Shriver, P. W. Atkins, and C. H. Langford, *Inorganic chemistry*, 2nd edn. Oxford University Press, Oxford (1994).

More-specialized inorganic topics

C P. A. Cox, *The elements*. Oxford University Press, Oxford (1989).

D W. E. Dasent, *Inorganic energetics*, 2nd edn. Cambridge University Press, Cambridge (1982).

J D. A. Johnson, *Some thermodynamic aspects of inorganic chemistry*, 2nd edn. Cambridge University Press, Cambridge (1982).

M A. G. Massey, *Main group chemistry*. Ellis Horwood, Chichester (1990).

PM R. V. Parish, *The metallic elements*. Longman, London (1977).

R H. Rossotti, *The study of ionic equilibria*. Longman, London (1978).

SM L. Smart and E. Moore, *Solid state chemistry*. Chapman & Hall, London (1992).

D and SM are particularly useful for clear presentations of important inorganic topics that are often given rather limited space in more general texts.

Data compilations

E J. Emsley, *The elements*, 2nd edn. Clarendon Press, Oxford (1991).

SC L. G. Sillén and A. E. Martell, *Stability constants of metal-ion complexes*, and *Supplement No. 1*. Chemical Society, London (1964, 1971).

There are also rich deposits of data in AP, D, and SAL.

'Get into shape' books for those who feel that their pre-university chemistry is inadequate or rusty.

AB P. W. Atkins and J. A. Beran, *General chemistry*. Scientific American, New York (1992).

Z S. S. Zumdahl, *Chemical principles.* D. C. Heath & Co., Lexington (1995).

And, finally, a book for you to recommend to non-chemists.

AEC P. W. Atkins, *Atoms, electrons and change.* Science American Library, Freeman, New York (1991).

Some of these books have been very valuable, even invaluable, sources of illustrations. Greenwood and Earnshaw (GE), in particular, has been a well-stocked and diverting hunting ground. Copyright material taken from any work in the list is denoted by its letter code (except for the isotope diagram[E] or the data table[D, E] for each element); and the source of material taken from an article is given in the caption. A list of copyright holders who have given permission to reproduce material is given on pp. 575–7, but it is more appropriate to thank them here, at the outset, as this book owes them so much.

Part I

Filling the 1s orbital
Hydrogen and helium

H Hydrogen

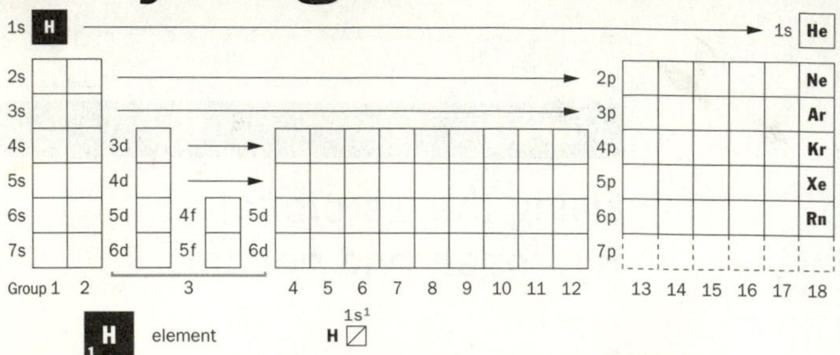

Name		**Hydrogen**
Symbol, *Z*		H, 1
RAM		1.00794
Radius/pm	**atomic**	78
	covalent	30
	ionic, H^+	(10^{-5})
Electronegativity (Pauling)		2.2
Melting point/K		14.01
Boiling point/K		20.28
$\Delta H^{\ominus}_{fus}$/kJ mol^{-1}		0.12
$\Delta H^{\ominus}_{vap}$/kJ mol^{-1}		0.46
Ionization energy for removal of *j*th electron/kJ mol^{-1}	***j* = 1**	1312.0
Electron affinity/kJ mol^{-1}		72.8
Bond energy/kJ mol^{-1}	**E–E**	453.6
	E–C	411
	E–N	390
	E–O	464
	E–F	566
	E–Cl	431
	E–Br	366
	E–I	299
$E^{\ominus}$/V for $H^+ + e^- \rightarrow \frac{1}{2}H_2(g)$		0 {H^+} = 1 mol dm^{-3} 0.828 {OH^-} = 1 mol dm^{-3}

Stable isotopes of hydrogen

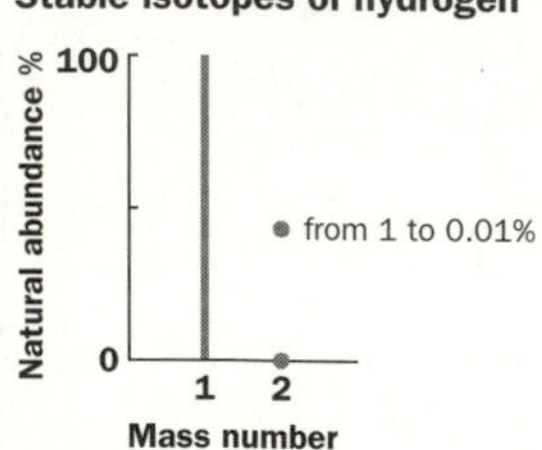

Promotion energy

$H(1s^1) \rightarrow H(1s^0 2s^1)$ 984 kJ mol^{-1}

Some bond strengths (/kJ mol^{-1}) for single bonds between like atoms [E]

H–H	B–B	C–C	N–N	O–O	F–F	Cl–Cl
453.6	335	348	160	146	159	242

Hydrogen is an amazing element. Its atom is the simplest possible, as its nucleus consists (usually) of just one proton whose single positive charge is balanced by the negative charge of the one associated electron. Since the hydrogen nucleus H^+ is a major component of stars, hydrogen has pride of place as the most abundant element in the universe; and the fusion of these nuclei is the first step in the life giving emission of energy from our local star, the sun. On Earth, hydrogen, combined with oxygen, provides water to fill not only oceans and rivers, but also living cells. In order of terrestrial abundance, hydrogen comes third after oxygen and silicon.

Hydrogen is a component of almost all organic substances, and, since it has a much more diverse inorganic chemistry than carbon, it has more known compounds than any other element. But its fascination is based not on a mere count of its compounds, but on the rich variety of its behaviour. The simple hydrogen atom is able to have such a wide range of chemical associates because it can interact with other atoms in a number of very different ways; and even in its elemental state, hydrogen appears in a surprising number of guises.

So how is this variety achieved? In an unexcited hydrogen atom,* its single electron is most likely to be found within a spherically symmetrical shell at a distance** of about 52.5 pm from the nucleus. This most stable atomic orbital,[A] labelled 1s, can, like any other orbital, hold up to two electrons (of spin $+\frac{1}{2}$ or $-\frac{1}{2}$). The next most stable orbitals, 2s and 2p, are of much higher energy, and so play no part in the normal chemical reactions of hydrogen; these involve only the 1s orbital, used in an impressive variety of ways.

By far the commonest form of elemental hydrogen[A] is the diatomic molecule H_2 in which the two electrons are likely to be found between the two protons. Both electrons are attracted to each nucleus, and both lessen the internuclear repulsion. It is tempting to recall the elementary idea that, by sharing the electrons, each atom has filled its 1s orbital. But the energy of each 1s atomic orbital is totally changed by the presence of a second nucleus nearby and the result is two non-identical *molecular orbitals*,[A] both of them associated with the *pair* of nuclei. One, the bonding[A] orbital, lies between the nuclei and, if occupied,

* Many of the ideas used in this book explained concisely in *Concepts in Physical Chemistry* by P. W. Atkins, Oxford University Press, 1995, and are denoted here by [A].

** Atomic distances are measured in picometres (pm). 1 pm is 10^{-9} m (one thousandth millionth of a metre).

Molecular orbitals in the dihydrogen molecule

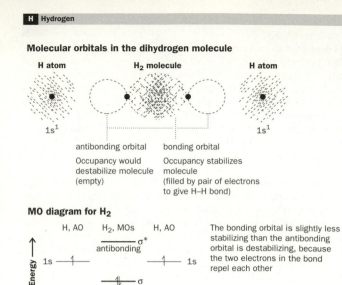

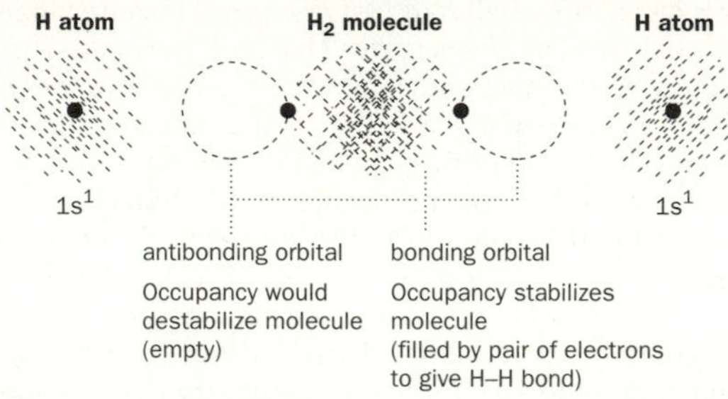

MO diagram for H_2

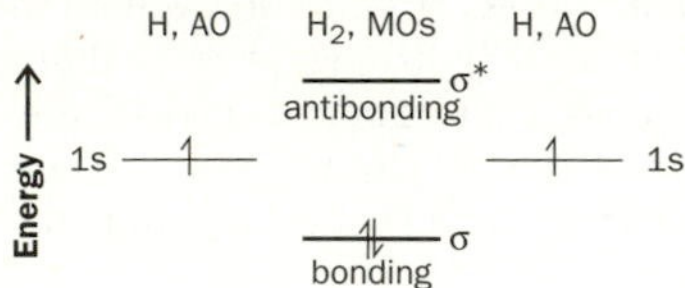

The bonding orbital is slightly less stabilizing than the antibonding orbital is destabilizing, because the two electrons in the bond repel each other

The hydrogen molecule-ion H_2^+

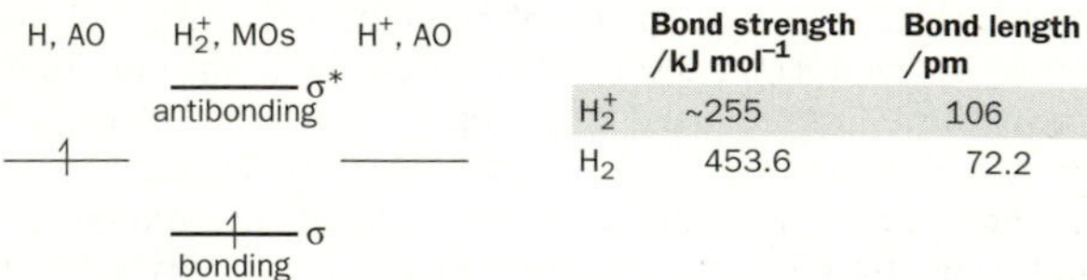

	Bond strength /kJ mol^{-1}	Bond length /pm
H_2^+	~255	106
H_2	453.6	72.2

Energy changes for the reaction of hydrogen with oxygen and with fluorine

(N.B. Sign convention: energy absorbed is positive, energy emitted is negative)

Reaction	$H_2(g)$	$+ \frac{1}{2} O_2(g) \rightarrow H_2O$ (l)	$H_2(g)$	$+ F_2(g) \rightarrow 2HF$ (g)
Bonds broken	1 × H–H	$\frac{1}{2}$ × O=O	1 × H–H	1 × F–F
(energy change/kJ mol^{-1})	453.6	249	453.6	159
Bonds formed	2 × O–H		2 × H–F	
(energy change/kJ mol^{-1})	–928		–1112	
Total energy change/kJ mol^{-1}	–225.4		–499.4	

decreases the repulsion between them and so lowers the energy of the arrangement. The other lies in two regions on the far sides of the nuclei and so does nothing to shield the repulsion between them. Since an electron in this higher energy region decreases the stability, the orbital is described as antibonding.[A] So, when two hydrogen atoms come close together, the two 1s atomic orbitals (AOs) form two molecular orbitals (MOs). The two electrons (with spins opposed and despite the electrostatic repulsion between them) both occupy the bonding orbital, and the dihydrogen molecule H_2 is formed, held together by a covalent[A] non-polar,[A] σ bond[A] ('sigma' bond). It is not surprising that dihydrogen, with molecules of such low mass, is a gas under normal conditions, although it can be liquefied and even frozen at very low temperatures (and for more bizarre behaviour see p. 35, footnote).

The H–H bond is the strongest known two-electron bond formed between identical atoms. When dihydrogen molecules are broken down by an electric arc discharge into atoms, the heat evolved when they recombine to form H_2 can be harnessed in the atomic hydrogen torch, which produces temperatures up to 4000°C, high enough to weld high-melting metals such as tungsten. Two protons can also be held together, though much more weakly, by a single electron in the bonding orbital, to give the simplest of all polyatomic species, the cation H_2^+, clumsily named the 'hydrogen molecule ion'.[A] A similarly weak bond between the two protons is found in the anion H_2^- in which the bond formed by two electrons in the bonding orbital is partially destabilized by the one extra electron in the antibonding orbital. Both these ions have been detected in electrical discharges, but are so unstable that they exist only fleetingly.

The H_2 molecule reacts with a large number of other elements. Hydrogen is so small that its 1s orbital can combine easily with a 2p orbital[A] of small atoms such as carbon, nitrogen, oxygen, and fluorine, to give molecular orbitals of low energy. As the atomic number of these elements increases from carbon to fluorine, the more strongly do their nuclei attract electrons away from the hydrogen atom. These polar bonds are often stronger than 2-electron bonds between identical atoms, and much energy may be given out. Indeed, hydrogen can react with the two most electronegative elements, fluorine and oxygen, with explosive violence.

Hydrogen does not, however, explode on immediate contact with air, even though the H–O bonds in a water molecule are of lower energy than an H–H bond and half an O=O bond. But if the mixture is sparked, the very high temperature in the tiny hotspot breaks down a few H_2 and O_2 molecules into single atoms, which are extremely reactive and set off a series of changes. Once started, the

The molecular shapes of some simple hydrides

showing (a) agreement with VSEPR theory, and (b) polarity

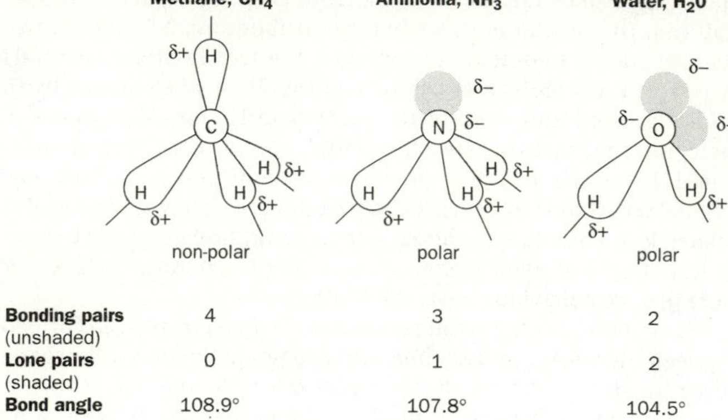

Bonding pairs (unshaded)	4	3	2
Lone pairs (shaded)	0	1	2
Bond angle	108.9°	107.8°	104.5°

Regular tetrahedral angle

Dipole moments, $\mu/(10^{-30}$ C m) for some simple hydrides [AP]

CH_4	NH_3	H_2O	HF
0	4.90	6.17	6.37

MO diagram for methane, CH_4 (simplified)

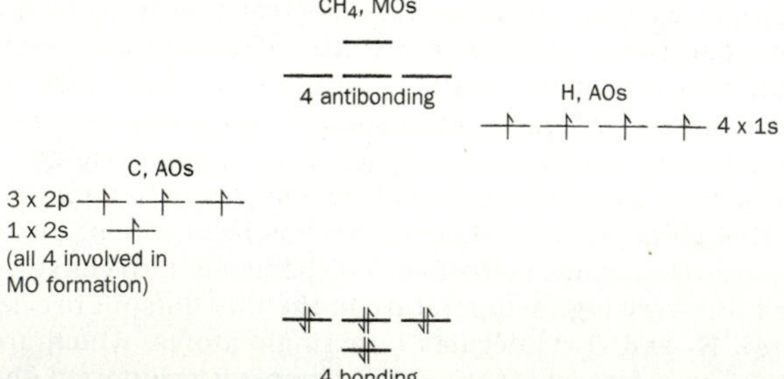

reaction generates enough heat to keep itself going. The speed of the change depends on the conditions. If we put a lighted match to the neck of a test-tube to test for hydrogen, we get a satisfying but restrained pop. A steady stream of hydrogen can, under carefully controlled conditions, burn steadily in air with a stable flame and it has been suggested that the energy supplied by this 'clean' reaction should form the basis of our industrial economy. Hydrogen can be burned in pure oxygen to give a high temperature for welding. But when the conditions are not controlled, mixtures of hydrogen and oxygen are very hazardous. In 1937 the hydrogen-filled airship *Hindenberg* exploded in flames when it grazed the tip of a wireless mast. Liquid hydrogen ignites spontaneously when it comes into contact with liquid oxygen, and the two are used together as a bipropellant for space rockets. At the other end of the speed range, a catalyst can be used to make hydrogen and oxygen combine so slowly that they can provide a steady supply of electric power to a space capsule.

The molecular shapes of methane, CH_4, ammonia, NH_3, and water, H_2O, agree surprisingly well with those predicted by the simple valence-shell electron-pair repulsion (VSEPR) theory, which supposes that the most stable structure is that which minimizes repulsion between the electron pairs on the central atom. In these molecules there are four electron pairs and the basic shape is tetrahedral. Not all electron pairs are involved in bonding but, since the lone, non-bonding pairs are nearer to the central atom than the bonding ones, they exert more repulsion. So, in ammonia, and even more in water, the bond angles at the central atom are slightly tighter in methane, in which the four bonds are directed to the points of a regular tetrahedron.

The tetrahedral structure of methane has long been explained by the mixing or *hybridization*[A] of the four $n = 2$ atomic orbitals (one 2s and three 2p) of carbon, to give four identical sp^3 orbitals, which form tetrahedral single bonds with four hydrogen atoms. In the molecular orbital treatment, the eight atomic orbitals (four from the carbon and one from each hydrogen) give four bonding and four antibonding orbitals. The eight outer electrons (again, four from the carbon and one from each hydrogen) are paired in the four bonding orbitals, leaving the antibonding orbitals vacant. We might guess that the four bonds would be equally strong but this may not be the case. We can now study energy levels in molecules by photoelectron spectroscopy[A] (or PES), and there is some suggestion that 'sp^3 hybridization' is not complete, and that one bond in methane has a somewhat lower energy than the other three.

MO diagram for ammonia, NH_3 (simplified) [SAL]

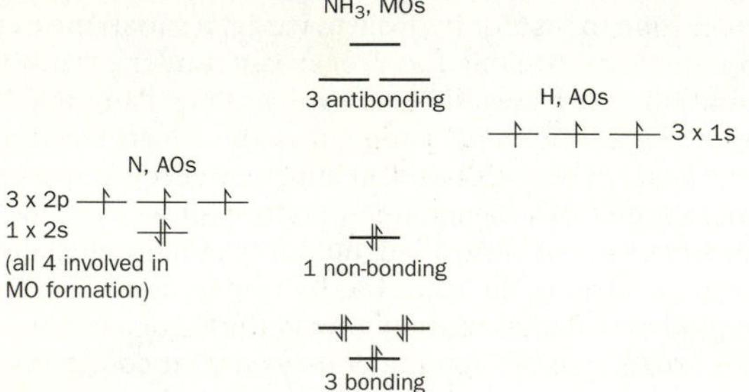

A full non-bonding orbital corresponds to a 'lone pair' and has little effect on stability

MO diagram for water, H_2O (simplified)

H_2O, MOs

σ_6^*

σ_5^*

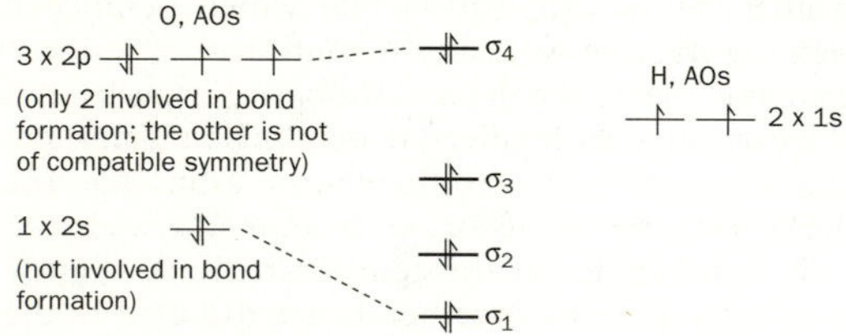

orbital: σ_1 mainly non-bonding, from 2s AO of oxygen
σ_2, σ_3 bonding, from 2 x 2p AOs of oxygen
σ_4 non-bonding, from 1 x 2p AO of oxygen
σ_5^*, σ_6^* antibonding (empty)

MO diagram for HF (simplified)

HF, MOs

H, AO

σ_4^* antibonding

1 x 1s

F, AOs

2 x σ_3

σ_2 non-bonding

3 x 2p (only 1 is of suitable symmetry to mix with 1s of H)

1 x 2s

σ_1 bonding

The other simple hydrides may be treated in the same way. The polar ammonia molecule has seven orbitals: three bonding (giving the N–H bonds on the nitrogen), three empty antibonding, and one mainly non-bonding (which is occupied by the lone pair). But PES shows that this is of only slightly higher energy than the bonding orbitals, and so probably it too contributes to the bonding. In water, there are four filled orbitals and two empty antibonding ones. The four occupied ones are all of different energies; the most stable one is mainly non-bonding (from the 2s atomic orbital of oxygen), the next two are bonding, and the highest energy orbital is again non-bonding (from a 2p orbital oxygen). Since the more electronegative[A] oxygen atom attracts electrons more strongly than nitrogen does, its lone pairs contribute less to the bonding; and water, with two tightly held lone pairs, is a far more polar molecule than ammonia.

Hydrogen fluoride is, of course, linear. The energy of its one filled bonding orbital is very similar to that of the 2s orbital in fluorine and the energy of its one empty antibonding orbital lies near that of the 1s orbital of hydrogen. The three lone pairs of the fluorine are in isolated non-bonding orbitals, derived mainly from the 2p atomic orbitals of fluorine, and the molecule is very polar. Like ammonia and water, hydrogen fluoride is highly associated (see p. 17–21).

So there is considerable diversity even for the two-electron bonds formed between hydrogen and the four adjacent elements from carbon to fluorine; and a very large number of compounds contain such bonds, including those in the vast realm of organic chemistry. But hydrogen is too small and its 1s orbital is too low in energy to form equally strong bonds with heavier non-metallic elements, which, partly for this reason, form hydrides of lower stability. Phosphine, for example, is much more reactive than ammonia and the hydrides of silicon are much fewer and less stable than methane and its heavier hydrocarbon analogues (see p. 151).

Much of the hydrogen used industrially is converted into C–H bonds (maybe also with O–H bonds) in organic substances such as methanol (from reaction with carbon monoxide using a cobalt catalyst) and edible fats (made by catalytic hydrogenation of unsaturated oils). A substantial amount of hydrogen is also converted into N–H bonds in ammonia by catalytic reaction with nitrogen (see p. 83). Metal catalysts often function by adsorbing hydrogen molecules on to their surface and so weakening the strong H–H bonds. Some hydrogen is directly combined with chlorine to give hydrogen chloride, another intermediate for use in the chemical manufacturing industry. The hydrogen is invariably obtained from

Methanol production

$$CO(g) + 2H_2(g) \xrightarrow[\text{catalyst}]{\text{Co}} CH_3OH(g)$$

Reduction of water to give hydrogen

Electrolysis: $4H_2O(l) \rightleftharpoons 4H^+(aq) + 4OH^-(aq)$

$+4e \downarrow$ $-4e \downarrow$

$2H_2(g)$ $O_2(g) + 2H_2O(l)$

Reaction with coke: $H_2O(g) + C(s) \xrightarrow{\text{heat}} H_2(g) + CO(g)$

$H_2O(g) + CO(g) \longrightarrow H_2(g) + CO_2(g)$

Reaction with hydrocarbons:

e.g. $3H_2(g) + C_3H_8(g) \xrightarrow[900°C]{\text{Ni}} 7H_2(g) + 3CO(g)$

propane

Lithium hydride, Li^+H^-

Reaction with water:

$Li^+H^-(s) + H_2O(l) \rightarrow Li^+(aq) + OH^-(aq) + H_2(g)\uparrow$

Electrolysis of melt:

$Li^+(l) + e^- \rightarrow Li(l)$ at cathode

$H^-(l) - e^- \rightarrow \frac{1}{2} H_2(g)\uparrow$ at anode

The radius[GE] of the hydride ion, H^-

Compound	MgH_2	LiH	NaH	KH	RbH	CsH
r_{H^-}/pm	130	137	146	152	154	152

The value 208 pm has been calculated for the free H^- ion.

some form of H_2O, usually by passing an electric current through water ('electrolysis'), but also from high temperature reaction of steam with coke or hydrocarbons.

Somewhat surprisingly, hydrogen can also form σ bonds to transition metals, provided that some other group is also present. Examples are compounds such as $HCo(CO)_4$ and $H_2Fe(CO)_4$ where the metal atom has the particularly stable arrangement of eighteen electrons in its valence shell (see p. 227).

In the simple uncharged molecules discussed so far, a hydrogen atom joins with another atom by means of a covalent bond resulting from a filled σ-bonding molecular orbital. But hydrogen has several other ways of forming compounds. With electropositive metals, such as lithium, sodium, or calcium, it forms salt-like compounds that react with water to give hydrogen. When molten, they conduct electricity, and, if electrolysed, give off hydrogen at the *anode*. These hydrides contain the anion H^-, which has two electrons in the 1s orbital of the hydrogen atom, in combination with a metal cation such as Li^+, Na^+, or Ca^{2+}. The energy needed to strip an electron off the metal is supplied partly by the energy that is lost when the hydrogen atom (like atoms of other elements) acquires an electron to form an anion, but is supplied mainly by the energy lost when the metal cations and hydride anions H^- pack together to form an ordered crystal, held together by forces that are at least partly electrostatic. However, particularly in lithium hydride, the small metal cation may deform the electron cloud of the hydride ion, which, with twice the number of electrons to protons, is surprisingly large and is so susceptible to outside influence that its radius depends markedly on the cation with which it is combined. There is probably some overlap between the orbitals of adjacent ions, giving the electrostatic forces some 'covalent character'. There is doubtless also considerable orbital overlap in the complex metal hydride ions, such as AlH_4^-, which is familiar as its lithium salt. Larger metal ions can accommodate more hydrogen atoms, as in FeH_6^{4-} and even ReH_9^{2-} (see p. 469).

Hydrogen, like other ns^1 elements ('alkali metals' such as sodium $3s^1$), can be stripped of its s electron to form a singly charged cation, although for hydrogen more energy is needed. But the hydrogen cation, H^+, being only a bare proton, has such a high concentration of positive charge that it polarizes, and combines with any suitable neighbour that has a non-bonding pair of electrons. The resulting σ bond differs from those discussed earlier (see p. 5) only because both electrons originated on one atom, which 'donates' them to the bonding orbital, as opposed to each electron being contributed by a different atom, as in hydrogen fluoride. With

Brønsted acids: some conjugate acid–base pairs

Protonated species ('acid')	H_3O^+	H_2O	NH_4^+	HF	H_2SO_4	HSO_4^-
Unprotonated species ('base')	H_2O	OH^-	NH_3	F^-	HSO_4^-	SO_4^{2-}

Amphoteric species, e.g. H_2O and HSO_4^- can act both as proton donors (acids) and acceptors (bases)

Dissociation of inorganic oxoacids

$$(O)_2S(OH)_{2(aq)} \rightleftharpoons H^+_{(aq)} + [(O)_2S(OH)O]^-_{(aq)}$$

The electron can be delocalized over three oxygen atoms

pK < 0
(strong acid)

$$Cl{-}OH_{(aq)} \rightleftharpoons H^+_{(aq)} + Cl{-}O^-$$

The electron is localized on the single oxygen atom

pK = 7.2
(weak acid)

$pK = -\log K_a \sim -\log [H^+][A^-]/[HA]$
[ ] = concentration of hydrated species

Strong acids have high values of K_a and low values of pK

Aquo–metal ions as Brønsted acids

Hydrolysis of iron(III):

Fe^{3+}(aq) in very dilute solution $\rightleftharpoons$ $FeOH^{2+}$(aq) + H^+(aq)

in other solutions $\rightleftharpoons$ $\frac{1}{2}$ Fe_2O^{4+}(aq) + H^+(aq)

or

$\frac{1}{2}$ $Fe_2(OH)_2^{4+}$(aq) + H^+(aq)

and similar species

Oxidation of metals by H^+(aq)

Reaction of Na with water:

$$Na(s) + H^+(aq) \rightarrow Na^+(aq) + \tfrac{1}{2} H_2(g)$$

The (thermodynamic) tendency for a metal to react with H^+(aq) is expressed by the standard electrode potential[A] $E^\ominus_{M^{Z+}, M}$ for the reduction of the ion M^{Z+}(aq) to the metal; so very reactive electro*positive* metals have high *negative* values of $E^\ominus_{M^{Z+}, M}$

M^{Z+}, M	$E^\ominus_{M^{Z+}}$/V, M	
K^+, K	−2.9	metals react with water
Na^+, Na	−2.7	reaction of metal with water energetically favoured, but prevented by oxide coating. (Na, Al)
Al^{3+}, Al	−1.67	
Zn^{2+}, Zn	−0.75	metals react with dilute aqueous acids
Fe^{2+}, Fe	−0.44	metal attacked by some concentrated oxidizing acids (Fe, Cu)
Cu^{2+}, Cu	+0.34	
Au^{3+}, Au	+1.50	metal resistant to most acids

Ease of reduction of M^{Z+} (increases downward); Stability of M in acid (increases downward); Reactivity of M in acid (increases upward)

water, for example, the proton forms H_3O^+, with ammonia NH_4^+, and with the fluoride and sulfate ions HF and HSO_4^-. These proton acceptors (or lone-pair donors) are, of course, familiar as *bases*,[A] while the proton, as a lone-pair acceptor, is an *acid*.[A] Indeed, under Brønsted's definition, but not under the more general Lewis one, an acid is defined as a *proton* donor. However, since naked protons do not exist under normal conditions, acid–base reactions involve transfer of protons between one base and another. In dilute aqueous sulfuric acid, the H_2SO_4(aq) molecules are largely dissociated into H_3O^+(aq), and HSO_4^-(aq), ions, since H_2O is a better proton acceptor (stronger base) than HSO_4^-(aq). Some of the HSO_4^-(aq) ions are further dissociated into more H_3O^+(aq) ions and SO_4^{2-}(aq), but many fewer, since SO_4^{2-}(aq) is a stronger base than H_2O. The hydroxyl ion is a still stronger base and so, when sodium hydroxide is added to sulfuric acid, it takes protons from all three acidic species, H_2SO_4(aq), H_3O^+(aq), and HSO_4^-(aq), and converts them to H_2O and unprotonated SO_4^{2-}(aq) ions.

Most inorganic acids generate their protons by rupturing an –O–H bond: common examples are oxyoacids of non-metals, such as sulfuric and hypochloric acids. Whilst H_2SO_4(aq) is a fairly strong acid, which is highly dissociated, HOCl (aq) is much weaker, even though there is little to choose between the strengths of the O–H bonds in the two molecules. The difference lies in the two anions: in HSO_4^- (better written as $HOSO_3^-$) the negative charge can be delocalized over three oxygen atoms, while in OCl^- it is more concentrated (see also p. 12). Many aquo metal ions also act as acids, liberating H_3O^+ by hydrolysis, which is most marked if the parent cation is small (like Be^{2+}, see p. 45) or highly charged (like Fe^{3+}, see p. 251). Protons can, of course, also be generated by breaking other types of bond. All the hydrogen halides are acidic in aqueous solution (see pp. 114and 184), as is $HCo(CO)_4$, which dissociates by rupturing a metal to hydrogen bond.

The hydrated proton [which from now on we shall call the hydrogen ion and write as H^+(aq)] is a mild oxidizing agent. Even at its very low concentration (10^{-7} mol dm^{-3}) in water, it reacts with the most electropositive metals such as sodium, and is itself reduced to gaseous hydrogen. Dilute solutions of strong acids containing about 0.1 mol dm^{-3} of H^+(aq) attack several less electropositive metals such as iron and zinc, although hydrogen comes off a zinc surface very sluggishly unless the metal is impure. Neither aluminium or gold is dissolved by dilute acids, although for quite different reasons. We should guess that the oxidation of the very elecropositive aluminium by H^+(aq) would liberate energy and hence that the metal might react vigorously with dilute acids; but it soon gets

The shape [GE] of the short-lived molecule-ion H_3^+

Reaction, in gas phase			ΔH/kJ mol^{-1}
$H + H + H^+$	→	H_3^+ (triangular)	–770
$H_2 + H^+$	→	H_3^+ (linear)	–337
$H + H_2^+$	→	H_3^+ (linear)	–515

Diborane B_2H_6

(a)

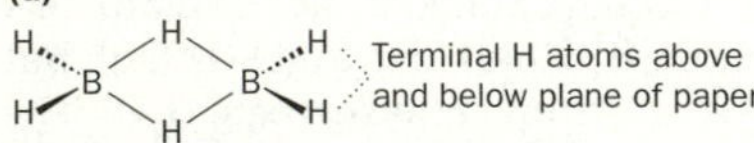

Bridge H atoms in plane of paper

(b) MO diagram for bridge

B_2H_6, MOs

antibonding

H, AOs

2 x 1s

available MOs from ($BH_{2,\ terminal}$)

2 x 2

non-bonding

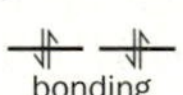

bonding

(c) 'banana' bonds in bridge

3-centre–2-electron bond

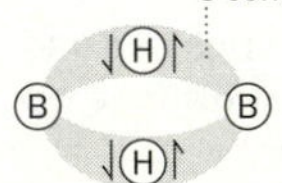

Beryllium hydride $(BeH_2)_x$

suggested structure

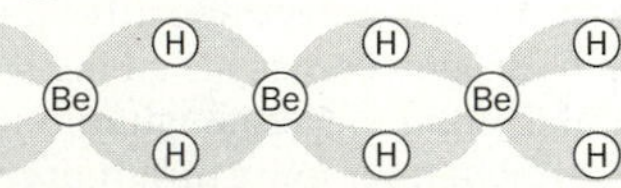

coated with a protective layer of oxide and hardly reacts at all. Gold, on the other hand, is unaffected by most acids as reaction with them would lead to an increase in energy, making the products less stable than the reactants. These examples are a reminder of the limitation of all predictions based on energy calculations. Whilst we can be sure from the laws of thermodynamics[A] that an unfavourable energy change WILL NOT occur spontaneously, all we can say about a favourable one is that it MIGHT happen; whether or not it occurs in practice depends on other factors such as the rate of the change or any interference by other reactions such as oxide formation.

Although the hydrogen compounds discussed so far show great variety, there are no major surprises. A hydrogen atom may combine with another atom by donating its electron to σ bond, or by losing an electron to form H^+, which then accepts a lone pair of electrons to form a σ bond. Alternatively, the atom may gain an electron to form the hydride ion H^-, which can combine with cations in an ionic lattice. But hydrogen has other tricks.

The triangular ion H_3^+ exists fleetingly in discharge tubes and is more stable than either of the species (H_2 + H^+) or (H + H_2^+) which would both be linear. So it seems that the three hydrogen atoms can be held together in some way by only two electrons. Hydrogen combines with boron to form a number of hydrides in which two boron atoms are bridged by one hydrogen atom. Since only two electrons are formally available for the B–H–B span, the bond is described as a 3-centre–2-electron ('3c–2e') bond, or more coloquially as a 'bent' or 'banana' bond. The simplest boron hydride is diborane, B_2H_6, in which two BH_2 groups are linked through two bridging hydrogen atoms (see p. 15). Each boron atom has three valence[A] electrons ($2s^22p^1$) in four atomic orbitals; it uses two electrons to form ordinary σ bonds with the two terminal hydrogen atoms, and so can contribute two orbitals and one electron to the central bridges. Each of the two bridging hydrogen atoms contributes one electron and one orbital, making a total of four electrons and six orbitals for building the bridge. Of the six molecular orbitals associated with the bridge, two are bonding, two non-bonding, and two antibonding. Each bonding orbital embraces one hydrogen atom and *both* boron atoms, and each accommodates one pair of electrons. The non-bonding and antibonding orbitals are empty, and, despite its departure from the traditional idea that the stability of the lighter elements requires a full valence shell ($1s^2$ for H and $2sp^3$ for B), the diborane molecule is not particularly unstable. 'Bent' three-centred bonds have also been suggested for BeH_2, a solid polymer in which the units are linked by the bridging

Structure of ice

A four molecule fragment showing H bonds (dotted)

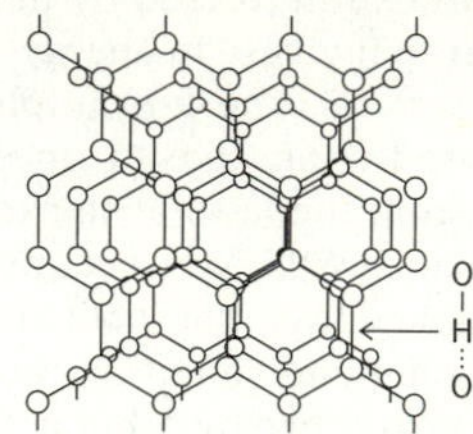

A larger fragment, showing open, hexagonal structure. The circles represent oxygen atoms and the lines O–H···O bonds SAL

Limiting ionic conductivity AP (λ) of H^+ and OH^- in water at 298 K

	H^+	OH^-	Na^+	Cl^-
λ (S cm^2 mol^{-1})	349.6	199.1	50.1	75.35

Grotthus mechanism for proton transfer in water SAL

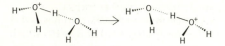

Hydrogen bonding in $(HF)_x$

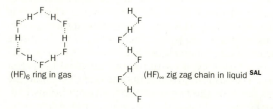

$(HF)_6$ ring in gas

$(HF)_\infty$ zig zag chain in liquid SAL

Hydrogen bonding in two organic compounds

The gas phase dimer of ethanoic acid contains two H bonds

Internal H bonding in salicylic acid

hydrogen atoms to form a chain (see p. 47). Similar hydrides are formed by aluminium, gallium, and zinc, which also have a high ratio of charge to size.

Many compounds of hydrogen with the lighter non-metals are markedly associated. Water is a liquid, despite the fact that its heavier analogue (hydrogen sulfide, H_2S, of almost twice its mass) is a gas. Ice floats on water because it has an open structure, so is less dense than the liquid. The very high ionic mobility[A] of the hydrogen ion in water, by the so-called Grotthus mechanism,[A] depends on the strong association in the liquid. Gaseous hydrogen fluoride forms hexameric $(HF)_6$ rings and is easily condensed to a liquid that contains HF_2^- and H_2F^+ ions, together with long $(HF)_x$ chains. Ammonia has a much higher boiling point and heat of vaporization than either methane or phosphine. These observations, together with many others from crystallography and spectroscopy, have long been interpreted in terms of the 'hydrogen bond'.[A] A molecule in which hydrogen is joined to an atom of a fairly light non-metal X, interacts through that hydrogen atom with a non-metal atom Y in another molecule. The two molecules involved may, but need not, be identical; and hydrogen bonds may also be formed within a single molecule. The non-metals X and Y are more electronegative than hydrogen and are most often nitrogen, oxygen, and fluorine; but weaker interactions also occur with sulphur and chlorine, and perhaps with even less electronegative elements such as carbon and bromine. Many hydrogen bonds are linear and unsymmetrical, the hydrogen being bonded to X and attracted, presumably electrostatically, to Y. A few slightly bent H bonds are known, and there may also be bent three-centre 'agostic' bonds in which a hydrogen atom lies between two metal atoms, or between one metal atom and a carbon atom.

Some of the strongest hydrogen bonds are, however, symmetrical, as in the ions $[F^-–H^+–F^-]$ and $H_2O–H^+–OH_2$. In these X_2H systems, there would appear to be appreciable 'covalent bonding' or orbital overlap over the whole $X–H^+–X$ group. We can describe these interactions as three-centre bonds but, unlike the 3c–2e bonds in diborane (p. 15), there are now four electrons spread over the three atoms. Three atomic orbitals would be involved, one (containing a lone pair) from each of the X atoms, together with the vacant 1s orbital from the H^+ ion. The resulting molecular orbitals would be one bonding, one non-bonding, and one antibonding. The four electrons would occupy the lower two orbitals, spread over all three atoms to form a 3c–4e bond and the antibonding orbital would be empty. Although the theory of the hydrogen bond may not yet be fully worked out, the importance of these interactions is

Some heats of vaporization ΔH_{vap} (at b.p.) [S]

similar graphs are obtained for melting and boiling points

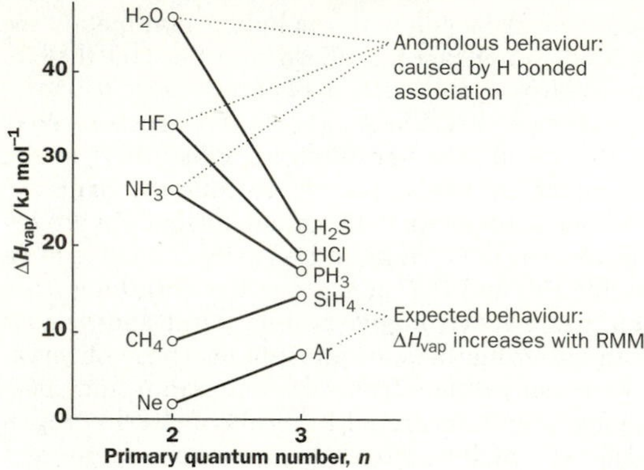

Strengths (kJ mol^{-1}) of some hydrogen bonds (…) and covalent bonds (–) [SAL]

		H bond (…)	Covalent bond (–)
Ammonia	H_2N–H … N	–17	–431
Water	HO–H … O	–22	–452
Hydrogen sulfide	HS–H … S	–7	–363
Hydrogen fluoride	F–H … F	–29	–568
Anion in K + HF_2^-	F … H … F (symmetrical)	–165	

MO diagram for H bonded $[F\cdots H\cdots F]^-$ [GE]

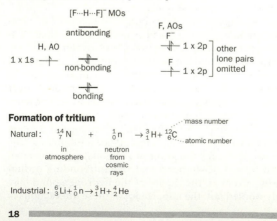

$[F\cdots H\cdots F]^-$ MOs

antibonding

H, AO

1 x 1s

non-bonding

bonding

F, AOs

F^- 1 x 2p

F 1 x 2p

other lone pairs omitted

Formation of tritium

Natural: $^{14}_{7}N + {}^{1}_{0}n \rightarrow {}^{3}_{1}H + {}^{12}_{6}C$

(N: in atmosphere; n: neutron from cosmic rays; superscript: mass number; subscript: atomic number)

Industrial: $^{6}_{3}Li + {}^{1}_{0}n \rightarrow {}^{3}_{1}H + {}^{4}_{2}He$

inestimable for the living world: without the hydrogen bond there would be no water as an external or internal environment, few biopolymers, and no genetic code.

Hydrogen forms a wide variety of compounds with those transition metals (including the lanthanides and actinides) that have rather few d electrons. Metals with fuller d shells seem to be unaffected by hydrogen, except for palladium (see p. 363), which for some reason absorbs large quantities of it. Hydrides of transition metals are often 'non-stoichiometric' (with bizarrely non-integral formulae, such as Th_4H_{15}) and may form several phases. Some have a stoichiometric but unexpected composition. Lanthanum, normally trivalent, forms LaH_2; but perhaps this is better written as $La^{3+}(H^-)_2^-$ (see p. 425). Since many of these hydrides are dark, shiny electrical conductors, they must contain at least some delocalized electrons. Such compounds, with their variable composition, were formerly classified as 'interstitial' because the small hydrogen atoms were envisaged primarily as slipping into the gaps between the metal ions; and our own understanding of their structure and bonding is not much further forward.

The variety shown by hydrogen is largely a result of the many ways it can use its 1s orbital or any molecular orbital derived from it, because those changes that we call 'chemical' are accompanied by rearrangement of electron clouds. But even further diversity is provided by the nucleus.

All hydrogen nuclei[A] have (by definition) a charge of + 1 and so contain a single proton; but they do not all have the same mass. In nature, about three in every 20 000 hydrogen nuclei are twice as heavy as the others, because they contain a neutron in addition to the proton. This heavy isotope of hydrogen is known as deuterium (D or ^{2}H).

An even heavier hydrogen isotope, tritium (T or ^{3}H) contains one proton and two neutrons. The tritium nucleus is unstable and decays, with a half-life of about 12.3 years, to an isotope of helium. Tritium is formed continuously in minute amounts in the atmosphere as cosmic rays collide with ^{14}N nuclei; and so natural waters always contain traces of it. In underground reservoirs, which are protected from cosmic rays, the tritium decays and so, by comparing the tritium content of underground and surface waters, we can estimate how long the water has been underground.

The two heavy isotopes of hydrogen are not, however, quite as heavy as we might expect: their nuclear masses are less than the sum of the masses of their component particles. The lost mass has been given out as energy, since on the subatomic scale these quantities are interconvertible though Einstein's famous equation. The

$^1_0n \rightarrow ^1_1p + ^0_{-1}e$

Radioactive decay of a neutron to give a proton and an electron

Tritium ^3_1H

1 proton ●] A=3
2 neutrons ○]
1 electron

Helium ^3_2He

2 proton ●] A=3
1 neutrons ○]
2 electrons

Einstein's equation

		Planck's constant		velocity of light
ΔE	=	$h\nu$	=	$\Delta m \times c^2$
energy emitted		frequency of radiation		mass defect

Nuclear binding energy per nucleon = mass defect/number of nucleons = $\Delta m/A$

Nuclear fusion energy

Energy evolved/kJ mol^{-1}

$^1_1H + ^1_0n \rightarrow ^2_1H$ 1.68×10^8

$^1_1H + 2\,^1_0n \rightarrow ^3_1H$ 8.73×10^8

(for the combustion of methane in air:

$CH_4(g) + 3O_2(g) \rightarrow CO_2(g) + 2H_2O$

energy evolved : ~ 900 kJ mol^{-1})

Isotopes of hydrogen

Atomic properties	^1H	$^2H(D)$	$^3H(T)$
Number of protons, Z	1	1	1
Number of neutrons (A–Z)	0	1	2
RAM	1.007825	2.014102	3.016049
Mass defect (amu)	0	1.84×10^{-3}	8.73×10^{-3}
Nuclear spin	$\frac{1}{2}$	1	$\frac{1}{2}$
Half-life	stable	stable	12.35 y
Properties of H_2	$^1H_2$	D_2	T_2
Melting point/K	13.96	18.73	20.62
Boiling point/K	20.39	23.67	25.04
Bond energy at 298 K/kJ mol^{-1}	436	443	447
Zero point energy/kJ mol^{-1}	25.9	18.5	15.1
Properties of H_2O	^1H_2O	D_2O	T_2O
RMM	18.0151	20.0276	22.0313
Melting point/°C	0.00	3.81	4.48
Boiling point/°C	100.0	101.42	101.51
Temperature of maximum density/°C	3.98	11.23	13.4
Density (25°C)/g cm^{-1}	0.99701	1.1044	1.2138
pK_w* (25°C)	14.0	14.7	~ 15.2
Symmetric O–H stretching/cm^{-1}	3657	2671	—

* $pK_w = -\log [H^+][OH^-]$

mass lost per nuclear particle is a measure of the stability of the nucleus, and shows us that deuterium has a more stable nucleus than hydrogen, and that tritium is more stable still. Huge changes are involved. The energy given out by fusion of a certain number of protons with twice as many neutrons is about one million times greater than that generated by fully burning the same number of methane molecules in air. Despite the immense gain in stability that accompanies these reactions, nuclear fusion is generally thought to occur only at the exceedingly high temperatures of the stars or of atomic (fission) explosions because a large amount of initiation energy is needed. We have not yet managed to produce controlled nuclear fusion for technological use.

Since all three hydrogen isotopes have identical nuclear charge and electronic structure, we rightly expect that they would differ little in chemical behaviour. The 100% and 200% increases in mass do, however, cause some differences. Molecules of 'heavy hydrogen' diffuse more slowly than $^1H_2$ and have slightly higher melting and boiling points. They also have lower energies at absolute zero temperature and this alters both the rates of reactions involving hydrogen, and the positions of chemical equilibrium. Bonds of H and D to a second element absorb infrared radiation at different frequencies, and this effect can be used to detect which part of an organic molecule has been deuterated. Since hydrogen and deuterium undergo the same reactions, but can easily be distinguished by mass spectroscopy, deuterium is widely used as a tracer to probe the mechanisms of chemical changes. Some physical properties of D_2O are appreciably different from those of H_2O. When water containing ^2H is electrolysed, $^1H_2$ comes off more readily than $^2H^1H$ or $^2H_2$, and so the concentration of ^2H in the remaining liquid is enriched; and, if the process is repeated many times, almost pure deuterium gas may be obtained. Since the deuterium nucleus captures neutrons less effectively than does the proton, D_2O can be used to moderate but not block the output of neutrons from a nuclear reactor. Tritium, which is used as a tracer and in nuclear fusion, is prepared by bombarding lithium-6 with neutrons.

The hydrogen nucleus provides yet other diversity. One peculiarity is that samples of isotopically pure $^1H_2$ can vary up to 50% in their thermal conductivity, because the two nuclei, which have non-zero spin (see p. 25), may combine with their spins either aligned (ortho-H_2)[A] or opposed (para-H_2). At absolute zero, the para form is the more stable, but at room temperature the favoured ratio is ortho:para, 3:1. Tritium, with the same nuclear spin, behaves similarly; but for deuterium, which has a different nuclear spin, the ortho form is favoured at absolute zero, while room temperature

Effect of deuteration on infrared absorbtion

The N–^{1}H and N–D peaks for the nickel(II) bis(oxamido) anions

```
        H       H   ⌉2–
OC–N\     /N–CO
 |      Ni      |
OC–N/     \N–CO
        H       H
```

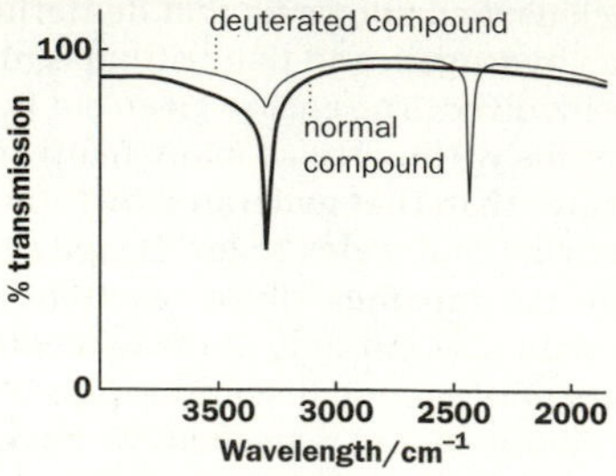

From P. X. Armendez and K. Nakomoto, *Inorg. Chem.*, **5**, 796 (1966).

gives a 2:1 ortho:para mixer. This nuclear spin isomerism has no appreciable effect on chemical behaviour, but is included for curiosity value.

Nuclear spin does, however, form the basis of an important research tool, nuclear magnetic resonance[A] spectroscopy, or NMR. If placed in a magnetic field, any nuclear particle with a non-zero spin can be oriented with the spins aligned with the field or against it; intermediate positions are unstable. The two allowed alignments differ slightly in energy, and if the sample is irradiated with electromagnetic radiation of a frequency that corresponds to the exact difference in energy, the magnetic nuclei can 'flip' from one alignment to the other, 'resonating' with the radiation. Naturally, the size of the energy gap depends not only on the nucleus, but also on the magnetic field; and, while this in turn depends largely on a powerful external magnetic, it depends also on the electronic environment of the nucleus (because electrons, too, have magnetic properties). So, protons in different chemical environments (as in the $-CH_3$ and $-OH$ group in methanol, CH_3OH), will resonate at different frequencies. And since the internal magnetic field depends also on the presence of other magnetic nuclei, NMR is an extremely sophisticated tool for unravelling the niceties of molecular structure. Much of the early work was carried out using proton resonance, and PMR, or ^1H NMR, is now also a powerful medical technique for detection of abnormalities using a body scanner.

Hydrogen does not, of course, have any monopoly on chemical diversity, as we shall see in the rest of the book; but the variety it can show, given only one proton, one electron, and maybe one or two neutrons to add weight, is impressive indeed.

For summary, see pp. 24–5.

Summary

Electronic configuration

- $1s^1$

Element*

- Low density diatomic gas, H_2, undetectable by human senses.
- Forms explosive mixture with air

Occurrence

- H^+ major component of stars; terrestrially, as water and in all organic matter

Extraction

- Reduction of water, electrolytically, or with coke or hydrocarbons

Chemical behaviour

- More compounds than any other element
- Wide variety of bonding
  - *Ionic*
    - electron loss to give H^+ ($1s^0$); too high charge/size ratio to exist naked; coordinated by bases, e.g. H_2O to give H_3O^+:
    - electron gain to give H^- ($1s^2$) in solids and melts with Na^+, Ca^{2+}, etc.
  - *Covalent:*
    - two-centre, bond order 1, e.g. H_2 (non-polar bond) and non-metal hydrides (polar bonds)
    - two-centre, bond order $\frac{1}{2}$, e.g. H_2^+ and H_2^- (transient)
    - three-centre, non-linear, e.g. H_3^+ (transient), B–H–B in B_2H_6 and derivatives, and M–H–M in polymeric hydrides, e.g. BeH_2
  - *'Hydrogen bonds':*
    - dipolar three-centre–four-electron bonds with electronegative atoms
    - fairly weak, unsymmetrical, usually linear, e.g. O–H···H
    - stronger symmetrical, e.g. $F^-\cdots H^+\cdots F^-$
  - *Interactions with conduction band*
    - hydrides of d and f metals, often non-stoichiometric, conducting, and intensely coloured

* At room temperature and pressure.

Uses

- Chemical
  - as a reductant
    - of C=C in oils to give CH–CH in fats (margarine) of CO, N_2, and Cl_2 to give HCHO, NH_3, and HCl for use in chemical industry
    - of O_2 for welding, as a rocket fuel, as a source of power for spacecraft, and (potentially) as basis of pollution free economy
  - in atomic recombination reaction for welding
- Nuclear (see below)
    - ^1H in NMR for research and medical diagnosis, in nuclear fusion reactions (energy produces stellar and, potentially, industrial power)
    - ^2H as tracer (detected by mass spectroscopy, and infra-red absorption)
    - ^3H for dating

Nuclear features

- ^1H naturally occurring (nuclear spin $\frac{1}{2}$)
- ^2H traces in nature; obtained from water by successive electrolysis (nuclear spin 1)
- ^3H minute trace in nature from cosmic ray bombardment (nuclear spin $\frac{1}{2}$)
- Nuclear spin isomerism (ortho and para)

He Helium

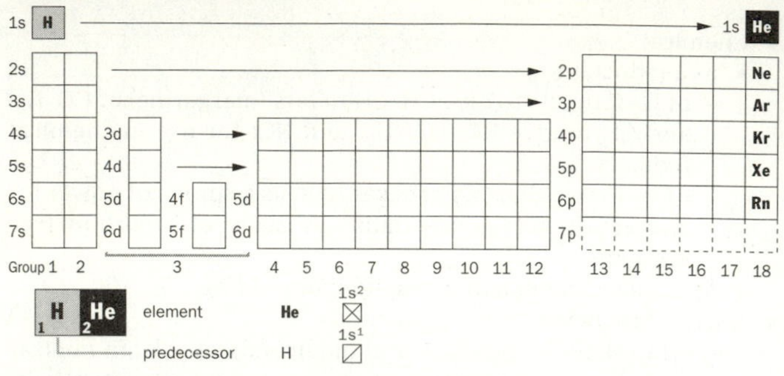

	Predecessor	Element
Name	**Hydrogen**	**Helium**
Symbol, *Z*	H, 1	He, 2
RAM	1.00794	4.002602
Radius/pm: atomic	78	126
Boiling point/K	20.28	4.216
$\Delta H^{\ominus}_{fus}$**/kJ mol**$^{-1}$	0.12	0.021
$\Delta H^{\ominus}_{vap}$**/kJ mol**$^{-1}$	0.46	0.082
Ionization energy for removal of *j*th electron/kJ mol^{-1} ***j*** **= 1**	1312.0	2372.3
j **= 2**		5250.4
Electron affinity/kJ mol^{-1}	72.8	0

Stable isotopes of helium

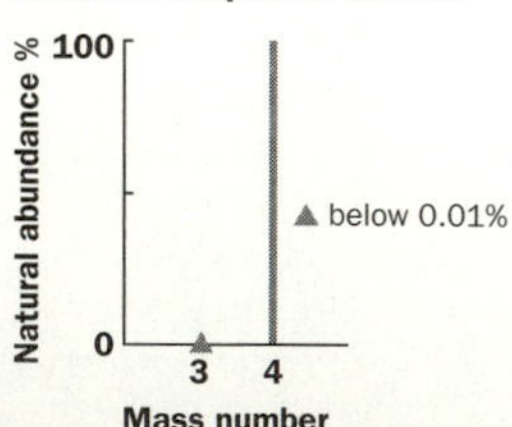

The second lightest element, helium, is also the second commonest in the universe, and was first detected spectroscopically in the sun: hence its name. Being a model of both nuclear and electronic stability, it provides a complete contrast to the chemical variety of hydrogen; indeed, much of its interest lies in its very inertness.

The two protons in the helium nucleus are almost always accompanied by two neutrons, although naturally occurring helium also contains about one part per million of the lighter isotope $^{3}_{2}$He. The high binding energy per particle of the usual nucleus $^{4}_{2}$He is about seven times that of deuterium, $^{2}_{1}$H, and the conglomerate of two neutrons with two protons is so stable that helium nuclei are generated (as if spat out whole) by spontaneous disintegration of many heavier nuclei. In this context, $^{4}_{2}$He nuclei are still called α-particles, the name they were given in early studies of radioactivity[A]. Helium is being continuously formed by radioactive decay of heavier nuclei, both within the earth's crust, where it is trapped in minerals, and in pockets of natural gas, from which it can be obtained by fractionation. A small quantity of helium is also present in the atmosphere but, as the gas is so light, much of that which reaches the surface escapes from the earth's gravitational field.

In the helium atom, the second electron, like the first, occupies the innermost atomic orbital. This $1s^2$ configuration is particularly stable, as the electrons are very near the nucleus and are strongly attracted to it; more energy is needed to remove an electron from helium than for any other element. Since the next lowest (2s) atomic orbital is of much higher energy, helium has no tendency whatever to gain an electron, or to share electrons with any other atom. Unsurprisingly, there are no known helium compounds, and for this reason, helium is known as a 'noble' or 'inert' gas. Helium cannot be trapped in the cage-like crystals that can contain some of its heavier analogues (see pp. 191, 319) as it can escape through the gaps in the lattice; and it even diffuses slowly out of a glass container.

The physical properties of helium are mainly a result of its very low mass and the fact that its small, tightly held, electron cloud is barely distorted by neighbouring atoms. So, forces between helium atoms are extremely weak. The low density of the gas and its chemical inertness makes helium ideal for filling airships and meteorological (and party) balloons. Gaseous helium is also used instead of nitrogen in deep sea diving, since it is less soluble in blood plasma and so causes fewer problems, such as the 'bends', on decompression. Liquid helium has the lowest boiling point of any known substance; it is used as a refrigerant in low temperature

Stability of the ^4He nucleus [c]

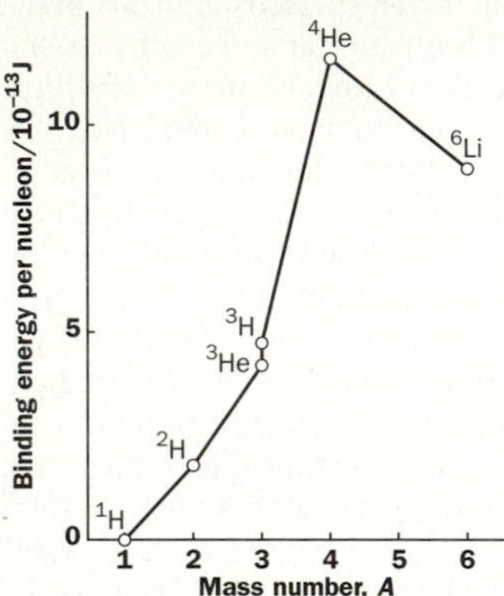

Some examples of α decay

$$^{235}_{92}U \xrightarrow[7.04 \times 10^8 \text{ y}]{t_{1/2}} {}^{231}_{90}Th + {}^4_2He$$

Successive α decay is often written as:

$$^{252}_{98}Cf \xrightarrow[65 \text{ y}]{\alpha} {}^{248}_{96}Cm \xrightarrow[3.4 \times 10^5 \text{ y}]{\alpha} {}^{244}_{94}Pu \xrightarrow[8.2 \times 10^7 \text{ y}]{\alpha} {}^{240}_{92}U$$

Half-lives can be quite brief:

$$^{261}_{105}Ha \xrightarrow[\sim 1 \text{ s}]{\alpha} {}^{257}_{103}Lr \xrightarrow[35 \text{ s}]{\alpha} {}^{253}_{101}Md$$

Phase diagram [AP] for ^4He

showing conditions for forming superfluid He-II

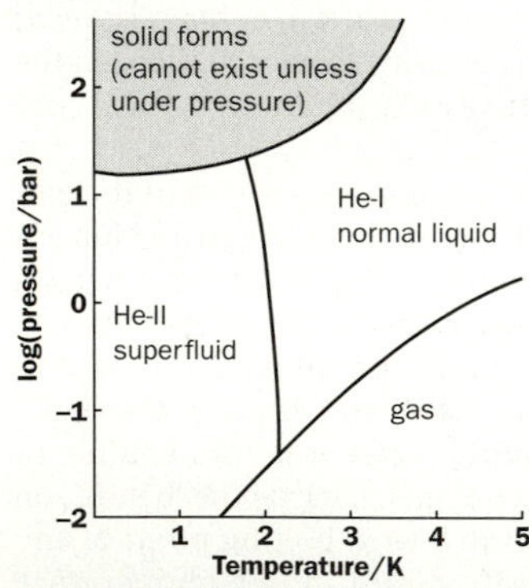

(Note the logarithmic pressure scale)

research. It cannot be solidified at any temperature unless pressure is applied.

So the properties of helium, with its light, Is^2 atoms appear to be boringly predictable—down to about 2.2 K, if the gas pressure is low. When the temperature is lowered even further, however, there is a total change of character, because liquid helium ^{4}He, is transformed from a normal liquid ('He–I') to a most extraordinary one. The low temperature form, known as He–II, conducts heat one million times as fast as He–I, and seems to have zero viscosity. Since it also has the peculiar ability to flow over any solid that is at or below 2.2 K, including vertically upwards, it is known as a 'superfluid'[A]. There is also a superfluid ^{3}He, but the isotopic forms of liquid helium-II differ more than we should expect merely from their masses. This may be because the different nuclear spins of ^{3}He and ^{4}He make themselves felt for such light atoms at very low temperatures. Until these various bizarre phenomena are satisfactorily explained, it is clearly inappropriate to dismiss the behaviour of helium as predictable, let alone boring.

Summary

Electronic configuration

- $1s^2$

Element

- Very light monatomic gas; not detectable by human senses; very difficult to liquefy.

Occurrence

- He^{2+} continuously formed in sun; terrestrially in natural gas and trapped in minerals; trace in atmosphere

Extraction

- Fractionation of natural gas

Chemical behaviour

- Extreme inertness; no known compounds; no clathrates (too small); less soluble in blood plasma than nitrogen

Uses

- Filling balloons
- Deep sea diving: replacement for nitrogen
- Refrigerant in low temperature research

Nuclear features

- ^{4}He has very high binding energy: nuclei emitted 'α-particles', in radioactive decay
- ^{4}He at 2.2 K becomes superfluid, highly conducting liquid 'He–II'

Part II

Filling the 2s and 2p orbitals

The first short period: lithium to neon

Li Lithium

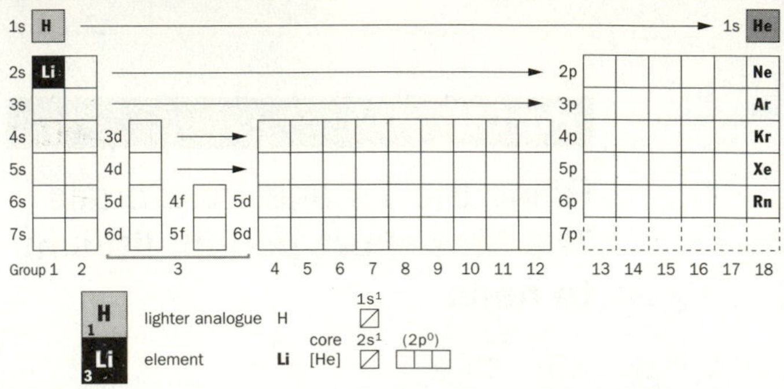

		Element	Lighter analogue
Name		Lithium	Hydrogen
Symbol, Z		Li, 3	H, 1
RAM		6.941	1.00794
Radius/pm:	atomic	152	78
	covalent	123	30
	ionic, M^+	78	(10^{-5})
Electronegativity (Pauling)		1.0	2.2
Melting point/K		453.69	14.01
Boiling point/K		1620	20.28
$\Delta H^{\ominus}_{fus}$/kJ mol^{-1}		4.60	0.12
$\Delta H^{\ominus}_{vap}$/kJ mol^{-1}		134.7	0.46
Density/kg m^{-3} (Temperature)		534 (293 K)	76.0 (S) (11 K)
Electrical conductivity/Ω^{-1}m^{-1}		1.17×10^7	
Ionization energy for removal of jth electron/kJ mol^{-1}	$j = 1$	513.3	1312.0
	$j = 2$	7298.0	
Electron affinity/kJ mol^{-1}		59.6	72.8
Dissociation energy of E_2(g)/kJ mol^{-1}		107.8	453.6
$E^{\ominus}$/V for $M^+(aq) + e \rightarrow M(s)$		−3.04	

Stable isotopes of lithium

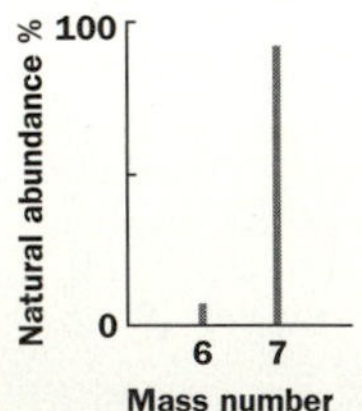

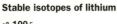

Lithium has three electrons and, since the 1s orbital can hold only two, it must also make use of the 2s orbital, of higher energy and further out from the nucleus. Those elements whose outer electrons occupy the four $n = 2$ orbitals make up the so-called first short period of the periodic table. Although lithium, $[He]2s^1$, has a higher nuclear charge than hydrogen or helium, it can lose its outermost (2s) electron much more easily than hydrogen loses its (1s) electron: partly because it is further away from the nucleus, but also because the two intervening 1s electrons 'screen' or '*shield*'[A] it from the full attractive force of the protons.

In an isolated atom of lithium, the 2s orbital is of slightly lower energy than the three 2p orbitals because it is more 'penetrating'. We know that no orbital has a fixed boundary, but that an electron in a specified orbital has a certain probability of occupying a particular volume of space. For example, there is a finite possibility that an electron from the $n = 2$ shell may be found, at any given instant, within the region also occupied by the 1s electrons, and it will then be more firmly held than if it were further away. This *penetration*[A] of an inner electron shell is more marked for an s than for a p orbital, and so the 2s orbital is slightly more stable than the 2p trio.

At room temperature and pressure, the most stable arrangement of lithium atoms is a close-packed three-dimensional (3D) array. A bulk-sized piece of solid lithium contains a large number* of atoms in close proximity to each other, and each one affects the orbital energy of the others, so that the previously sharp 2s and 2p energy levels of the individual atoms broaden and overlap to form a broad 'band'[A] of permissible energies, spread over the entire 3D array. Since the electrons (one per atom) that occupy this band are free to move throughout it, the array is a good conductor of electricity. The sample of lithium is solid, and behaves in many ways as if it were a regular lattice of Li^+ ions vibrating within a delocalized cloud of electrons, which acts as a barrier to any light falling on it. The lithium is totally opaque and, if smooth, reflects the light and so looks shiny. If the temperature is raised, the vibrations become more active and increasingly interfere with the flow of electrons.

So, lithium is a metal: that is, it has high electrical conductivity which decreases with increased temperature, and also a high thermal conductivity and shiny appearance. Lithium is the lightest element to exist in a metallic form under normal pressures and

* A cube of lithium of side 1 cm contains about 4×10^{22} atoms.

The formation of a conduction band in lithium [M]

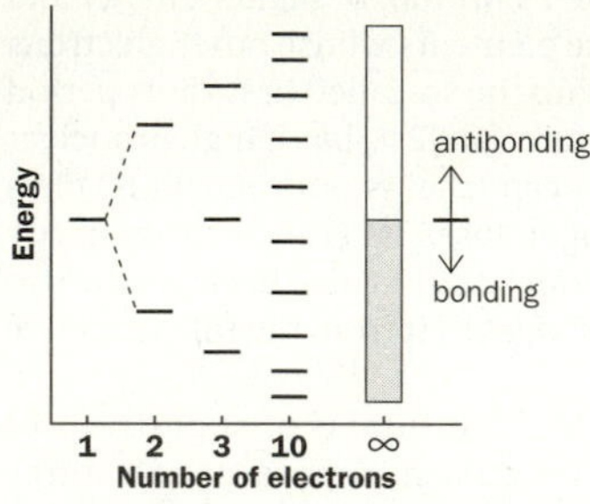

Bonding and antibonding orbitals formed from a line of atoms

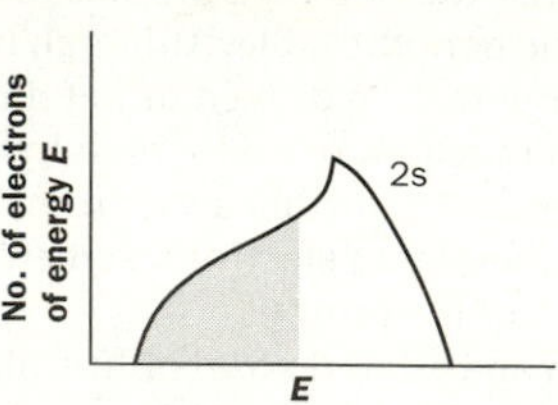

The 2s band in Li can accommodate two electrons per atom

Densities of some light metals at 20°C

Density/kg m^{-3}	Li 534	Be 1847.7	
	Na 971	Mg 1738	Al 2698

it is also by far the least dense metal*: not only do the atoms themselves have low mass but, being singly charged and each contributing only one delocalized electron, the forces within the lithium lattice are weaker than in metals that have ions of higher charge, and therefore a higher concentration of electrons in the cloud (or 'conduction band'). Lithium can be added to other metals such as magnesium and aluminium to give alloys that are both tougher and lighter than the pure components.

The reactions of lithium are entirely due to the 2s shell, because the second ionization energy is so high that the $1s^2$ helium core always remains intact. The possibilities of bond formation for lithium ($2s^1$) might be expected to mirror those for hydrogen ($1s^1$), but the similarities are mainly formal.

Sigma bonding is much less important for lithium than for hydrogen. The 2s orbitals are too diffuse to overlap effectively and the gaseous species Li_2 readily dissociates. Some rather unstable σ bonds are also formed between lithium and aromatic organic groups to give low melting compounds that dissolve in organic solvents.

The Li^- anion, $2s^2$, is formed when the metal dissolves in liquid ammonia, but, unlike the hydride ion H^-, it does not occur in ionic solids. As anticipated from the low value of the first ionization energy, most compounds of lithium contain the Li^+ ion, which, in common with all other singly charged ions except H^+, is large enough to be considered an independent chemical entity. The Li^+ ion occurs in nature in aluminosilicate minerals, such as $LiAlSi_2O_6$, which can be converted first to lithium chloride and then to metallic lithium. Much energy is needed to convert the stable Li^+ ion, tightly held in an ionic lattice, into the reactive metal, so the reduction must be carried out by electrolysis, rather than by chemical reduction, and in a molten salt solution (lithium and potassium chlorides), instead of an aqueous one.

Lithium is a fairly soft metal, which reacts slowly with water to give hydrogen and an alkaline solution containing hydroxyl ions. For this reason, lithium and heavier ns^1 elements are known as 'alkali metals'. Lithium has a greater thermodynamic incentive to

* Elements that are non-metallic at normal pressures may become more metallic if their atoms are forced together by increasing the pressure (see, for example, iodine, p. 407). Calculations show that the core of the planet Jupiter probably contains liquid metallic *hydrogen*. But the conditions there are extreme: over 4 million atmospheres pressure and a temperature of 10 000 K.

Thermodynamic incentive

The tendency for a reaction to happen, and the position of equilibrium reached, depend on the value K; and for 'high incentive' with equilibrium far to the right:

K is large $\quad$ K = equilibrium constant

A large value of K implies large values of:

$-(G_{products} - G_{reactants}) = -\Delta G = RT \ln K$ $\quad$ G = 'Gibbs free energy'

$-\Delta H + T\Delta S = -\Delta G$

H = 'heat content' or 'enthalpy'

S = 'entropy' (usually envisaged as disorder)

So, reaction is favoured if it:

(a) is exothermic (ΔH negative)
(b) increases disorder (ΔS positive)
} but these factors may oppose each other

Two special cases

Acid dissociation

$HA(aq) \rightleftharpoons H^+(aq) + A^-(aq)$ $\quad$ $K > 1$ implies pK negative

Reduction of H^+ by metal in acidic solution, $[H^+] = 1$ mol dm^{-3}

e.g. $M(s) + H^+(aq) \rightleftharpoons M^+(aq) + \frac{1}{2} H_2(g)$ $\quad$ $K > 1$ implies $E^{\ominus}_{M}$ negative

Electrode potentials of lithium and sodium

For the reaction

$$M(s) + H_2O(l) \rightarrow M^+(aq) + OH^-(aq) + \tfrac{1}{2}H_2(g)$$

$E^{\ominus}_{Li} = -3.04$ V and $E^{\ominus}_{Na} = -2.71$ \

A high value of $(-E^{\ominus}_{M})$ is favoured by:

	Li	Na
(1) low ΔH_{atom}, for $M(s) \rightarrow M(g)$	139.30	**91.68**/kJ mol^{-1}
(2) low IE, for $M(g) \rightarrow M^+(g)$	513.3	**495.8**/kJ mol^{-1}
(3) high $(-\Delta H_{hyd})$ for $M^+(g) \rightarrow M^+(aq)$	**–502**	–406/kJ mol^{-1}
(4) low $(-\Delta S_{hyd})$ for $M^+(g) \rightarrow M^+(aq)$	–119	**–89**/J K^{-1} mol^{-1}

The smaller Li^+ ion is more strongly hydrated than the Na^+ ion, giving it a more favourable enthalpy change ΔH_{hyd}; it also has a greater ordering effect on the water molecules and so has a more unfavourable entropy change ΔS_{hyd}. The large difference in the values of ΔH_{hyd} for Li^+ and Na^+ outweighs the differences in terms (1), (2), and (4), even though these all combine to oppose it. The values in bold face indicate the greater contribution to a high value of $-E^{\ominus}_{M}$

Ionic lattices

For the reaction

$M^{z+}(g) + X^{z-}(g) \rightarrow MX(s)$ $\quad$ $-\Delta H \sim U$ $\quad$ lattice energy

For ionic lattices of same formula (here MX)

$U \propto z_+z_-/(r_+ + r_-)$ **Kaputstinskii's equation**

r_+, r_- are ionic radii

react with water than almost any other metal, despite the relatively large amounts of energy needed both to vaporize the metal and to ionize the atoms formed. The thermodynamic driving force that produces the high value of $-E^{\ominus}$ (see p. 12) comes mainly from the energy given out when the Li^+ ions are hydrated; although they are only singly charged, their small size gives them a high surface density of positive charge, which makes them strongly attracted to regions of negative charge, such as the oxygen atoms of the dipolar molecules of water. The reaction between lithium and water, although energetically highly favoured, is quite restrained in practice because the metal surface becomes partially coated with lithium hydroxide, which is only sparingly soluble in water.

The Li^+ ion is small (almost the same size as Mg^{2+}) and this is the key to much of the chemistry of lithium. The hydrated ion $Li(H_2O)_6^+$ can replace $Mg(H_2O)_6^{2+}$ in silicate minerals. The anhydrous lithium ion forms stable lattices with small or highly charged anions; when the metal burns in air or when the nitrate is heated, lithium forms the oxide $(Li^+)_2O^{2-}$ rather than $(Li^+)_2O_2^{2-}$ or $Li^+O_2^-$, which contain larger anions. Like magnesium, lithium reacts with nitrogen to form a nitride Li_3N. Many of these observations are in keeping with Kaputstinskii's equation, which states that the energy evolved on forming an ionic crystal from its constituent ions increases with the product of the charges of the anion and the cation, the number of ions per formula unit, and the inverse of the interionic distance. So, for two substances of the same general formula and charge type, the one with the smaller ions will form the more stable lattice. On this simple electrostatic model, we should therefore expect ions such as F^-, OH^-, O^{2-}, and N^{3-} to form stable, largely ionic, compounds with the Li^+ ion; although this may also deform ('polarize') the electron clouds of the anions to give some overlap of orbitals which further strengthens the lattice. If there is strong overlap the electrostatic model is inadequate and the structure may depart from an infinite 3D array. Lithium nitride, for example, has a layer structure rather than a continuous ionic lattice.

Indeed, the very labels 'ionic' and 'covalent' refer only to idealized situations. In a purely ionic solid two spherically symmetrical ions of opposite charge interact purely electrostatically; while in a perfect covalent bond the region of maximal electron density is centrally placed between the two atoms. Although purely covalent bonds are formed between like atoms (as in H_2), there is some electron displacement ('partial ionic character') in most of the bonds we describe as 'covalent', just as there is some disturbance of electron clouds ('partial covalent character') in many high melting and primarily 'ionic' solids.

Solubilities of ionic solids

The reaction

$M^+X^-(s) + H_2O(l) \rightarrow M^+(aq) + X^-(aq)$

is favoured by:

(1) low lattice energy, U i.e. low $\dfrac{1}{(r_+ + r_-)}$

(2) high heat of hydration ΔH_{hyd}, i.e. high $(\frac{1}{r_+}) + (\frac{1}{r_-})$

(3) low $(-\Delta S_{hyd})$ of hydration, i.e. low $(\frac{1}{r_+}) + (\frac{1}{r_-})$

Since the unfavourable entropy change on hydration is usually much smaller than the favourable heat change, the solubility depends mainly on a balance of factors (1) and (2).

When one ion is much smaller than the other, hydration is often dominant, and solubility high.

When the ions are more similar in size, the lattice energy becomes relatively more important, and solubility decreases.

Solubilities of some lithium compounds

Values[J] of $-\Delta G^\circ$/kJ mol^{-1} for the solution of some anhydrous lithium compounds in water. (The observed solubility is complicated by the formation of hydrates, but in general, the higher the value, the higher the solubility.)

Anion	F^-	Cl^-	Br^-	I^-	OH^-	NO_3^-	CO_3^{2-}	SO_4^{2-}
r_-/pm	133	180	196	218	140	189	185	230
$-\Delta G^\circ$/kJ mol^{-1}	–14	41	57	78	8	15	–17*	10*

[r_+ for Li = 74 pm]

*The marked difference between two salts of the same charge type, both with large anions, shows how delicately the energy terms are balanced.

Lithium ions with large anions

(a) Polarization

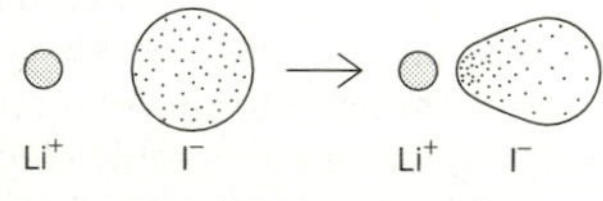

The small Li^+ ion deforms the easily polarized iodide ion, and so strengthens the lattice

(b) Repulsion between anions

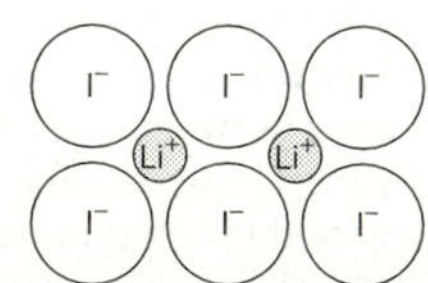

When r_+/r_- is low, the lattice is weakened by repulsion between the anions

Some compounds such as LiF and LiOH (both with small anions) and Li_2CO_3 and Li_3PO_4 (both with highly charged anions) have high lattice energies and are only sparingly soluble in water. But there are other, often opposing, factors at work. The small cation, and small or highly charged anions, are strongly hydrated, causing a large release of heat (which favours solubility) but also a marked ordering of the water molecules (which opposes it). As is frequently the case, actual chemical behaviour is determined by small differences between large energy terms. The lithium ion is the only singly charged cation to form a nitride, M_3N, although analogous compounds, M_3N_2, are formed by all $([\text{core}]ns^0)^{2+}$ metal ions; and it also combines with the anion AlH_4^-. Lithium salts of large oxoanions, such as carbonate and nitrate, often decompose at fairly low temperatures because, although smaller anions usually form stronger lattices than larger ones, this effect is most marked with small cations.

With large, singly charged anions such as iodide and perchlorate (ClO_4^-), however, the Li^+ ion forms lattices in which the ionic contribution is less than that predicted by Kaputstinskii's equation, because the cation is too small to keep the bulky anions away from each other; and the repulsion between them weakens the crystal; but this effect may be offset by covalent contributions to the lattice stability (see above).

Electrostatic forces between negative dipoles[A] and positive ions are stronger for the compact Li^+ ion than for larger singly charged cations; for example, with substances that contain charged or partially charged oxygen atoms. Not only do Li^+ ions have a high hydration energy, but some lithium salts are soluble in polar organic solvents such as 2-propanone ('acetone') and diethyl ether. Lithium ions also form adducts such as $Li(NH_3)_4^+$ with dipolar compounds containing nitrogen.

Low size can, however, also decrease interactions for geometric reasons, and the Li^+ ion appears to be too small to replace K^+ in alums or to be held in the crown ethers that enclose larger M^+ ions (see p. 128); for Li^+, a tighter 'hole' is needed.

It seems that the bonding in lithium compounds is fairly predictable. Apart from logically plausible but energetically unlikely 2s–2s overlap in Li_2 and the one-off $2s^2$ ion Li^- in liquid ammonia, most lithium compounds contain Li^+, relatively tightly bound. Like lithium alloys, some compounds of lithium are exploited for their low density: in small, light batteries or in space capsules (where Li_2O_2 is used as a source of oxygen and LiOH as an absorbent for carbon dioxide). Some lithium compounds are used because they are tougher than their sodium or potassium counterparts;

Solvation of Li^+ by organic solvents

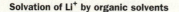

$$\delta+ \quad (H_3C)_2C{=}\ddot{O}: \delta- \qquad \delta+ \quad (H_3C{-}CH_2)_2\ddot{O}: \delta-$$

'acetone' (propan-2-one) — diethyl ether

The negative dipoles on the oxygen atoms are attracted to the small positively charged Li^+ ions

'Alum' $K_2SO_4 \cdot Al_2(SO_4)_3\ 24H_2O$
no lithium analogues known

The structure of the lithium methyl tetramer

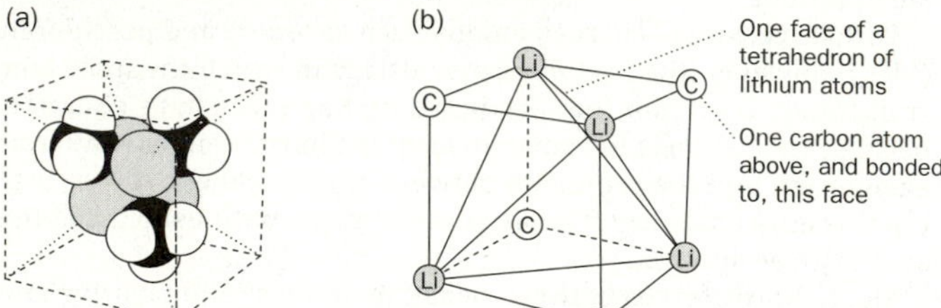

The structure of $(CH_3Li)_4$ (a) Showing the tetrahedral Li_4 unit with the CH_3 groups located symmetrically above each face of the tetrahedron. [Adapted from E. Weiss and E. A. C. Lucken, *J. Organmet. Chem.*, 1964, **2**, 197.] The structure can also be regarded as derived from a cube (b)

Li_2F_2 and Li_2Cl_2

X–Li–X–Li ring (X bridging the two Li atoms): almost square

^6Li 'milking'

$${}^6_3Li + {}^1_0n \rightarrow {}^3_1H + {}^4_2He$$

tritium

examples are lithium carbonate in enamels and lithium stearates in grease. Lithium also features as the counter ion of the versatile organic reductant, $LiAlH_4$. Alkyls and aryls of lithium are more covalent and less violently reactive than those of heavier ns^1 metals.

The chemistry of lithium is not, however, without its surprises. The compound $Li_4(CH_3)_4$ is a nearly cubic cluster around a tetrahedron of lithium atoms. We can envisage each methyl group to be attached by delocalized (4-centre–2-electron) bonding to the three metal atoms bounding one face; but there is also marked interaction both between the lithium atoms within a cluster and between adjacent molecules. The gaseous dimeric fluoride, Li_2F_2, and chloride, Li_2Cl_2, have planar diamond-shaped structures and are thought to contain 3-centre–2-electron bonds similar to those in diborane (p. 15).

The lithium ion, administered as lithium carbonate, is used as a mood stabilizer in some types of manic depression. (The level of lithium in the body is monitored by measuring the intensity of the crimson colour produced when a sample of blood is injected into a flame. The red light is emitted, not by excited lithium ions or atoms, but by the molecule LiOH.)

A further peculiarity of lithium is the fact that different samples vary appreciably in relative atomic mass. This is, however, a result of human vagaries rather than to chemical irregularity. Natural lithium contains, in addition to ^{7}Li, a trace of a lighter stable isotope ^{6}Li which can be converted by neutron bombardment into tritium, which is used in nuclear fusion reactions. Samples of natural lithium are 'milked' of ^{6}Li and may later be sold without their history being declared (and this is unhelpful for those chemists who need to know the exact mass of a lithium compound they should dissolve to make a lithium solution of a particular molar concentration).

For summary see p. 43.

Summary

Electronic configuration

- $[He]2s^1$

Element

- Metallic: very light, relatively hard, fairly reactive, crimson flame

Occurrence

- In aluminosilicate minerals

Extraction

- By electrolytic reduction of molten $(Li^+ + K^+)Cl^-$

Chemical behaviour

- (1) Most compounds contain Li^+ ($2s^0$) and are predominantly ionic; small size of Li^+ results in:
  - high enthalpy of hydration, and hence high negative electrode potential (despite fairly high energy needed for sublimation and ionization of Li)
  - stable lattices with small or highly charged anions; metal burns in air to Li_2O (*not* Li_2O_2) and Li_3N
  - easy thermal decomposition of salts with oxoanions
  - low aqueous solubility of salts with small or highly charged anions
  - high aqueous solubility of salts of large anions
- (2) Li^- ions ($2s^2$) formed when metal dissolves in liquid ammonia
- (3) Organometallic compounds partially covalent, e.g. alkyls, aryls, and $Li_4(CH_3)_4$ clusters

Uses

- Manufacture of hard light alloys, thick greases, and light electric batteries
- Treatment of manic depression
- ^6Li source of tritium for nuclear fusion (see below)

Nuclear features

- RAM varies, as ^6Li may be 'milked' off main natural isotope ^7Li.

Be Beryllium

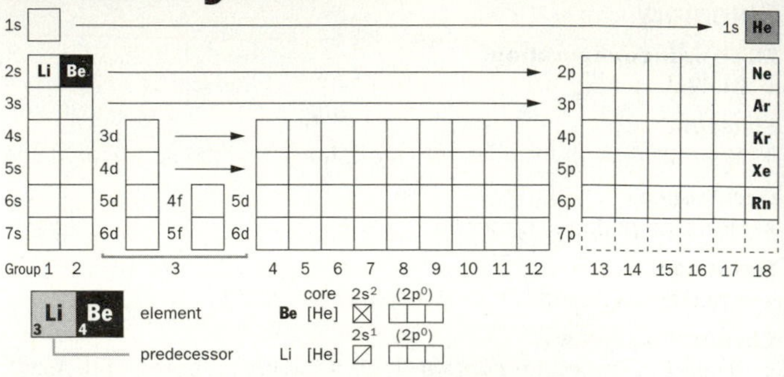

	Predecessor	Element
Name	**Lithium**	**Beryllium**
Symbol, *Z*	Li, 3	Be, 4
RAM	6.941	9.01218
Radius/pm: atomic	152	113.3
covalent	123	89
ionic, M^{2+}		34
Electronegativity (Pauling)	1.0	1.6
Melting point/K	453.69	1551 ± 5
$\Delta H^{\ominus}_{fus}$/kJ mol^{-1}	4.60	9.80
$\Delta H^{\ominus}_{vap}$/kJ mol^{-1}	134.7	308.8
Density (at 293 K)/kg m^{-3}	534	1847.7
Electrical conductivity/Ω^{-1}m^{-1}	1.17×10^7	2.5×10^7
Ionization energy for removal of *j*th electron/kJ mol^{-1} *j* = 1	513.3	899.4
***j* = 2**	7298.0	1757.1
***j* = 3**		14 848
Electron affinity/kJ mol^{-1}	59.6	−18
Bond energy/kJ mol^{-1} E–H		226
E–O		523
E–F		615
E–Cl		293

Stable isotopes of beryllium

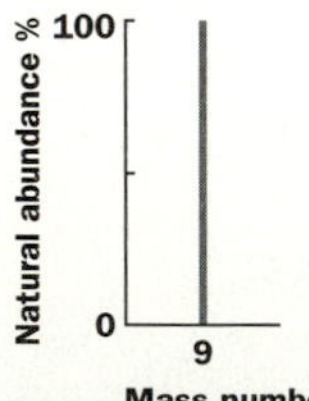

The outer occupied orbital, 2s, is the same in beryllium as in lithium, but the nuclear charge is a third as much again; so we should guess that the beryllium atom, [He]$2s^2$, would be very small. We should expect beryllium to have a higher value of the first ionization energy than lithium, but a much lower value of the second one. Since the 2s shell is full, no dimers would be expected in the vapour; but if the 2s band in the solid overlapped the 2p band (as is likely since they are derived from orbitals of similar energy), electrons would still be free to move if a potential were applied. The Be^{2+} ion, if formed, would be so small and so highly polarizing that its compounds would have marked covalent character.

On the whole, beryllium fulfils our expectations. The element is indeed a metal. Its small atomic size and the presence of two delocalized electrons per atom lead to tight bonding in the metal, which has a high melting point and heat of sublimation, and is very dense for its atomic mass. It is used for the windows of X-ray machines. (Transparency to X-rays decreases with increased electron density around the nucleus, and hence with increasing atomic number. Since beryllium, unlike lithium, is stable to damp air, it is the obvious choice for the job.) Other uses of the metal include neutron technology and alloy manufacture. It is added to copper to increase its strength, and is used to produce non-sparking alloys for use in fire-sensitive situations such as oil refineries. As expected, the metal vapour is monomeric.

The ion Be^{2+} seems to be too highly polarizing to exist on its own in an ionic lattice, although it is stabilized when surrounded by negatively charged ions such as F^- or negative dipoles such as in water. Only four such binding groups (or 'ligands') can pack round the small, central beryllium ion, and they are almost always in a tetrahedral arrangement, which minimizes repulsion between them. Examples are the complex ions $Be(OH_2)_4^{2+}$, $Be(OH)_4^-$, and BeF_4^{2-}, and in each of these the four ligands contribute a total of eight electrons to the beryllium, which then has a complete octet ($2s^2 2p^6$) in its $n = 2$ shell.

The strong polarizing power of beryllium causes deformation of the bound water in the aquo-cation $Be(OH_2)_4^{2+}$, which releases protons in solution, unless acid is added (see p. 13). This hydrolysis produces mainly polynuclear hydroxy complexes such as $Be_3(OH)_3^{3+}$ in neutral or mildly acidic solutions. Dilute acids dissolve beryllium powder, forming the hydrated cation and hydrogen, although concentrated nitric acid gives the beryllium a passive coating of oxide. In alkaline solutions the main species is the simple anion $Be(OH)_4^{2-}$, which is also formed, together with hydrogen

The formation of a conduction band in beryllium [M]

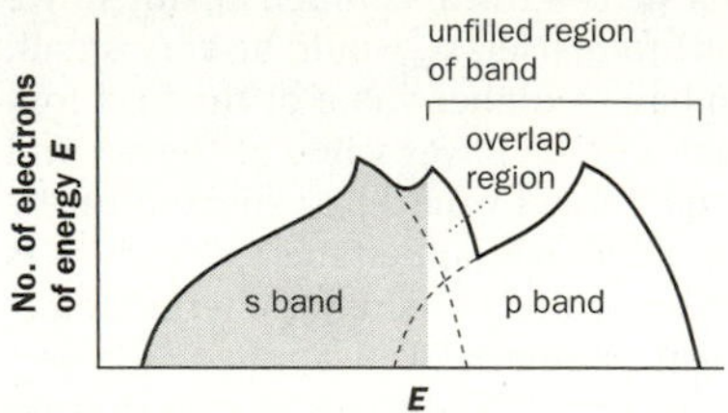

Metallic beryllium conducts electicity because electrons can move into the unfilled region

Some BeX_2 chain polymers [S]

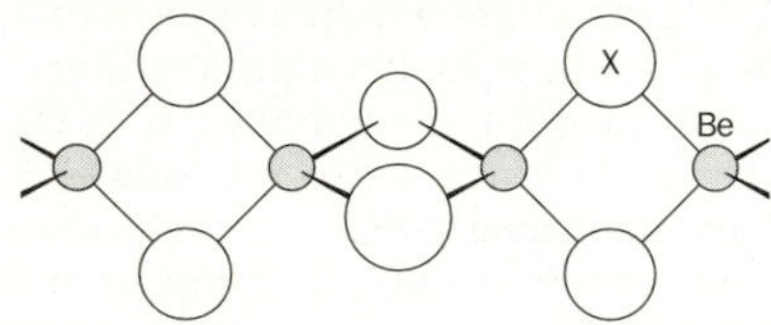

In the *halides* BeX_2, each Be–X–Be unit contains four electrons, an average of 1/2 from each beryllium, and 3 from the halogen atom (2 lone pairs on each halogen are non-bonding). There are two bonding orbitals (1/2 a 2s orbital from each Be and one p orbital from the halogen), which accommodate the four bridge electrons.

In the *hydride* BeH_2 and *dimethyl* $Be(CH_3)_2$, the bridging group also contributes one orbital, but only one electron, to give a 3-centre–2-electron bridging bond

'Basic beryllium acetate' [GE]

$Be_4O(OOCCH_3)_6$ with only one of the bidentate ligands $H_3C{-}CO_2^-$ shown in full

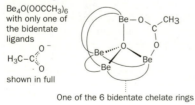

One of the 6 bidentate chelate rings

arranged tetrahedrally around the central oxygen atom. So the carboxylate group is bi*dentate*, since it forms a ring using two donor atoms, acting like teeth. (The term 'dentate' comes from the same Latin root as 'dentist'.) Each beryllium is joined to all the three others by a carboxylate group, which acts as an O–C–O bridge, forming a six-membered 'chelate' ring (chelate comes from the Greek for a crab's claw, and the 'ch' is pronounced hard as in 'orchestra'). Beryllium forms similar basic complexes with nitrate and carbonate ions, which also chelate through two oxygen atoms; the 'basic' carbonate $Be_4O(CO_3)_6^{6-}$ is, of course, a complex ion rather than an uncharged molecule.

Beryllium (like aluminium) forms a carbide that contains isolated carbon atoms and hydrolyses with water to give methane. In Be_2C the beryllium atoms are again in a tetrahedral environment.

Even beryllium, however, can, on occasion, be lured to sit in the centre of a planar ring, bound by four nitrogen atoms. With rigid ligands, such as phthalocyanine, which have several 'donor' atoms able to supply electron pairs, strong interaction between the metal ion and donors more than compensates for some increased electron–electron repulsion caused by changing from tetrahedral to square planar bonding.

Beryllium interacts strongly with some yet more complex organic molecules, as we know to our cost, for it is extremely toxic (particularly if inhaled). It is thought to interfere with phosphate metabolism, which involves enzymes activated by the magnesium ion. Maybe the smaller beryllium ion ousts magnesium by binding more strongly at the same site. Since beryllium does not occur widely or abundantly, there has been little biological impetus for the evolution of any beryllium-tolerant form of life.

- Very toxic: may replace Mg^{2+} in enzymes
- Compounds of special interest
  - halides $(BeX_2)_x$: covalent polymers, utilizing filled halogen orbitals
  - hydride $(BeH_2)_x$ and dimethyl $[Be(CH_3)_2]_x$: covalent polymers with 3-centre–2-electron bonds
  - complexes containing the tetrahedral OBe_4 group chelated by six bidentate O donors (e.g. carboxylates)
  - carbide Be_2C hydrolyses (like Al_4C_3) to methane

Uses

- For X-ray windows
- In nuclear technology

Nuclear features

- Natural Be almost isotopically pure ^9Be

B Boron

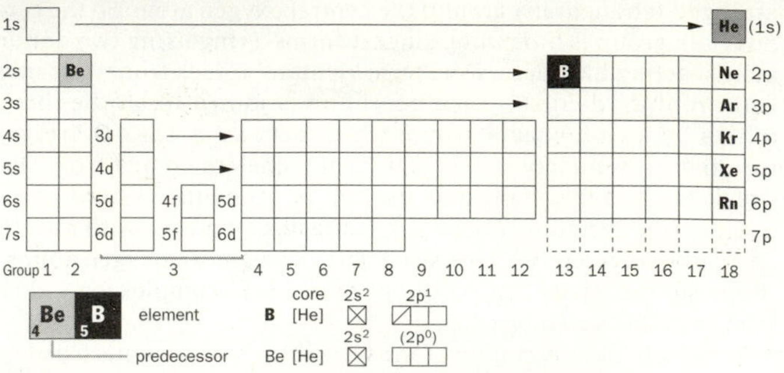

		Predecessor	Element
Name		**Beryllium**	**Boron**
Symbol, Z		Be, 4	B, 5
RAM		9.01218	10.81
Radius/pm:	**atomic**	113.3	83
	covalent	89	88
	ionic M^{3+}		23
Electronegativity (Pauling)		1.6	2.0
Melting point/K		1551±5	2573
Boiling point/K			3931
$\Delta H^{\ominus}_{fus}$ **/kJ mol^{-1}**		9.80	22.2
$\Delta H^{\ominus}_{vap}$ **/kJ mol^{-1}**		308.8	538.9
Density (at 293 K)/kg m^{-3}		1847.7	2340
Electrical conductivity/Ω^{-1}m^{-1}		25×10^{6}	5.55×10^{-5}
Ionization energy for removal of *j*th electron/kJ mol^{-1}	***j*=1**	899.4	800.6
	***j*=2**	1757.1	2427
	***j*=3**	14 848	3660
	***j*=4**		21 006
Electron affinity/kJ mol^{-1}		–18	26.7
Bond energy/kJ mol^{-1}	**E–E**		335
	E–H	226	381
	E–C		372
	E–O	523	523
	E–F	615	644
	E–Cl	293	444

Stable isotopes of boron

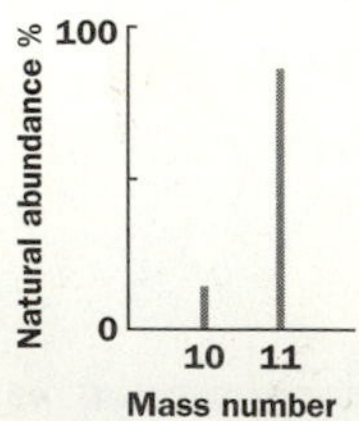

Since the s subshell is full at beryllium ($2s^2$), the outermost electron of the next element, boron, must go in to one of the three 2p orbitals. These are less penetrating than the 2s orbital (see p. 33), and so the outermost electron in boron is slightly easier to remove than that in beryllium, despite the higher nuclear charge.

So, boron, with its electronic configuration of $[He]2s^22p^1$, has three electrons in its four outer orbitals. How easy is it to predict its chemistry? A triply charged cation B^{3+} seems extremely unlikely, since much energy would be needed to remove all three electrons in the $n = 2$ shell, and the resulting ion would be so small and so highly polarizing that the tendency for covalent bond formation would be even higher than for beryllium. This effect would be lessened by increasing the size of the ion by surrounding it by large uncharged ligands, such as pyridine (py). Indeed, Bpy_4^{3+} is the only known cation of this type.

We may think of the anions BF_4^- and $B(OH)_4^-$ as being derived from B^{3+} and four negatively charged ligands, but this is a pure formality. It is more realistic to consider that the covalent molecules BF_3 and $B(OH)_3$ (see later) are acting as Lewis acids by adding a fluoride or a hydroxyl ion, and so filling the four outer orbitals of boron to give it the stable octet configuration $2s^2p^6$. Elementary bonding theory of this type might lead us to guess that boron would form dimeric halides (as does aluminium; see p. 145), where each boron forms single bonds with three halogen atoms and completes its octet by accepting a lone electron from a fourth bridging halogen atom. We might also suppose that it would not form a hydride, since hydrogen has no lone pair of electrons to donate in a similar way. Both guesses would be wrong: boron forms monomeric halides BX_3 and a dimeric hydride B_2H_6 (see p. 15), which is the first member of a vast realm of hydrides and their derivatives. Apart from the fact that the halides are Lewis acids, and complete their octet by accepting a lone pair from a Lewis base such as ammonia, the complete octet has little place in the chemistry of boron.

We have seen (p. 8) that we can visualize diborane as being held together by two 3-centre–2-electron bonds, and, as a result of its liberation from conventional electronic bookkeeping, boron forms a wide range of compounds, mostly of alarming complexity and apparently implausible bonding. Here, we shall give only the briefest indication of the varied chemistry of boron, since any attempt to do justice to its richness would be out of place in this introduction to atomic diversity.

Boron occurs naturally as mineral borates, which vary in isotopic composition, containing roughly 4:1 ratios of the two stable nucleides ^{11}B and ^{10}B: borates from Turkey being slightly higher in

The B^{3+} ion

can exist if surrounded by large ligands

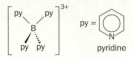

$$\left[B(py)_4 \right]^{3+} \qquad py = \text{pyridine}$$

Preparation of elemental boron

$$B_2O_3 + 3Mg \xrightarrow{\text{heat}} 2B + 3MgO$$

$$2BCl_3 + 3H_2 \xrightarrow{\text{heat}} 2B + 6HCl$$

$$2BI_3 \xrightarrow{\text{heat}} 3B + 3I_2$$

Hardness, on Moh scale

boron 11 diamond 10

The B_{12} icosohedron [M]

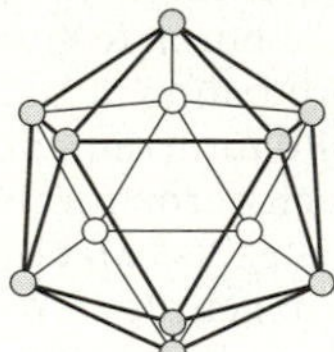

Each of its twenty faces is an equilateral triangle, and all sites are identical (but those marked ○ are not visible from the front)

^{10}B than those from California. So, the RAM of boron cannot be quoted with great precision.

The element is obtained by heating the oxide, B_2O_3, with magnesium; the chloride, BCl_3, with hydrogen; or the iodide, BI_3, on its own, and here the complications start. The different methods may produce different forms of boron that are fiendishly difficult to obtain pure or to investigate. They are very hard and high melting materials, but of low density and non-metallic: the B^{3+} ion, if it existed, would be too highly polarizing to allow delocalization of three electrons, even though a bond derived from one s and three p orbitals, being only partially occupied, might, in principle, give rise to electronic mobility. The simplest structural unit is an almost regular icosahedron in which twelve boron atoms are arranged in a cluster with twenty nearly equilateral triangles as faces. The B_{12} clusters have other clusters, which may or may not be identical, associated with them, and we can envisage the conglomerates as held together by a mixture of single covalent (two-centre) bonds and 3-centre–2-electrons bonds. The cavities in the structure account for both the low density and the difficulty of purification.

The B_{12} cluster is also the basis of 'boron carbide', $B_{13}C_2$, in which the icosahedra are linked by C–B–C groups. It, too, is hard and is used as an abrasive. Boron combines with a large number of metals to form borides in which the ratio of metal to boron ranges from 5:1 to 1:66, and which may contain more than one type of metal. In the metal-rich borides, single boron atoms may be accommodated in the metal structure, while those with a higher proportion of boron may contain boron atoms linked in pairs, chains (simple, branched, or double), nets (planar or puckered), or clusters (B_6 octahedra or the familiar B_{12} icosahedra). Over two hundred metal borides are known and many of these are used as toughening fibres in composite materials.

We have seen (p. 15) how in diborane two hydrogen atoms can bridge two boron atoms by means of two 3-centre–2-electron bonds, by filling only the lowest energy ('bonding') molecular orbitals. Higher boranes, which all contain similar hydrogen bridges, are also known: there are over 25 neutral ones, and an even greater number of borane anions. They are unstable, reactive, toxic substances, with melting and boiling points similar to hydrocarbons of the same molecular mass. Boranes are named by classical prefixes that classify their shape as cages, birds' nests, spiders, nets, or composites, the last category being two or more of the others joined together in, of course, a great variety of ways. The complexity by no means ends here, because a B–H group in a borane can be replaced by one of a number of other groups. The

Diborane B_2H_6

One of four B–H σ bonds (bond energy 381 kJ mol^{-1})

Two 3-centre–2-electron B–H–B 'banana' bonds (bond energy 439 kJ mol^{-1})

Some complex boranes SAL

○ boron ○ hydrogen

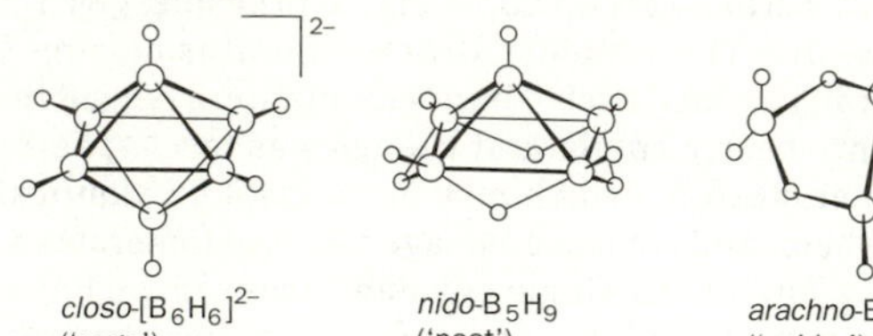

closo-$[B_6H_6]^{2-}$ ('cage') *nido*-B_5H_9 ('nest') *arachno*-B_4H_{10} ('spider')

Some 'cage' carboranes SAL

● carbon ○ boron ○ hydrogen

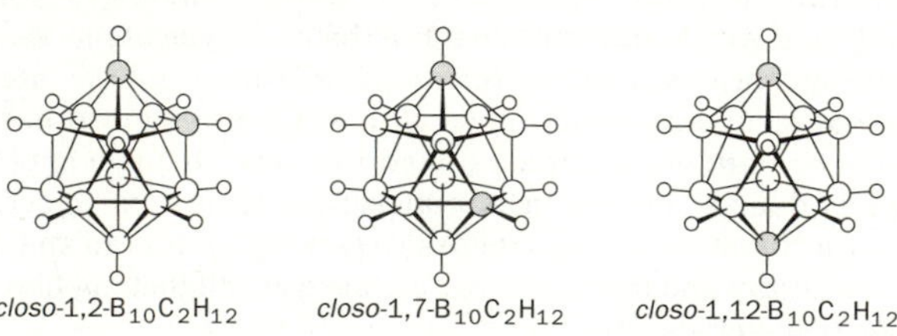

closo-1,2-$B_{10}C_2H_{12}$ *closo*-1,7-$B_{10}C_2H_{12}$ *closo*-1,12-$B_{10}C_2H_{12}$

Some groups that replace B–H in boranes

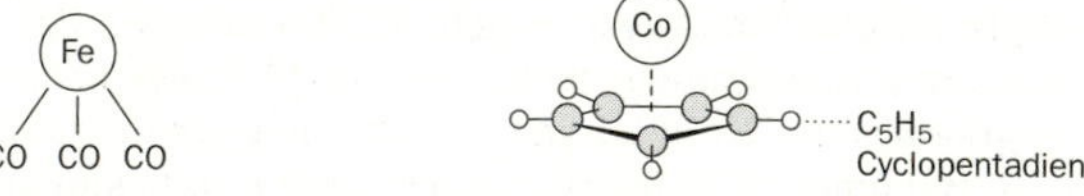

On Fe, there are 14 electrons: eight from the metal atom ($3d^64s^2$) and two from each CO molecule.

On Co, there are 14 electrons: nine from the metal atom ($3d^74s^2$) and five from the ligand.

boron atom in such a group has four electrons (three from itself and one from the hydrogen), which is not only four more than the noble gas helium ($1s^2$) but also four less than the next noble gas, neon ($2s^22p^6$); and it can often be replaced by another group that is also four electrons short of an inert gas configuration, provided that the orbitals are of the same shape and symmetry and of similar energy. Suitable groups are the carbon atom and the ions N^+ and P^+. Carbon, in particular, can replace a B–H, giving a carborane. Starting with a single borane, we can vary both the number of boron atoms replaced by carbon, and the positions at which the substitution occurs. So a huge range of carboranes is possible.

This type of substitution is not, however, limited to single atoms or ions of the lighter elements. Many complexes of transition metals in low oxidation states are four electrons short of a very favourable structure, such as that of the noble gas krypton ($3d^{10}4p^64s^2$), with 18 outer electrons. Examples of such 14-electron groups are the fragments $Fe(CO)_3$ (the iron atom has eight outer electrons and each of the carbon monoxide molecules contributes two) and $Co(C_5H_5)$ (the cobalt atom has nine outer electrons, while the cyclopentadienyl ring contributes five). Since the examples given are only two of a very large number of fragments with outer orbitals that can replace one or more B–H groups, again in a choice of positions, there is immense scope for the formation of these metalloboranes.

Compounds between boron and oxygen are yet another widely varied group. A boron atom may be surrounded by three oxygen atoms in a plane, or by four in a tetrahedron, and one or the other, or both, of these two borate units may be linked to form chains or rings, with or without H^+ ions to balance negative charges on oxygen atoms. It is no wonder that when heated these substances often give a glassy form of B_2O_3. Like glass itself (see p. 153), this is an amorphous solid with little more internal order than a liquid in which movement of molecules from place to place has been much restricted by supercooling. When it is heated, it does not melt at a fixed temperature, but gradually softens until it becomes fluid.

Natural borax, $Na_2B_4O_7.10H_2O$, contains the $B_4O_5(OH)_4^{2-}$ ion. Its use in the ancient world in making glasses, glazes, and enamels extends today to the large-scale manufacture of fibreglass, glass insulation, and hard heat-resistant borosilicate glasses. Not surprisingly, in acidic solution boron exists as the fully protonated boric acid, $B(OH)_3$ and in alkaline solution as the $B(OH)_4^-$ ion. At intermediate values of pH, we find the 'borax' ion together with other anions containing three or five boron atoms.

Borate units [s]

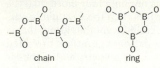

chain

ring

The 'borax' ion [s], $B_4O_5(OH)_4^{4-}$

The perborate ion, $B_2O_2(OH)_4^{2-}$

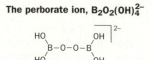

Boric acid $B(OH)_3$ [s]

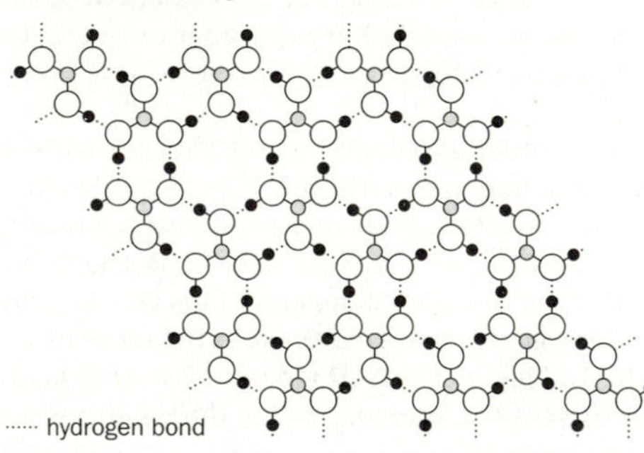

A further variation occurs in the perborates, where two boron atoms are joined by an –O–O– group instead of by sharing one or more oxygen atoms. The perborate $Na_2B_2(O_2)(OH)_4$ is used as a mild bleaching agent in laundry detergents since it decomposes in water to give hydrogen peroxide.

The monomeric trihalides of boron might seem to be one of the few types of boron compound that are not immensely complicated; but even so, there are some questions to be answered. Can we explain why they are monomeric? (Maybe there would be excessive repulsion between electrons in a dimeric bridged system involving halogen atoms together with so small an atom.) Any explanation must take into account two unexpected observations: that the boron to halogen bonds are shorter than expected, particularly in the fluoride; and that although fluorine is the most electronegative halogen, pulling the electron clouds away from the boron, ammonia adds to BI_3 more easily than to BF_3.

It is thought that the three boron to halogen σ bonds are reinforced by π bonding. Of the one s and three p orbitals in boron, there is one p orbital that is not used in σ bonding, and is perpendicular to the plane of the molecule. Parallel to this empty orbital are filled p orbitals on the three halogen atoms. These p orbitals of boron and the halogen can overlap sideways on, above and below the plane of the molecule, forming a π orbital, which is delocalized over the whole molecule. The π orbital can accept one pair of electrons, strengthening and shortening the boron to halogen bond. The π bond both stabilizes the monomer relative to the (non-existent) dimer, B_2X_6, and somewhat opposes the formation of an adduct with ammonia. The formation of a π bond by sideways overlap is much more sensitive to distance than is the head-on encounter in a σ bond; so π bonds will be stronger between smaller atoms and will be more important in BF_3 than in BI_3. It is likely that the bonds in all the boron trihalides have some double bond character, which is greatest in the fluoride (see p. 58). Boron also forms some unstable lower halides, B_2X_4, in which a halogen atom in BX_3 is replaced by a $-BX_2$ group; and further short chains may be built up in the same way.

There is yet another most interesting group of boron compounds: the nitrides and their derivatives. The pair of atoms B–N has the same number of electrons as a singly bonded pair of carbon atoms; and it also has the same bond length. Moreover, the electronegativity of carbon is midway between that of boron and nitrogen. The most stable form of boron nitride $(BN)_x$, like that of elemental carbon, is a layered structure containing planar sheets of hexagonal rings, B_3N_3. But boron nitride differs from graphite in

Boron trifluoride, BF_3 MM

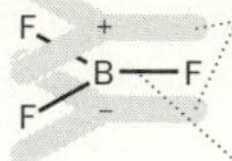

One delocalized π bond above and one below the plane of the molecule

One of three B–F σ bonds

Pauling electronegativities

boron 2.04

carbon 2.55

nitrogen 3.04

Boron nitride $(BN)_x$ S

showing vertical B–N–B stacking

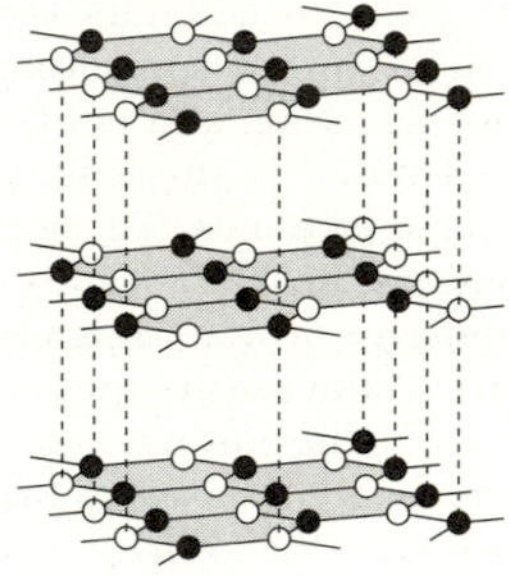

Borazine $B_3N_3H_6$

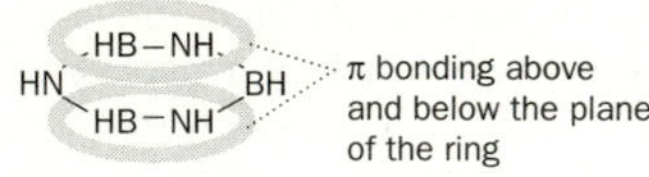

that the hexagons are stacked exactly one above the other, so that each boron has a nitrogen above and below it, and vice versa. Weak interaction between the boron and nitrogen in different planes prevents them from sliding over each other, and so $(BN)_x$ cannot be used, as oxidized graphite is, for lubrication. At very high pressures, the most stable form of $(BN)_x$ is borazon, an extremely hard substance with the same structure as diamond; but very high temperatures are again needed if the hexagonal form is to be converted to borazon at a practicable rate. The similarity between B–N and C–C even extends to the hydrogenated compound borazine, $B_3N_3H_6$, a liquid in which the molecules are planar rings of equally spaced atoms and which resembles benzene in its physical properties. But since boron is more electronegative than nitrogen it attracts negatively charged species, with the result that borazine is more reactive than benzene.

The lively chemistry of boron is indeed one of the reasons why the element is treated so briefly in books of elementary chemistry but at such length in advanced treatises. Apart from the metal borides and the borates, many boron compounds are rather unstable and have to be coaxed into existence. With only three outer electrons, and those under the very firm control of the nucleus, boron cannot release them into a delocalized metallic band, nor form a simple triply charged cation that could exist in an ionic lattice, nor contribute enough electrons for the formation of the four single bonds needed to fill its $2sp^3$ orbitals. Instead, the lowest energy options involve electron pair donation or the formation of 3-centre–2-electorn bonds. Although departure from conventional bonding greatly increases the range of possible compounds, the 3-centre–2-electron bonds are not very stable and are prone to chemical attack. So, it is no surprise that although impure boron was isolated nearly two centuries ago, and the first boron hydrides were made as early as 1912, much of boron chemistry is less than one-third of a century old, and it is still growing excitingly fast.

For summary see p. 61.

Summary

Electronic configuration

- $[He]2s^22p^1$

Element

- Hard, high melting, non-metallic
- Various forms, containing B_{12} and other B_x clusters, with some 3c–2e bonds

Occurrence

- As borates

Extraction

- By reduction of B_2O_3 (with Mg), BCl_3 (with H_2), or BI_3 (by heat)
- Very difficult to purify

Chemical behaviour

- B^{3+} does not exist (would be very small and highly polarizing): formal derivates Bpy_4^{3+}, BF_4^-, $B(OH)_4^-$
- Covalent (3 σ bonds do not complete octet): halides, BX_3, stabilized by
  - π donation from halogen atom (greatest for F)
  - bonding with Lewis base, e.g. ammonia to complete octet (greatest for I)
- Other compounds: extremely complicated, owing to
  - 3c–2e bonds rather than octet completion, e.g. hydrides (B_2H_6 and many higher) and derivatives with BH replaced by C (in carboranes) or [core-4] fragments, e.g. $Fe(CO)_3$
  - mixture of three and four coordination, e.g. borates, often polymerized and glassy [e.g. $B_4O_5(OH)_2^-$ in borax]
  - many boron cluster structures, e.g. metal borides, have atoms, chains, layers, clusters of B accommodated in the metal: boron carbide $B_{13}C_2$ has B_{12} icosahedra and C–B–C links
  - 'aromatic-type' π delocalization, e.g. nitrides. $(BN)_x$ has graphite and diamond structures; $B_3N_3H_6$ similar to benzene

Uses

- Metal borides to toughen metals
- Borates in glass manufacture
- Perborates as bleaches

Nuclear features

- Natural isotopes ^{11}B and ^{10}B about 4:1 but vary with location, so RAM imprecise

C Carbon

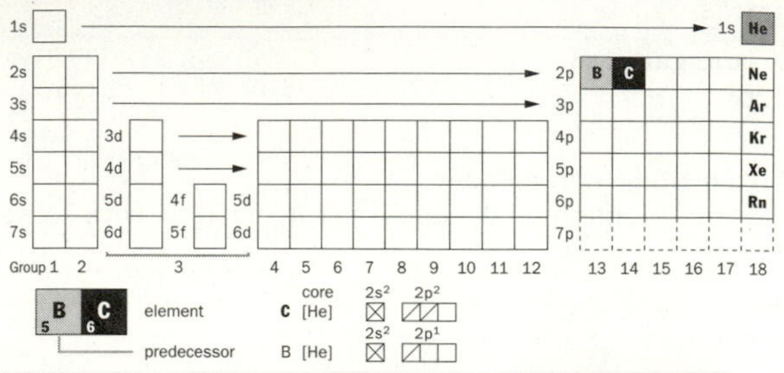

		Predecessor	Element
Name		**Boron**	**Carbon**
Symbol, *Z*		B, 5	C, 6
RAM		10.81	12.011
Radius/pm: covalent		88	77 (C–C)
			67 (C = C)
			60 (C ≡ C)
Electronegativity (Pauling)		2.0	2.6
Melting point/K		2573	~ 3820 (diamond)
$\Delta H^{\ominus}_{fus}$/kJ mol^{-1}		22.2	150.0
$\Delta H^{\ominus}_{vap}$/kJ mol^{-1}		538.9	710.9
Density (at 293 K)/kg m^{-3}		2340	3513 (diamond)
			2260 (graphite)
Electrical conductivity/Ω^{-1}m^{-1}		55.55×10^{-5}	1×10^{-11} (diamond)
			7.27×10^{4} (graphite)
Ionization energy for removal of *j*th electron/kJ mol^{-1}	***j* = 1**	800.6	1086.2
	***j* = 2**	2427	2352
	***j* = 3**	3660	4620
	***j* = 4**	21 006	6222
	***j* = 5**		37 827
Electron affinity/kJ mol^{-1}		26.7	121.9
Bond energy/kJ mol^{-1}	**E–E**	335	348
	E–H	381	411
	E–C	372	348
	E–N		305
	E–O	523	358
	E–Cl	447	325
	C = C		614
	C = N		615
	C = O		745
	C ≡ C		839
	C ≡ N		891
	C ≡ O		1074

Carbon, [He]$2s^22p^2$, is even less likely than its predecessor, boron, to form simple ionic compounds. A great deal of energy would be needed to remove the four outer electrons; and the small quadruply charged C^{4+} cation would be so highly polarizing that it would distort orbitals of any neighbouring anion to such an extent that the bonding would be far more covalent than ionic. The chemistry of carbon is indeed almost entirely based on covalent bonds formed by orbital overlap; but it is, none the less, of breath-taking variety and complexity, which owes as much to biological evolution as to laboratory ingenuity. Almost all the carbon atoms in non-living matter, be they in chalk cliffs, coal seams, paper, food, oilfields, pharmaceuticals, natural gas, synthetic polymers, limestone mountains, or new organic compounds, have at one time been part of a green plant, and before that of a molecule of atmospheric carbon dioxide. And this is surely part of the romantic appeal that carbon exerts over many chemists.

Naturally occurring carbon is mainly ^{12}C, together with about 1% of the non-radioactive isotope ^{13}C. Living plants, however, contain a trace (about 10^{-10} %) of the radioactive ^{14}C, which is continually formed by cosmic-ray bombardment in the upper atmosphere, and absorbed as carbon dioxide during photosynthesis. When the plant dies, the ^{14}C is no longer taken in and the (very low) radioactivity of the wood or other material falls off exponentially, with a half-life of 5730 years, enabling archaeologists to estimate the age of the material. (Corrections for the small variations of ^{14}C in the atmosphere can be made by combining radioactive measurements with data from growth rings in trees.)

The chemical diversity of carbon owes much, but by no means all, to the single covalent bond. It does not take much energy to promote one of the paired 2s electrons in the $2s^22p^2$ ground state of a carbon atom up to the empty 2p level, and it needs only a little

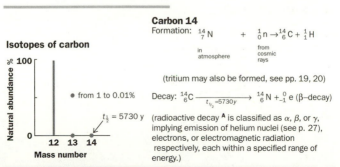

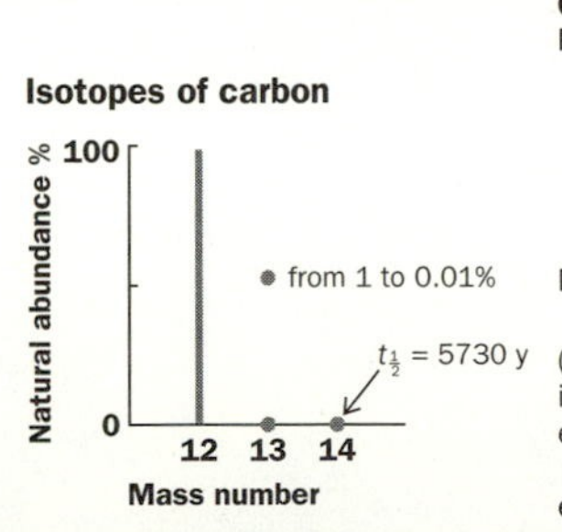

Carbon 14

(tritium may also be formed, see pp. 19, 20)

Decay: $^{14}_{6}C \xrightarrow[t_{1/2}=5730y]{} ^{14}_{6}N + ^{0}_{-1}e$ (β–decay)

(radioactive decay **A** is classified as α, β, or γ, implying emission of helium nuclei (see p. 27), electrons, or electromagnetic radiation respectively, each within a specified range of energy.)

Combustion of methane

$$CH_4(g) + 2O_2(g) \rightarrow CO_2(g) + 2H_2O(l)$$
$$\Delta H^{\ominus} = -890 \text{ kJ mol}^{-1}$$

Although the reaction is so exothermic, methane does not catch fire spontaneously when mixed with air. It will burn only if ignited.

Some hydrocarbons with C–C single bonds

CH_4
methane
(parent)

iso-C_4H_{10}
iso-butane

n-C_4H_{10}
butane

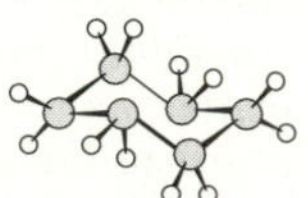

cyclo-C_6H_{12}
cyclo-hexane

The strongest X–X bonds [D]

Bond	H–H	C–C	C–C	S–S	Cl–Cl
Found in	$H_2(g)$	diamond	C_2H_6	$S_8(g)$	$Cl_2(g)$
Enthalpy/kJ mol^{-1}	436	356	331	266	242

Some derivatives of methane, CH_4

Replacement of –H by:

–CH_3 gives a higher paraffin: C_2H_6, ethane
(with successive substitution giving straight-chain or branched analogues, e.g. C_4H_{10}

–NH_2 gives a primary amine: CH_3NH_2, methylamine

–OH gives a primary alcohol: CH_3OH, methanol

X gives a haloalkane, e.g. CH_3Cl, chloromethane

Some derivatives of propane, $CH_3CH_2CH_3$

Replacement of the central –CH_2-group by:

–NH– gives a secondary amine: CH_3–NH–CH_3

–O– gives an ether: CH_3–O–CH_3

Diamond

an infinite 3D array of tetrahedrally linked carbon atoms

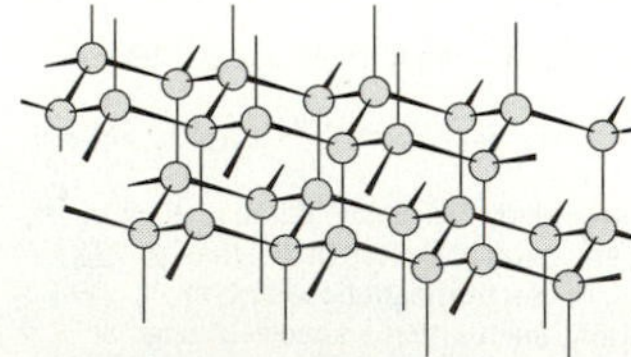

more energy to promote the most stable configuration $2s^1 2p^3$ (all electron spins aligned) into a more excited state in which the spins of the electrons are random. If this excited species forms a single bond with each of four neighbouring groups, as in methane (see p. 6), the energy given out greatly exceeds that needed for excitation and so there is a net gain in stability. Each bound group contributes one orbital and one electron. In a molecule CX_4 such as methane, the electrons are paired in the four tetrahedral bonding orbitals, while the four antibonding orbitals are empty.

Many carbon compounds are unaffected by air and unhydrolysed by water, but their apparent stability is a consequence of kinetic sluggishness. All are thermodynamically unstable in air, relative to the carbon dioxide and water that would be formed if they burned; but the energy needed to initiate combustion is too high for it to happen spontaneously. A hotspot is needed for ignition, partly because of the rather large energy gap between the highest occupied and lowest unoccupied orbitals (see p. 6). This in turn arises because carbon is the only element (except hydrogen) that has the same number of valence orbitals as electrons; it has neither a vacant orbital to accommodate electrons from an attacking group, nor an electron pair able to attack another species. So, although reactions such as combustion are energetically favourable, they need a large initiation energy.

The hydrogen atoms in methane may be replaced by any group with a half-filled orbital if this is of a symmetry suitable for σ bond formation, and of a size and energy compatible with the molecular orbitals provided by carbon. Substitution by successive methyl (CH_3) radicals gives hydrocarbon chains (straight or branched) or puckered rings in which the repeat unit is $-CH_2-$. After H–H, the C–C bond is one of the strongest known single bonds between two like atoms, and this favours chain formation ('catenation') by carbon.

All four hydrogen atoms in methane may be replaced by carbon atoms, each of them surrounded tetrahedrally by other carbon atoms, themselves similarly placed, and so on, to give a 3D array that is transparent, highly refractive, and very hard: the largest sample yet found (the Cullinan diamond) contained over 3×10^{25} atoms (over 50 moles) of carbon. Hydrogen atoms in methane and heavier hydrocarbons may also be replaced by $-NH_2$ or –OH groups, or by halogen atoms, to give primary amines, alcohols, or halo derivatives, while a $-CH_2-$ link may be replaced by –NH– or –O– to give secondary amines or ethers. These compounds too, are thermodynamically unstable in air, and also in water; but they are often

Ethene: the carbon–carbon link

The C_2H_4 molecule has six bonding orbitals, six antibonding orbitals and 12 electrons.

4 bonding orbitals and 8 electrons form 4 C–H σ bonds.

1 bonding orbital and 2 electrons form 1 C–C σ bond.

1 bonding orbital (formed by sideways overlap of the p_y orbitals of the C atoms) accomodates the remaining 2 electrons to form a π bond above and below the C–C σ bond.

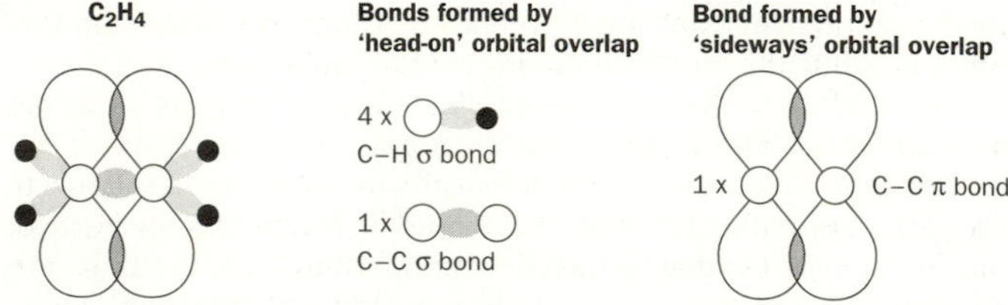

The 'double bond' written as C=C contains one σ bond and one π bond.

Ethyne (acetylene): the carbon–carbon link

The 'triple bond' written as C≡C contains

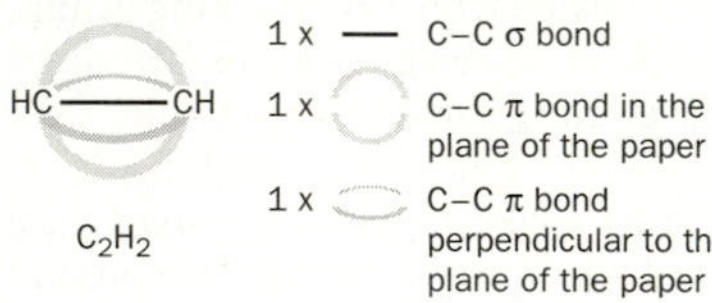

Some delocalized π bonds

H₂C=… butadiene: H, H, C–CH–CH–C, H, H — C_4H_4 butadiene

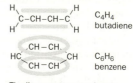

HC, CH–CH, CH, CH–CH — C_6H_6 benzene

The lines represent σ bonds, and the shaded areas represent conjugated π systems

Graphite [s]

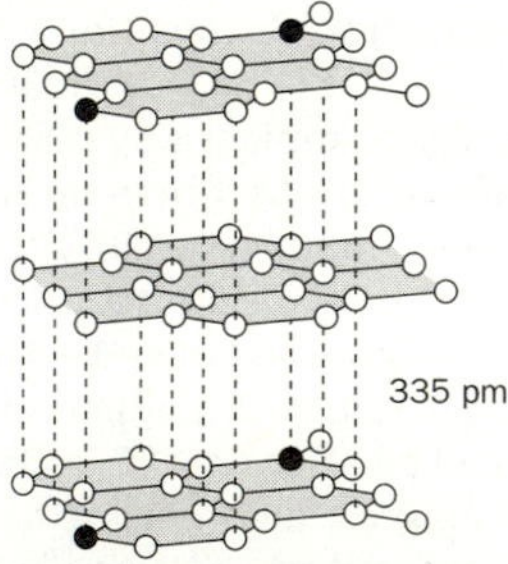

Alternate layers (see marked atoms) are directly above each other

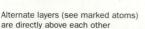

–C–C–C–C–C–C–C–C–C– π σ π

vdW

C–C–C–C–C–C–C–C–C π σ π

vdW

–C–C–C–C–C–C–C–C–C– π σ π

Within the layers: –C< σ bonding

Above and below the layers: π bonding

Between the layers: van der Waals (vdW) forces

unreactive through kinetic sluggishness. Tetrachloromethane, for example, does not react with water, although hydrolysis is energetically favourable. (The energetics of silicon tetrachloride hydrolysis are very similar, but reaction is instantaneous, see p. 151).

The variety that carbon shows in its singly bonded compounds, although remarkable, does scant justice to the diversity of its chemistry. Carbon has other tricks up its sleeve. The group CH_2, for example, has six valence orbitals and six outer electrons. Two orbitals and four electrons are involved in the two C–H σ bonds, leaving four orbitals and two electrons that can interact with other groups. A pair of CH_2 groups for example, can join together by forming a double bond between the carbon atoms, to give ethene, C_2H_4. There is one bonding σ orbital between the two nuclei, and one bonding π orbital above and below the σ bonding orbital. The analogous σ and π antibonding orbitals are empty. Since π bonding may be envisaged as a sideways interaction, it is most likely to occur if the atoms are small and can get sufficiently near to their neighbours to allow appreciable overlap.

Ethyne (or acetylene) C_2H_2 consists of two CH groups joined by one σ bond and two π bonds, perpendicular to each other. As π bonds, with their far less efficient overlap, are weaker than σ bonds, ethene is more reactive than its saturated analogue ethane, C_2H_6, and ethyne yet more so.

If we remove one hydrogen atom from each of two ethene molecules, and join up the two $H_2C{=}CH^{\cdot}$ fragments, we get butadiene, sometimes written as $H_2C{=}CH{-}CH{=}CH_2$. More often, however, it is represented by $H_2C{\overset{\ldots}{-}}CH{\overset{\ldots}{-}}CH{\overset{\ldots}{-}}CH_2$ to emphasize that there are not two localized π bonds between the two end pairs of carbon atoms. Instead, the p orbitals on all four carbon atoms interact to give a composite π orbital which can accommodate four electron pairs extending the whole length of the chain and keeping the molecule linear. In the cyclic molecule of benzene, C_6H_6, the π system forms a ring embracing all six carbon atoms. The C–C distances are all the same, and the π overlap above and below the σ bonds holds the ring planar. Many long-chain and cyclic organic molecules (some containing several rings) have delocalized (or 'conjugated') π systems of this type.

A familiar conjugated array of carbon atoms is graphite, which occurs in nature. It is the most stable form ('allotrope') of the element and consists of infinite layers of C_6 hexagons. Each carbon atom is joined to three others by σ bonds within the plane and by conjugated π bonds above and below it. The delocalized π electrons give graphite semi metallic properties: it is black, shiny,

Graphite fluoride [S]

Puckered rings of $(CF)_n$

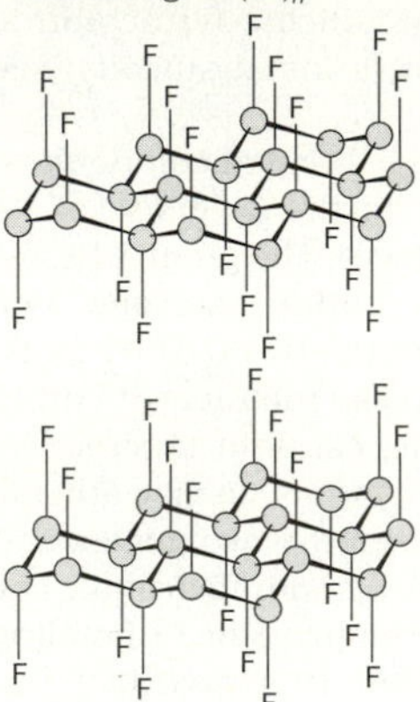

Soot [AEC]

A possible structure for a speck of soot

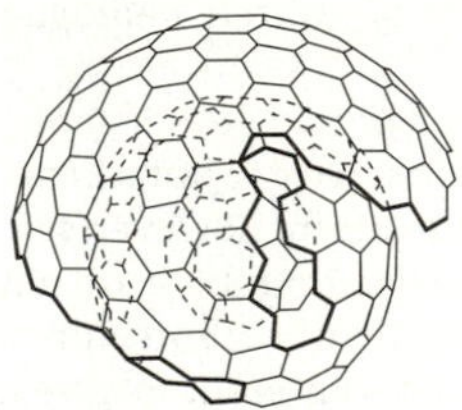

Buckminsterfullerine, C_{60} [AB]

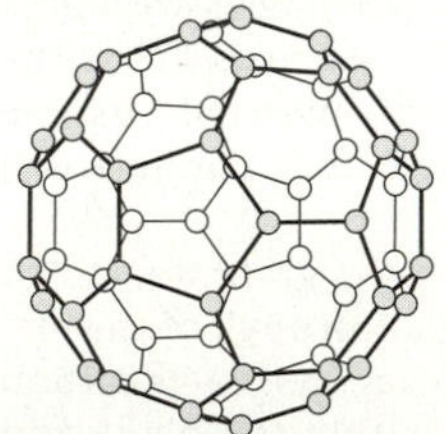

Each carbon atom is part of one of the 12 pentagons and of two of the twenty hexagons

and a weak electronic conductor used to make arcs and electrodes. Since the layers are rather widely spaced, reactants can get in between them to form intercalation compounds. For example, the π system is attacked by strong oxidizing agents to form graphite oxide, which is a valuable lubricating agent as the layers are held together less strongly than in graphite and can slip over each other. Cleavage between layers also allows graphite to be rubbed off on to flat surfaces. It is used in pencil 'leads' (see p. 347) and its name is related to a host of English words ending in '-graph' or '-graphy' and implying drawing, writing, or recording. Graphite can also be fluorinated to give $(CF)_n$ in which fluorine atoms are bound by σ bonding above and below the carbon hexagons. As the π bonds have been broken, the hexagons are now puckered and no metallic character remains. Graphite can combine with potassium, which transfers some electronic charge to the π bonds and greatly increases the electrical conductivity. Crystalline graphite normally has alternating planes exactly on top of each other, although in a rarer form every third plane is 'in phase'; in the compound C_8K, however, all the graphite planes are in phase, but are much further apart. Graphite also reacts with bromine and some metal halides, which withdraw some charge from the π bonds and again increase the space between the layers.

Graphite is slightly less dense than diamond and can be forced into a tetrahedral structure under extreme conditions. As with boron nitride (see p. 58), very high pressures are needed to make the change energetically possible, while very high temperatures must be used in order to excite the graphite enough to undergo the transformation. Despite the expense, small diamonds (of up to 0.2 g) are made artificially for use in tools for cutting and grinding. In several other types of carbon, the hexagonal graphitic structure persists. Very strong fibres of hexagons may be made by carbonization of regular polymers and may be spun, woven, or used in composite materials. Soot consists of irregular rolls of mainly hexagonal fragments. The unpaired electrons around the edges absorb all visible light and readily take part in bonding. Soot has been a well-known black pigment since ancient times, while 'active charcoal' is used for adsorbing gases and for catalysing a variety of reactions.

If carbon vapour condenses at high temperature and more slowly than is possible in a flame, it may form one of a number of hollow, nearly spherical structures that are more regular and stable than a haphazard spiral roll. The commonest of these is buckminsterfullerine, C_{60}, first made in 1983 but not characterised until 1990. The structure resembles a soccer ball, stitched from 20 hexagons and 12 pentagons; each hexagon touches three other hexagons and

Carbon dioxide

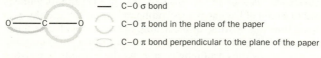

O——C——O

— C–O σ bond

C–O π bond in the plane of the paper

C–O π bond perpendicular to the plane of the paper

The group >C = O in organic compounds

$R_1R_2C=O$

R_1	R_2	Compound(s)
H	H	Formaldehyde
R	H	Aldehydes
R	R′	Ketones
R	NH_2	Amides
R	OH	Carboxylic acids

The carbonate ion, M CO_3^{2-}

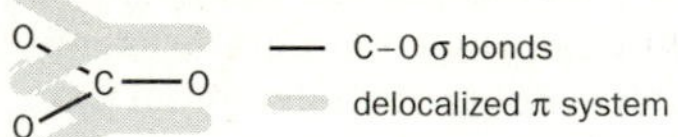

three pentagons, while each pentagon is surrounded by five hexagons. All the carbon atoms are in equivalent positions, connecting two hexagons with one pentagon; and no two pentagons are in contact. The other fullerines, or 'buckyballs', also have hollow molecules but they are less regular spheroids. Those smaller than C_{60} (with 32, 44, 50, or 58 atoms) have a higher ratio of pentagons, some adjacent to each other, and are slightly indented; the larger ones (with 70, 240, 540, or 960 atoms) have a lower ratio of pentagons and have small peaks on their surface. Fullerines seem to be chemically stable, doubtless because they have no 'loose edges' but a continuous slightly curved π band. But potassium atoms can slip into the spaces between the spheres in a fullerine crystal in much the same way as they can intercalate between the layers in graphite (see p. 69).

The formation of carbon chains can be used to produce a wide range of man-made polymers, or 'plastics'. Many have the skeleton $-CHR-CH_2-$ formed by polymerization of ethene derivatives, $H(R)C{=}CH_2$, using a suitable catalyst (see p. 147), often followed by some linking of neighbouring chains to give added strength and rigidity. By careful choice of both R and the degree of cross-linking, we can make polymers with a wide variety of properties. Examples are clear, flexible polythene (R = H) for wrapping; tougher polypropylene ($R = CH_3$) for ropes; fairly rigid polystyrene ($R = C_6H_5$) for household utensils; acrylic (R = CN) for wool-like textiles, and PVC (R = Cl) for clear coatings. The backbone, too, can be varied. In synthetic rubbers, chains contain double bonds, which make the polymer elastic but liable to chemical attack. (Natural rubber has more double bonds and is more elastic but even more perishable.)

Carbon can, of course, form multiple bonds not only with itself but also with nitrogen, and, in particular, with oxygen, as in the linear molecules of carbon dioxide O=C=O and tricarbon dioxide, O=C=C=C=O, where each σ bond is reinforced by one π bond*. The σ and π bonded >C=O group is a component of the functional groups of many organic compounds, such as aldehydes (RHCO), ketones (RR^1CO), amides ($RCONH_2$), acids (RCOOH), and esters (RCOOR). In the planar, triangular carbonate ion CO_3^{2-}, three C–O σ

* The role of CO_2 as the ultimate building block for biological storage is made possible by the 'moderately high' strength of the C=O bond. The molecule is stable eno ugh not to be destroyed by living matter, but sufficiently reactive to be able to take part in photosynthesis: it is as if the π-overlap conferred both durability and dormant reactivity, but with no tendency to interfere toxically in any way.

MO diagram for carbon monoxide, CO

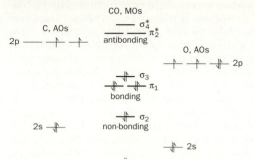

CO, MOs

C, AOs

σ_4^*

π_2^*

2p

antibonding

O, AOs

2p

σ_3

π_1

bonding

σ_2

2s

non-bonding

2s

σ_1

non-bonding

orbitals: 2 x π_1 and σ_3 form the triple bond
2 x π_2^* and σ_4^* (all empty) are antibonding

'Ellingham plot'

showing variation of oxide stability with temperature

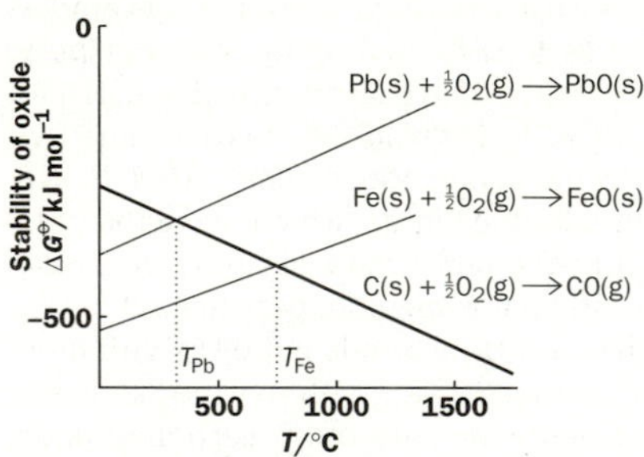

At temperatures above T_M, the products of the reaction

$MO(s) + C(s) \longrightarrow M(s) + CO(g)$

become thermodynamically more stable than the reactants, and the oxide can, in principle, be reduced to the metal with carbon

Calcium carbide, CaC_2 SAL

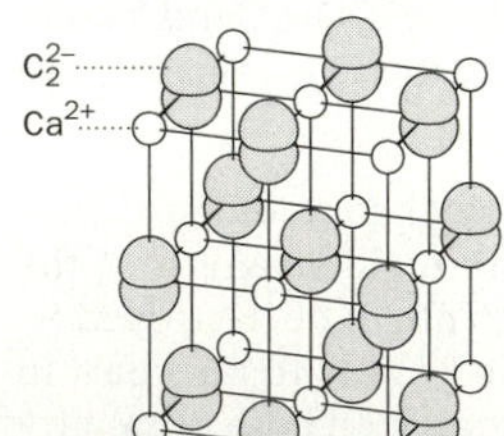

The lattice is like that of NaCl, but elongated to accommodate the non-spherical C_2^{2-} ions

bonds are strengthened by an electron pair occupying a π orbital that is delocalized over the central carbon and all three oxygen atoms (compare boron trifluoride, p. 58). The carbonate group can be inserted into the backbone of polymer chains, to give polycarbonates which contain –O–C–O– units as well as hydrocarbon ones. They resemble unbreakable glass and have many uses such as roofing, motor cycle helmets, lenses, marine propellers, and babies' bottles. A triple bond, similar to that in ethyne, C_2H_2, is found in carbon monoxide C≡O and in the cyanide ion $C{\equiv}N^-$, which is isoelectronic with it. The strong C≡O bond, and the increasing stability of CO with temperature, allow many metals (such as iron) to be extracted from their oxide ores with very hot carbon (usually coke).

Carbon also shows great diversity in the way in which it interacts with metal atoms and ions. The binary compounds of many electropositive metals are ionic and are often thought to contain the anion $(C{\equiv}C)^{2-}$. They give ethyne if treated with water: the calcium salt CaC_2 was used to fuel the acetylene lamps of early bicycles and cars, and more recently of fishing boats. Beryllium and aluminium carbides, Be_2C and Al_4C_3, however, react with water to give methane, and contain isolated carbon anions (but probably not carrying a full charge of –4). Many other metals accommodate carbon atoms in spaces within the metal lattice. A metal may form several such interstitial carbides, of varying composition and structure. As these are harder than the parent metal, they are of great technological importance: it is unlikely that the age of iron would ever have persisted (from the beginning of the Iron Age in about 1000 BC to the present day) without that $1 \pm \frac{1}{2}$ per cent of carbon that converts soft, brittle iron into hard, tough steel.

A vast and expanding branch of chemistry deals with organometallic substances in which a carbon compound interacts with a metal atom, often with a luxurious variety of bonding. One such compound, $Ni(CO)_4$, was made as early as 1890. Each carbon atom supplies two electrons and forms a σ bond with the nickel atom, making use of the low lying empty orbitals on the metal (see p. 74); simultaneously, some electron charge is transferred from the nickel atom to the low lying, empty π antibonding orbitals of the C≡O group. So the carbon to metal bond is formed at the expense of the carbon to oxygen bond (which is certainly strong enough to survive some seepage of charge). A wide variety of metal carbonyls have been made, some containing several metal atoms bridged by CO groups (see p. 74).

Alkyl groups, such as $-CH_3$ and $-C_2H_5$, can bind to metal atoms, sometimes seeming to form 'bent' multicentre bonds, as with

Metal carbonyls [M]

(a) A M–CO bond

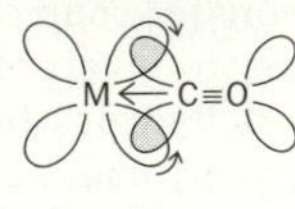

M← C σ donation from C to M [vacant *n*d and (*n*+1)s,p orbitals]

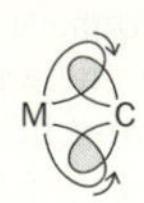

π donation from M (filled *n*d orbitals) to π* orbitals of CO

(b) A simple carbonyl, $Ni(CO)_4$

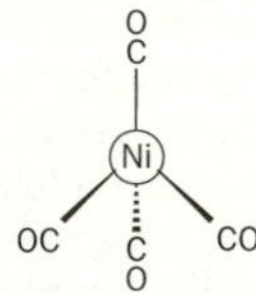

(c) A bridged carbonyl, $Fe_2(CO)_9$

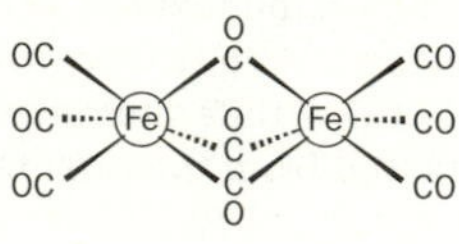

'Ferrocene' $Fe(C_5H_5)_2$ or '$Fe(Cp)_2$'

The Fe atom is sandwiched between two cyclopentadienyl, C_5H_5, rings and bound equally to all ten C atoms.

The rings are not exactly one above the other ('eclipsed') but are offset by 9°

lithium in $Li_4(CH_3)_4$ (see p. 41) and with aluminium in $Al_2(CH_3)_6$ (see p. 146). Chains of two or more carbon atoms joined by multiple or conjugated bonding often interact as a whole with a metal atom, rather than using a single carbon atom to form a σ bond. Ethene, for example, often binds perpendicularly to metals (such as platinum). Iron can be similarly 'sandwiched' between two cyclopentadienyl, C_5H_5 rings, and chromium between two benzene rings, in such a way that all the carbon atoms interact equally with the metal atom. However, in some conjugated systems, unequal interaction may occur. Cyclo-octatetraene (C_8H_8), for example, can interact with a metal atom using all eight of its carbon atoms, and in addition can act as a chelating agent and a bridging agent. The rich variety of organometallic chemistry again emphasizes the amazing variety that carbon achieves by having the same number of electrons as it has accessible outer orbitals, provided of course that the orbitals are of the right size to allow good sideways overlap. If further illustration were needed, we need look no further than our own bodies to see carbon compounds in action: all but carbon dioxide are thermodynamically unstable in air, but none the less so inert kinetically that species with carbon–carbon links form the great majority of our molecular components. For example, in the form of proteins, they help to make the protective coverings provided by skin, hair, and nail; the mechanical filaments in muscle; and compounds for the storage and transport of energy. The proteins are major units in growth, repair, and reproduction. And in the form of enzymes, they act as catalysts in the myriad steps of the reaction systems that any organism needs in order to remain alive.

For summary see p. 76.

Summary

Electronic configuration

- $[He]2s^2 2p^2$

Element

- Various solids
  - diamond: infinite array of tetrahedral σ bonds, transparent, very hard insulator
  - graphite: sheets of hexagons, σ and π bonds, shiny black (weak) electronic conductor, forms intercalation compounds
  - fullerines: hollow near-spherical clusters (e.g. C_{60} 'football') of graphite-like hexagons and pentagons
  - active charcoal, soot, etc.: planar and spiral fragments, edge electrons give adsorbant sites for gases

Occurrence

- Graphite and diamond native
- Chalk, limestone
- All natural organic matter, e.g. wood, coal, oil
- Carbon dioxide in atmosphere

Extraction

- Partial combustion of wood, etc.

Chemical behaviour

- No compounds of C^{4+} or C^{4-} (too much energy needed to form ions)
- Little energy needed to excite atom to 'valence' form of four degenerate sp^3 orbitals, each with one electron; stabilized by variety of bonding, e.g.
  - four tetrahedral σ bonds, e.g. in diamond, and to, –H, $-CH_2-$ to give chains and rings, $-NH_2$, –OH and halogen derivatives, and higher analogues

- σ bonds and localized π bonds, e.g. in –CO– or –C(=NH)–; –C≡C, CO, and the very stable O=C=O and $C{\equiv}N^-$.
- σ bonds and delocalized π bonds, e.g. in graphite, C_{60}, benzene, and in the carbonate ion CO_3^{2-}

- Compounds (except CO_2) thermodynamically unstable in air and water, but do not react because they are kinetically inert
- Metal carbides: electropositive metals form acetylides containing $C{\equiv}C^{2-}$; others give hard interstitial carbides, e.g. iron (to give steel) and tungsten
- Organometallic compounds: unsaturated molecules can bind to metal atoms, e.g. $Ni(CO)_4$, $Fe(C_5H_5)_2$

Uses

- Essential components of living matter and of all organic substances: biological and energy trapping (photosynthesis) and liberation (respiration)
- Fossil fuels (coal, oil, natural gas)
- Diamond, and carbides of silicon and tungsten, for cutters and grinders
- Graphite for electrodes, and as component of lubricants and pencil 'leads'
- Coke for reduction of steam and metal oxides
- 'Active charcoal' as catalyst
- Fibres in composites

Nuclear features

- Natural carbon main stable: ^{12}C with some ^{13}C
- Some ^{14}C (radioactive), formed by cosmic rays and used for dating organic materials

N Nitrogen

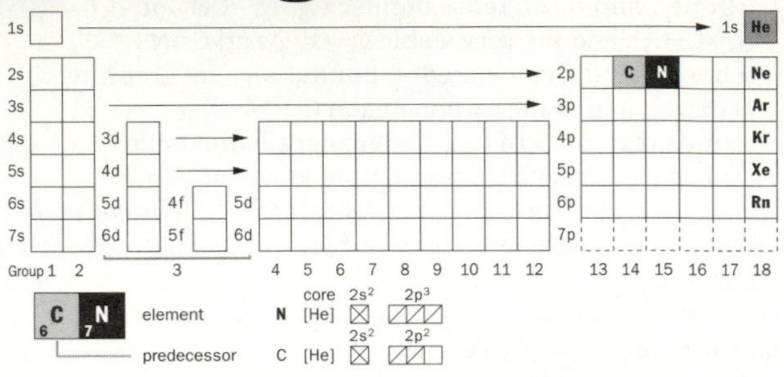

		Predecessor	Element
Name		**Carbon**	**Nitrogen**
Symbol, *Z*		C, 6	N, 7
RAM		12.011	14.00674
Radius/pm:	**atomic**		71
	covalent	77	70
	ionic X^{3-}		171
Electronegativity (Pauling)		2.6	3.01
Melting point/K		~ 3820 (d)	63.29
Boiling point/K			77.4
ΔH°_{fus}**/kJ mol^{-1}**		150.0	0.72
ΔH°_{vap}**/kJ mol^{-1}**		710.9	5.577
Ionization energy for removal of *j*th electron/kJ mol^{-1}	***j* = 1**	1086.2	1402.3
	***j* = 2**	2352	2856.1
	***j* = 3**	4620	4378.0
	***j* = 4**	6222	7474.9
	***j* = 5**	37 827	9440.0
	***j* = 6**		53 265.6
Electron affinity/kJ mol^{-1}		121.9	–7
Bond energy/kJ mol^{-1}	**E–E**	348	160
	E–H	411	390
	E–C	348	305
	E–N	305	160
	E–O	358	200
	E–F		272
	E–Cl	325	193
	E = E	614	415
	E ≡ E	839	946

After carbon we break new ground: nitrogen [He]$2s^22p^3$ has more electrons than it has orbitals in its outer occupied shell. Since the unoccupied 3s orbital is of much higher energy than the $n = 2$ orbitals, any species in which the bonding involved $n = 3$ orbitals would be extremely unstable. The rich and varied chemistry of nitrogen, like that of its predecessors, carbon and boron, is due solely to changes that occur within the $n = 2$ shell.

Nitrogen gas was first identified as that component of the air which did not support respiration or combustion and was given such names as '*azote*' (without life) in French and '*stickstoff*' (suffocating substance) in German. Elemental nitrogen is still obtained from the atmosphere, but by fractional distillation of liquid air. The gas is used mainly to provide an inert environment, at high pressure if need be, and the liquid is used as a refrigerant. Elemental nitrogen, N_2, is indeed extremely stable but once a nitrogen atom can be persuaded to combine with anything other than a second nitrogen atom, its chemical behaviour is far from lifeless. Nitrogen compounds are, in fact, essential for living matter, and the element accounts for 2.5% of our own mass. In the laboratory, nitrogen can be coaxed to show all integral formal oxidation states* from +V to –III and some non-integral ones, reflecting a lively variety of bonding.

The atom is small, with a high first ionization energy and a fairly high effective nuclear charge. Classical valence bond theory would suggest that the nitrogen atom could combine with other atoms by acquiring, or by having a share in, three additional electrons, to achieve the stable $2s^22p^6$ configuration of neon, as in the N^{3-} ion (see p. 83). Not surprisingly the N^{5+} ion ($1s^2$) is unknown. To achieve a formal oxidation state of V, nitrogen *shares* electrons with more electronegative atoms, rather than shedding them (see p. 80).

In elemental dinitrogen, each atom contributes five electrons and four orbitals to the $n = 2$ valence shell. Six of the ten electrons are

* The '*formal oxidation state*' of an element is the average number of electrons it would need to lose if, in that species, all bonds between unlike atoms were completely ionic. The oxidation state is zero in molecules containing atoms of a single element; and it is negative if the element would gain electrons. For monatomic ions, and for simple covalent species without any bonds between like atoms, the oxidation state is numerically equal to the *valence number*, or the number of electrons that an atom loses, gains, or shares equally with another atom.

Some species containing nitrogen in different oxidation states

based on imaginary structures with fully polarized bonds

O^{2-}, O^{2-}, O^{2-} around N^{5+}	$N^0{\equiv}N^{2+}{-}O^{2-}$	$N^0{-}N^0$	H^+, H^+, H^+ around N^{3-}	$(N^0{-}N^0{-}N^0)^-$
N^V in NO_3^-	N^I in N_2O	N^0 in N_2	N^{-III} in NH_3	oxidation state in N_3^- is $-\frac{1}{3}$

Oxidation states of nitrogen compounds

Latimer diagrams [SAL]

Oxidation state	+5	+4	+3	+2	+1	0	−1	−2	−3

Acid solution pH = 0

$$NO_3^- \xrightarrow{0.803} N_2O_4 \xrightarrow{1.07} HNO_2 \xrightarrow{0.996} NO \xrightarrow{1.59} N_2O \xrightarrow{1.77} N_2 \xrightarrow{-1.87} NH_3OH^+ \xrightarrow{1.41} N_2H_5^+ \xrightarrow{1.275} NH_4^+$$

Basic solution pH = 14

$$NO_3^- \xrightarrow{-0.86} N_2O_4 \xrightarrow{0.887} NO_2^- \xrightarrow{-0.46} NO \xrightarrow{0.76} N_2O \xrightarrow{0.94} N_2 \xrightarrow{-3.04} NH_2OH \xrightarrow{0.73} N_2H_4 \xrightarrow{0.1} NH_3$$

Values of $E^\ominus$/V are given above the arrows

Any species which is *followed* by a high *positive* number is a strong *oxidising* agent

Any species which is *preceeded* by a high *negative* number is a strong *reducing* agent

These relationships are better illustrated by a simpler system, such as Tl

$$Tl^{3+} \xrightarrow{1.25} Tl^+ \xrightarrow{-0.34} Tl$$

where Tl^{3+} is strongly oxidising, and the metal strongly reducing.

Any species which has a higher value of $E^\ominus$ to the right than to the left is thermodynamically unstable relative to the two adjacent species and may *disproportionate* into products of higher and lower oxidation state e.g. in basic solution

$$\underset{(-I)}{4NH_2OH} \longrightarrow \underset{(0)}{N_2} + \underset{(-II)}{N_2H_4} + 4H_2O$$

Further disproportionation may occur

$$\underset{(-II)}{2N_2H_4} \longrightarrow \underset{(0)}{N_2} + \underset{(-III)}{2NH_3}$$

For Tl^+ above the value of $E^\ominus$ to its right is lower than that to the left, showing that it is stable to disproportionation

Oxidation state diagram [SAL]

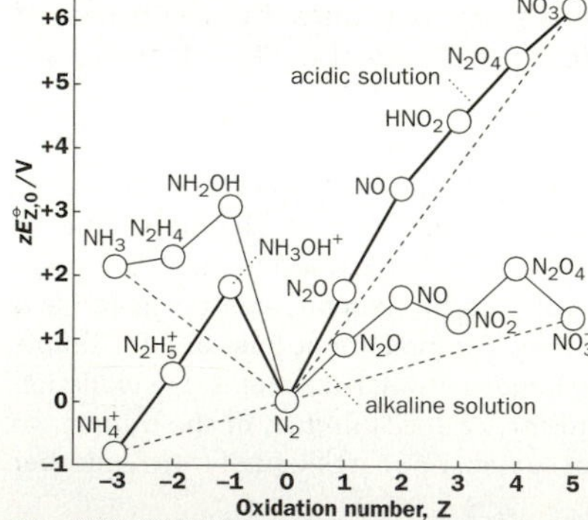

$E^\ominus_{Z,0}$ is the standard electrode potential for conversion of the species of oxidation state Z to the element by transfer of *z* electrons.

Any species which appears above the line joining two others is unstable with respect to them.

In both acidic and alkaline solution, all species except N_2 are thermodynamically unstable relative to oxidation states (V) (NO_3^-) and (−III) (NH_4^+ or NH_3)

accommodated in the three lowest energy orbitals (all of σ symmetry): one is strongly bonding, one weakly bonding, and one weakly antibonding. Since the effects of the last two orbitals cancel out, those six electrons between them contribute one σ bond. The remaining four electrons occupy the two π bonding orbitals, which are degenerate[A] (i.e. have the same energy). One gives overlap on either side of the σ bond, and the other overlaps above and below it. So the nitrogen atoms are joined by a triple bond, exactly analogous to that in ethyne but stronger, because the smaller nitrogen atoms allow even better π overlap than is possible for carbon. The three unfilled orbitals (two π and one σ) are all antibonding.

The very strong N≡N bond is not the only cause of the lack of reactivity of dinitrogen. Even for thermodynamically favourable changes, nitrogen may react very slowly because the N_2 molecule is non-polar and so is not attracted into transient combination with polar species; and since the lowest unoccupied orbitals are of much higher energy than the highest occupied ones, they have no great tendency to accept electrons from other species they may encounter. The sluggish behaviour of dinitrogen has advantages for those who are using isotopically enriched nitrogen compounds

MO diagram of N_2 [MM]

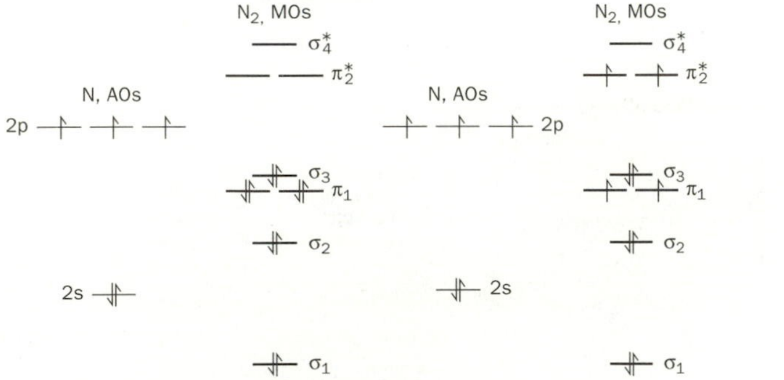

N, AOs | N_2, MOs

σ_4^*

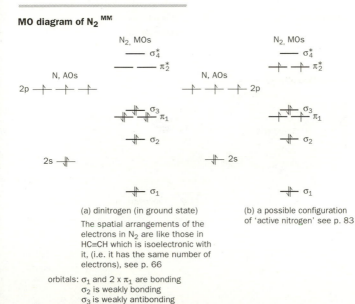

π_2^*

2p

σ_3

π_1

σ_2

2s

σ_1

(a) dinitrogen (in ground state)

The spatial arrangements of the electrons in N_2 are like those in HC≡CH which is isoelectronic with it, (i.e. it has the same number of electrons), see p. 66

orbitals: σ_1 and 2 x π_1 are bonding
σ_2 is weakly bonding
σ_3 is weakly antibonding
2 x π_2^* and σ_4^* (all empty) are antibonding

N, AOs | N_2, MOs

σ_4^*

π_2^*

2p

σ_3

π_1

σ_2

2s

σ_1

(b) a possible configuration of 'active nitrogen' see p. 83

Stable isotopes of nitrogen

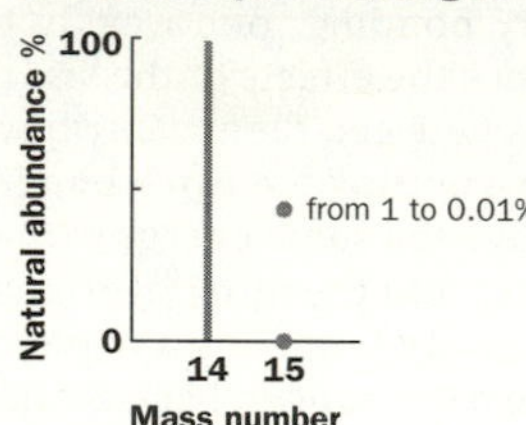

Enthalpies/kJ mol^{-1} of some triple bonds [E]

N≡N	C≡C	C≡N	C≡O
946	839	891	1073

Bonding of N_2 to a metal atom

M ← N ≡ N
end-on

M ← N ≡ N → N
end-on bridge

N ||| with arrows to M N M
side-on bridge

[M ← N≡N]
side-on (not yet confirmed)

A simplified nitrogen cycle [GE]

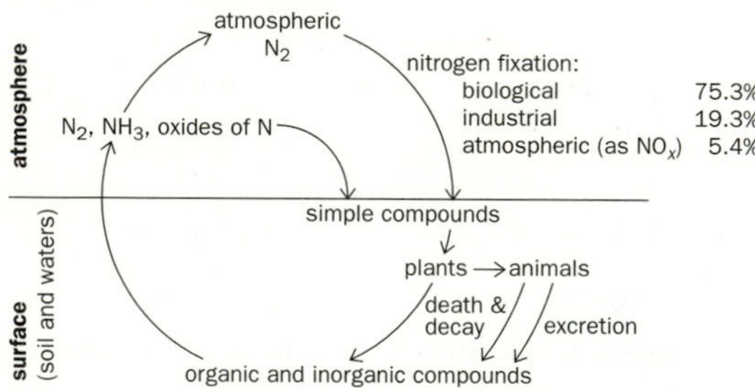

because there is no appreciable exchange between the ^{15}N in the labelled species and the $^{14}N_2$ in the atmosphere. (Work with species containing labelled hydrogen and oxygen atoms is, however, complicated by rapid exchange between the isotopically enriched species and any water and dioxygen that may be present.)

An unexpectedly exciting property of N_2 is its ability to replace some (but seemingly never all) of the ligands coordinated around some transition metal ions, as in the complex $[Ru(NH_3)_5(N_2)]^{2+}$. The dinitrogen may either join (end on) to a metal ion, or act as a bridge between two of them. The bonding is similar to, though weaker than, that in analogous metal carbonyls. Dinitrogen has the same number of electrons as carbon monoxide and forms bonds by a combination of σ donation from ligand to metal, and π donation from metal to ligand (see p. 173). It is thought that coordination of dinitrogen to a molybdenum atom in the enzyme nitrogenase is one step in the reduction of atmospheric nitrogen to ammonia by some algae and bacteria (e.g. those in the root nodules of leguminous plants). This 'fixation' of nitrogen is the first step in the natural 'nitrogen cycle' in which ammonia is oxidized to nitrates, which are absorbed by the plants and converted to the proteins that control the functioning of living systems (see p. 87). Excretion, death, and decay eventually produce small inorganic species, some of which return to the atmosphere as ammonia, oxides of nitrogen, or dinitrogen itself.

Despite both its stability and its kinetic sluggishness, dinitrogen is not totally unreactive. With light electropositive metals such as lithium, and magnesium (see pp. 37 and 133), it forms partially ionic nitrides. At higher temperatures, dinitrogen reacts with a variety of metals, such as nickel, which are hardened by assimilating nitrogen atoms into the metal lattice in much the same way as iron absorbs carbon to form steel.

Dinitrogen is more reactive when its triple bond is weakened by energetic vibration. At high temperatures nitrogen will combine with oxygen (see p. 91). An electric discharge (as in a stroke of lightning) may provide enough energy to separate the atoms, which then recombine; but the 'active' dinitrogen that is formed is often excited, perhaps with two unpaired electrons in the π antibonding orbitals, and only a double N=N bond. These high-energy species can attack even such stable species as H_2O and CO_2. The reactivity of triply bonded dinitrogen can be increased by using a catalyst to provide a less energetic pathway for reaction (see p. 84): in the Haber process for manufacturing ammonia, a finely divided form of iron enables nitrogen to combine with hydrogen at high temperature and pressure. The metal provides adjacent sites for contact

Haber synthesis of ammonia

Starting materials: natural gas (CH_4),
air ($O_2 + 4N_2$)
steam (H_2O)

Main processes:

$$CH_4 + H_2O \xrightarrow{\text{Ni, 750°C}} H_2,\ CO_2,\ CO$$

$$H_2 + \text{air} \xrightarrow{\text{1100°C}} H_2O,\ N_2$$

$$CH_4 + H_2O \xrightarrow[\text{(b) copper, 200°C}]{\text{(a) iron oxide, 400°C}} \underbrace{CO_2,\ H_2}_{\text{Ni, }\downarrow\text{ 325°C}},\ CO(\text{trace})$$

$$CH_4,\ H_2O$$

Removal of by-products:

CO_2 with K_2CO_3 as $KHCO_3$
H_2O by use of excess CH_4
CO removed as above (if present it could poison the Fe catalyst)

Synthesis:

$$N_2 + 3H_2 \xrightarrow[\text{200 atm}]{\text{Fe catalyst, 400°C}} 2NH_3 \quad \text{(exothermic)}$$

The equilibrium is favoured by high pressure and low temperature; but the temperature must be high enough to allow the (catalysed) reaction to occur.

Ammonia complexes [R] of Cu^{2+}

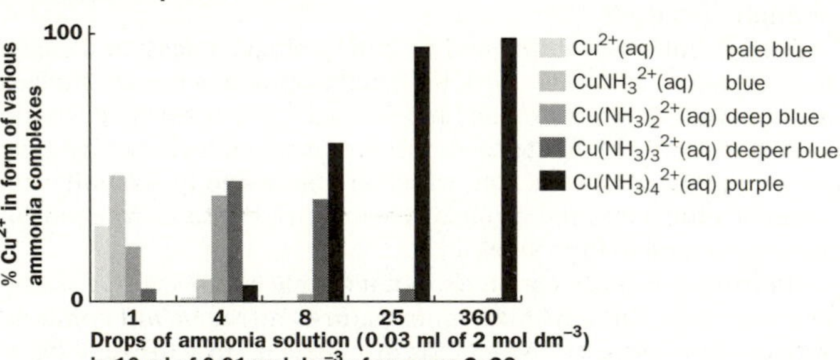

Self-ionization of ammonia [PIC]

		K at 25°C
$2NH_3 \rightleftharpoons NH_4^+ + NH_2^-$	[in NH_3(l)]	5×10^{-27}
$2H_2O \rightleftharpoons H_3O^+ + OH^-$	[in H_2O(l)]	1×10^{-14}

Bond enthalpies/kJ mol^{-1}

N–N 160	C–C 348

between the two sorts of molecules and may also withdraw electrons from, and so weaken, the N≡N bond. The ammonia is used to make vast quantities of agricultural fertilizers, and also explosives, synthetic polymers, dyestuffs, and pharmaceuticals; indeed, the chemical industry synthesizes more molecules of ammonia than of any other substance.

We have seen that the ammonia molecule with its lone pair of electrons (see p. 9), is a (Brønsted or Lewis) base that reacts with the Lewis acids BX_3 (see p. 51) to form $H_3N.BX_3$, and with the proton to give the ammonium ion NH_4^+ (see p. 12). Ammonium salts are very similar to those of potassium, doubtless because the two cations are almost the same size. Metal ions can also act as Lewis acids (see p. 84), often binding ammonia more strongly than water, which can be replaced, forming complex ions such as the series of royal blue cations $Cu(NH_3)_n^{2+}(aq)$.

The lone pair of electrons is also responsible for the electrical polarity of the ammonia molecules, and hence for the hydrogen bonding between them. Like water, ammonia is a liquid at room temperature and can dissolve many ionic solids. As it can both gain and lose a proton, liquid ammonia self-ionizes (though to a lesser extent than water does) to give NH_4^+ and NH_2^-.

We can replace one or more of the hydrogen atoms in ammonia (or indeed in the ammonium ion) by a large variety of groups or atoms that can contribute one electron to a σ bond. Halogen atoms with configuration [core]ns^2np^5 might seem suitable; but NF_3 is the only stable trihalide of nitrogen. Electrostatic repulsion is lowest between the small electronegative fluorine atoms. The higher halides, NCl_3, NBr_3, and $NI_3.NH_3$ (which is apparently somewhat stabilized by N–I–N bonding) are explosively unstable.

Replacement of a hydrogen atom by an $-NH_2$ group gives hydrazine H_2N-NH_2; but we cannot repeat the process to make stable compounds containing chains of N–N–N bonds. The N–N bond has only about half the strength of the C–C bond, probably because of the repulsion between the lone pairs on the small nitrogen atoms. Only a few chain compounds of nitrogen have been made and these are short, unstable, and usually contain multiple bonds.

Hydrazine itself is readily oxidized to water together with such species as dinitrogen, hydrazoic acid HN_3, or ammonia. Liquid hydrazine, often together with liquid N_2O_4, is used as a rocket propellent: the two liquids are hypergolic, i.e. they ignite spontaneously on mixing. Less dramatically, hydrazine is used to prevent corrosion by removing dissolved oxygen from aqueous solutions (see p. 86).

Hydrazine

Configuration

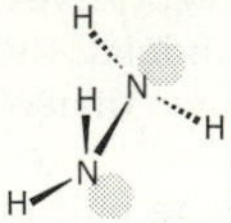

H, N, H / H, N, H (structural diagram)

As a rocket propellant

$2N_2H_4(l) + N_2O_4(l) \longrightarrow 3N_2(g) + H_2O(g)$

(methyl derivatives of hydrazine are also used) — dinitrogen tetroxide (other oxidants are HNO_3, O_2, H_2O_2 and F_2)

As an antoxidant

$2N_2H_4(aq) + O_2(g) \longrightarrow N_2(g) + 2H_2O(l)$

Hydroxylamine

Configuration

relative orientation of the two groups may vary

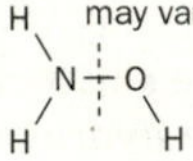

H, N, O, H, H (structural diagram)

Basic strength

$B(aq) + H_2O(l) \rightleftharpoons BH^+ + OH^-(aq)$

base B	*K*
ammonia NH_3	1.8×10^{-5}
hydrazine N_2H_4	8.5×10^{-7}
hydroxylamine NH_2OH	6.6×10^{-9}

Redox behaviour

reduction $NH_2OH + 2H^+ + 2e^- \longrightarrow NH_3 + H_2O$

$2NH_2OH + 2H^+ + 2e^- \longrightarrow N_2H_4 + 2H_2O$

oxidation $2NH_2OH + 2OH^- - 2e^- \longrightarrow N_2 + 2H_2O$

disproportionation $4NH_2OH \longrightarrow N_2H_4 + N_2 + 4H_2O$

DNA [GE]

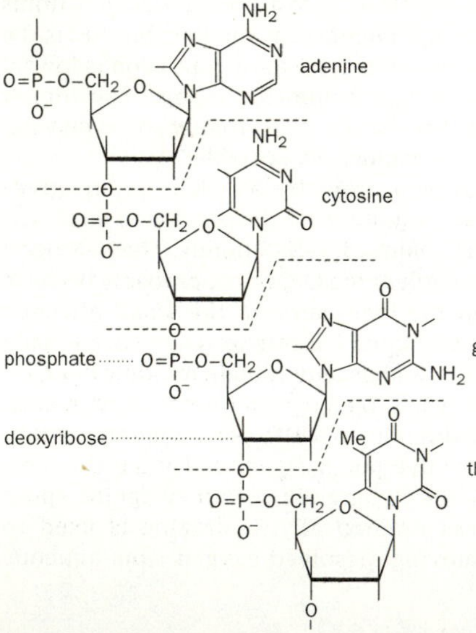

adenine

cytosine

guanine

thymine

phosphate

deoxyribose

The $-NH_2$ group can also be joined to an $-OH$ group to give hydroxylamine, NH_2OH, which is a weaker base than ammonia or hydrazine. Hydroxylamine has a roughly equal tendency to be reduced (to give hydrazine or ammonia) and to be oxidized (to dinitrogen), and so will often disproportionate to undergo both processes simultaneously (see opposite). Its ability to both lose or gain electrons makes it useful for carrying out the rather delicate oxidation–reduction reactions needed in photography and dyeing.

Substitution of the hydrogen atoms in ammonia by organic groups gives a huge range of organic nitrogen compounds such as primary (RNH_2), secondary (RR′NH) and tertiary (RR′R″N) amines, and quaternary ammonium ions ($RR'R''R'''N^+$), where the groups R, R′, etc. may be alkyl or aryl. We might have expected that a tetrahedral structure NRR′R″ with four different vertices (including the lone pair) would exist in different left- and right-hand forms, as do the tetrahedral compounds RR′R″R‴C derived from methane (see p. 64). But in ammonia and its derivatives the molecules can easily invert (like an umbrella blowing inside out), so that the nitrogen atom can oscillate between positions above and below the plane of the bonded groups. The dynamic equilibrium between the two mirror forms is therefore so rapid that they cannot be separated.

Substituted ammonia molecules, RNH_2, provide wide diversity. In aniline (see p. 88), R is a benzene ring. When R is the carboxylic acid group $-CH_2COOH$, we have glycine, H_2NCH_2COOH, one of twenty naturally occurring amino acids that can link together to give a wide variety of 'peptides' such as $H_2N-CH_2-CO-NH-CH_2COOH$. The proteins, essential for biological growth and reproduction, are made from many hundreds of amino acids joined together by the –CO–NH– peptide link to form a linear sequence, like beads on a string. This necklace is then folded to give each amino acid unit a very precise orientation. Bound nitrogen is also a component of the four cyclic 'bases' of the DNA double helix and it is the arrangement of these bases that stores information in an organism so that particular proteins are made by selecting the right amino acids opposite bases in a particular order. This genetic code helps to determine the precise characteristics of an organism as it develops.

For the great majority of living organisms, which cannot directly assimilate atmospheric dinitrogen for the building of proteins,

'Inversion' of ammonia

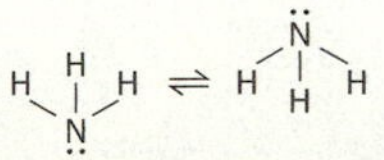

H H H N̈ (pyramid, lone pair below) ⇌ H H H N̈ (pyramid, lone pair above)

Delocalized π orbitals in the NO_3^- ion [M]

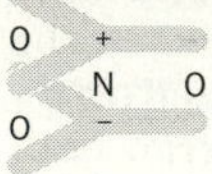

Starting materials for dyes

$C_6H_5NH_2$ — aniline

$C_6H_5N{=}N^+ \ Cl^-$ — benzene diazonium chloride

Oxides of nitrogen [SAL]

N–N–O	N–O	O_2N–NO	NO_2	O_2N–NO_2	O_2N–O–NO_2
dinitrogen oxide	nitrogen monoxide	dinitrogen trioxide (planar)	nitrogen dioxide	dinitrogen tetroxide (planar)	dinitrogen pentoxide (planar)

intake of nitrogen starts with the absorption of the nitrate ions from natural waters by plants (see p. 82).

The nitrate ion is by far the commonest of the many known combinations of nitrogen and oxygen, in all of which the nitrogen atom departs from the simple σ bonding found in derivatives of ammonia. Since there is such strong π overlap in N_2, it is not surprising that nitrogen can also form multiple bonds with carbon and oxygen atoms, which have a similarly low radius. The nitrate ion NO_3^- is isoelectronic with the carbonate ion CO_3^{2-} and has the same flat triangular structure, stabilized by a delocalized π orbital (see p. 73). The cyanide ion $C{\equiv}N^-$ is isoelectronic with N_2 but its triple bond is slightly weaker. Partly because it is polar, it binds much more strongly than N_2 to metal ions and probably owes its extreme toxicity to its ability to complex with the metal ion present in many enzymes.

Rupture of one of the two π bonds in dinitrogen gives the diazogroup –N=N– which, when attached to a benzene ring in aniline derivatives, is the basis of many dyestuffs. A single nitrogen atom, $2s^2 2p^3$, is isoelectronic with the group –CH= and can replace it in many organic compounds, particularly in those aromatic rings such as benzene. A large ring containing –CH= and –N= may, if it is of the right size and shape, accommodate a metal ion; and if the inner enclosure is surrounded by other aromatic rings, the result may be a very stable metal porphyrin complex. Many biologically important molecules are of this type, including haemoglobin (see iron, p. 255) and chlorophyll (see magnesium, p. 134).

Nitrogen can also be incorporated into a rich and bizarre collection of compounds in which it forms (often multiple) bonds to phosphorus or to sulphur, the elements directly and diagonally below it in the periodic table (see pp. 168 and 177).

With oxygen, nitrogen forms a remarkable collection of compounds that are of similar energy, although all are thermodynamically unstable relative to a mixture of N_2 and O_2. They exist merely because they are too sluggish to decompose. All the oxides of nitrogen have some degree of multiple bonding, the details of which are still not fully known.

Dinitrogen oxide (or 'nitrous oxide'), N_2O, is produced, together with dinitrogen and 'nitric oxide' NO, by carefully heating ammonium nitrate. Previously used as a dental anaesthetic ('laughing gas'), its solubility in fats now makes it a good whipping and propelling agent for foods such as ice cream, but in quantities too low to produce mirthful intoxication. With sodamide, $NaNH_2$, dinitrogen oxide gives the azide ion N_3^- which is the conjugate base of

pK values of nitrogen oxoacids

HONO 3.3 $HONO_2$ −1.4

Bonding of NO_2^- and ONO^- ions to metal ions GE

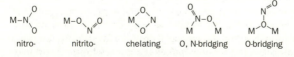

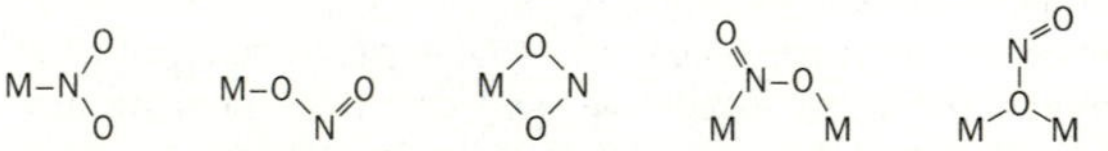

M–N(O)(O) nitro-

M–O–N=O nitrito-

M(O)(O)N chelating

O=N–O, M M O, N-bridging

N=O, M–O–M O-bridging

the very toxic weak acid HN_3. Azides of heavy metals such as lead are so unstable that they explode if hit, and are used as detonators.

Nitrogen oxide (or 'nitric oxide'), NO, is formed by reaction between oxygen and nitrogen during thunderstorms, in the internal combustion engine, and by catalytic oxidation of ammonia during nitric acid manufacture. Despite its odd electron, it dimerizes only partially in the liquid phase and not at all as a gas. It decomposes at high temperatures to dinitrogen and dioxygen, and, under other conditions, to various other oxides of nitrogen. Like dinitrogen, nitrogen oxide can complex with a number of transition metal ions in the presence of other ligands. It may contribute three electrons to the metal ion to give a 'straight' (180–160°) M–N–O bond, or it may provide only one electron to give a bond which is decisively bent (140–120°). The molecule may also form a bridge with the nitrogen atom attached (sometimes symmetrically, sometimes not) to two metal ions. Nitric oxide is environmentally unfriendly. It reacts with atmospheric oxygen to form the brown irritant gas NO_2, visible in the photochemical smog that may blankets cities like Los Angeles and Athens; and since it is oxidized also by ozone, it contributes to the depletion of the ozone layer.

Gaseous nitrogen dioxide, NO_2, is in equilibrium with N_2O_4, its colourless dimer. The liquid contains only N_2O_4, which is almost unionized, although it can react with other species to give products containing the ions NO^+ and NO_3^-.

Dinitrogen pentoxide N_2O_5 is a very unstable solid that sublimes and decomposes just above room temperature to give NO, NO_2, and O_2. The short-lived nitrogen trioxide, NO_3, is probably formed as an intermediate. Dinitrogen pentoxide is obtained by dehydrating nitric acid and reacts readily with water to regenerate it.

Nitrogen forms a number of oxoacids and their salts. All except the somewhat esoteric hyponitrous acid HO–N=N–OH contain the –N=O group, together with the hydroxy group HO– or, in peroxoacids, the group H–O–O–. The most important oxoacids of nitrogen are nitrogen(III) acid, HONO, and, in particular, nitrogen(V) acid, $HONO_2$, which we shall call by their traditional names of nitrous and nitric acids. Both anions contain N–O σ bonds strengthened by one π bond delocalized over the entire ion. Nitrous acid, HONO, is a weak acid that exists only in solution but its sodium and potassium nitrites can be made by heating the nitrates. The nitrite ion ONO^- can combine with a metal ion in a variety of ways: through a single donor atom, either N or O; through both O atoms to form a chelate complex; by use of one O atom to form a bridge between two metal ions; or by joining two metal ions through an O–N bridge. It forms a pink complex with haemoglobin and so is used (despite its mild

Thermal decomposition of some nitrates

$$\begin{aligned}
&NH_4NO_3(s) &&\rightarrow && N_2(g) + 2H_2O(g) + \tfrac{1}{2}\,O_2(g)\\
&NaNO_3(s) &&\rightarrow && NaNO_2(s) + \tfrac{1}{2}\,O_2(g)\\
&2KNO_3(s) &&\rightarrow && K_2O(s) + N_2(g) + \tfrac{5}{2}\,O_2(g)\\
&Pb(NO_3)_2(s) &&\rightarrow && PbO(s) + 2NO_2(g) + \tfrac{1}{2}\,O_2(g)\\
&AgNO_3(s) &&\rightarrow && Ag(s) + NO_2(g) + \tfrac{1}{2}\,O_2(g)
\end{aligned}$$

Detonation of TNT

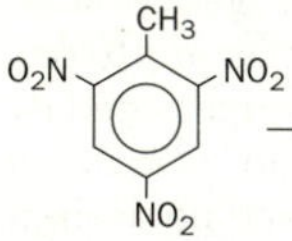

CH_3, O_2N, NO_2, NO_2 $\rightarrow$ CO_2, H_2O, N_2

trinitrotoluene 'TNT'

For complete combustion of 1 mol TNT, 8.5 mol oxygen is needed

toxicity and suspected carcinogenic influence) to improve the appearance of meat. The human body may also be at risk from nitrites derived from the use of nitrate fertilizers.

Nitric acid, $HONO_2$ is made (together with NO) when NO_2 reacts with water. It is produced industrially on a large scale, and much of it is converted to ammonium nitrate for use as an agricultural fertilizer. Precautions are needed to prevent the explosive decomposition of NH_4NO_3, which releases not only dinitrogen and water vapour, but also dioxygen, ready to react with any combustible material present. Its instability is exploited by mixing it with fuel oil and using it as a blasting explosive.

The nitrates of sodium and potassium also give dioxygen on heating, and are used as a source of oxygen in gunpowder and other pyrotechnic materials. Lead and silver nitrates, on the other hand, give mainly nitrogen dioxide, which again illustrates the delicate energy balances amongst compounds of nitrogen with oxygen.

If chemical behaviour were determined only by the energy differences between reactants and products, the energetically favourable reaction between dinitrogen, dioxygen, and water would turn all natural waters into dilute solutions of nitric acid; but, happily, the change is very unfavourable kinetically and so does not occur.

Nitric acid is a strong acid and its salts are usually soluble in water. The large, singly charged planar nitrate ion has no point of high electron density at which a proton may be firmly anchored, nor does it form strong lattices with metal ions. It is a strong oxidizing agent which dissolves relatively 'noble' metals like copper and is used as an oxidant in rocket propellants and in the synthesis of organic polymers. It can also substitute the nitro group, $-NO_2$, for a hydrogen atom in a wide variety of organic substances, sometime introducing several $-NO_2$ groups into one molecule, as in trinitrotoluene (TNT). Such species are so unstable that they decompose very rapidly with evolution of a lot of hot gas (N_2, H_2O vapour, and CO_2) and they are powerful explosives. The instability of such compounds must be largely owing to the extreme stability of one of the decomposition products: the non-polar, triply bonded dinitrogen molecule. In a way, it is that very quality that earned nitrogen its lifeless 'azo' label that confers so active and varied a chemistry on its compounds.

For summary, see p. 95.

Summary

Electronic configuration
- $[He]2s^2 2p^3$

Element
- Diatomic gas with very stable N≡N triple bond (one σ and two π)

Occurrence
- Major component of air, nitrates in soil, proteins in living matter

Extraction
- Distillation of liquid air

Chemical behaviour
- Anion N^{3-} in ionic nitrides of (Li^+, Mg^{2+}, Al^{3+}); nitrides of other metals often interstitial
- Often forms three σ bonds
  - ammonia, NH_3, roughly tetrahedral; lone pair causes Lewis basicity with H^+ and M^{z+}, and H bonding; derivatives include N_2H_4, H_2NOH, many organonitrogen compounds, and natural products e.g. amino acids, peptides, proteins, and DNA bases
  - halides NX_3: only NF_3 stable
- Many multiply bonded species involving p_π–p_π overlap with other small atoms, e.g. $C{\equiv}N^-$, N_3^-, NO, NNO, NO_2 (dimerizes)
  - oxides of nitrogen have delicate redox interelationships
  - oxoacids and derivatives, $HONO_2$ (nitric acid) strong; nitrates soluble in water and decompose on heating (to oxides or nitrites); organic nitrocompounds by substitution of $-NO_2$ for –H, polynitrocompounds explosive; HONO (nitrous acid) only in solution

Uses
- N_2, gas as inert atmosphere, liquid as coolant
- NH_3 and $HONO_2$ as bulk chemicals in manufacturing industries
- NH_4NO_3 as fertilizer
- Proteins essential for enzymes, reproduction, growth, etc.
- Nitrocompounds as explosives

Nuclear features
- Mainly ^{14}N with some ^{15}N

O Oxygen

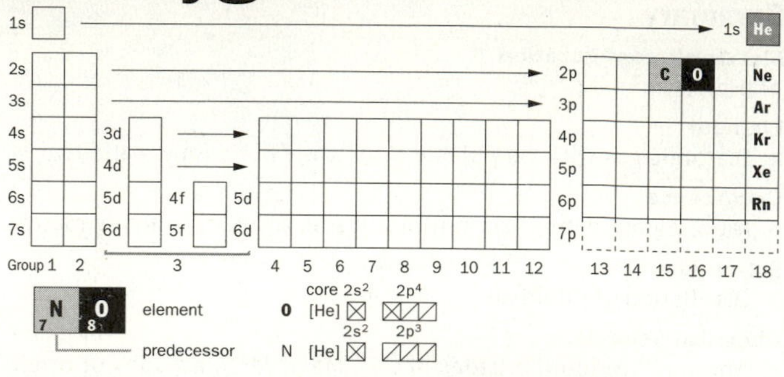

	Predecessor	Element
Name	**Nitrogen**	**Oxygen**
Symbol, *Z*	N, 7	O, 8
RAM	14.00674	15.9994
Radius/pm: covalent	70	66
ionic, X^+		22
X^2		132
Electronegativity (Pauling)	3.01	3.4
Melting point/K	63.29	54.8
Boiling point/K	77.4	90.188
$\Delta H^{\ominus}_{fus}$/kJ mol^{-1}	0.72	0.444
$\Delta H^{\ominus}_{vap}$/kJ mol^{-1}	5.577	6.82
Ionization energy for removal of *j*th electron/kJ mol^{-1} *j* = 1	1402.3	1313.9
***j* = 2**	2856.1	3388.2
***j* = 3**	4378.0	5300.3
***j* = 4**	7474.9	7469.1
***j* = 5**	9440.0	10 989.3
***j* = 6**	53 265.6	13 326.2
***j* = 7**		71 333.3
Electron affinity/kJ mol^{-1}	–7	141
Bond energy/kJ mol^{-1} E–E	160	146
E–H	390	464
E–C	305	358
E–N	160	200
E–O	200	141
E–F	272	190
E–Cl	193	206
$E^{\ominus}$/V for $O_2 + 4H^+ + 4e^- \rightarrow 2H_2O$		1.23

Electronegativity (Pauling)

O	3.4	F	4.0

Oxygen is the most abundant element in the earth's surface, and we need look no further than its natural manifestations for many examples of its chemical diversity. Its electronic configuration $[He]2s^22p^4$ would suggest an atom smaller than its predecessor, nitrogen. Like other $n=2$ elements, it cannot make use of its $n=3$ orbitals as they are of much higher energy than the $n = 2$ ones. We should expect it to be more electronegative than nitrogen; indeed, oxygen has the highest electronegativity of any element except fluorine. The first ionization (from a filled, rather than a half-filled, orbital) is slightly easier than for nitrogen, as is electron addition. We might guess that the energy needed to add a second electron to complete the $n=2$ shell might be handsomely rewarded by the high lattice energies of solids containing the small doubly charged oxide ion O^{2-}, particularly if the accompanying cation is small or highly charged. We might also, rightly, predict that covalent bonding could be achieved by two single, covalent σ bonds to different atoms, or a double (one σ plus one π) bond to one other atom with two half-filled orbitals compatible with $2p_{y,z}$. With its two donatable lone pairs, the oxygen atom seems a likely Lewis base; and when joined to the more electropositive hydrogen atom, the partial negative charge on the oxygen could give rise to hydrogen bonding between molecules (see pp. 13 and 17). Other delocalized bonding with an average of one and a bit interatomic pairs of electrons is also possible, as in the carbonate and nitrate ions (see pp. 71 and 89).

The earth's rocks contain about 46% by weight of oxygen, much of it σ bonded to silicon (see p. 149), some as oxide ion lattices containing metal ions, and some bound to carbon as carbonates, in which the carbon–oxygen σ bonds are strengthened by a delocalized π system. Roughly 86% of the oceans is oxygen, σ bonded to hydrogen to form H_2O molecules which are hydrogen bonded to each other to give water, and, in polar regions, ice. In dioxygen, which accounts for nearly 23% of the atmosphere's mass, two oxygen atoms interact by σ and π bonding (see p. 99); in the stratosphere and above the atmosphere, some of the dioxygen is destroyed by sunlight, oxygen atoms are joined up in threes, by σ bonds and a 3-centre–2-electron π system, to give ozone (see p. 101). The amount of dioxygen in the atmosphere has probably been roughly constant for the last 500 million years. It seems that the first green plants capable of trapping the energy of sunlight to build up energy-storing sugars from carbon dioxide and water, evolved over 2500 million years ago; and the by-product of their photosynthesis, dioxygen, gradually built up in our atmosphere. Although many simple organisms were then forced to live in anaerobic environments, some evolved to tolerate the new pollutant, and

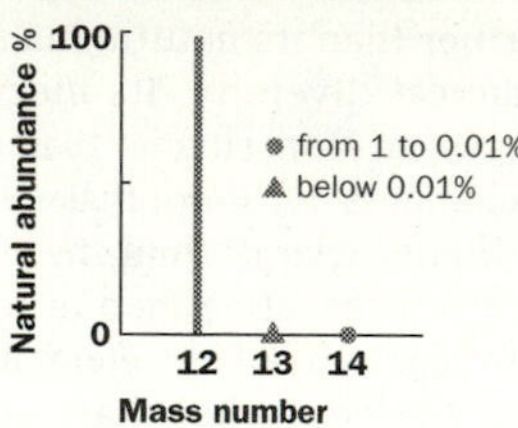

Preparation of oxygen gas

(i) Fractional distillation of liquid air

(ii) Electrolysis of water:

at anode $2H_2O(l) - 2e^- \rightarrow O_2(g) + 4H^+(aq)$

(iii) Chemical decomposition

$$2\ KClO_3(s) \xrightarrow{\text{heat}} 3O_2(g) + 2KCl(s)$$

potassium
chlorate

$$2H_2O_2(l) \xrightarrow{\text{catalyst}} O_2(g) + 2H_2O(l)$$

hydrogen
peroxide

Generation of oxygen by green plants

Photosynthesis: overall reaction

$$6CO_2 + 6H_2O \xrightarrow[\text{chlorophyll}]{\text{sunlight}} C_6H_{12}O_6 + 6O_2$$

glucose

Liberation of energy by oxygen

Overall reactions

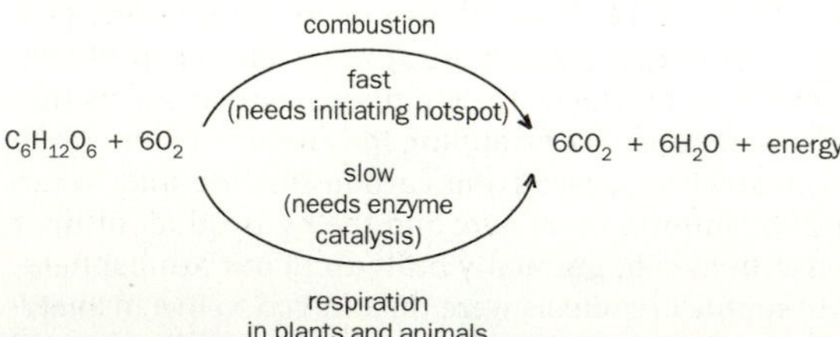

to exploit it for useful energy. Now it is a necessity for the great majority of living organisms on our planet. Dioxygen is sufficiently soluble in water, including sea water, to support life.

Dioxygen is extracted from natural sources by just those methods we should expect: fractional distillation of liquid air and electrolysis of water. On a smaller scale, pure oxygen can be obtained in the laboratory from unstable oxygen-rich products: for example, by heating the chlorate or permanganate of potassium, or by the catalytic decomposition of hydrogen peroxide. Many uses of dioxygen involve the reliberation of the solar energy once trapped by plants; such reactions are the basis of both respiration and combustion. We need commercial oxygen for breathing apparatus in medicine, in aircraft, and for deep sea diving. As an oxidant for combustion, it is used for producing very high temperatures for furnaces and welding, and for rocket propulsion. Its ready combination with carbon, and with small highly charged metal ions, makes it widely used in chemical industries, such as steel making, plastics manufacture, and the production of titanium dioxide. Although oxygen is essential for almost all terrestrial life, in excess it is toxic to living matter, and it is also used to destroy bacteria in sewage. The fact that the evolution of life on earth has resulted in such an intricate relationship with oxygen suggests that the atom can take part in reactions in which there is a very delicate balance of energy changes. Much research is directed at unravelling the stages by which dioxygen is reduced, and energy liberated, by living organisms.

Dioxygen itself has twelve electrons, and eight orbitals in which to accommodate them. The σ bonding and σ antibonding molecular orbitals, derived from the 2s atomic ones, are full, as are the one σ bonding and two π bonding orbitals derived from the two 2p orbitals. The two remaining electrons are unpaired, one in each of the π^* antibonding orbitals, which are degenerate[A] (that is, of identical energy). The last orbital is σ^* antibonding, of much higher energy. In the most stable form of dioxygen the two unpaired electrons have the same spin, but there is also a more reactive form in which these electrons have different spins, and a still higher energy state ('singlet' oxygen) in which two outer electrons are paired in only one of the two π^* antibonding orbitals. The presence of two unpaired electrons in degenerate antibonding orbitals of ordinary dioxygen accounts for its (previously inexplicable) magnetic properties, the blue colour of liquid and solid oxygen, the relatively high boiling point of the liquid, and, more significantly for the living world, its reactivity. But many oxidation reactions happen slowly, partly because of the energy needed to excite the dioxygen

MO diagram of O_2

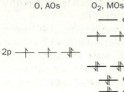

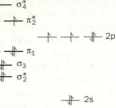

O, AOs	O_2, MOs	O, AOs	O_2^*, MOs
	σ_4^*		σ_4^*
	π_2^*		π_2^*
2p		2p	
	π_1		π_1
	σ_3		σ_3
	σ_2^*		σ_2^*
2s		2s	
	σ_1		σ_1

Ground-state ('triplet') dioxygen
Paramagnetic[A] with two unpaired electrons*

Excited ('singlet') dioxygen
Diamagnetic[A] with all electrons paired*

In both forms, the two lowest orbitals (σ_1 and σ_2^*, derived from the 2s AO) are self-canceling. In the higher set (derived from the 2p AOs) there are six electrons in bonding orbitals (1 x σ_3 and 2 x π_1) and two in the π_2^* antibonding orbitals, giving a double O=O bond. 'Singlet' oxygen is less stable than the 'triplet' form because the two electrons in the same π_2^* orbital repel each other.

In O_2 (unlike N_2, p. 81) the third highest σ orbital is more stable than the π_p orbital; as it is derived from an atomic s orbital, it is more penetrating, and so more influenced by the nucleus, than the π orbital. So, the energy difference between the s and p electrons increases with increasing nuclear charge, lowering the possibility of molecular orbitals having mixed s and p origins.

Bonding of O_2 to metal atoms [MM]

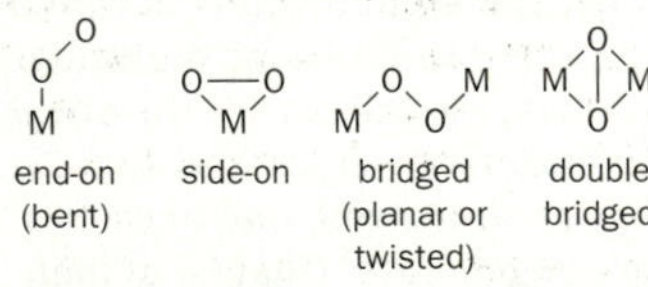

(see also p.263)

Bonding in O_3

: lone pairs: two on end O atoms, one on central O atom

— O–O σ bonds

delocalized 3-centre–4-electron π bond

Ozone, O_3 has the same structure as the (isoelectronic) NO_2^- ion

* A negligible magnetic moment (diamagnetism) shows that electrons are paired while a non-zero moment (paramagnetism) indicates one or more unpaired electrons.

molecule; if happenings were governed by thermodynamics rather than by kinetics (that is by overall energy changes, rather than by intermediate ones), all living matter would be killed by oxidation and all fuel would catch fire spontaneously in air.

Dioxygen can combine in a number of ways with the metal atom in some transition metal complexes and can also act as a bridge between two metal atoms. The haemoglobin in our own blood (see p. 155) acts as an oxygen carrier because the iron atom at its centre can readily accept, and release, dioxygen. The molecule is held in an 'end on, bent' position and it may be that the O=O bond is weakened by its attachment to the metal atom, making the dioxygen more reactive.

A strong oxidizing agent can remove one of the π^* electrons from dioxygen to give the O_2^+ ion, which can form a stable lattice with large polyfluoroanions such as BF_4^- and PF_6^-. An electron can also be added to one of the half full π^* orbitals to give the superoxide ion O_2^- which forms a stable lattice with large M^+ and M^{2+} cations, for example in CsO_2. Addition of a second electron to the other π^* orbital gives the peroxide ion $^-O{-}O^-$, for example in Na_2O_2, which is obtained by heating sodium in oxygen.

Elemental oxygen can also exist as ozone, O_3, a bent molecule that can be pictured as containing a 3-centre–4-electron bond. It gets its name from the same Greek root as osmium (see p. 473): we can smell it. Under usual conditions, ozone is gaseous and decomposes slowly to dioxygen; but if liquid or solid, it explodes. Since the ozone layer above the atmosphere absorbs ultraviolet radiation from the sun, any depletion (by reaction with, for example, the chlorofluorinated carbon compounds, 'CFCs', used as refrigerants and aerosol solvents) could allow a harmful increase in the ultraviolet rays reaching the surface of the earth. Not surprisingly, ozone is an oxidizing agent: with an unsaturated organic compound it destroys the C=C double bond, forming ozonides by addition, whilst with electropositive metals, particularly the heavier ones, it forms ionic ozonides such as $Cs^+O_3^-$.

Oxygen, like carbon and nitrogen, forms a great variety of bonds with other elements. Amongst the lighter ($n = 1$ or $n = 2$) elements, its strongest single bonds are formed with boron, followed by

---

Ozonides GE

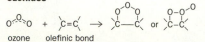

ozone + olefinic bond → (ring ozonide) or (alternative ozonide)

Ozonides are useful unstable intermediates in organic syntheses

Some properties[GE] of H_2O

	H_2O	Ar
RMM	18	18
T_{fus}/°C	0	–189.37
T_{vap}/°C	100	–185.86
$\Delta H^{\ominus}_{vap}$/kJ mol^{-1}	44.02	6.52
$\Delta S^{\ominus}_{vap}$/J K^{-1} mol^{-1}	118.8	74.8

Decomposition of water

Oxidation, e.g. $2H_2O(l) + 4Ce^{4+}(aq) \rightarrow O_2(g) + 4H^+ + 4Ce^{3+}$

$$\left[\begin{array}{l} \text{for } O_2(g) + 4H^+(aq) + 4e^- \rightarrow 2H_2O(l)\ E^{\ominus}/V = 1.23 \\ \quad Ce^{4+}(aq) + e^- \rightarrow Ce^{3+}(aq) = 1.76 \end{array}\right]$$

Reduction, e.g. $2H_2O(l) + 2Li(s) \rightarrow H_2(g) + Li^+(aq) + OH^-(aq)$

$$\left[\begin{array}{ll} \text{for } 2H^+(aq) + 2e^- \rightarrow H_2(g) & E^{\ominus}/V = 0 \\ \quad Li^+(aq) + e^- \rightarrow Li(s) & E^{\ominus}/V = -3.04 \end{array}\right]$$

hydrogen and then by carbon; single bonds to nitrogen, fluorine, or to another oxygen atom are weaker, doubtless because of the repulsion between the lone pairs, which more than counteracts the advantage of better σ overlap with these smaller atoms. The weakness of the O–O single bond contributes to the reactivity (oxidizing power) of peroxides, and to the absence of an oxygen analogue of S_8 (see p. 171).

Oxygen forms a great number of binary compounds with other elements. The most familiar of all oxides is surely water (see p. 9), containing the strong σ bonds from oxygen to the two hydrogen atoms. Extensive and fairly strong hydrogen bonding between the molecules affects all phases, including the vapour in which small clusters $(H_2O)_x$ may be formed. Ice is reminiscent of silica (SiO_2, p. 153) in that one of the many solid phases is amphorous, like glass. The most usual form, however, contains $(H_2O)_6$ hexagons (see p. 16) and at 0°C is less dense than water. The influence of strong hydrogen bonding in the liquid is indicated by the high values of many physical constants, such as boiling and freezing points, $\Delta H^{\ominus}$ and $\Delta S^{\ominus}$ of vaporization, surface tension, and viscosity. There is appreciable self-ionization (see p. 16) to the acidic ion, H^+(aq), and the basic OH^-(aq), with high ionic mobility by the proton switch mechanism.

Since water has a high dielectric constant and can bond both to metal cations and to oxoanions, it is an excellent solvent for many ionic, and polar organic, components. It is also held in many crystals: in clathrates, in which it is surrounded by cage-like molecules; in zeolites,* where it is held in cavities or channels of silicates; and in hydrated salts where some water molecules are usually attached by electron pair donation from the oxygen atom to the cation, but others may be associated with the anion, or be equally distant from either ion. In aqueous solution, hydrated metal ions may dissociate (see p. 13), transferring a proton to water and leaving a hydrolysed species in which an hydroxy ion is attached to the metal ion (or acts as a bridge between two metal ions).

The strong H–O bonds in water can be broken by a powerful oxidizing agent such as cerium(IV), which liberates oxygen from it, or by a strong reducing agent, such as lithium or sodium, which liberates hydrogen (which is also obtained when steam is passed over red hot iron).

* 'zeo' comes from the Greek and means to seethe, which is what the rock seems to do when it is heated in a geologist's blowlamp and the water inside it boils.

Hydrogen peroxide

Structure [GE]

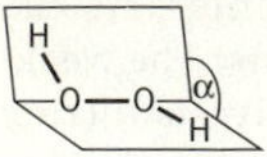

variation in angle α

$H_2O_2(g)$	111.5°
$H_2O_2(s)$	90.2°
$M_2C_2O_4{\cdot}H_2O_2$	~102° (M = K, Rb)
$M_2C_2O_4{\cdot}H_2O_2$	180° (M = Li, Na)

Redox properties [GE]

Latimer diagrams $E^\ominus$/V

In acidic solution: $O_2 \xrightarrow{0.695} H_2O_2 \xrightarrow{1.763} H_2O$

In alkaline solution: $O_2^- \xrightarrow{0.065} HO_2^- \xrightarrow{0.867} OH^-$

In both acidic and alkaline solution H_2O_2 is unstable to disproportionation, but decomposition occurs only if catalysed

H_2O_2 as oxidizing agent

In acid: $2Fe^{2+} + H_2O_2 + 2H^+ \longrightarrow 2Fe^{3+} + 2H_2O$

In alkali: $2Mn^{2+} + 2H_2O_2 \longrightarrow 2MnO_2 + 2H_2O$

H_2O_2 as reducing agent

In acid: $2MnO_4^- + 5H_2O_2 + 6H^+ \longrightarrow 2Mn^{2+} + 8H_2O + 5O_2$

$HOCl + H_2O_2 \longrightarrow H_3O^+ + Cl^- + O_2^*$

In alkali: $2Fe(CN)_6^{2+} + H_2O_2 + 2OH^- \longrightarrow 2Fe(CN)_6^{4-} + 2H_2O + O_2$

$Cl_2 + H_2O_2 + 2OH^- \longrightarrow 2Cl^- + 2H_2O + O_2^*$

The O_2^* is excited 'singlet' oxygen, which glows red

Acid dissociation [GE]

For: $H_2O_2 + H_2O \rightleftharpoons H_3O^+ + OOH^-$ $K = 1.78 \times 10^{-12}$

[For: $2H_2O \rightleftharpoons H_3O^+ + OH^-$ $K = 1.01 \times 10^{-14}$]

Dioxygen difluoride O_2F_2

Similar to H_2O_2 in shape, but with very different bond lengths[GE]

	$O_2F_2(g)$	$H_2O_2(g)$	$O_2(g)$	$OF_2(g)$
O–O distance/pm	121.7	147.5	120.8	–
O–X distance/pm	157.5	95.0	–	140.5
Angle between X–O–O planes	87.5°	111.5°	–	–

Hydrogen peroxide, also a colourless tasteless liquid, has a non-planar molecule, in which the angle between the two H–O–O planes varies with its surroundings. In both acidic and alkaline solution it can act as a mild oxidizing agent (being reduced to water), but it can also act as a reductant and be oxidized to oxygen. The pure liquid was used as an early missile propellant (as a reductant of potassium permanganate). The H_2O_2 molecule is amphoteric. As an acid it can lose one or even both of its protons, or it can gain a proton and act as a base; but, compared with water, it is a stronger acid and a weaker base. It is unstable relative to water and oxygen, but is kinetically sluggish and does not decompose appreciably unless a catalyst is present.

Many other elements also form more than one binary compound with oxygen. Those of the non-metals are often molecular, and the lighter ones are gaseous at room temperature. Some, such as O=C=O, C≡O and N=O, involve p_π–p_π bonding. Oxygen fluorides, OF_2 (analogous to H_2O) and O_2F_2 (geometrically similar to H_2O_2, but with a much shorter O–O bond), are strong oxidizing and fluorinating agents. Many oxides of non-metals, such as CO_2, SO_2 and SO_3, dissolve in water to give oxoacids, $(HO)_x\ EO_y$ (see p. 106). The strength of the acid is often found to increase with the number y of unprotonated oxygen atoms attached to the element E and, for acids of analogous formula, with the electronegativity of E. One or more hydroxyl groups in these acids may be replaced by other groups; and the units –C(H)=O, HO–C=O, and –C(OH)=O, in particular, are of immense importance in organic chemistry. A few non-metal oxides, such as CO and NO, do not react with water.

Electropositive metals react readily with oxygen. Those with high ratios of charge to size (z/r_+) form 'normal' oxides such as $Li_2^+O^{2-}$, $M^{2+}O^{2-}$ ($M=Mg^{2+}$, Ca^{2+}, Sr^{2+}), and $Al_2{}^{3+}(O^{2-})_3$. Sodium and barium, with lower ratios of z/r_+, form the peroxides $(Na^+)_2O_2^{2-}$ and $Ba^{2+}O_2^{2-}$, while the large, singly charged cations of potassium, rubidium, and caesium form the superoxides $M^+O_2^-$. Oxides of very electropositive metals are basic: they react with acids to form ionic solutions, and, if soluble in water, form solutions of the metal hydroxide. Some metal oxides and hydroxides, such as those of zinc, are amphoteric. They not only dissolve in dilute acids to give metal ions, but are also soluble in aqueous alkali to give metal oxoanions [e.g. ZnO_2^{2-}(aq)]. The acidity of an oxide usually increases both across a period in the periodic table and, for the same element, with oxidation state: for example, from basic MnO to very acidic Mn_2O_7. The oxide ion is larger than many cations, but, as it is smaller than the sulfide ion, oxide lattices are often more stable than sulphides. Moreover, the oxide ion, unlike the sulfide ion, is not oxidized by

Oxide acidity

Element	Oxide	
Electropositive metal	Basic,	e.g. $Li_2O + H_2O \rightarrow 2Li^+ + OH^-$
Less electropositive metal	Amphoteric,	e.g. $ZnO + 2HCl \rightarrow Zn^{2+} + H_2O + Cl^-$ $ZnO + 2NaOH \rightarrow ZnO_2^{2-} + 2Na^+ + H_2O$
Non-metal	Acidic,	e.g. $SO_3 + H_2O \rightarrow H_2SO_4$
Non-metal	Non-reactive with water,	e.g. CO, NO

Strengths [GE] of oxoacids $EO_y(OH)_x$

Electro-negativity of E	$y = 0$	pK	$y = 1$	pK	$y = 2$	pK*	$y = 3$	pK**
3.2	Cl(OH)	7.2	ClO(OH)	2.0	$ClO_2(OH)$	–1	$ClO_3(OH)$	(–10)
3.0	–		NO(OH)	3.3	$NO_2(OH)$	–1.4	–	
2.96	Br(OH)	8.7	–	–	–			
2.66	I(OH)	10.6	–	–				
2.6	–		–		$SO_2(OH)_2$	< 0		
2.2	–		$PO(OH)_3$	2.1	–		–	
2.04	$B(OH)_3$	9.2	–	–	–			
	weak		intermediate		strong*		very strong**	

It is difficult* or very difficult** to obtain precise values of pK when dissociation is almost complete.

For discussion, see p. 13.

metals in high oxidation states. Indeed, oxygen (like fluorine) is sufficiently electronegative to bring out the highest oxidation state of an element, as in $BaFe^{VI}O_4$, $KMn^{VII}O_4$, and even $Os^{VIII}O_4$.

The stability of many oxide lattices is reflected in the flammability of magnesium ribbon and of powdered metals such as iron, aluminium, and zinc, which are used in firework manufacture. Metal oxide formation is of enormous economic significance. Huge resources are invested to prevent (or, more realistically, to delay) corrosion, much of which is a result of reaction with oxygen: rust (a hydrated oxide) forever flaking off iron has been notorious since biblical times. But for some metals, such as aluminium and titanium, the ready formation of a thin, close-packed oxide film is a great advantage as it protects the metal from further attack.

Solid oxides show a wide range of structures: molecular, chain, layer, and 3D arrays. These may be ionic, interstitial (for cations of $r_+ < 50$ pm), 'covalent' (for those of high formal charge), and either (for example, SiO_2) crystalline or amorphous. Their composition may vary slightly, or widely, from the stoichiometric ratio of the formula, depending on the partial pressure of oxygen over the solid. Either the metal ($Zn_{1+x}O$) or the oxygen (VO_{2+x}, $Cu_{2-x}O$) may be in excess. The irregularities or 'defects' in the lattice may themselves be regular or disordered. The defects may arise in different ways: in $Fe_{1-x}O$ a few of the Fe^{2+} ions are replaced by Fe^{3+} ions. In stoichiometric titanium dioxide TiO_2 chains, of TiO_6 octahedra, joined through their apices, share corners with similar chains; but non-stoichiometric Ti_nO_{2n-1} (where $n = 4, 5, 6, 7, 8, 9, 10$, or infinity) can be produced if some corner sharing is replaced by edge sharing. One type of metal atom may be replaced by one or more other types. Mixed oxides of transition metal ions often have interesting electrical and magnetic properties. Many are used in information technology, while others are being developed in the hope of finding commercially viable high-temperature superconductors. Research on the earth's most abundant element, whether aimed at biochemistry or technology, is increasing rapidly.

For summary see p. 109.

Summary

Electronic configuration
- [He]$2s^2 2p^4$

Element
- Diatomic paramagnetic gas O_2 with filled σ and two π bonding orbitals and unpaired electrons in two π^* antibonding orbitals
- Triatomic gas, ozone O_3; bent molecule with 3c–4e bond; absorbs UV light; strong oxidizing agent.

Occurrence
- O_2 in atmosphere, from photosynthesis
- O_3 in upper atmosphere
- On earth's surface as H_2O and oxoanions in rocks

Extraction
- Fractional distillation of liquid air
- Electrolysis of water

Chemical behaviour
- Oxidizing agent, often slow acting, in combustion, respiration and corrosion of metals
- With strongly electropositive metals forms basic oxides, usually based on the O^{2-} ion (but sometimes also O_2^- or O_2^{2-})
- With more polarizing cations, e.g. Zn^{2+}, forms amphoteric oxides
- Forms strong single bonds with H (to give amphoteric, H bonded water) and with B and Si (to give oxides and wide variety of oxo-anions)
- With most non-metals forms acidic oxides (some, e.g. NO, CO, are neutral)
- Oxoanions of $n = 2$ elements, e.g. CO_3^{2-}, NO_3^- planar, with one delocalized 4c–2e π bond: of $n = 3$ elements, e.g. PO_4^{3-}, SO_4^{2-}, some p_π–d_π bonding
- X=O in organic compounds
- O–O and F–O bonds weak (peroxides strong oxidizing agents)

Uses
- Oxidation, e.g. in respiration, combustion, and as bleaching agent and germicide
- Many useful oxygen-containing compounds listed under other elements

Nuclear features
- Mainly ^{16}O, some ^{18}O and ^{17}O

F Fluorine

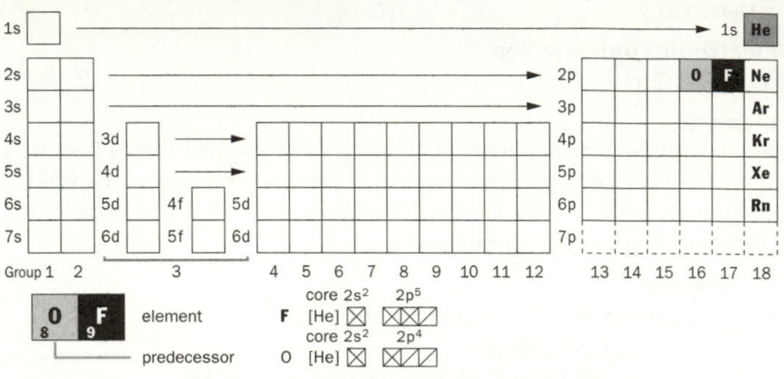

		Predecessor	Element
Name		**Oxygen**	**Fluorine**
Symbol, *Z*		O, 8	F, 9
RAM		15.9994	18.9984032
Radius/pm	**atomic**		70.9
	covalent	66	58
	ionic X^-		133
Electronegativity (Pauling)		3.4	4.0
Melting point/K		54.8	53.53
Boiling point/K		90.188	85.01
$\Delta H^{\ominus}_{fus}$/kJ mol^{-1}		0.444	5.10
$\Delta H^{\ominus}_{vap}$/kJ mol^{-1}		6.82	6.548
Ionization energy for removal of *j*th electron/kJ mol^{-1}	***j* = 1**	1313.9	1681
	***j* = 2**	3388.2	3374
	***j* = 3**	5300.3	6050
	***j* = 4**	7469.1	8408
	***j* = 5**	10 989.3	11 023
	***j* = 6**	13 326.2	15 164
	***j* = 7**	71 333.3	17 867
	***j* = 8**		92 036
Electron affinity/kJ mol^{-1}		141	328
Bond energy/kJ mol^{-1}	**E–E**	146	159
	E–H	464	566
	E–N	200	272
	E–O	141	190
	E–F	190	159
	E–Cl	206	257
	E–B		645

The chemistry of fluorine* is simpler and more predictable than that of its immediate predecessors. With an electronic configuration of [He]$2s^22p^5$, it can accept only one electron into the 2p shell; and it can make no use of the orbitals of the $n = 3$ shell which are of much higher energy. The extra nuclear charge makes the fluorine atom smaller than oxygen and even more electronegative, with a greater tendency to add an electron and a lower tendency to lose one (despite the opposing effect of repulsion between the seven electrons crowded into the nearly full $n = 2$ shell of a small atom). We should guess that such an atom would be highly oxidizing, and would form both ionic compounds containing the F^- ion, and covalent ones containing singly bound fluorine. We should also expect strong hydrogen bonding between fluorine atoms and hydrogen compounds. With four lone pairs of electrons, the F^- ion is likely to be a Lewis base and may also act as a bridging ligand. Although fluorine cannot form multiple bonds, it shows great chemical diversity in other ways. It is the most reactive of all the elements, and forms compounds (often several of them) with all other elements except helium, neon, and argon.

Obviously, so reactive an element does not occur free in nature. It is usually found as the F^- ion in combination with calcium ions, as fluorite* CaF_2 (see p. 201), together with calcium and phosphate as fluorapatite, $Ca_5(PO_4)_3F$, or as a complex ion with aluminium in cryolite, Na_3AlF_6. The difficult preparation of fluorine gas, F_2, by oxidation of F^- has to be carried out electrolytically, using much the same method as was used when Moissan first made it in 1886: the electrolysis of a solution of potassium fluoride in anhydrous hydrofluoric acid HF. Since hydrofluoric acid is extremely corrosive, and difluorine can react explosively with its co-product dihydrogen or even with the graphite electrode, modern developments have concentrated on safety. Compounds of fluorine are needed for a range of industrial processes. Fluorination of organic compounds provides polymers such as polytetrafluoroethylene ('Teflon') and the notoriously ozone unfriendly refrigerants such as the Freons, CCl_2F_2, and other 'CFCs'. Synthetic cryolite is used for the extraction of aluminium (p. 139), and uranium hexafluoride is used to obtain uranium enriched with ^{235}U for nuclear reactors. Addition of small amounts of fluoride ion to drinking water, and of tin fluoride

* The name 'fluorine' comes from that of the mineral 'fluorite', derived from the Latin root 'fluor' meaning flow, because CaF_2 melts easily in a blowlamp.

'Teflon' (polytetrafluoroethylene)

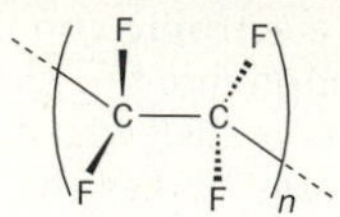

Stable isotopes of fluorine

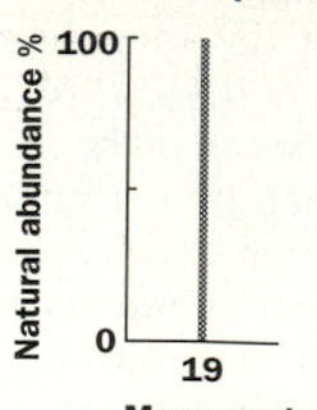

Bond enthalpies/kJ mol^{-1}

F–F	H–H	Cl–Cl
159	453.6	242

Anion radii / pm

F^-	O^{2-}
133	132

MO diagram of F_2

F, AOs | F_2, MOs | F, AOs

σ_4^*

π_2^*

2p | | 2p

π_1

σ_3

σ_2^*

2s | | 2s

σ_1

orbitals: σ_1 and σ_2^* are self cancelling
$2 \times \pi_1$ and $2 \times \pi_2^*$ are self cancelling
σ_3 is bonding
σ_4^* (empty) is antibonding

The relative stabilities of the F_2 MOs resemble those in O_2 (see p. 100)

to toothpaste, makes our teeth more resistant to decay by converting some of the calcium phosphate to the tougher fluoroapatite.

The molecular orbital diagram of difluorine resembles that of dioxygen (see p. 100), but with both π^* antibonding orbitals filled. There is no possibility of the multiple bonding found in dinitrogen and dioxygen, and we can guess that the single F−F σ bond is weakened by the repulsion between the lone pairs of electrons on the two closely spaced fluorine atoms. So the molecule can dissociate, without the expenditure of too much energy, into atoms that react extremely vigorously with anything they encounter. The fluorine atoms may remove an electron, forming a fluoride ion that is hard to polarize, since the $n = 2$ electrons are firmly under the control of the nucleus. Although the fluoride ion is only singly charged, it is almost the same size as the oxide ion, and with many metals it forms stable ionic fluorides with structures similar to those of the metal oxides.

Non-metallic elements form covalently bonded molecular fluorides. A number of the small fluorine atoms can often fit around a central atom, attracting negative charge to themselves, and so reducing interelectronic repulsion; so fluorine can often bring out the highest possible oxidation state of the central element. Iodine, for example, forms the fluorides IF, IF_3, IF_5, and IF_7, and the highest fluoride of sulphur is SF_6; the highest chlorides of these elements are ICl_5 and SCl_4. Fluorine (like oxygen but unlike chlorine) can react with the heavier noble gases to form, for example, XeF_2, XeF_4, and XeF_6 (see p. 411). Very unstable oxofluorides can sometimes be made by substituting an oxygen atom for two fluorine atoms in higher interhalogen compounds or xenon fluorines. Difluorine, like strong oxygen-containing oxidizing agents (but here unlike dioxygen), can also break down the σ bonding system in graphite (see p. 69). Covalent molecular fluorides are also formed by some metals in high oxidation states: UF_6 is a gas. The fluoride ion can act as a Lewis base, donating a lone pair of electrons to form a variety of fluoroanions such as BF_4^-, ClF_2^-, AlF_6^{3-}, AsF_6, XeF_8^{2-}; and by donating two pairs of electrons, it can act as a bridging ligand, as in $Xe_2F_3^+$.

The strength of the single covalent bond from fluorine to another $n = 2$ element decreases from boron to fluorine (being greatest for the greatest difference in electronegativity and for the lowest interelectronic repulsion). The ability of difluorine to oxidize water is, like other aspects of its reactivity, a result, in part, of the low F−F bond strength, together with the (fairly) high electron affinity of the fluorine atom; but it is reinforced by the high favourable enthalpy (and only low unfavourable entropy) of hydration for the small fluoride ion, in much the same way as the very negative electrode

Some fluorides of iodine GE

IF$_3$ IF$_5$ IF$_7$

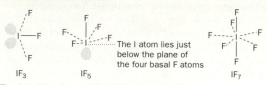

The I atom lies just below the plane of the four basal F atoms

The dotted bonds are slightly longer than the full ones.

Lone pairs are shaded

Some properties GE of HF

	HF	H_2O
Melting point/°C	–83.4	0
Boiling point/°C	19.5	100.0
Self-ionization at 298 K	$[H_2F^+][HF_2^-] \sim 8 \times 10^{-12}$	$[H_3O^+][OH^-] = 1.008 \times 10^{-14}$

Dissociation of HF in water D

Acid dissociation

$$HX(aq) \rightarrow H^+(aq) + X^-(aq).$$

for HF	for HCl
$K_a = 6 \times 10^{-4}$	$K_a \sim 10^8$
$pK = 3.22$	$pk \sim -8$

Dissociation is favoured by:

		F	Cl	
(1)	low $(-\Delta H)$ for $HX(g) \rightarrow HX(aq)$	–48	**–18**	/kJ mol^{-1}
(2)	weak H–X bond	567	**431**	/kJ mol^{-1}
(3)	high EA for $X(g) \rightarrow X^-(g)$	328	**349**	/kJ mol^{-1}
(4)	high $(-\Delta H^{\circ}_{hyd})$ for $X^-(g) \rightarrow X^-(aq)$	**–524**	–378	/kJ mol^{-1}
(5)	low $(-\Delta S^{\circ}_{hyd})$ / high $(+\Delta S^{\circ}_{hyd})$ } for $X^-(g) \rightarrow X^-(aq)$	–14	**57**	/J K^{-1} mol^{-1}

Terms (1), (2), (3) and (4) combine to make both dissociations exothermic, but for HF ($\Delta H^{\circ} = -16$ kJ mol^{-1}) less favourable than for HCl ($\Delta H^{\circ} = -57$ kJ mol^{-1}). The values in bold face indicate the greater tendency to dissociate.

The smaller F^- ion is more strongly hydrated than Cl^-, and while this contributes favourably to the heat change of dissociation, the much greater ordering of water molecules opposes the change and outweighs the (modestly) favourable heat change for dissociation of HF.

potential of lithium, which is in part a result the high heat of hydration of its small cation (see p. 37).

Fluorine forms a strong bond with hydrogen (the bond enthalpy of H–F is between that of B–F and C–F), and hydrogen fluoride is the product of many explosive reactions (such as oxidation by fluorine of hydrogen or of hydrazine, N_2H_4) which have been used in rocket propulsion. In some ways, anhydrous hydrogen fluoride resembles water. At (low) room temperature, it is a colourless but more volatile liquid, self-ionized (to H_2F^+ and HF_2^-) to much the same extent as water, and a good solvent both for many ionic fluorides and for biochemical material. It is, however, extremely corrosive: it even reacts with the silica in glass (to form SiF_6^{2-}) and is used in etching. Burns from hydrofluoric acid on animal tissue heal very slowly, because the calcium ions necessary for new growth are removed as insoluble CaF_2. As expected, hydrogen fluoride is strongly hydrogen bonded but, with only one hydrogen atom per molecule, association can take place only in two dimensions. In both the solid and the anhydrous liquid, planar zigzag chains are formed. In the gas phase, single HF molecules are in equilibrium with linear hydrogen bonded $(HF)_2$ dimers and with the main product of association, the unlikely looking 12-membered hydrogen bonded $(HF)_6$ rings (see p. 17). In 'not too dilute' aqueous solution, the symmetrical anion $[F...H...F]^-$ is formed.

Hydrofluoric acid, unlike the other halogen acids, is only weakly dissociated in water. Compared with the other halogens, fluorine has a stronger H–X bond and stronger halide to water bonds. The heat change on dissociation, although favourable, is too small to overcome the unfavourable ordering effect of the small fluoride ion on the water molecules. Hydrogen fluoride is widely used for making both organic and inorganic fluorine compounds, because, despite being so corrosive and toxic, it is less aggressive than difluorine. Many of the inorganic products, often compounds of oxygen, fluorine, and one or more other non-metals, are, however, very unstable.

Hydrogen bonding to fluorine is not restricted to acids. As the potassium and ammonium ions are almost the same size, it is not surprising that the salts of the two cations with the same anion usually have the same structure. But the fluorides are an exception. Potassium fluoride has, as we should expect, an ionic lattice, with the NaCl structure. Ammonium fluoride, however, has the wurtzite lattice, in which the fluoride ions are aligned through hydrogen bonds with the tetrahedrally dispersed hydrogen atoms on the ammonium ions.

Fluorine, condemned to restrict its behaviour to filling a single vacancy in its valence shell and to exerting a strong nuclear pull, may have a more predictable chemistry than some of its predecessors. But what with tough teeth and Teflon, rocket fuels and glass etching, xenon compounds, gaseous uranium hexaluoride, and hexameric hydrogen fluoride vapour, it seems to have exploited richly its limited possibilities for diversity. Predictable (perhaps) it may be; boring it is not.

Summary

Electronic configuration

- $[He]2s^2 2p^5$

Element

- Diatomic gas, very reactive: F–F bond weakened by repulsion of three non-bonding pairs on each (small) atom

Occurrence

- With Ca^{2+} or Al^{3+} (sometimes with PO_4^{3-})

Extraction

- From molten cryolite (Na_3AlF_6) by electrolysis

Chemical behaviour

- F (small and very electronegative) brings out high coordination numbers (IF_7, UF_6^+, XeF_8^{2-}) and oxidation state, and maximal ionic character, e.g. (AlF_3)
- F^- formed readily (despite repulsion from lone pairs), r_- small, stable ionic lattices, e.g. with Ca^{2+} in teeth; Lewis base, e.g. BF_4^-, AlF_6^{3-}, SiF_6^{2-} (in glass etching)
- σ bonds to non-metals, and replaces H from organic C–H bonds (refrigerants and polymers)
- Strong, polar F–H bond in HF: strong H bonds give cyclic $(HF)_6$ in gas phase, zigzag chains in liquid and symmetrical $[F...H...F]^-$ in concentrated aqueous solution. HF_{aq} is a weak acid.
- Oxides unstable, e.g. F_2O_2

Uses

- 'CFCs' as (former) refrigerants (ozone unfriendly)
- 'Teflon' polymers
- Strengthening teeth
- Glass etching

Nuclear features

- Isotopically pure ^{19}F

Ne Neon

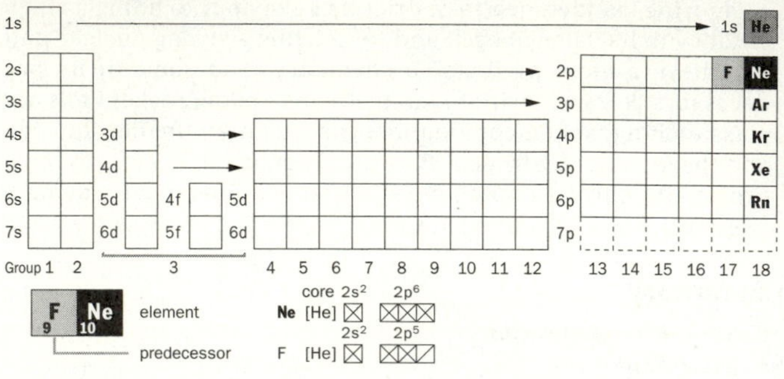

	Predecessor	Element
Name	**Fluorine**	**Neon**
Symbol, *Z*	F, 9	Ne 10
RAM	18.9984032	20.1797
Melting point/K	53.53	24.48
Boiling point/K	85.01	27.10
$\Delta H^{\ominus}_{fus}$**/kJ mol**$^{-1}$	5.10	0.324
$\Delta H^{\ominus}_{vap}$**/kJ mol**$^{-1}$	6.548	1.736
Ionization energy for removal of *j*th electron/kJ mol^{-1} ***j* = 1**	1681	2080.6
***j* = 2**	3374	3952.2
***j* = 3**	6050	6122
***j* = 4**	8408	9370
***j* = 5**	11 023	12 177
***j* = 6**	15 164	15 238
***j* = 7**	17 867	19 998
***j* = 8**	92 036	23 069
***j* = 9**	106 432	115 377
Electron affinity/kJ mol^{-1}	328	−29 (calc)

Stable isotopes of neon

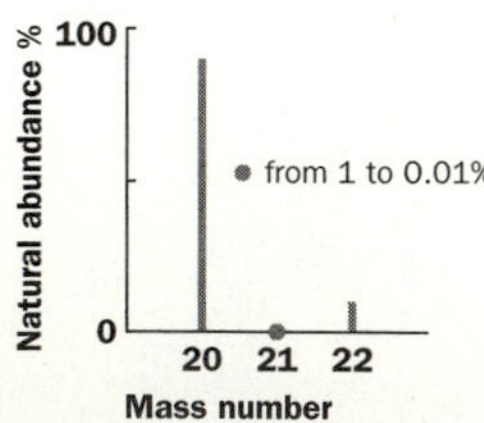

Neon, [He]$2s^22p^6$, with its six 2p electrons, completes the $n = 2$ shell. It is the second lightest member of group 18, now called the noble gases and previously known as the 'rare' or the 'inert' gases. (The first member is helium, $1s^2$, which also has a filled outer electron shell, see p. 6.) Although neon is present in our atmosphere in only very small quantities (about 0.0018%), it is not as rare as many other elements; but it is certainly inert.

Its high first ionization energy for total removal of an electron from the filled $n = 2$ shell is in line with the high energy needed to promote an outer electron into the lowest unoccupied orbital. It is unlikely that any chemical reaction would result in bonds strong enough to overcome this unfavourable rupture of the s^2p^6 arrangement, and the only evidence of a bond formed by neon is in the very transient ion Ne_2^+ (of bond order $\frac{1}{2}$), detected spectroscopically after a high energy electric discharge.

The combination of low size and high effective nuclear charge is unfavourable even to van der Waals interactions and so neon has very low heats of fusion and vaporization. It remains gaseous to very low temperatures and solidifies at only a few degrees below its boiling point. As neon is found (in three isotopic forms) only as the uncombined gas, our main source of it is the atmosphere, from which it can be obtained by fractional distillation of liquid air.

Since neon is heavier than helium, it diffuses less rapidly and escapes more slowly through a small orifice. It is, however, too small to be contained in the molecular cages that have been used to entrap atoms of heavier noble gases (see p. 191).

The high first promotion energy of neon implies that its atoms do not interact with relatively low energy, visible light, so the gas is colourless; and since it does not react with other substances, it has no effect on our senses of taste and smell. Some excited species do, however, emit visible light in the discharge tubes that produce the familiar red of neon signs, and of sodium lights that have not yet 'struck'. Different colours are produced by mixtures of neon with other noble gases such as argon. The red emission of the Ne^+ ion is now also exploited as a laser.

For summary see p. 121.

Summary

Electronic configuration

- [He]$2s^22p^6$

Element

- Monatomic gas, difficult to liquefy

Occurrence

- In air (7 ppm)

Extraction

- From liquid air by distillation

Chemical behaviour

- No known compounds: ionization very difficult, large energy gap between 2p and 3s, and negligible electron affinity; too small for clathrates.
- Transient species: Ne_2^+ in discharge tubes

Uses

- In lighting
- For lasers

Nuclear features

- ^{20}Ne, $^{22}Ne \sim 10{:}1$, with some ^{21}Ne

Part III

Filling the 3s and 3p orbitals

The second short period: sodium to argon

Na Sodium

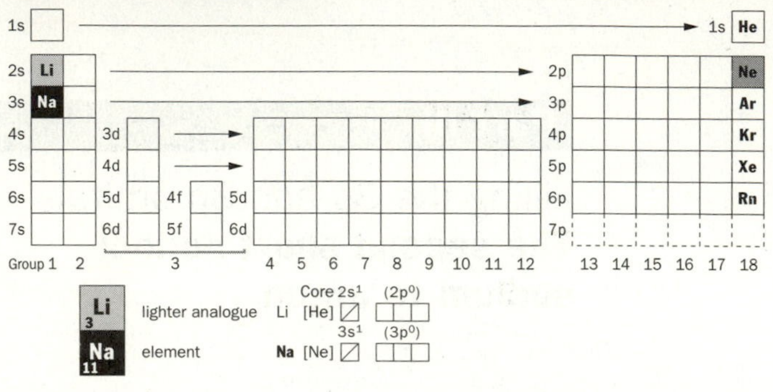

	Element	Lighter analogue
Name	**Sodium**	**Lithium**
Symbol, *Z*	Na, 11	Li, 3
RAM	22.989768	6.941
Radius/pm atomic	153.7	152
ionic, M^+	98	78
Electronegativity (Pauling)	1.0	1.0
Melting point/K	370.96	453.69
Boiling point/K	1156.1	1620
$\Delta H^{\ominus}_{fus}$/kJ mol^{-1}	2.64	4.60
$\Delta H^{\ominus}_{vap}$/kJ mol^{-1}	89.04	134.7
Density (at 293 K)/kg m^{-3}	971	534
Electrical conductivity/Ω^{-1}m^{-1}	23.81×10^6	1.17×10^7
Ionization energy for removal of *j*th electron/kJ mol^{-1} $j = 1$	495.8	513.3
$j = 2$	4562.4	7298.0
Electron affinity/kJ mol^{-1}	52.9	59.6
Dissociation energy of E_2(g)/kJ mol^{-1}	73.2	107.8
$E^{\ominus}$/V for $M^+(aq) + e \rightarrow M(s)$	–2.713	–3.04

Stable isotopes of sodium

Natural abundance %

100

0

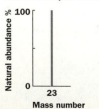

23

Mass number

Yellow sodium flame

$Na^+(g) + e^-(g) \rightarrow Na^*(g) \rightarrow Na(g) +$ yellow light

$Na^+(g)$	$Na^*(g)$	$Na(g)$
	[Ne]$3p^1$	[Ne]$3s^1$
ion	excited atom	ground state atom

Although sodium, of configuration [Ne]$3s^1$, breaks new ground with occupancy of the $n = 3$ shell, it predictably shows little variety in its everyday chemistry. It readily loses its one 3s electron, and in nature it occurs only in the form of the Na^+ ion ([Ne]$3s^0$)$^+$, in company with a number of anions such as carbonate, nitrate, sulphate, and borate, but most commonly with chloride; if all the 'common salt' dissolved in the oceans were extracted and shaped into a cube, an edge would stretch 325 km, as far as from London to Manchester.

Not surprisingly, sodium ([Ne]$3s^1$) behaves in many ways like lithium ([He]$2s^1$); but the sodium atom is larger and its outer ($3s^1$) electron is held less firmly than the $2s^1$ electron of lithium because its higher nuclear charge is more than outweighed by the increased distance from the nucleus, together with the extra screening by the eight $n = 2$ electrons. So, sodium has an even lower ionization energy than lithium. In the solid, the loosely held 3s electron is delocalized into the conduction band and so sodium too is a metal, but is softer and more readily disrupted than lithium. Its low melting point makes it useful as a heat exchanger in nuclear reactors. Reduction of the very stable Na^+ ion to sodium metal can be carried out only by electrolysis (of a molten mixture of sodium and calcium chlorides). With its ready loss of the 3s electron, sodium is used as a reducing agent, to obtain other metals such as potassium and titanium from their chlorides. Dissolved in mercury as 'sodium amalgam', it is also a useful reductant in organic chemistry. Sodium chloride is used in snow clearance, in human and animal foods, and as a starting point for the manufacture of many key substances in the chemical industry, e.g. sodium hydroxide, chlorine, and sodium carbonate. In sodium vapour, containing widely separated atoms, the loosely held 3s electrons can readily be excited, and when they fall back into more stable orbitals they emit energy in the visible part of the spectrum. At low pressure, the light is almost monochromatic (in fact it consists of two wavelengths of very similar energies) and is used in the familiar bright yellow street lamps; and the same colour may be obtained by dropping common salt into a gas flame. If the atoms are much closer together, their energy bands are broadened, although less so in the vapour than in the metal. Light is emitted over a wider range of energies, and high pressure sodium lamps give a whitish apricot light.

Despite its lower energies of atomization and ionization, sodium has a slightly less negative standard electrode potential than lithium, partly because, even though the larger sodium ion can accommodate up to six groups around it, less heat is given out when it is hydrated (see p. 36). Sodium, however, reacts with water more violently than lithium does, because its hydroxide is very

Combustion in air

$2\,Na(s) + O_2(g) \rightarrow Na_2^+\,(O{-}O)^{2-}\,(s)$

$\left[\begin{array}{l} 2\,Li(s) + \frac{1}{2}O_2(g) \rightarrow Li_2^+\,O^{2-}\,(s) \\ 3\,Li(s) + \frac{1}{2}N_2(g) \rightarrow Li_3\,N\,(s) \end{array}\right]$

Thermal decomposition of nitrate

$2NaNO_3\,(s) \rightarrow 2NaNO_2(s) + O_2(g)$

$[2LiNO_3\,(s) \rightarrow Li_2O(s) + 2NO_2(g) + \frac{1}{2}O_2(g)]$

Solubility of sodium halides in water [D]

$NaX(s) \rightarrow Na^+(aq) + X^-(aq)$

Heat changes ΔH°/kJ mol^{-1}	**F**	**Cl**	**Br**	**I**
(1) for $NaX(s) \rightarrow Na^+(g) + X^-(g)$ strongly endothermic	930	788	752	704
(2) for $Na^+(g) + X^-(g) \xrightarrow{H_2O} Na^+(aq) + X^-(aq)$ strongly exothermic	–929	–784	–753	–713
Heat change on solution, ΔH_s is (1) + (2)	1 (endothermic)	4 (endothermic)	–1 (exothermic)	–9 (exothermic)
Solubility/mol NaX(kg H_2O)$^{-1}$	0.987	6.14	9.19	12.26
	Sparingly soluble	Soluble	Soluble	Soluble
Entropy changes, ΔS/J K^{-1} mol^{-1}				
$T\Delta S_s^\circ$ for solubility of NaX at 298, kJ mol^{-1}	–2	13	18	23
$\Delta G_s^\circ = (\Delta H_s^\circ - T\Delta S_s^\circ)$	3	–9	–19	–32
Observed order of solubility is that of $-\Delta G_s^\circ$.				
Favourable ΔH_s term?	No	No	Yes	Yes
Favourable ΔS_s term?	No	Yes	Yes	Yes

NaCl is soluble in water because $T\Delta S^\circ$ (favourable) > ΔH_s° (unfavourable).

Main anion-dependent factors favouring solubility

	F	**Cl**	**Br**	**I**
(1) low ΔH for $NaX(s) \rightarrow Na^+(g) + X^-(g)$/kJ mol^{-1}	930	788	752	**704**
(2) high ($-\Delta H$) for $X^-(g) \rightarrow X^-(aq)$/kJ mol^{-1}	**524**	378	348	308
(3) low ($-\Delta S$) for $X^-(g) \rightarrow X^-(aq)$/J K^{-1} mol^{-1}	–160	–96	–80	**–62**
r_-/pm	2.01	2.57	2.75	3.03

In practice, terms (1) and (3) outweigh term (2), and solubility increases from fluoride to iodide. The values in bold face indicate the greatest tendency to dissolve.

soluble, and so the reaction is not slowed down by the formation of a protective coating (see p. 37). Indeed, the reaction proceeds rapidly, giving out enough heat to melt the sodium (and, on occasion, to cause explosion of the surrounding mixture of hydrogen and air). The subtly different behaviour of lithium and sodium with water illustrates both the delicate balance between various energy terms and the importance of kinetic factors in determining what actually happens.

Sodium, like lithium, burns in air; but unlike lithium [(which gives $(Li^+)_3N^{3-}$ and $(Li^+)_2O^{2-}$)], sodium forms no nitride and gives mainly the peroxide $(Na^+)_2O_2^{2-}$. Two other oxides of sodium can also be made: the monoxide $(Na^+)_2O^{2-}$ and the superoxide $Na^+\ O_2^-$ (see p. 105). Many of the differences in the ionic behaviour of sodium and lithium are in accord with the Kapustinskii equation (p. 36) which predicts that increase in lattice energy with decreased anion size is greater for small cations. So, sodium, unlike lithium, forms a hydrogen carbonate, $Na^+\ HCO_3^-$; and sodium hydroxide, carbonate, and peroxide are more stable to heating than are their lithium counterparts. Sodium nitrate decomposes on heating to the nitrite and dioxygen, while lithium nitrate gives the monoxide.

Almost all sodium salts (unlike lithium salts of small or highly charged anions) are very soluble in water, and this might seem to be yet another predictable consequence of the lower lattice energy predicted for the larger cation. But solubility in water depends on the stability of the hydrated ions as well as on the strength of the ionic lattice. Since the larger sodium ion has a less favourable heat of hydration, only slightly offset by a less unfavourable entropy change (see p. 38), the increase in solubility on changing from lithium to sodium is not as great as the decrease in lattice energy might suggest.

Many sodium salts (including sodium chloride) dissolve endothermically in water because the heat liberated by hydration of the ions is inadequate to break down the lattice. The driving force for the breakdown of the NaCl crystal in water is the favourable entropy change, which reflects the large increase in disorder that occurs when the ions become mobile; the effect is somewhat offsest by the order they impose on the solvent molecules. Although both the heat of solution and the entropy of hydration oppose dissolution, these are (for sodium chloride) small quantities, and are overridden by the large favourable entropy change accompanying the release of ions from the crystal. This last term is, of course, similar for all the sodium halides. The variation in solubility with anion size is a result, as is commonly the case, of a delicate balance of energy terms. A large anion forms a weaker

Some crown ethers with different 'hole sizes' GE

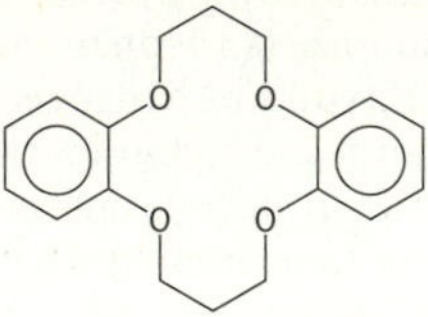

dibenzo-14-crown-4

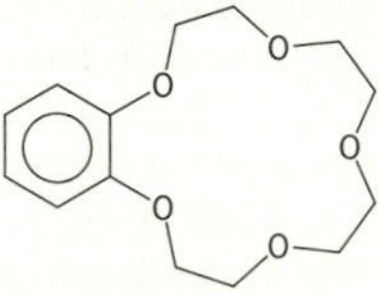

benzo-15-crown-5

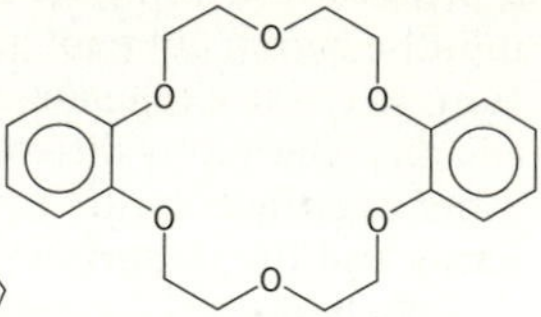

dibenzo-18-crown-6

Comparison of ionic diameters and crown ether 'hole sizes'

Cation	ionic diameter/pm	polyether ring	'hole size'/pm
Li^+	152	14-crown-4	120–150
Na^+	204	15-crown-5	170–220
K^+	276	18-crown-6	260–320
Rb^+	304	21-crown-7	340–430
Cs^+	334	—	—

Limiting ionic conductivity AP, $\lambda/(\mathrm{S\ cm^2\ mol^{-1}})$

In water at 25°C Na^+ 50.1 [Li^+ 38.7]

Sodium in liquid ammonia

$$Na(s) + NH_3(l) \rightarrow Na^+(solv) + e^-(solv) \xrightarrow{Na} Na^+(solv) + Na^-(solv)$$

In presence of crown or crypt, L:

$Na(s) + NH_3(l) + L \rightarrow NaL^+(solv) + e^-(solv)$

or $\rightarrow NaL^+, e^-(solv)$ (unstable)

or $\rightarrow NaL^+Na^-(solv)$ (less unstable)

lattice and has a lower ordering effect on the water molecules. Both factors favour solubility and for sodium halides are dominant, although they are opposed by less exothermic ionic hydration. So solubility is low for the fluoride but increases with the size of the anion.

The sodium ion forms weak complexes with a number of oxygen donor ligands. It is held to strong acid cation exchange resins more strongly than lithium, because, although it has a larger crystal (naked) radius, it has for this very reason a smaller hydrated radius (and also a higher ionic conductivity). For complex formation with ligands such as crowns and crypts, which surround or enclose it, the sodium ion naturally needs a larger hole than the lithium ion. These size-dependent complexes are of particular interest to biochemists because they model the forms in which living organisms control the flow of sodium and potassium ions across cell membranes.

Sodium ions also form complexes with some simple organic groups. These are much less stable than the covalent lithium alkyl clusters such as $Li_4(CH_3)_4$ and appear to be ionic. Less reactive ionic compounds such as $Na^+C_5H_5^-$ are formed between sodium and some π bonded systems, where the 3s electron from sodium can readily be accepted into a delocalized orbital of the ligand.

So far, the chemistry of sodium seems to be that of a very reactive metal and its stable, singly charged cation, which behaves in a predictably ionic way. But the sodium atom has one 3s electron, and could perhaps take part in σ bonding, although the 3s orbital is more diffuse than the 2s and less able to form bonds involving concentrated electron overlap. Some σ bonded Na_2 dimers are, none the less, formed in sodium vapour. More exciting, however, is the formation of the Na^- anion, in which the 3s atomic orbital is filled. Sodium metal, like lithium, dissolves in liquid ammonia to give a blue solution, which contains solvated cations and solvated electrons and which is widely used to reduce organic compounds. As the concentration of sodium in the ammonia is increased, the additional metal atoms mop up the solvated electrons to form solvated Na^- ions. Sodium metal (here, unlike lithium) is oxidized when it dissolves in solutions of some crowns or crypts, L, in ether. The ousted electron may merely be solvated, and form a (very unstable) electride complex NaL^+ e^-(solv); or it may react with another sodium atom to give a more stable alkalide, NaL^+Na^-. So, even sodium can surprise us by forming some compounds that make use of the newly entered $n = 3$ shell.

For summary see p. 131.

Summary

Electronic configuration

- [Ne] $3s^1$

Element

- Soft, low melting, very reactive metal; soluble in mercury to give amalgam
- Na_2 and Na in vapour; yellow emission spectrum

Occurrence

- As $Na^+(aq)$ in oceans, and in deposits of NaCl

Extraction

- By electrolysis of $Na^+(Cl^- + Ca^{2+})$ melts

Chemical behaviour

- Mainly cationic; Na^+ very stable, salts usually soluble in water (except NaF)
- Maximal coordination number of six, e.g. $Na(H_2O)_6^+$
- Burns in air to form $(Na^+)_2O_2^{2-}$, but no nitride (unlike Li)
- Na^+ complexes with oxygen donors (polycarboxylates, crowns, and crypts) and electron-accepting hydrocarbons, e.g. $C_5H_5^-$. Na^+ ions 'pumped' out of living cells by complexing?
- Na in liquid ammonia gives $Na^+(NH_3)_x$ and $e^-(NH_3)_x$ (blue); forms anion $Na^-(NH_3)_x$ at higher concentration in liquid ammonia, and as NaL^+Na^- in cryptates

Uses

- Salt in food, ice clearance, and chemical industry
- Liquid sodium as heat exchanger in nuclear reactors
- Vapour in street lights
- Solid, and Na/NH_3, in organic syntheses

Nuclear features

- Isotopically pure ^{23}Na

Mg Magnesium

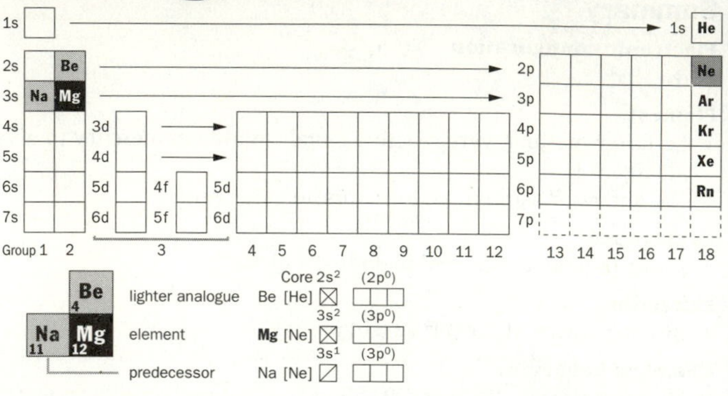

		Predecessor	Element	Lighter analogue
Name		**Sodium**	**Magnesium**	**Beryllium**
Symbol, *Z*		Na, 11	Mg, 12	Be, 4
RAM		22.989768	24.3050	9.01218
Radius/pm	**atomic**	153.7	160	113.3
	covalent	98	136	89
	ionic, M^{2+}		78	34
Electronegativity (Pauling)		1.0	1.3	1.6
Melting point/K		370.96	922.0	1551 ± 5
Boiling point/K		1156.1	1363	
$\Delta H^{\ominus}_{fus}$/kJ mol^{-1}		2.64	9.04	9.80
$\Delta H^{\ominus}_{vap}$/kJ mol^{-1}		89.04	128.7	308.8
Density (at 293 K)/kg m^{-3}		971	1738	1847
Electrical conductivity/Ω^{-1}m^{-1}		23.81×10^6	22.83×10^6	2.5×10^7
Ionization energy for removal of *j*th electron/kJ mol^{-1}	***j* = 1**	495.8	737.7	899.4
	***j* = 2**	4562.4	1450.7	1757.1
Electron affinity/kj mol^{-1}		52.9	–21	–18
$E^{\ominus}$/V for $M^{2+} \rightarrow M$			–2.356	

Stable isotopes of magnesium

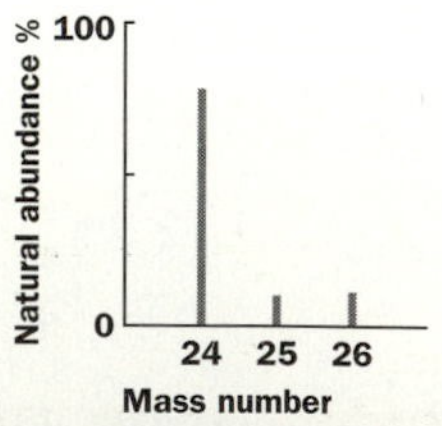

One of the joys of indoor fireworks, or of elementary practical chemistry, is the brilliant white flame we see when magnesium burns in air. The energy needed to form Mg^{2+} and O^{2-} from magnesium, $[Ne]3s^2$, and oxygen is as nothing compared with the energy given out when these doubly charged ions combine to the very stable ionic oxide, $Mg^{2+}O^{2-}$. Since the Mg^{2+} ion is about the same size as Li^+, and equally underformable, magnesium behaves like lithium in several ways; the heat and light given out by burning magnesium is the result of the formation not only of the oxide, but also of a stable ionic nitride, Mg_3N_2.

As expected, magnesium, with its higher nuclear charge, is more difficult to ionize than sodium, but, as predicted by the Kaputstinskii equation, its cation forms stronger ionic lattices. Magnesium fluoride and hydroxide are less soluble in water than their sodium counterparts, despite the higher hydration energy of the cation. Again (see p. 127), the lattice energy seems to be the dominant factor. Magnesium occurs in nature mainly as almost insoluble minerals such as talc and asbestos (which are sodium magnesium silicates) and as dolomite, ($MgCa(CO_3)_2$). But it is also found as the more soluble sulfate and it is present in natural waters. Many salts of magnesium resemble those of lithium rather than their sodium analogues: the carbonate and nitrate of magnesium decompose fairly readily on heating to give the oxide; and neither lithium nor magnesium forms compounds of HCO_3^- or O_2^{2-} under normal conditions. Magnesium, here like both lithium and sodium, gives a carbide ($Mg^{2+}C_2^{2-}$) when heated with carbon at 500°C; but at higher temperatures it behaves idiosyncratically to give a compound written as $(Mg^{2+})_2C_3^{4-}$, although it is most unlikely to contain C^{4-} ions. Magnesium is not quite as electropositive as lithium or sodium. Although the metal is sometimes obtained from its salts by electrolysis, the magnesium ion can also be reduced chemically (by coke at 2000°C or, more usually, by ferrosilicon). Clean magnesium breaks down steam to give hydrogen, but does not react with cold water as it becomes protected by a surface layer of oxide. It is, however, sufficiently electropositive to be used both as a reducing agent and as a sacrificial electrode to protect other metals from corrosion. The dramatic combustion of powdered magnesium in air is exploited by manufacturers of flares and fireworks.

Magnesium metal, with two 3s electrons in the conduction band, is much harder and stronger than sodium. Its low density and high strength, combined with easy machinability, makes magnesium and its alloys invaluable for making a wide range of objects such as aircraft, luggage, and pocket cameras (which, if made out of steel of the same strength, would weigh three to four times as much).

Heat of formation [AP] of MgO/kJ mol^{-1}

MgO, –569.43 [NaCl, with the same structure, –384.14]

Heat of hydration of Mg^{2+}/kJ mol^{-1}

Mg^{2+}, –1920 [Na^+, –405]

Sacrificial corrosion

O_2(aq) H^+(aq) ······ Mg $E^\ominus_{Mg^{2+}, Mg} = -2.36$ V

······ Fe $E^\ominus_{Fe^{2+}, Fe} = -0.44$ V

$E^\ominus_{Fe^{3+}, Fe} = -0.04$ V

Any oxidizing agent, e.g. H^+(aq), or O_2 reacts preferentially with the metal with the lowest (most negative) $E^\ominus$ value, so Mg corrodes before Fe.

Chlorophyll *a* [S]

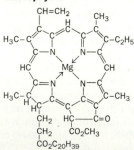

CH=CH$_2$, CH$_3$, H$_3$C–C, C–C$_2$H$_5$, HC, Mg, CH, H$_3$C–C, C–CH$_3$, CH$_2$, HC—C=O, CH$_2$, CO$_2$CH$_3$, CO$_2$C$_{20}$H$_{39}$

Magnesium exhibits almost entirely 'metallic' character. Unlike the very small beryllium ion, Be^{2+}, the charge to size ratio of the magnesium Mg^{2+} ion, and its consequent deforming power, is not high enough to coerce neighbouring anions into covalent overlap. The hydrated ion $Mg(H_2O)_6^{2+}$, unlike $Be(H_2O)_4^{2+}$, is not acidic and $Mg(OH)_2$, unlike $Be(OH)_2$, does not dissolve in aqueous alkali. The Mg^{2+} ion combines with a range of oxygen donating ligands, such as oxalate $(COO)_2^{2-}$. Its inorganic compounds hardly deviate from 'purely ionic' character (although the chloride, bromide, and iodide crystallize in layered rather than continuous lattices, and the ion $MgCl^+$ has been detected in the molten chloride).

The main interest in magnesium chemistry does not, however, lie in its fairly predictable inorganic behaviour, nor in its manifold technical applications. It is in its organic and biochemical contexts that magnesium is at its most impressive. With alkyl halides, RX, it forms a series of Grignard reagents, RMgX, which play a crucial role in organic syntheses. Here, the magnesium is far from ionic and is presumably bound by two σ bonds derived from its 3s orbitals. The wide range of different products that can be made using Grignard reagents is a result of the complexity of the equilibria that occur in the solutions: careful control of solvent conditions allows the position of equilibrium to be shifted almost at will to favour the production of whatever reactant is needed. So we may find: ionic dissociation of RMgX into RMg^+ and X^-; formation of dimers bridged through X, either symmetrically to give $RMg(X_2)MgR$ or asymmetrically to give $R_2Mg(X_2)Mg$; and the former may ionize to RMg^+ and $RMgX_2^-$, while the asymmetric dimer may dissociate to MgR_2 and MgX_2. Such complexity allows the organic chemist great scope for designing syntheses.

In biochemistry, magnesium plays an even more significant role. The element is a major component of life on earth for two reasons. The Mg^{2+} ion sits at the middle of the porphyrin ring that forms the skeleton of chlorophyll molecules; and it is these substances that enable green plants to trap the energy from sunlight and convert it into sugars, which are the starting point of the food chain for all terrestrial life. The magnesium ion gives the ring a rough fourfold symmetry, and the complex absorbs the energy needed for further reactions. Each magnesium ion is also attached to a fifth group, which is joined, through hydrogen bonding and proteins, to other molecules of chlorophyll, and so on, forming a stack of rings through which energy from light can be quickly transmitted. Organisms also need magnesium in most phosphate enzymes: for example, those in carbohydrate metabolism, and for nerve and muscle action. The element is also essential in DNA and related

substances, and in the enzymes that control their replication. But we do not yet know what factors make the magnesium ion so particularly suitable for its vital roles. It may be that it was favoured by evolution not only for its size and charge, which allow it to be held in the right place at the right strength, but also because it is so available. Unlike many heavier metals with apparently more subtle chemical behaviour, magnesium cannot interfere with electron transfer by adopting another oxidation state, nor can it waste energy by causing fluorescence. It may be that the factors that make its chemistry (fairly) predictable are just those that made magnesium a safe bet for natural selection.

Summary

Electronic configuration

- $[Ne]3s^2$

Element

- Strong, light, fairly reactive metal (but passive surface layer of oxide)

Occurrence

- With oxygen in silicates and carbonates and as the sulfate; in natural waters

Extraction

- Reduction of Mg^{2+} in ores by electrolysis or with coke or ferrosilicon

Chemical behaviour

- Mg^{2+} predominant (charge:size ratio similar to Li^+); stable ionic lattices; metal burns in air to MgO and Mg_3N_2; fluoride, carbonate, oxide of low aqueous solubility
- $Mg(H_2O)_6^{2+}$ unhydrolysed
- Halides (except F^-) have layered structures
- Mg^{2+} complexes with O donors and with 4-N porphyrin ring of chlorophyll
- Covalent bonding in Grignard reagents, RMgX, with alkyl halides RX

Uses

- Light alloys
- Sacrificial corrosion
- Flares and fireworks
- Grignard reagents for organic synthesis
- Essential for photosynthesis and many animal processes

Nuclear features

- ^{24}Mg: ^{25}Mg: $^{26}Mg \sim 8:1:1$

Al Aluminium

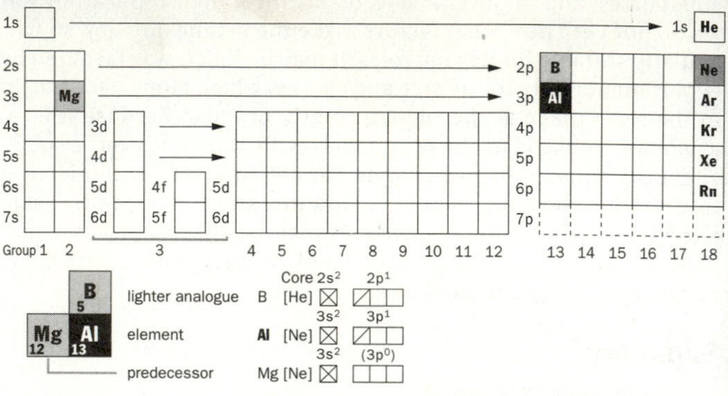

	Predecessor	Element	Lighter analogue
Name	**Magnesium**	**Aluminium**	**Boron**
Symbol, *Z*	Mg, 12	Al, 13	B, 5
RAM	24.3050	26.98154	10.81
Radius/pm **atomic**	160	143.1	83
covalent	136	125	88
ionic, M^{3+}		57	23
Electronegativity (Pauling)	1.3	1.6	2.0
Melting point/K	922.0	933.52	2573
Boiling point/K	1363	2740	3931
$\Delta H^{\ominus}_{fus}$**/kJ mol^{-1}**	9.04	10.67	22.2
$\Delta H^{\ominus}_{vap}$**/kJ mol^{-1}**	128.7	293.72	538.9
Density (at 293 K)/kg m^{-3}	1738	2698	2340
Electrical conductivity/Ω^{-1}m^{-1}	22.83×10^6	37.67×10^6	5.55×10^{-5}
Ionization energy for removal of *j*th electron/kJ mol^{-1} ***j* = 1**	737.7	577.4	800.6
***j* = 2**	1450.7	1816.6	2427
***j* = 3**		2744.6	3660
Electron affinity/kj mol^{-1}	−21	44	
Bond energy/kg mol^{-1} **E–E**		~ 200	
E–H		285	
E–C		225	
E–O		585	
E–F		665	
E–Cl		498	

'Thermit' process

$Cr_2O_3(s) + Al(s) \rightarrow Al_2O_3(s) + Cr(s) \quad \Delta H^{\ominus} \sim -330$ kJ mol^{-1}

Aluminium is the third most common element (after oxygen and silicon) in the earth's crust. With an electronic structure of [Ne]$3s^23p^1$ it is not surprising that, despite the increased nuclear charge, the first electron of aluminium is lost (from the 3p shell) more easily than the first electron (from the 3s shell) of magnesium; but the energy required to remove the first *two* electrons is greater for aluminium. Aluminium (like magnesium [Ne]$3s^2$, but unlike boron [He]$2s^22p^1$) is, predictably, a metal but, with all three $n = 3$ electrons in its valence band, its metallic bonding is somewhat stronger than in magnesium. The energy given out when the Al^{3+} ion forms compounds, whether these are finally 'ionic' or 'covalent', is sufficiently large to ensure that aluminium exhibits only the oxidation number of + 3.

The small, highly charged 'hard' Al^{3+} ion (unlike the even more highly polarizing B^{3+}) can exist, provided that it is combined with similarly undeformable 'hard' anions such as F^-, OH^-, and O^{2-}, or forms complexes with donor oxygen or fluoride atoms, as in $Al(OH_2)_6^{3+}$ and AlF_6^{3-}, where the charge is much less concentrated than on the naked ion. But 'softer' atoms, such as bromine, are so deformed by the highly concentrated charge on the Al^{3+} ion that covalent overlap can occur.

However, the chemical behaviour of aluminium, particularly that outside the laboratory, is dominated by its very strong interaction with oxygen. It is found as many aluminosilicates, such as mica and feldspar, and as bauxite [(a hydrated oxide of formula which varies between AlOOH and $Al(OH)_3$]. The metal is extracted by electrolysis of bauxite in molten cryolite (Na_3AlF_6), since the Al_2O_3 lattice is too strong for chemical reduction except by very electropositive metals such as sodium (which are expensive and hazardous). Indeed, aluminium combines so powerfully with oxygen that the powdered metal can be used (as in the 'Thermit' process) for the extraction of other metals, such as chromium and manganese, whose oxides are too stable to be reduced with coke.

There are many other examples of the strong affinity of aluminium for oxygen. When powdered it combines explosively with liquid oxygen and has been used as a rocket propellant. Aluminium amalgam reduces water to hydrogen; the hydration of the Al^{3+} ion to $Al(OH_2)_6^{3+}$ supplies more than enough energy to drive the unfavourable conversion of the metal to its gaseous cation. Solid aluminium dissolves in moderately dilute acids for the same reason. It also dissolves in concentrated alkali to give hydrogen and an oxoanion, probably AlO_2^-. The preference of aluminium for oxygen or fluorine is demonstrated by the impossibility of using aqueous solutions to make any aluminium salts of weak acids (except

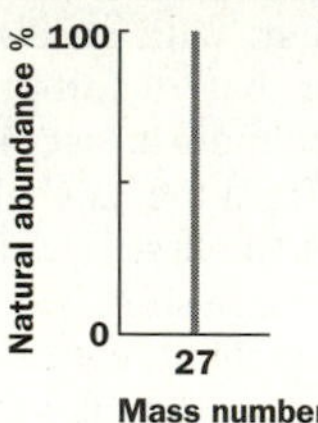

Reaction of Al^{3+}(aq) with aqueous anions

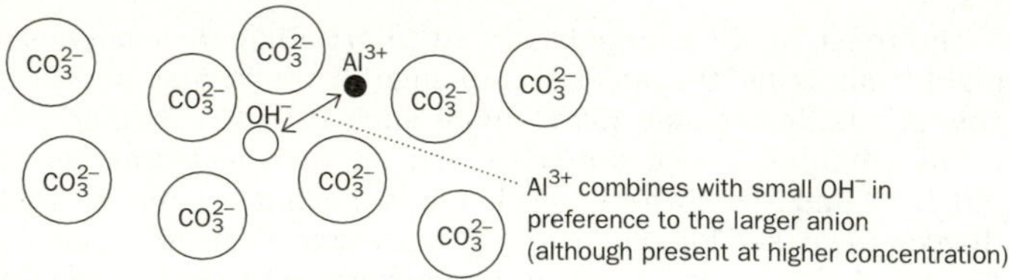

Hydrolysis of Al^{3+}(aq)

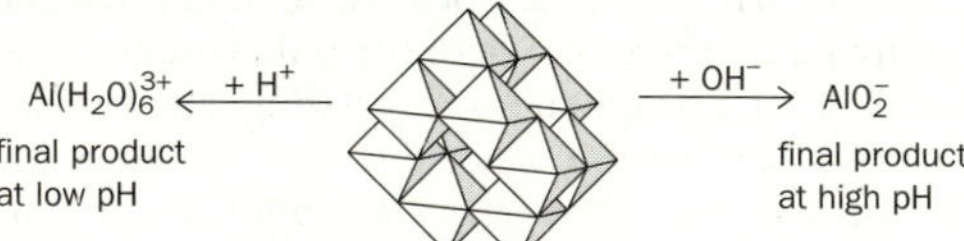

Arrangement of twelve AlO_6 octahedra round the central AlO_4 tetrahedron in $[AlO_4Al_{12}(OH)_{24}(H_2O)_{12}]^{7+}$

[See *Journal of the Chemical Society, Dalton Transactions*, 1347 (1988)]

the fluoride). When salts such as sodium sulphide, carbonate, or cyanide are dissolved in water, some anions will become protonated, and so will generate hydroxyl ions. If aluminium ions are added, they will combine with the hydroxyl ions to precipitate the hydroxide, rather than join up with the larger anions, even though these are present in much higher concentrations. Aluminium hydroxide is amphoteric, and dissolves both in dilute acids to form $Al(OH_2)_6^{3+}$ and in alkalis to give AlO_2^-, at intermediate values of pH the composition of the solutions is very sensitive to concentration and contains a range of polymeric hydroxo complexes. In hydrolysed, but mildly acidic, solutions, an important species is $AlO_4Al_{12}(OH)_{24}(H_2O)_{12}^{7+}$, in which twelve AlO_6 octahedra are grouped round a tetrahedral AlO_4 unit. However, we know less about those species that occur in mildly alkaline solutions. It is not surprising that such an electropositive metal reacts even with damp air, to give Al_2O_3; but the oxide is so close packed a structure that it forms a very thin continuous layer over the surface of the metal and protects it from further attack. Thicker layers of oxide may be produced by oxidizing the metal either with concentrated nitric acid or by using it as the anode in electrolysis.

Aluminium is produced on a vast scale for a multitude of uses, mainly in the construction of aircraft and in the aerospace and building industries. It is light, strong, can be polished, is very easily worked, and, thanks to its coating of oxide, is not at risk from corrosion. A thin oxide film can be given an attractive weatherproof finish and a thicker anodically oxidized layer can even be coloured since the surface of the oxide provides sites for the strong attachment of molecules of dyes. Weight for weight, aluminium has about twice the electrical conductivity of copper and so is much used for cables.

The strong interaction between aluminium and oxygen produces a number of interesting and useful compounds. Alumina itself, Al_2O_3, has several structural manifestations. The most closely packed structure, α-corundum and its impure form, emery, is nearly as hard as diamond and is used as an abrasive. It is occasionally found coloured by various metallic impurities such as chromium (often red, but see p. 234), and iron(II) with titanium(IV) (blue). These gemstones, ruby and sapphire, can now also be produced synthetically. There is also a softer form of Al_2O_3, together with several layer and chain structures of AlO.OH and $Al(OH)_3$, which involve hydrogen bonding. Fibres of Al_2O_3 may now be made commercially and woven into heat-resistant fabric or used, like carbon fibres, to strengthen metals.

Mixed oxides containing a second type of metal ion include 'β-alumina', which in fact also contains sodium ions. These are

Spinel, $MgAl_2O_4$ MM

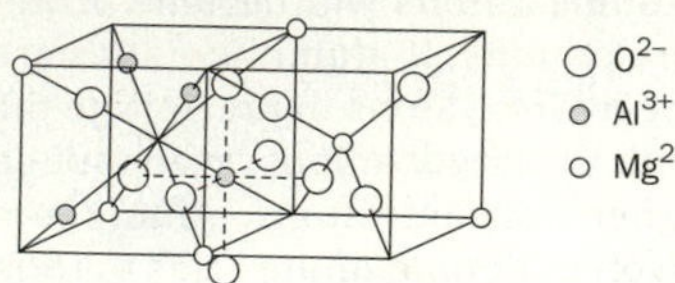

The cyclic anion, $Al_6O_{18}^{18-}$ in Portland Cement GE

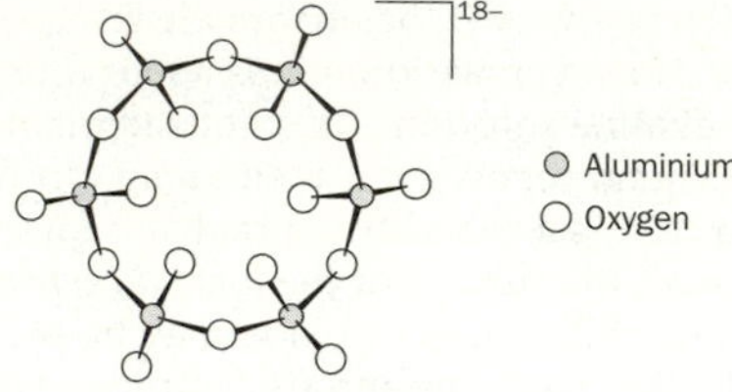

accommodated rather sparsely in separate layers where they are free to move, and so make the material a good electrical conductor. Spinel, $MgAl_2O_4$, has an interesting structure because the Al^{3+} ions are surrounded (octahedrally) by six O^{2-} ions while the Mg^{2+} ions, with a lower charge, are surrounded (tetrahedrally) by only four, an arrangement that maximizes the stability of the lattice. The calcium salt of the cyclic anion $[Al_6O_{18}]^{18-}$ is a component of Portland cement which sets under water, partly because of the ready formation of hydrates within the very open structure.

It is thought that aluminium plays some part in the nervous disorders of people who suffer from Alzheimer's disease in old age. The element seems to do no harm to healthy tissue, and, indeed, the hydroxide is often used to treat indigestion by lowering the acidity of the stomach. But, if some defect is already present, aluminium may then cause trouble, perhaps by interacting with phosphate groups, which play a large role in our metabolism. This would be yet another example of the affinity of aluminium for charged oxygen atoms.

A metal as electropositive as aluminium naturally reacts also with elements other than oxygen; indeed, it forms alloys with most metals and also a number of interesting compounds in which it is linked to a non-metallic element. Although aluminium is, of course, larger than either boron (with the same outer electronic structure s^2p^1) or beryllium (with an ion of similar charge:size ratio), it resembles both these elements in being unable to complete its ns^2p^6 octet merely by σ bond formation. But aluminium can sometimes accommodate more than four electron pairs in its valence shell, as in AlF_6^{3-}. This 'octet expansion' (see p. 151) is not possible for the $n = 2$ elements.

Like beryllium but unlike boron, aluminium forms no gaseous hydrides, but a polymerized solid hydride $(AlH_3)_x$; in one form the metal atom is surrounded by six hydrogen atoms, all envisaged as forming 3-centre–2-electron bridges to other aluminium atoms. With Lewis bases, such as $N(CH_3)_3$, the polymer disintegrates to give the tetrahedral, complete-octet compound $H_3AlN(CH_3)_3$, and this can add another molecule of base to give $H_3Al(N(CH_3)_3)_2$ which has a trigonal bipyramidal structure, with an electronic configuration expanded beyond the octet. Reaction of aluminium with hydrogen in the presence of sodium or lithium gives $MAlH_4$, containing the group AlH_4^-. Like the simple hydride, these compounds react with traces of water, but the lithium salt is stable in dry ether and is widely used as a versatile reducing agent.

Amongst other binary compounds, the sulfide Al_2S_3 can be made by direct combination (but not by reaction in water, see

Aluminium trihalides: melting points/°C

AlF_3 1290 $AlCl_3$ 192.4 $AlBr_3$ 97.8 AlI_3 189.4

Aluminium trihalide dimers

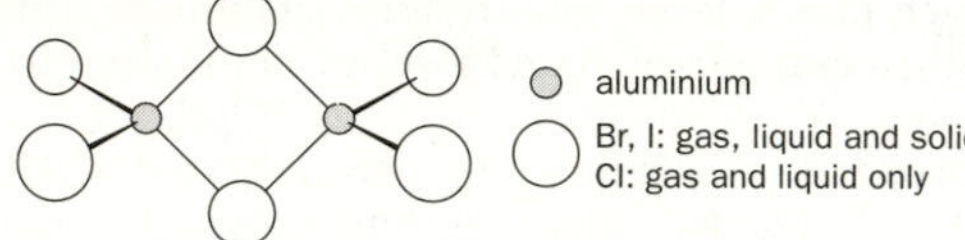

p. 141). There is also a high melting nitride AlN. Only four sulphide or nitride ions can be fitted round the small Al^{3+} ion. The carbide, AL_4C_3, gives methane on hydrolysis, but, none the less, is unlikely to contain C^{4-} anions, particularly in the presence of a small highly charged cation such as Al^{3+}.

Aluminium forms all four trihalides, whose properties vary most pleasingly with the size and polarizability of the halogen atom or ion. The fluoride, which has a very high negative enthalpy of formation, is clearly ionic with a high melting point and a coordination number of six, as we see also in the complex ion AlF_6^{3-} in cryolite and in aqueous fluoride solutions. The bromide and iodide are much less stable, low melting, volatile dimers in which two AlX_2 groups are linked by two bridging halogens giving the metal atom a coordination number of four. We may consider that each AlX_3 molecule acts both as a Lewis base and a Lewis acid, donating and accepting an electron pair. The chloride shows intermediate behaviour. Much less stable than the fluoride, somewhat more so than the bromide and iodide, it has a six-coordinate metal atom in the solid, but it melts and vaporizes readily, and exists as the dimer in both the melt and the vapour. Above 200°C, some dimer molecules dissociate into planar $AlCl_3$ monomers similar to BCl_3. The ability of $AlCl_3$ to act as a Lewis acid is exploited for catalysing Friedel–Crafts organic syntheses. Reaction of an organic halide, RX for example, may give $AlCl_3X^-$ and an organic cation, R^+, which undergoes further reaction.

Aluminium also forms monohalides, AlX, which readily disproportionate at room temperature to the metal and the trihalide. They are interesting as our first example of the 'inert pair effect' in which an element forms a compound by using its np^x electrons, but not its ns^2 electrons (see p. 503).

It might seem surprising that although boron and aluminium have analogous ns^2np^1 configurations, aluminium trichloride dimerizes but boron trihalide does not. However, the mainly σ bonding in BCl_3 is thought to be strengthened by some π bonding, which involves donation of electrons from the filled 3p orbital of the chlorine into vacant 2p orbitals of the boron (see p. 57). Multiple bonding is less common for $n = 3$ than for $n = 2$ elements, and so aluminium would be less likely to take part in π bonding than boron, probably because 3p–3p overlap would be less efficient. For each aluminium atom, the most stable arrangement involves four σ bonds (with no 3p–3p π bonding), unlike the smaller boron which favours three σ bonds with some 2p–3p π reinforcement.

Aluminium, unlike boron (and also unlike the heavier ns^2np^1 elements, with $n \geqslant 4$), forms trialkyls and triaryls that are also dimeric

Bridging group angles [GE] in some Al_2X_6 compounds

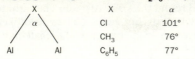

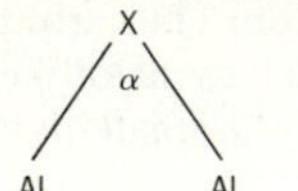

X	α
Cl	101°
CH_3	76°
C_6H_5	77°

Summary

Electronic configuration

- [Ne]$3s^2 3p^1$

Element

- Strong, light, easily worked metal, good electrical conductor, protected by oxide coating

Occurrence

- With oxygen (as hydrated oxide and aluminosilicates) and fluorine

Extraction

- From molten cryolite, Na_3AlF_6, by electrolysis

Chemical behaviour

- Small ('hard') Al^{3+} has strong affinity for O and F, and forms mainly ionic compounds; maximum coordination number (CN) six (e.g. AlF_6^{3-} and $Al(H_2O)_6^{3+}$)
- Al_2O_3 very stable; $Al(OH)_3$ amphoteric; Al^{3+} (aq) strongly hydrolysed to polynuclear cations and anions
- $(AlH_3)_x$ solid (3c–2e bridges); with Lewis base, B, depolymerizes $AlH_3.B$ and $AlH_3.(B)_2$
- Halides:
  AlF_3 high melting ionic solid (CN6)
  $AlCl_3$ (s) similar, but Lewis acid and volatilizes to Al_2Cl_6 (CN 4, with Cl bridging by lone pair donation)
  Al_2Br_6 and Al_2I_6 unstable, dimeric in all phases,
  unstable AlX
- Dimeric Al_2R_6 trialkyls and triaryls (3c–2e)

Uses

- Metal in construction industries, and for electrical cables; as a reductant for rocket fuels; pyrotechnics
- Al_2O_3 in fibres; $(Al_6O_{18})^{18-}$ in Portland cement
- In organic syntheses: $Li^+AlH_4^-$ as reductant, $AlCl_3$ and Al_2R_6 as catalysts

Nuclear features

- Isotopically pure ^{27}Al

and, presumably, contain two 3-centre–2-electrons bonds. The bond angle at the bridging group is much tighter than in the dimeric trihalides, which have bridging atoms larger than carbon. The aluminium alkyls are unstable but extremely useful. They react with and reduce titanium tetrachloride, with shuffling of alkyl and chloride groups, to give solids that catalyse the polymerization of olefines. These Ziegler–Natta, and related, catalysts are important because different ones can be used to make polymers of different structure, such as a highly ordered form of polythene with most of the methyl groups on the same side of the carbon chain.

Aluminium resembles boron by forming compounds containing four- or six-membered rings of alternate R−Al < and R′−N < groups; and these rings may be fused into clusters. It seems that organometallic chemists can cajole aluminium into a much more varied behaviour than would be suggested by the strong preference for oxygen that it shows outside the laboratory.

Si Silicon

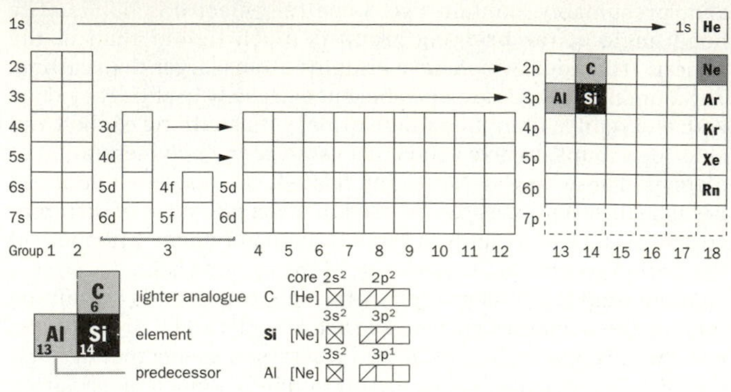

	Predecessor	Element	Lighter analogue
Name	**Aluminium**	**Silicon**	**Carbon**
Symbol, *Z*	Al, 13	Si, 14	C, 6
RAM	26.98154	28.0855	12.011
Radius/pm atomic	143.1	117	
covalent	125	117	77
Electronegativity (Pauling)	1.6	1.9	2.6
Melting point/K	933.52	1683	~3820 (d)
Boiling point/K	2740	2628	
$\Delta H^{\ominus}_{fus}$**/kJ mol**$^{-1}$	10.67	39.6	150.0
$\Delta H^{\ominus}_{vap}$**/kJ mol**$^{-1}$	293.72	383.3	710.9
Density (at 293 K)/kg m^{-3}	2698	2329	3513 (d)
Electrical conductivity/$\Omega^{-1}m^{-1}$	37.67×10^6	1×10^3	1×10^{-11} (d)
Energy/of band gap in solid/kJ mol^{-1}	0	106.8	~ 580 (d)
Ionization energy for removal of *j*th electron/kJ mol^{-1} ***j* = 1**	577.4	786.5	1086.2
***j* = 2**	1816.6	1577.1	2352
***j* = 3**	2744.6	3231.4	4620
***j* = 4**		4355.5	6222
Electron affinity/kj mol^{-1}	44	133.6	121.9
Bond energy/kg mol^{-1} **E–E**	~ 200	226	348
E–H	285	326	411
E–C	225	301	348
E–O	585	452	
E–F	665	582	
E–Cl	498	391	614

Silicon is the second most common element (27.2% w/w) in the earth's surface and always occurs in combination with the commonest one, oxygen. We should expect silicon, with an electronic configuration of [Ne]$3s^2 3p^2$, to be a non-metal and to form only covalent compounds. Like carbon, [He]$2s^2 2p^2$, silicon could promote one of the s electrons into the empty p orbital and use the four half-filled sp^3 orbitals to form four tetrahedral σ bonds. Like its predecessor aluminium, [Ne]$3s^2 3p^1$, but unlike carbon, it would be able to accommodate more than four pairs of electrons in its valence shell. With its higher nuclear charge, silicon would be smaller and more electronegative than aluminium. But silicon would, of course, be larger and less electronegative than carbon, and so form a weaker bond than carbon does to hydrogen and a slightly weaker one to carbon; bonds with these elements are not very polar and better overlap is achieved between smaller atoms, provided that there is no overcrowding. With other non-metals, however, the greater ionic character of the bond with silicon would make for a greater bond strength and silicon predictably forms its strongest bonds with electronegative atoms: oxygen, fluorine, and chloride. Moreover, for larger atoms such as chlorine, there is more space around silicon than around carbon. Increased size is, however, an overriding disadvantage for the formation of double or triple bonds; the 3p orbitals are too extended to allow for good p_π–p_π overlap as in the strong multiple bonds formed by the $n = 2$ elements carbon, nitrogen, and oxygen. We shall see that, in many ways, silicon behaves as predicted.

Silicon can be extracted from naturally occurring silica, SiO_2, by heating it with coke (keeping the silica in excess to prevent the carbide SiC from being formed). It has a structure like diamond, but, since the bonding is weaker, it is more volatile. There is a smaller energy gap between the occupied and unoccupied orbitals in silicon than in diamond and some of the outer electrons, which are less tightly held than in carbon, may be promoted from the full valence band to the empty conduction band, particularly at higher temperatures. Silicon is therefore a semiconductor. Its conductivity may be increased by adding small amounts of impurities, for example phosphorus ([Ne]$3s^2 3p^3$) which has one more electron than silicon, or aluminium ([Ne]$3s^2 3p^1$) which has one fewer. These give the material an excess of either *n*egative or *p*ositive charge and so are known respectively as *n*-type or *p*-type semiconductors[A] (see p. 295). Junctions between the two types of material are the basis of the transistor and of the 'silicon chip' in computer memories.

At room temperature, silicon, like aluminium, has a very thin protective coating of oxide, and so reacts with no other element except fluorine. At higher temperatures it reacts with the other

Stable isotopes of silicon

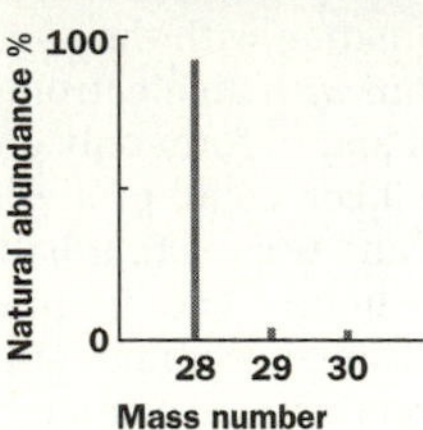

Silicon tetrafluoride, SiF_4

Tetrahedral molecule, melts –90°C

[AlF_3: extended lattice, sublimes 1291°C]

Trisilylamine	**Trimethylamine**
$N(SiH_3)_3$	$N(CH_3)_3$
planar	tetrehedral

A possible scheme for d orbital involvement GE

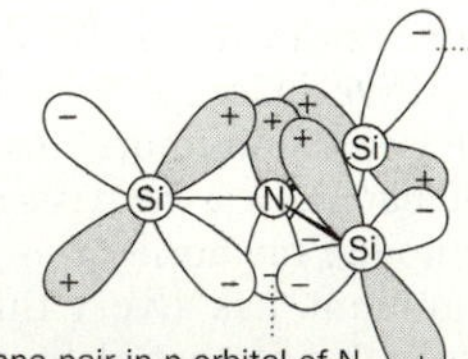

Empty d orbital of Si

Overlap is possible only when the orbital lobes have the same sign

Lone pair in p orbital of N

halogens, and at about 1000°C will also combine with oxygen, nitrogen, phosphorus, and sulfur. When molten, it is even more reactive and readily acts as a reducing agent, being itself converted to the very stable oxide SiO_2. Silicon combines with carbon to form carborundum, SiC, which has a diamond-type structure, and since SiC is extremely hard and resistant to heat and chemical attack, it is used in heater elements and as an abrasive.

Silicon reacts with many transition and pre-transition metals to form silicides which, like borides (see p. 63), show a wide variety of structures and formulae, ranging from M_6Si to MSi_6. Silicides of the pre-transition metals often react with water, sometimes to give hydrogen, but more often to give a hydride of silicon. For example, alloys of silicon and aluminium give SiH_4, and Mg_2Si gives Si_2H_6; but $CaSi_2$ gives hydrogen and polysilene $(SiH_2)x$. Higher straight chain silanes, Si_nH_{2n+2}, are known for valves of n up to 8, and a few cyclic, branched, and even unsaturated silicon hydrides have been synthesized since 1980. The higher hydrides are very unstable in air, but even the lighter ones are more reactive than their carbon analogues. The silicon atom is more prone to attack than carbon (see p. 65), as it is large enough to accommodate more than four atoms around it. This of course means that it has more than four electron pairs in its valence shell, and is showing 'octet expansion' or 'hypervalence'. Since any attacking group contributes an orbital, as well as an electron pair, it is not necessary to assume that silicon makes use of its 3d orbitals, although, since they are of only slightly higher energy than the 3p, it may well do so. The bond formed by hydrogen to silicon is more polar than that to carbon and this also makes the silicon compounds more attractive to attacking groups.

The silicon halides differ markedly from those of carbon. Silicon tetrachloride, unlike its kinetically inert carbon analogue tetrachloromethane (see p. 65), is hydrolysed by water, doubtless again because of octet expansion. The tetrafluoride is a volatile molecular compound, here resembling tetrafluoromethane rather than the high melting ionic aluminium fluoride. In its fluorides and chlorides, silicon shows its greatest tendency for catenation, forming silicon–silicon chains in halides Si_nX_{2n+2} up to $n = 16$ for the fluorides and $n = 6$ for the chlorides. It may be that the σ bonds from silicon to the electronegative halogen are reinforced by π bonds formed by donation from the full p orbitals of the halogens to the empty 3d orbitals of the silicon.

Some evidence for p_π–d_π bonding in silicon compounds is provided by the shapes of the molecules $N(SiH_3)_3$ (planar) and H_3SiNCO and H_3SiNCS (linear). (In the carbon analogues, there can be no π

Silicate ions — a small selection

Simple ions [GE]

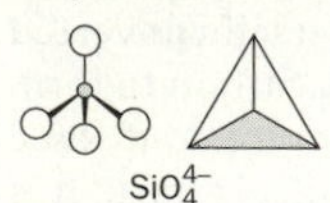

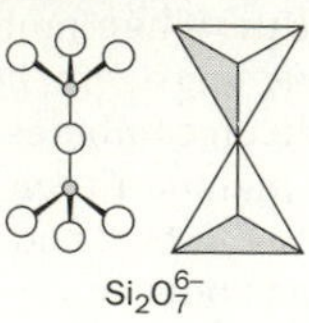

SiO_4^{4-} $Si_2O_7^{6-}$

Single strand chains [GE]

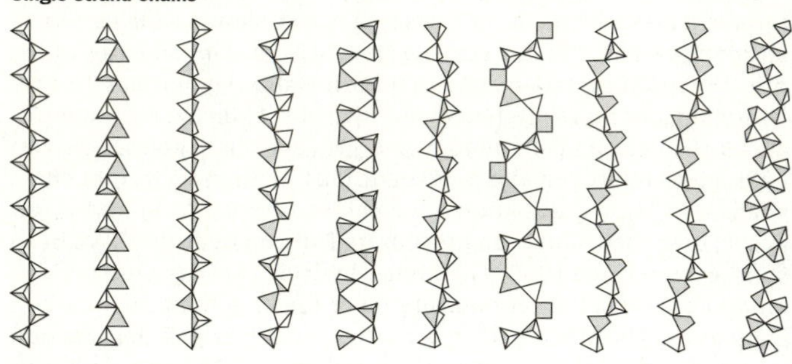

Rings [GE]

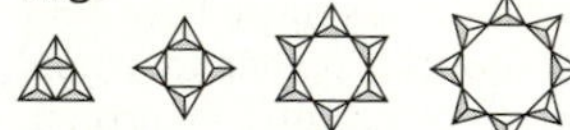

Double strands [GE]

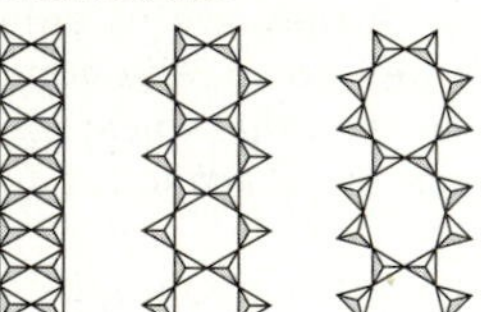

Layers [PIC]

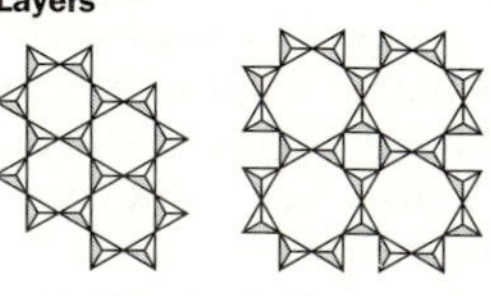

Cages [SAL]

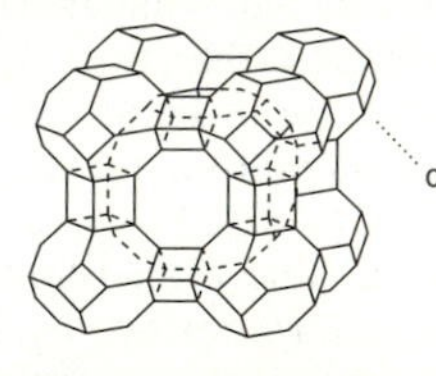

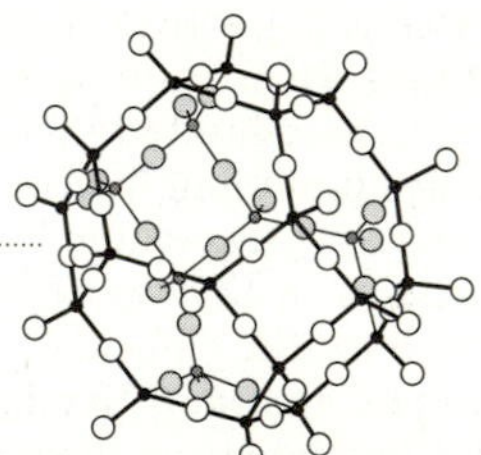

a type-A zeolite 'supercage'

bonding from nitrogen to methyl and so trimethylamine is tetrahedral whilst methyl isocyanate and isothiocyanate are bent.)

Silicon, like aluminium, forms a few unstable 'inert pair' compounds such as SiO and $SiCl_2$, which are stable at higher temperatures. The discussion so far might give the impression that the chemistry of silicon is a pale, distorted, and fragmentary reflection of organic chemistry. But the element shows its true richness elsewhere: in its oxygen-containing compounds. Since any silicon to oxygen p_π–p_π bonds would be very weak, silicon invariably binds oxygen with four tetrahedral σ bonds. The number of ways in which these tetrahedra can be linked up is the basis of the immense variety of silicate chemistry. With four oxygen atoms per group, the possibilities exceed even those for joining up the BO_3^{3-} groups to form borates (see p. 55).

The parent compound silica, SiO_2, can exist in several forms. cristobalite has a diamond-like structure with an oxygen atom in between each silicon; quartz contains spirals and so can form optically active crystals; synthetic quartz which is used in clocks and watches is asymmetric and piezoelectric (i.e. pressure produces a gradiant of charge across the crystal); vitreous silica is a glass with no long-range order and is used in laboratory apparatus because it is fairly inert to chemical attack, except by fluorine, hydrogen fluoride, and concentrated alkalis.

A hydrated form of silica, prepared by acidifying aqueous sodium silicate, has some of the Si–O–Si links broken and replaced by two Si–OH end groups. A confusing mixture of variously condensed and hydrated silicate ions and 'silicic acids' seems to be formed. Removal of water gives the finely divided 'silica gel' which reabsorbs water avidly and is used as a drying agent. Mixtures of silica fused with alkali metal oxides, such as Na_2O, give glasses on cooling, and traces of oxides of transition metal or of p block elements can be added to provide a colour or opacity. These vitreous materials provide not only glass, but also pottery glazes and enamels.

Silica and the silicates form about 95% of the earth's rocks and their erosion products: sands, soils, and clays. The SiO_4 tetrahedra may (occasionally) be discrete. More often they are joined: in pairs, into small or larger closed rings, in chains (straight or variously undulating, single or doubly stranded), into sheets, or into continuous 3D arrays. The properties of the minerals sometimes give us a strong hint as to their structure. The fibrous asbestos is a chain silicate, while mica, with its peeling flakes, has a layer structure. As we have seen (p. 139), the silicon atoms may be partially replaced by aluminium, and some of these may be displaced by magnesium.

From silicates to silicones GE

When the terminal O^- groups (shaded) in silicates are replaced by e.g. $-CH_3$ and $-C_6H_5$, the result is a siloxane or 'silicone'

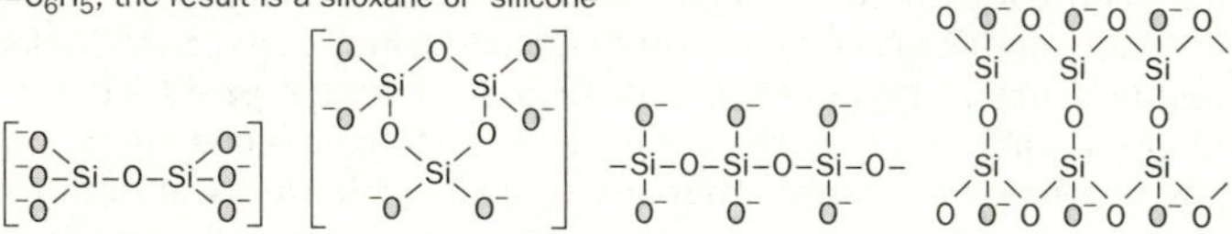

simple ion

ring

single strand chain (gives oily silicones)

double strand chain 'ladder' (gives resinous silicones)

Since the SiO_4 groups are negatively charged, natural silicates must also contain cations; and, even with the requirements of size and overall electrical neutrality, here too there is much scope for variety. Silicate minerals may also hold various, and sometimes vast, amounts of water, as in the vermiculites, where layers of water can be accommodated between complex magnesium alumino-silicate sheets. Aluminosilicates can also form 3D polyhedral porous frameworks containing cavities. These cavity structures may themselves be linked by channels to form fibres, sheets, or 3D arrays. It is possible to synthesize these zeolites with particular sizes of pore, channel, and cavity for use as 'molecular sieves', to remove water and other small molecules; as ion exchange resins, to separate straight chain from branched hydrocations; and as catalysts, where the active site may be at the intersection of two channels.

Chemists have exploited the richness of the silicates to create and extend another very varied region of silicon chemistry. Although silicon bonds somewhat less strongly to carbon than to oxygen, it is possible to replace up to three of the oxygen atoms in the SiO_4 group by methyl groups. The $(CH_3)_nSiO_{(4-n)}$ ($n \geqslant 3$) tetrahedra are linked together by sharing the remaining oxygen atoms to form silicones, which consist of chains, rings, or ladders of alternate silicon and oxygen atoms, with each silicon carrying enough methyl groups to complete its octet with a total of four σ bonds. Silicone polymers are water repellent; non-toxic; stable to chemical attack, extremes of temperature, UV radiation; and they are widely used as oils, greases, emulsifiers, polishes, rubbers, and resins. (The analagous carbon compounds, $R_2C{=}O$ are ketones, which do not polymerize; but it is no surprise that oxygen participates in π bonding more readily with carbon than with silicon.)

Silicon can also be surrounded tetrahedrally by four sulfur atoms. Silicon disulfide, SiS_2, consists of chains of such groups joined by sharing not just one sulfur atom (as might be guessed from the silicates) but two bridging atoms. However, it is instantly decomposed by water.

The chemistry of silicon emphasizes how much variety can be achieved with apparently rather little scope. Almost all its compounds are built on a framework of four tetrahedral σ bonds, joining silicon either strongly to oxygen or fairly strongly to carbon. It is perhaps this constancy that excludes it from any apparent physiological role, at least in vertebrates (although it does do physiological damage, which is often lethal, to the lungs of those who have inhaled dust from some silicates such as asbestos). Some unicellular organisms such as diatoms do, however, use silicates as an

external structural framework, and vast deposits of the mineral kieselguhr are the remains of their prehistoric forebears. And if the sharp edge of a blade of grass or the sting of a nettle should cut or prick like a glass splinter, our feelings do not greatly deceive us: on the leaves of both plants are deposits of silica, in the case of the nettle shaped like the hollow needle of a syringe through which it injects its acid.

Summary

Electronic configuration

- $[Ne]3s^2 3p^2$

Element

- Solid semiconductor (diamond-type structure with smaller band gap), protected by oxide

Occurrence

- As oxide SiO_2 (silica) or silicated in most rocks, in vast variety

Extraction

- From silica by reduction with coke

Chemical behaviour

- Forms no ionic Si^{4+}, and negligible Si=Si (too large)
- Many compounds have four σ bonds (from four half-full orbitals), Si–O strong, e.g. in SiO_2 and SiO_4^{4-} in great variety of minerals
- Halides volatile, tetrahedral, readily hydrolysed (octet expansion)
- Wide variety of metal silicides; SiC similar to diamond
- Hydrides of lower stability and of less variety than carbon
- Wide variety of synthetic silicones $R_nSiO_{(4-n)}$
- Unstable Si^{II} compounds, e.g. $SiCl_2$ and SiO

Uses

- Silicon in electronics
- Silicon carbide as abrasive
- Quartz for time keeping
- Sand in the building industry
- Silicates in glass, glazes, and enamels
- Aluminosilicates for ion exchanges and molecular sieves
- SiF_6^{2-} in glass etching
- Silicones as resins, lubricants, polishes, etc.

Nuclear features

- Mainly ^{28}Si, some ^{29}Si and ^{30}Si

P Phosphorus

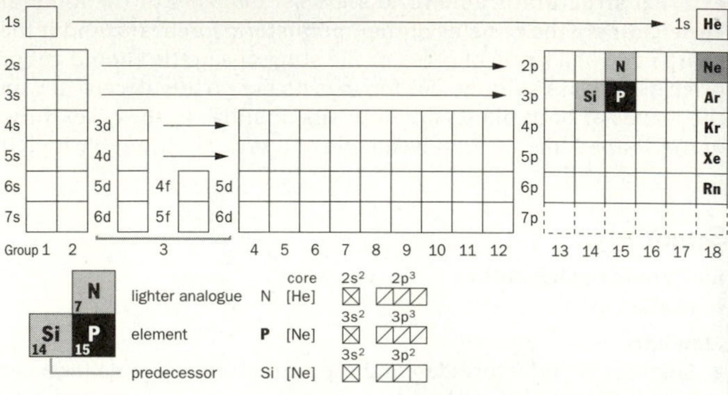

		Predecessor	Element		Lighter analogue
Name		**Silicon**	**Phosphorus**		**Nitrogen**
Symbol, *Z*		Si, 14	P, 15		N, 7
RAM		28.0855	30.973762		14.00674
Radius/pm	**atomic**	117	93 (white); 115 (red)		71
	covalent	117	110		70
	ionic, X^{3-}		212		171
Electronegativity (Pauling)		1.90	2.2		3.0
Melting point/K		1683	317.3	P_4	63.29
Boiling point/K		2628	553	P_4	77.4
$\Delta H^{\ominus}_{fus}$/kJ mol^{-1}		39.6	2.51	P_4	0.72
$\Delta H^{\ominus}_{vap}$/kJ mol^{-1}		383.3	51.9	P_4	5.577
Density (at 293 K)/kg m^{-3}		2329	1820	P_4	
Electrical conductivity/Ω^{-1}m^{-1}		1×10^3	1×10^{-9}	P_4	
Ionization energy for removal of *j*th electron/kJ mol^{-1}	***j* = 1**	786.5	1011.7		1402.3
	***j* = 2**	1577.1	1903.2		2856.1
	***j* = 3**	3231.4	2912		4378.0
	***j* = 4**	4355.5	4956		7474.9
	***j* = 5**		6273		9440.0
Electron affinity/kj mol^{-1}		133.6	72.0		−7
Bond energy/kg mol^{-1}	**E–E**	226	209		160
	E–H	326	328		390
	E–C	301	264		305
	E–O	452	407		200
	E–F	582	490		272
	E–Cl	391	319		193

Strength of P = O bond in $X_3P = O$

X	F	Cl	Br
Bond enthalpy/kJ mol^{-1}	544	511	498

Phosphorus seems a lively element, which glows in the dark (if air is present) and is essential for biological energy transfer and for heredity. Indeed, it was first isolated from material of recent animal origin: human urine.

With an electronic configuration of [Ne]$3s^2 3p^3$, phosphorus, like nitrogen, can complete its octet by forming three σ bonds, to give compounds such as PX_3 which use the lone pair of electrons to act as Lewis bases. Unlike nitrogen, however, but like silicon, phosphorus can also expand its octet (see p. 151).

As with compounds of silicon, σ bonds formed by electron pair donation from $:PX_3$ to a Lewis acid might be strengthened by some flow of electric charge back from the acid into the empty 3d orbitals of the phosphorus. Phosphorus would then be acting both as a σ donor (Lewis base) and as a π acceptor (Lewis acid) towards the same atom.

Another more obvious manifestation of octet expansion is the existence of compounds such as the pentahalides PX_5 in which phosphorus forms five σ bonds. There are of course no analogous compounds of nitrogen, which, like all the $n = 2$ elements, is too small to accommodate more than four (small) atoms round it. Pentavalent phosphorus compounds can often act as simple Lewis acids, accepting a lone pair of electrons from Lewis bases such as halide ions, to form species such as PCl_6^-.

Predictably, phosphorus is slightly smaller than silicon and slightly more electronegative. The strengths of the P–H and Si–H bonds are about the same (that with phosphorus having the closer approach but forming an almost non-polar bond), but the P–P bond is weaker than Si–Si, and phosphorus forms somewhat weaker σ bonds than silicon to carbon, oxygen, fluorine, and chlorine.

Since much energy would be needed to make the P^{3-} anion, and the radius of such an ion would be high (and so would not lead to very favourable values of lattice energy), there are no phosphorus analogues of the partially ionic nitrides such as Li_3N and Mg_3N_2. However, the ability of phosphorus to expand its octet and to act as both donor and acceptor provides scope for a wide variety of covalent chemistry.

Although phosphorus (like silicon but unlike nitrogen) is too large for successful np_π–np_π bonding, it forms a strong double bond with oxygen by sharing its two non-bonding electrons with two electrons from the half-filled 2p orbitals of oxygen, and, like silicon, occurs in nature only in combination with oxygen. There are about 200 orthophosphate rocks, such as the apatites $Ca_5(PO_4)_3X$, where X is usually F, Cl, or OH (see p. 201). The element can be volatilized off phosphate rock by heating with sand and

Stable isotopes of phosphorus

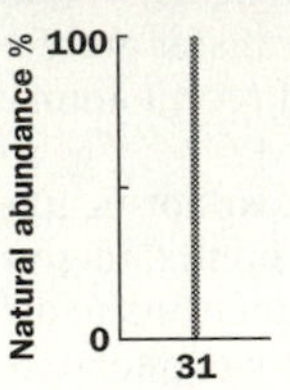

Extraction of phosphorus

Overall reaction

$$2Ca_3(PO_4)_2 + 6SiO_2 + 10C \xrightarrow{\sim 1450^\circ C} 6CaSiO_3 + 10CO + P_4$$

$2Ca_3(PO_4)_2$	$6SiO_2$	$10C$	$6CaSiO_3$	$10CO$	P_4
crushed rock	sand	coke	slag sold for hard-core	used for supple-mentary heating	condensed in water jet

Impurities

F Some $\rightarrow SiF_4$ (collected and used for fluorination): rest absorbed by slag

Fe $\rightarrow Fe_2P$, sinks below slag and is run off. (Various technological uses depend on high density ~ 6600 kg m^{-3})

Allotropes of phosphorus

P_4 tetrahedra in white phosphorus and in vapour

P≡P

some P_2 dimers in vapour above 800°

—P<P|P>P—P<P|P>P—P<P|P>P—

suggested structure for red phosphorus [GE]

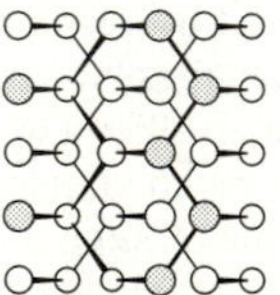

layers of puckered hexagons in black phosphorus [SAL]

coke. Calcium phosphate also provides the support structure of bones and teeth, while phosphate groups play a number of vital roles in metabolism (see p. 165).

There are several solid forms of phosphorus, all containing single bonds. The low melting, white allotrope, which is soluble in carbon disulphide, consists of P_4 tetrahedra, in which each atom forms three σ bonds. It is very toxic and very reactive, perhaps because the bond angles are too acute for kinetic stability in air. The P_4 tetrahedra are also present in the vapour, but decompose at high temperatures to the triply bonded dimers P≡P. White phosphorus is oxidized by air (emitting light) at room temperature and ignites spontaneously at about 35°C. Red phosphorus is formed by heating the white form in the absence of air; it is denser, harder, much higher melting, and is less toxic and less reactive. Although red phosphorous is normally amorphous, heat treatment can produce various crystalline forms which probably contain rejoined fragments of P_4 tetrahedra. Heating the white form at high pressure gives a still denser and more stable black form, which is a semiconductor, consisting of puckered sheets of hexagons reminiscent of graphite but with non-planar layers.

Phosphorus forms a great variety of binary compounds with almost all elements, often by direct action; the exceptions are antimony, bismuth, and the noble gases. Like boron and silicon, phosphorus may form a number of different binary compounds with a given element: at least eight phosphides of nickel are known. Phosphides that are rich in phosphorus may be semiconductors, while many of the metal-rich ones show metallic conductivity. Phophides of electropositive metals often have some ionic character and may contain Zintl anions such as P_7^{3-} and P_{11}^{3-}. In others, phosphorus atoms may be present in pairs, or as P_4 clusters, or as chains or layers.

Phosphorus forms a series of hydrides, P_nH_{n+2}, of which only the first member, phosphine, PH_3, is stable. As phosphorus is less electronegative than nitrogen, there is little hydrogen bonding in phosphine, which, despite its greater mass, is more volatile than ammonia. The PH_3 molecule has the same shape as NH_3 but a tighter apical angle and a much higher energy barrier to inversion. Liquid phosphine self-ionizes less than ammonia and has little tendency to lose or gain a proton, although it does act as a Lewis base, donating its lone pair to boron trifluoride and also stabilizing BH_3 as $H_3P.BH_3$. Pure phosphine is stable in air, but traces of the second hydride, P_2H_4, often make an impure sample ignite spontaneously; and the higher hydrides ($n \leqslant 3 \leqslant 6$) are very unstable indeed.

Some properties of phosphine

	PH_3	NH_3
Average energy of E–H bond/kJ mol^{-1}	322	391
Bond angle for H–E–H/°	93.7	106.6
Boiling point/°C	–87.7	–33.35
Inversion energy/kJ mol^{-1}	155	24
Proton affinity in gas/kJ mol^{-1}	770	866

Decomposition of solid phosphonium halides, $PH_4^+X^-$

$PH_4^+X^-(s) \rightarrow PH_3(g) + HX(g)$ (ΔS favourable)

decomposition favoured by	most favourable for	size of effect
(i) low lattice energy of $PH_4^+X^-$	$I^- > \ldots > F^-$	small (PH_4^+ large)
(ii) low electron affinity of X	$I^- > Br^- \sim F^- > Cl^-$	small
(iii) strong H–X bond	$F^- > \ldots I^-$	moderate

The dominant factor here is (iii) and the most stable member is $PH_4^+I^-$ with decomposition disfavoured by the weak H–I bond.

Non-existent PH_5; real PF_5

$PX_5(g) \xrightarrow{?} PX_3(g) + X_2(g)$ (ΔS favourable)

$PX_5(g)$	$PX_3(g)$	$X_2(g)$
2 P–X bonds broken	P–X bonds less crowded	1 X–X bond formed

Non-existent PH_5.

P–H bonds in PH_5 would be weakened by crowding	H–H bond very strong (433 kJ mol^{-1})

Although the total number of bonds is decreased, the overall free energy of decomposition is favourable and PH_5 is not formed.

Real PF_5

P–F bonds less crowded than P–H bonds	F–F bond weak (158 kJ mol^{-1})

(F more electronegative than H)

Here, the unfavourable ΔH on breaking two P–F bonds to form one F–F bond more than counterbalances the favourable entropy of decomposition; so PF_5 can exist.

Solid phosphonium halides, $PH_4^+X^-$, can be made but only the iodide is stable above 0°C. (The relative thermal stability of $PH_4^+I^-$ is partly a result of the fact that the H–I bond is weak, and so decomposition to form HI is not very favourable. Other factors that vary with the size of the halide ion have a much smaller effect: chlorine has the largest electron affinity, which would favour anion formation, but differences between the halogens are small; the small fluoride ion would give the strongest $PH_4^+X^-$ lattice, but with so large a cation, the differences again would be small.) There is no pentahydride PH_5: a mixture of phosphine and hydrogen is favoured by the increased disorder. Moreover, dissociation is more likely for PH_5 than for PCl_5, partly because of the high strength of the H–H bond, and possibly also because the shortness of the P–H bond would cause high interelectronic repulsion if five hydrogen atoms were crowded round a central phosphorus.

Phosphorus forms all twelve possible halides PX_3, PX_5, and P_2X_4; pseudohalides PY_3 for Y=CN, CNO, and CNS; and also some mixed derivatives. All are hydrolysed by water to oxyoacids (see p. 164). The reactive oxygen derivatives OPX_3 and SPX_3 are tetrahedral with short O=P and S=P bonds. The structures of the pentahalides vary interestingly with the size of the halogen. The pentafluoride is a trigonal bipyramid; fluorine, being larger than hydrogen and very electronegative, can fit five atoms round phosphorus and dissociation of PF_5 is discouraged by the weakness of the F–F bond. The pentachloride has a similar structure in the gas phase but is ionic in the solid, with the PCl_5 molecule acting as a Lewis acid to accept a chloride ion. The most stable form is $[PCl_4^+][PCl_6^-]$, but $[PCl_4^+]_2[PCl_6^-]Cl^-$ may also be formed. In solution, the pentachloride may be molecular or ionic depending on the solvent. The solid pentabromide is $[PBr_4{}^+]Br^-$, since there would presumably not be room to fit six bromine atoms round the phosphorus. The structure of the more recently prepared pentaiodide is probably similar.

Phosphorus forms an enormous number of oxides, sulfides, and oxosulfides, but happily all the molecular structures seem to be

Pentahalides of phosphorus

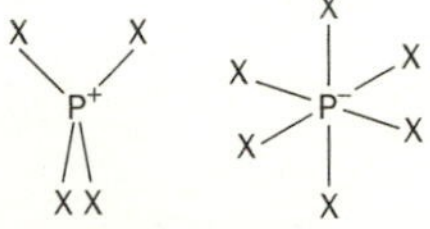

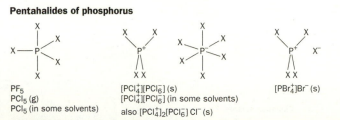

PF_5
PCl_5 (g)
PCl_5 (in some solvents)

$[PCl_4^+][PCl_6^-]$ (s)
$[PCl_4^+][PCl_6^-]$ (in some solvents)
also $[PCl_4^+]_2[PCl_6^-]$ Cl^- (s)

$[PBr_4^+]Br^-$ (s)

Oxides and sulfides of phosphorus

'peak'

'base'

P_4 framework
(no direct P–P bonding)

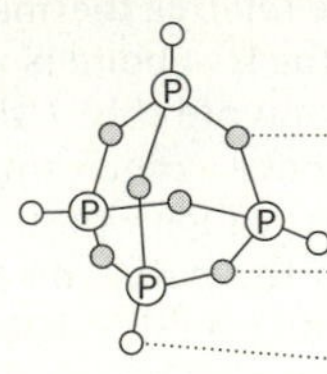

Compound	Occupied sites T O	T S	B_b O	B_b S	B_p O	B_p S
P_4O_{10} (see right) [SAL]	4	–	3	–	3	–
P_4O_6 (see right) [SAL]	–	–	3	–	3	–
P_4O_7	1	–	3	–	3	–
P_4S_3	–	–	–	–	–	3
$P_4O_4S_6$	4	–	–	3	–	3
$P_4O_6S_4$	–	4	3	–	3	–
P_4S_6	–	1	–	2	–	3

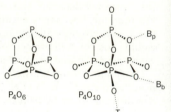

P_4O_6 P_4O_{10} B_p B_b T

Redox behaviour of phosphorus [SAL]

simplified to exclude species containing more than one P atom

Latimer diagrams $E^{\ominus}/V$

Acidic solution pH = 0: $H_3PO_4 \xrightarrow{-0.276} H_3PO_3 \xrightarrow{-0.499} H_3PO_2 \xrightarrow{-0.508} P \xrightarrow{-0.063} PH_3$

Alkaline solution pH = 14: $PO_4^{3-} \xrightarrow{-1.12} HPO_3^{2-} \xrightarrow{-1.57} H_2PO_2^- \xrightarrow{-2.05} P \xrightarrow{-0.89} PH_3$

Oxidation state diagram

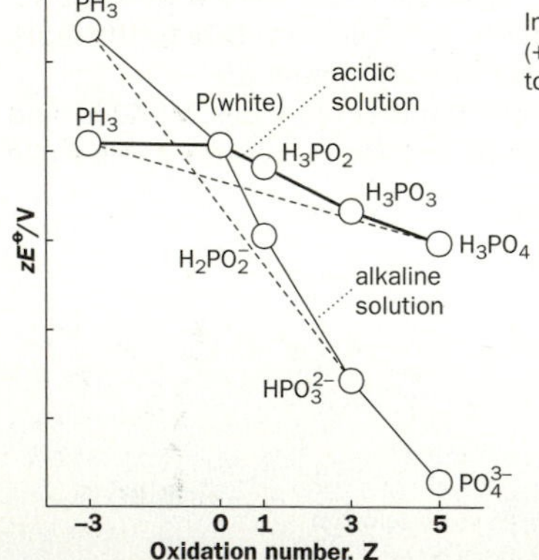

In both media, the element and the (+I) species are markedly unstable to disproportionation

built on the same model. This is always based on a tetrahedron of four phosphorus atoms; and it is convenient to envisage one of them forming the peak of the tetrahedron with the other three making the base. Between the three basal phosphorus atoms are three sites, B_b, for bridging atoms of oxygen or sulphur. There are three more bridging sites, B_p, between each of the basal phosphorus atoms and the peak; and there are four terminal sites, T, one on each phosphorus atom. In the commonest of these compounds, P_4O_{10}, all ten sites are occupied, and in the lower oxide, P_4O_6, the six bridging sites are all full and the four terminal ones are empty. These oxides react avidly with water to form oxoacids; and P_4O_{10} is used as a dehydrating agent and a desiccant.

In the sulfide P_4S_3 (used in strike-anywhere matches) only the B_p sites are filled, while in the oxosulfides $P_4O_4S_6$ and $P_4O_6S_4$ the element contributing six atoms occupies bridge sites while that contributing four takes up the terminal positions. The compounds are not necessarily symmetrical: oxides $P_4O_n (n \leqslant 6 \leqslant 10)$ and sulfides P_4S_n ($n = 3, 4, 5, 7, 9$, and 10) are known. In addition, there is a polymeric oxide consisting of PO_4 groups fused into sheets, and also a vitreous amorphous form.

Phosphorus shows even richer diversity in its oxoacids, forming more than any other element except for silicon. All contain the group $O{=}\overset{/}{\underset{\backslash}{P}}{-}O^-H^+$; the variety is provided by the two other groups on the phosphorus atom. When two hydroxyl groups are attached, we have phosphoric acid, H_3PO_4, which is used mainly to make fertilizers, but also in the manufacture of colas, paint strippers, detergents, and anti-rust (which was first used on corset stays). Addition of two hydrogen atoms to the $O\overset{/}{\underset{\backslash}{P}}{-}OH$ unit gives H_3PO_2, a moderately strong *mono*basic acid which, like white phosphorus itself, is thermodynamically unstable with respect to phosphine and one of the higher oxoacids (depending on pH). Further variety may be obtained by filling one of the sites with –O–OH to give a peroxoacid. Alternatively, two or more OPOH groups may be joined together, often with –P–O–P– links but also with –P–P– and –P–O–O–P– links. End-groups may be either H– or HO–. Polyphosphoric acids with chains containing up to 17 links are known, together with a very highly polymerized 'meta' phosphoric acid. The wide variety of chains formed by joining PO_4^{3-} tetrahedra reminds us of the linear silicates (see p. 152). Small rings are also formed. The cyclic $Na_5P_3O_{10}$ is used as a water softener as it binds

Some simple oxoacids of phosphorus [SAL]

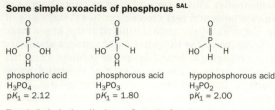

phosphoric acid	phosphorous acid	hypophosphorous acid
H_3PO_4	H_3PO_3	H_3PO_2
$pK_1 = 2.12$	$pK_1 = 1.80$	$pK_1 = 2.00$

The similarity in the pK values reflects the fact that when a proton is lost anions of all three acids have two oxygen atoms over which the charge can spread (see p. 13)

Some polyphosphate ions

$HP_3O_8^{4-}$	$P_4O_{10}^{4-}$	$P_6O_{12}^{6-}$
an anion with P–P and P–O links and a terminal H atom	an cyclic anion with P–P and P–O links	an cyclic anion with only P–P links

Some polyphosphate chains [GE] formed by joining PO_4 tetrahedra through P–O–P links

calcium and magnesium ions. The phosphates bound to bases and sugar derivatives play a major role in physiology: they are part of the DNA molecule used to carry genetic information (see p. 86); tri- and di-phosphates can be hydrolysed to provide energy for living processes; and a substance carrying the $-PO_4^{2-}$ group and the $-PO_2-O-PO_2-$ link is involved in the photosynthetic build-up of carbohydrates from carbon dioxide.

Phosphorus also forms a wide variety of compounds in which it is joined to nitrogen. We may think of some of these compounds as formal analogues of the oxoacids in which a group containing oxygen or hydrogen has been replaced by another that contains nitrogen and which is isolobal to it (having orbitals of the same shape, symmetry, and occupancy). So $-P-NH_2$ and $-P-NR_2$ groups can replace $-P-OH$ or $-P-H$; $-P{=}NH$ and $-P{=}NR$ are similarly isolobal with $-P{=}O$; and the link $-P-N(H \text{ or } R)-P-$ is interchangeable with $-P-O-P-$ (see p. 168).

The P–N bonds are short and strong, and many of the compounds are quite stable. There is a great tendency to form chains and rings, some of which are planar, even though, like $(R_3P-R'N)_2$, they contain no multiple bonds. Many phosporus–nitrogen compounds do, however, contain multiple bonds of some sort: a well-known example is $(NPCl_2)_x$ which can exist as a planar six-membered cyclic trimer, as various larger non-planar rings, or as an infinite polymer. All P–N distances are very similar and shorter than expected for a single σ bond, so there seems to be electron delocalization, although we do not know the details of the bonding. Since the polymeric phosphazene unit $-N{=}P-$ is isoelectronic with the silicone unit $-O-Si-$, it is not surprising that there is a range of phosphazene polymers that resemble the silicones in being water repellant, flexible, and stable to heat and solvents.

There are also a vast number of compounds in which phosphorus is bonded to carbon, since it can replace both nitrogen atoms and CH groups in a wide range of organic compounds. Although organophosphorus chemistry is outside the scope of this book, it is yet more evidence of the scope of compounds that phosphorus is able to achieve with its three unpaired 3p electrons, its $3s^2$ lone pair, and its ability to expand its octet.

For summary see p. 169.

Replacement of >O in P_4O_6 by >NR

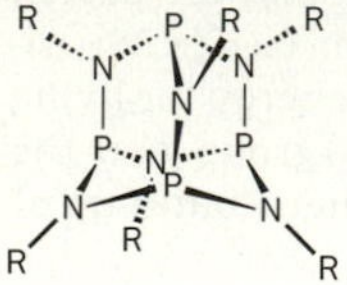

As in P_4O_6, there are three B_p bridges and three B_b bridges

$P_4(NR)_6$ (a phosphazene) SAL

Two phosphazine chlorides SAL

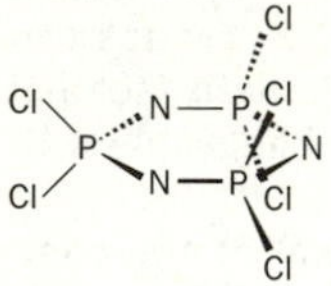

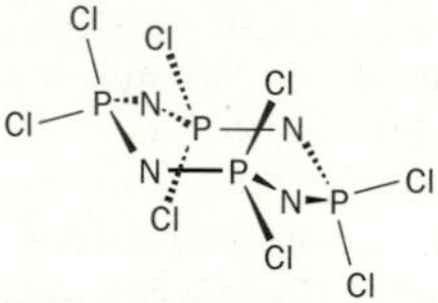

$(NPCl_2)_3$ planar

$(NPCl_2)_4$ puckered

In both compounds, all the P–N bonds are of very similar length

A silicone analogue SAL

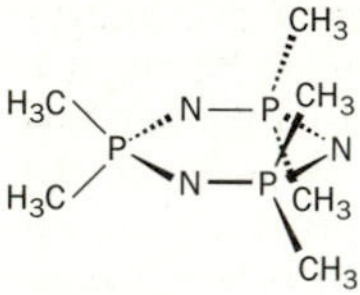

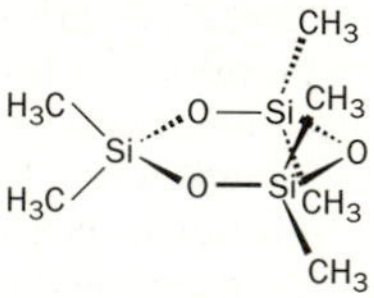

cyclic $[(CH_3)_2PN]_3$
a phosphazene

cyclic $[(CH_3)_2SiO]_3$
a silicone

Summary

Electronic configuration

- [Ne]$3s^2 3p^3$

Element

- Various singly bonded allotropes: white, P_4 tetrahedra, very reactive; red, less reactive, amorphous structure based on P_4 tetrahedra; black semiconductor, puckered layer structure

Occurrence

- In phosphate rocks

Extraction

- By heating phosphate minerals with coke and sand

Chemical behaviour

- No P^{3-} and no p_π–p_π bonding
- Wide variety of phosphides: with electropositive metals forms Zintl anions P_x^{3-}
- Phosphine, PH_3, only stable hydride; little H bonding, self-ionization or affinity for proton (but forms $PH_4^+I^-$ and acts as Lewis base)
- Halides PX_3, PX_5, P_2X_4, pseudohalides, and oxides OPX_3; octet expansion, e.g. in PF_5, PCl_5, and [PCl_6^-]
- Wide range of oxides, P_4O_x, based on P_4 tetrahedra, P–O–P bridges, and terminal P=O groups; similar sulfides and oxosulfides
- Wide variety of oxoacids based on OP(X, X′) O^-H^+ (X, X′ = H or OH), and polyacids joined through –P–P–, –P–O–P–, or –P–O–O–P–; some very highly polymerized; others small rings
- Analogous nitrogen compounds where –OH or –H replaced by –N and =O by =N–, forming rings and chains, similar to silicones; $PNCl_2$ six membered ring or infinite chain

Uses

- P_4S_3 in matches
- Phosphoric acid and phosphates as fertilizers, paint strippers, anti-rust coatings, water softeners, and in soft drinks
- In biological energy transfer and in DNA

Nuclear features

- Isotopically pure ^{31}P

S Sulfur

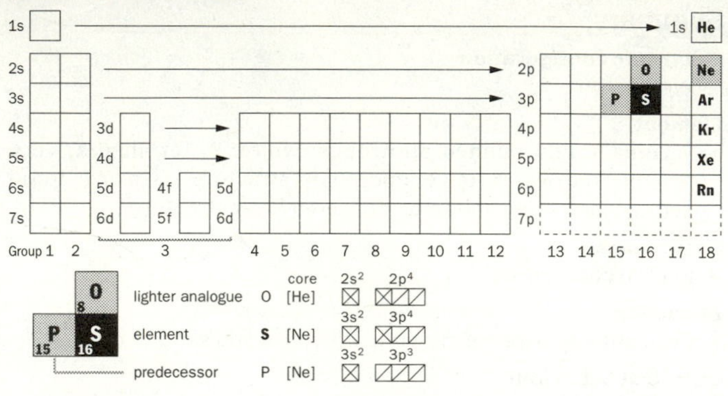

		Predecessor	Element	Lighter analogue
Name		**Phosphorus**	**Sulfur**	**Oxygen**
Symbol, *Z*		P, 15	S, 16	O, 8
RAM		30.973762	32.066	15.9994
Radius/pm	**atomic**	93 (white); 115 (red)	104	
	covalent	110	104	66
	ionic, X^{2-}		184	132
Electronegativity (Pauling)		2.2	2.6	3.4
Melting point/K		317.3	386.0 (α)	54.8
Boiling point/K		553	717.824	90.188
$\Delta H^{\ominus}_{fus}$/kJ mol^{-1}		2.51 (P_4)	1.23	0.444
$\Delta H^{\ominus}_{vap}$/kJ mol^{-1}		51.9	9.62	6.82
Density (at 293 K)/kg m^{-3}		1820	2070 (α)	
Electrical conductivity/Ω^{-1}m^{-1}		1×10^{-9}	5×10^{-16}	
Ionization energy for removal of *j*th electron/kJ mol^{-1}	***j* = 1**	1011.7	999.6	1313.9
	***j* = 2**	1903.2	2251	3388.2
	***j* = 3**	2912	3361	5300.3
	***j* = 4**	4956	4564	7469.1
	***j* = 5**	6273	7013	10 989.3
	***j* = 6**		8495	13 326.2
Electron affinity/kj mol^{-1}		72.0	200.4	141
Bond energy/kg mol^{-1}	**E–E**	209	226	146
	E–H	328	347	464
	E–C	264	272	358
	E–O	407	265	141
	E–F	490	328	190
	E–Cl	319	255	206

We know that the way in which an element occurs in nature often gives us an insight into its chemistry. Although both silicon and phosphorus behave with rich variety in the laboratory, in nature they are always surrounded by oxygen atoms. Sulfur may also be found combined with oxygen, as sulfates, either in rocks or dissolved in sea water; but it occurs in many other forms as well: as the free element, in crude oil as organic sulfur compounds, in natural gas as hydrogen sulfide, and in rocks as sulfides. Its electronic structure of $[Ne]3s^23p^4$ differs from that of phosphorus in having two lone pairs, instead of one, and so it has two sites that can accept a Lewis acid, such as an oxygen atom. As sulfur also has two half-filled orbitals, it forms two σ-bonds, as in H_2S. Moreover, unsubstituted sulfur atoms can join up without multiple bonding, 'catenating' to form chains or rings. Since sulfur needs only two electrons to complete its octet, it can, unlike phosphorus, form a monatomic anion; the ion S^{2-} is found in sulfides of very electropositive metals, although it can readily be distorted into covalent overlap by more highly polarizing cations. Also possible are singly charged anions with one σ-bond (such as HS^-) or singly charged cations with three (such as R_3S^+, which is formally analogous to R_4P^+).

Like silicon (see p. 151) and phosphorus, sulfur can expand its outer valence shell to accommodate more than eight electrons; in SF_6 there are 12. Although it is often assumed that octet expansion involves d orbitals, this need not be so. Certainly, the $n = 2$ elements, which have no d orbitals, do not expand their octets; but this is probably simply because their atoms are too small to accommodate more than four neighbours. In sulfur hexafluoride, sulfur contributes four orbitals and six electrons, while each fluorine contributes one orbital and one electron. Calculations have shown that of the ten orbitals, four are bonding, two non-bonding, and four antibonding. So the 12 electrons can be accommodated in the bonding and non-bonding regions, leaving all antibonding orbitals empty.

Since sulfur is more electronegative than phosphorus (though much less so than oxygen), it can take part in (weak) hydrogen bonding. As expected, sulfur forms stronger bonds than phosphorus to the more electropositive elements, hydrogen and carbon, but weaker ones to oxygen and the halogens, which are more electronegative. Its higher charge makes it somewhat smaller than phosphorus, and so, like oxygen, sulfur can form dimers, S_2, in which there is p_π–p_π bonding.

Sulfur demonstrates its diversity in many aspects of its chemistry. By catenating through σ bonds, sulfur produces more allotropes than any other element, with widely varying bond lengths and angles. The stable orthorhombic form, which is found in naturally occurring sulfur, contains the puckered S_8 ring, as does

Stable isotopes of sulfur

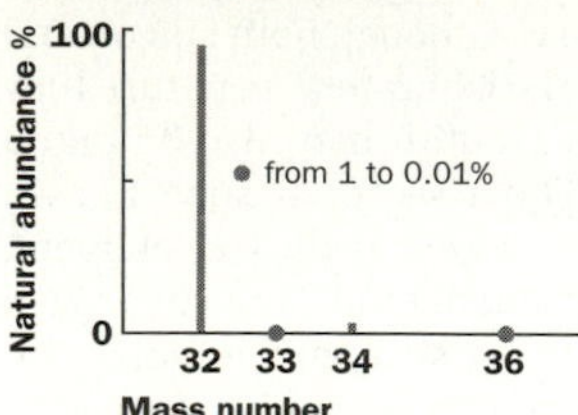

Some allotropes of sulfur

non-cyclic

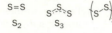

S=S — S_2; S_3; $(S_2)_{100\,000}$ — and many in between

cyclic [M]

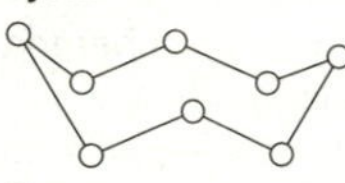

S_8
the most common form

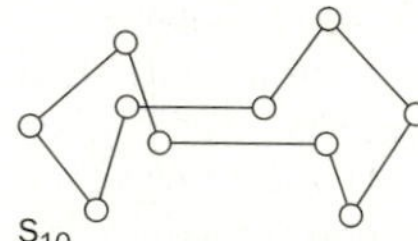

S_{10}
(there are similar S_n rings for n = 6 to 12, and 20)

Some properties of hydrogen sulfide [GE]

	H_2S	H_2O
Bond length E–H/pm	133.6	95.7
Bond angle H–E–H/°	92.1	104.5
Melting point/°C	–85.6	0
Boiling point/°C	–60.3	100
pK for $H_2E(aq) \rightleftharpoons HE^-(aq) + H^+(aq)$	6.9	14

Cubane structure of Fe_4S_4

S

Fe (or Mo)

* The structure is similar to that of NaCl (see p. 184) but with the bulky anions set skew between the cations

monoclinic sulfur, although this form has a slightly disordered stacking within the crystal. But there are other *cyclo*-S_n rings, with $6 \geqslant n \geqslant 12$ and $n = 20$: Solids S_n, for which $n \geqslant 5$, contain linear diradicals. Catena-S_n (or 'plastic' sulfur), which is obtained by pouring liquid sulfur into cold water, has $n = 2 \times 10^5$ if made from the melt at 180°C, but $n = 100$ when made from a 600°C melt. The chains form helices, either clockwise or anticlockwise. The low temperature vapour may contain S_8 rings and other species S_n ($2 \geqslant n \geqslant 10$). The trimer is bent, resembling ozone but cherry red, while S_2 is relatively stable with a short S=S bond.

Sulfur reacts with most other elements (except the noble gases), but heat must often be supplied in order to break the S–S bonds. Some sulfur molecules, S_n (e.g. $n = 4, 8$), may be oxidized by anhydrous sulfuric acid to cations S_n^{2+}; but these molecules have a greater tendency to be reduced to anions, which may then act as ligands to metal ions. The disulfide ion S_2^{2-}, present in the (partly) ionic lattice* of 'iron pyrites' FeS_2, can also act as a bridging ligand, while anions S_n^{2-} with longer chains can chelate.

Non-catenated sulfur occurs in hydrogen sulfide, which (unlike the very strongly hydrogen bonded water) is gaseous at room temperature and is a weak acid in aqueous solution. (Sulfanes H_2S_n ($2 \geqslant n \geqslant 8$) do exist but are reactive.) Since the sulfide ion is larger, less electronegative, and more polarizable than the oxide ion, the lattices of metal sulfides are often weaker and more covalent than the corresponding oxides. There is considerable variety: electropositive metals form mainly ionic lattices; zinc forms two stongly bound, probably partially covalent sulfide structures ZnS (zinc blende or wurtzite); many sulfides, such as FeS, have non-stoichiometric alloy-like structures and are often non-stoichiometric; layer structures as in NiAs (see p. 304) are common, as in MoS_2, which is used as a lubricant (see p. 347); sulfides MS_2, for metals like iron which do not readily form M^{4+}, contain the S_2^{2-} ion. There is equal diversity in electrical and magnetic properties of metal sufides, which range from insulators[A] to superconductors[A] and from diamagnetics[A] to ferromagnetics.[A]

Oddities abound. Iron and molybdenum form cubane sulfides, M_4S_4, with atoms alternately at the corners of a cube. This structure probably occurs in the biochemically important ferredoxins, where there may be electron exchange between iron (II) and sulfur(–II). Nitrogen and arsenic also form sulfides, Y_4S_4, of similar structure. It is thought that in one complex carbonyl of chromium a sulfur atom bridges to two metal atoms, donating three electrons to each.

There are seven fluorides of sulfur, of which the most important is the octahedral SF_6 which is very stable and a gas at room tem-

S_2F_2 and S_2F_6 GE

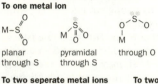

Bond lengths/pm	S_2F_2	S_8	S_2F_6
S–F	163.5	–	156.4
S–S	189	206	–

Binding of SO_2 to metal ions GE

To one metal ion

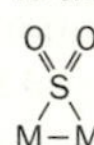

planar through S | pyramidal through S | through O | chelated through O and S

To two seperate metal ions | **To two joined metal ions**

Manufacture of sulfuric acid GE

$S + O_2 \rightarrow SO_2$ $\quad \Delta H^\ominus = -297\ \text{kJ mol}^{-1}$

also by roasting sulfide ores.

$SO_2 + \frac{1}{2}O_2 \xrightarrow{\text{catalyst}} SO_3$ $\quad \Delta H^\ominus = -9.8\ \text{kJ mol}^{-1}$

$SO_3 + H_2O$ (in 98% H_2SO_4) $\rightarrow H_2SO_4$ $\quad \Delta H^\ominus = -130\ \text{kJ mol}^{-1}$

The oxidation catalyst is V_2O_5, used between 400 and 620°C.

Formation of 'acid rain'

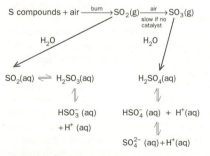

S compounds + air $\xrightarrow{\text{burn}}$ $SO_2(g)$ $\xrightarrow[\text{slow if no catalyst}]{\text{air}}$ $SO_3(g)$

$SO_2(g)$ + H_2O → $SO_2(aq) \rightleftharpoons H_2SO_3(aq) \rightleftharpoons HSO_3^-(aq) + H^+(aq)$

$SO_3(g)$ + H_2O → $H_2SO_4(aq) \rightleftharpoons HSO_4^-(aq) + H^+(aq) \rightleftharpoons SO_4^{2-}(aq) + H^+(aq)$

perature. One of the lower fluorides, S_2F_2, resembles O_2F_2 (see p. 104) and has a very short S–S bond. The main chlorides are SCl_2 and S_2Cl_2, but dichlorides S_nCl_2 are know up to at least $n \geqslant 8$. The bromides and iodides, being far less stable, have been studied less. Although fluorine is the only halogen small enough to form SX_6, two oxochlorides, $OSCl_2$ and O_2SCl_2, are well established.

When sulfur and many of its compounds are burned in air, sulfur dioxide is formed as a pungent, irritant gas. It can be condensed to a liquid, which is a good solvent, although (if pure) not a self-ionizing one. Many solutions in liquid sulfur dioxide may, however, be considered to contain SO^{2+} or SO_3^{2-} ions analagous to the H^+ and OH^- ions produced by hydrolysis of species such as Fe^{3+} and CN^- in aqueous solution. Gaseous sulfur dioxide dissolves in water to form an acidic solution containing, amongst other species, SO_2 (aq), H^+ (aq), and HSO_3^- (aq). The bent SO_2 molecule can coordinate to a metal atom, or act as a bridging ligand between two atoms, in at least six different ways. Its main importance, however, is that it can be oxidized to form sulfur trioxide, which can be hydrolysed to sulfuric acid. A catalyst is needed for oxidation of sulfur dioxide at an industrially useful rate. But, as slow oxidation takes place even in the air, any industrial exhaust gases that contain the already toxic sulfur dioxide will become further contaminated by sulfur trioxide, which is hydrolysed to give sulfuric acid when it comes into contact with water, either in the atmosphere or on the surface of living cells. This naturally does a great deal of harm when it falls as 'acid rain' into lakes, on forests, or on stone buildings, and may cause serious lung damage if it is inhaled, particularly by those who are very young, very old, or who already have respiratory problems. So there is a strong case for trying to minimize pollution from oxides of sulfur. Sulfur trioxide can associate cyclic trimers $(SO_3)_3$ in the gas and in solution, but as a solid forms chains $(SO_3)_n$.

There are at least eleven other much less important oxides of sulfur: S_nO_y (where $n = 6$, 7, or 8 and $y = 1$ or 2) where the oxygen atoms are attached to an S_n ring.

There are also many oxoacids of sulfur, though fewer than those of phosphorus. Some, like 'sulfurous' acid, exist only as their salts. If we take this hypothetical acid $(HO)_2\ \ddot{S}{=}O$ as our starting point, we may form other monosulfur acids by replacing the lone pair with =O, =S, –OH, or –OOH. The most important of these is sulfuric acid $(HO)_2S({=}O)_2$, a viscous corrosive liquid, which self-ionizes to give $H_3SO_4^+$ and HSO_4^- and is an ionizing solvent. It reacts avidly and dangerously exothermically with water: because the heat liberated by the hydration of the proton may vaporize the water and make the concentrated acid spit upwards. When we want to dilute sulfuric

Some sulfite derivates

Simple oxanions of (non-isolatable) 'H_2SO_3'

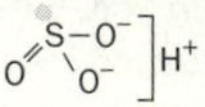

$[O{=}S(O^-)(O^-)]H^+$ hydrogen sulfite HSO_3^- $O{=}S(O^-)(O^-)$ sulfite SO_3^{2-}

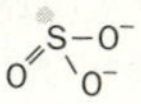

Anions of non-existent acids with S–S links

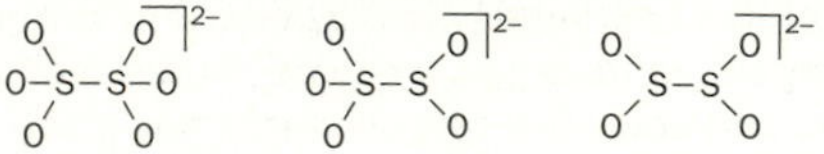

$[O_3S{-}SO_3]^{2-}$ $[O_3S{-}SO_2]^{2-}$ $[O_2S{-}SO_2]^{2-}$

Hydration of sulfuric acid[GE]

$H_2SO_4(l)$ + $H_2O(l)$ → very dilute solution containing $H^+(aq)$,

HSO_4^- (aq), SO_4^{2-} (aq), and H_2SO_4 (aq)

BEWARE! $\Delta H = -880$ kJ mol^{-1}

Some oxoacids of sulfur [GE]

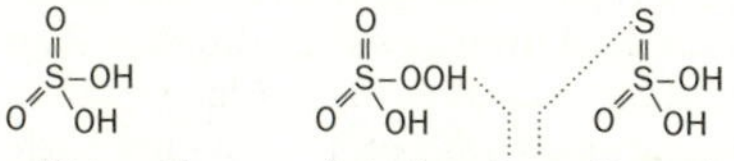

$O{=}S(=O)(OH)OH$ sulfuric acid

$O{=}S(=O)(OOH)OH$ $S{=}S(=O)(OH)OH$ derivatives formed by replacing –OH or =O by simple group

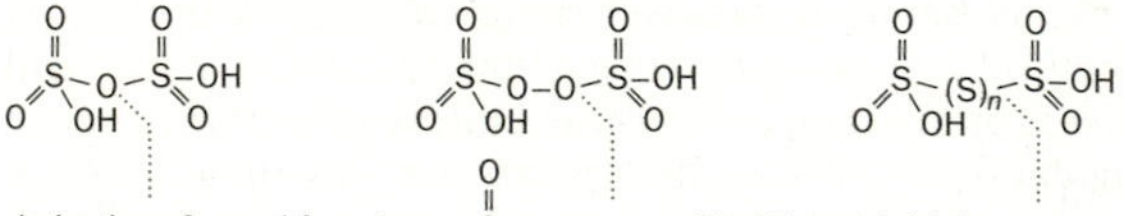

$HO(O{=})_2S{-}O{-}S(=O)_2OH$ $HO(O{=})_2S{-}O{-}O{-}S(=O)_2OH$ $HO(O{=})_2S{-}(S)_n{-}S(=O)_2OH$

derivatives formed from two $O{=}S(=O)(OH){-}$ groups with different bridging groups

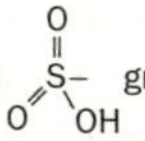

Latimer diagram, E°/V SAL

Acid solution pH = 0 $HSO_4^- \xrightarrow{0.158} H_2SO_3 \xrightarrow{0.500} S \xrightarrow{0.144} HS^-$

Alkaline solution pH = 14 $SO_4^{2-} \xrightarrow{-0.936} SO_3^{2-} \xrightarrow{-0.659} S \xrightarrow{-0.476} HS^-$

* The Greek root 'thio' is used as a label for sulfur compounds, often for those in which the sulfur replaces oxygen, as in thiosulfate ($S_2O_3^{2-}$) or a thiol (R–SH). Rather surprisingly, it is related to 'theo' which refers to a god (as in 'theology'). It seems that the two terms come from the same word, which meant 'smoke' and was used in connection both with burnt offerings to a god and with sulfur burning at the mouths of volcanoes.

acid we must always ADD ACID TO WATER and not the other way round. Concentrated sulfuric acid is used as a dehydrating agent, a desiccant, and a strong oxidizing agent; and it is a component of many large-scale industrial processes.

Sulfur, like phosphorus and silicon, forms a confusing array of polyacids, although not in nearly such rich variety. Those of sulfur consist only of three pieces. Two are 'acid units', derived by removing a hydroxy group from a molecule of one of the monosulfur (VI) acids (such as sulfuric) mentioned above. They are linked through a bridging group, which may be oxygen (–O–), peroxide (–O–O–), or catenated sulfur ($-S_n-$). Anions, but not the parent acid, can sometimes be formed by direct linkage of two 'acid units' (see opposite). It is not surprising that the redox behaviour amongst this plethora of species is extremely subtle.

Other bizarre compounds include those formed between sulfur and nitrogen (see p. 178). Since these two elements do not differ widely in electronegativity, the S–N bonds would be expected to be mainly covalent, but, like those formed between phosphorus and nitrogen, they cannot be described in terms of any simple theory of bonding. Although many of these compounds decompose to the elements if struck or heated, this does not imply particular weakness of the S–N bonds, but rather the unusual stability of the products. The N≡N bond is exceptionally strong, while S–S is stronger than other single bonds between like atoms, except for H–H and C–C.

There is no monomeric sulfur analogue of NO, although the best known nitrides of sulfur are indeed of formula $(SN)_x$. Tetrasulfur tetranitride, S_4N_4, although unstable relative to N_2 and S_8, is stable to air and water. Aqueous alkali can hydrolyse more of the S–N bonds: how many depends on the pH. The S_4N_4 molecule can be visualized as a boat-shaped eight-membered ring of alternate nitrogen–sulfur atoms, puckered so that the nitrogen atoms almost form a planar square. The sulfur atoms are in two pairs, possibly weakly bonded together, one each side of the plane of nitrogen atoms. (The structure of S_4As_4 is similar, except that the sulfur atoms now occupy the four central sites.) When S_4N_4 is heated, it gives S_2N_2, a cyclic molecule that is unstable and almost square planar. This slowly polymerizes to polythiazyl,* $(SN)_x$, which is, somewhat unpredictably, a 'one-dimensional metal': a solid with a bronze lustre, and a high electrical conductivity that decreases with increasing temperature, it clearly contains delocalized electrons. The atoms are linked alternately in wavy chains with two (not very) different N–S bond lengths. If these are formally but very approximately represented by double and single bonds, each sulfur atom would have one unpaired electron to contribute to the con-

Sulfides, S_4X_4 of nitrogen and arsenic [S]

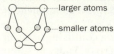

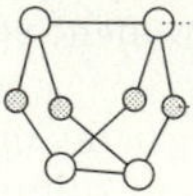

larger atoms
smaller atoms

Covalent radii/pm

N	70
S	104
As	121

in N_4S_4 ◉ = N ○ = S
in As_4S_4 ◉ = S ○ = As

Some other $(SN)_x$ compounds [SAL]

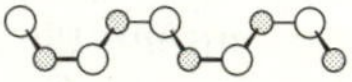

S—N
| |
N—S ⟶ ◉N ○S

S_2N_2
almost square
unstable

$(SN)_x$
'one dimensional metal'
bond lengths alternately 159 and 163 pm

Bonding of thiocyanate ion to metal ions

M ← NSC^- for lighter and for less polarizing ions,

e.g. Cu^{2+} ← NCS^-

La^{3+} ← NCS^-

M ← SCN^- for heavier, more polarizing ions,

e.g. Hg^{2+} ← SCN^-

$(H_3N)_5Co^{3+}$ can bind either ← NCS^- or ← SCN^-

thiocyanate can act as bridging ion

Co^{2+}–NCS^-–Hg^{2+}

duction band. Other sulfides have even more complex structures, sometimes with two rings, which may either be linked together by one S atom or by an S_x chain, or be fused sharing an N–S–N bond. Some nitrides of sulfur readily lose or gain electrons to form ions such as $S_4N_4^+$ and $S_3N_3^-$.

One important species of the many that contain sulfur, nitrogen, and a third element is the thiocyanate ion SCN^-, which can complex through nitrogen, to those metal ions (such as the lanthanides, see p. 429) that favour ionic bonding, and through sulfur to more polarizing ones, such as Hg^{2+}.

It is not surprising that sulfur can replace oxygen in a large number of organic compounds, subtly altering their acid–base, redox, and complexing properties by virtue of its lower electronegativity and increased size. These thio compounds, together with the stability and complexing power of the cross-linking S–S group, and the ability of sulfur compounds to take part in electron transfer under mild conditions, doubtless contribute to the varied roles that sulfur plays in biochemical processes: in group transfer reactions, in building of proteins (such as those in hair), directly in catalysis, and in retaining atoms of heavy metal such as copper, zinc, and iron. Reactions involving sulfur enable penicillin to inhibit the reproduction of bacteria.

For summary see p. 181.

Summary

Electronic configuration

- $[Ne]3s^2 3p^4$

Element

- σ bonded S_8 rings; $-(S)_n-$ diradicals; S_n in gas phase

Occurrence

- S_8 native; and as SO_4^{2-}, and S_2^{2-} in minerals

Extraction

- Mined from S_8 deposits

Chemical behaviour

- Negligible $p_\pi - p_\pi$ bonding; electronegativity O > S > P >, so stronger bonds than P to H and C, weaker to O and halogens
- Octet completion by two σ bonds (including catenation), two-electron gain (to S^{2-}), or S_2^{2-} formation
- Sulfides more covalent than oxides, many layer structures
- H_2S gaseous (no strong H bonding); weak acid in aqueous solution
- Cubane metal sulfides, M_4S_4 for M = Fe or Mo
- Wide variety of halides, by octet expansion up to SX_6 (for F only)
- Oxides SO_2 (ligand to metal ions) or SO_3, give acid solutions in water; H_2SO_4 strong oxidizing acid and ionizing solvent
- Thio acids formed by replacing O by S in oxoacids; two (but not more) sulfate-based groups can be joined by −O−, −O−O− or $-(S_n)-$
- Variety of nitrides: S_4N_4 (cubane type), S_2N_4 (squarish), $(SN)_x$ (polymeric, one-dimensional metal); and more complex structures

Uses

- H_2SO_4 for fertilizers and many industrial processes
- HSO_3^- as paper bleach
- Vulcanization of rubber, and many smaller scale uses throughout industry
- Important in many biochemical reactions

Nuclear features

- Mainly ^{32}S, with some ^{34}S and traces ^{33}S and ^{36}S

Cl Chlorine

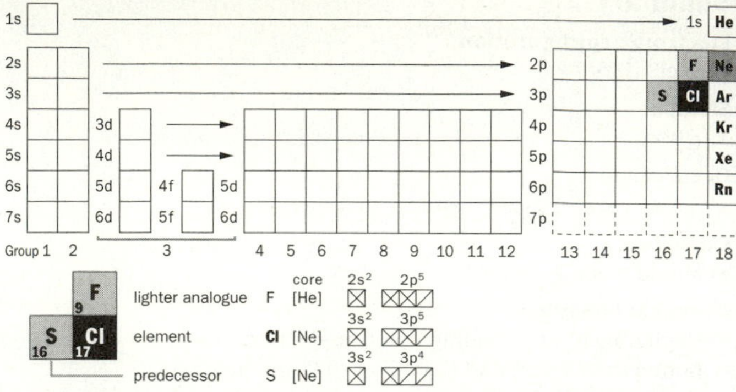

	Predecessor	Element	Lighter analogue
Name	**Sulfur**	**Chlorine**	**Fluorine**
Symbol, *Z*	S, 16	Cl, 17	F, 9
RAM	32.066	35.4527	18.9984032
Radius/pm covalent	104	99	58
ionic, X^-		181	133
Electronegativity (Pauling)	2.6	3.2	4.0
Melting point/K	386.0 (α)	172.17	53.53
Boiling point/K	717.824	239.18	85.01
$\Delta H^\ominus_{fus}$/kJ mol^{-1}	1.23	6.41	5.10
$\Delta H^\ominus_{vap}$/kJ mol^{-1}	9.62	20.4033	6.548
Ionization energy for removal of *j*th electron/kJ mol^{-1} *j* = 1	999.6	1251.1	1681
***j* = 2**	2251	2297	3374
***j* = 3**	3361	3826	6050
j = 4	4564	5158	8408
***j* = 5**	7013	6540	11 023
***j* = 6**	8495	9362	15 164
***j* = 7**		11 020	17 867
Electron affinity/kj mol^{-1}	200.4	349.0	328
Bond energy/kg mol^{-1} E–E	226	242	159
E–H	347	431	566
E–C	272	325	
E–N		193	272
E–O	265	206	190
E–F	328	257	159

Chlorine, with its electronic configuration [Ne]$3s^2 3p^5$, has a more limited chemistry than its immediate predecessors. One electron short of the octet, we expect, and find, a tendency to form either the chloride ion, Cl^-, or a single covalent bond. There are three lone pairs each of which could be donated to, for example, an oxygen atom, and octet expansion might be possible. The fairly high ratio of nuclear charge to size makes chlorine moderately electronegative, but we should expect that the 3s and 3p electrons, even when pulled in towards the nucleus by its higher charge, would experience less inter-electronic repulsion than do the crowded $n = 2$ electrons in fluorine.

In nature, chlorine occurs only as the chloride ion, in sea water and in some inland salt lakes and deposits. Salt was valued for our diet as early as 3000 BC. Chlorine gas is obtained by oxidation of chlorine (–I), industrially by electrolysis of aqueous or molten sodium chloride, or on a laboratory scale by adding a strong oxidizing agent (such as potassium permanganate) to concentrated hydrochloric acid. It is used in the manufacture of PVC (polyvinylchloride), of chlorinated derivatives of methane and ethane, and of aluminium trichloride for catalysis; and also in metallurgical processes (see, for example, titanium, p. 211 and vanadium p. 223). Chlorine has a very high first ionization potential and is never found as a simple cation (see, however, p. 185). Its electron affinity (greater than that of fluorine) and the strength of the Cl–Cl bond (stronger than F–F, but weaker than S–S) suggests that the inter-electronic repulsion in chlorine is much less than in fluorine, but slightly more than in sulfur. Chlorine also lies between fluorine (see p. 111) and sulfur in reactivity, as would be expected both from the stabilities of the binary compounds of the three elements, and from the instabilities of the elements themselves; although dichlorine dissociates less readily than difluorine, the chlorine atom, with a single vacancy in its 3p shell, lacks sulfur's ability to make stable rings by forming covalent bonds to two other atoms.

Chlorine forms a polar molecular hydride, which (unlike the strongly hydrogen bonded HF) is an unassociated gas at room temperature. However, the weaker hydrogen bonds acting within the solid are stronger than those that are possibly present in solid hydrogen sufide, and, with large cations, hydrogen bonded anions $[Cl-H \ldots Cl]^-$ analogous to the HF_2^- ion can be prepared from the anhydrous liquid. In water, hydrogen chloride (unlike hydrogen fluoride, see p. 115) dissociates completely to hydrochloric acid which has a vast number of uses in industries as varied as steel and food.

As the chloride ion is more polarizable than the smaller fluoride, the two chloride structures that are quoted as being typical ionic

Isotopes of chlorine

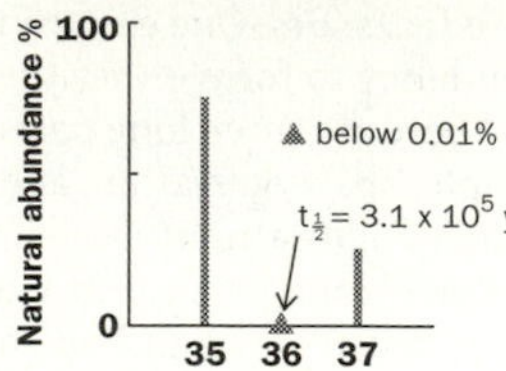

Preparation of chlorine gas

Oxidation of Cl^-

Industrial scale: electrolysis $2Cl^-[\text{in } Na^+Cl^-(l)] \xrightarrow[\text{at anode}]{-2e^-} Cl_2(g)$

Laboratory scale: chemical oxidation

e.g. $2Cl^-$ (in conc. aq. HCl) $- 2e^- \rightarrow Cl_2(g)$

$$\tfrac{2}{5}MnO_4^- + \tfrac{4}{5}H^+ + 2e^- \rightarrow \tfrac{2}{5}Mn^{2+} + \tfrac{2}{5}H_2O$$

Some properties of hydrogen chloride GE

	HCl	HF
Boiling point/°C	–84.2	19.5
Main component in vapour on boiling	HCl	H_6F_6
Main components in aqueous solution	$H^+(aq)$, $Cl^{-1}(aq)$	HF(aq) (some HF_2^-, $H^+(aq)$ if concentrated some $H^+(aq)$, $F^-(aq)$ if dilute)
p*K* (see p. 12)	~ –7	2.95

Some ionic metal chlorides S

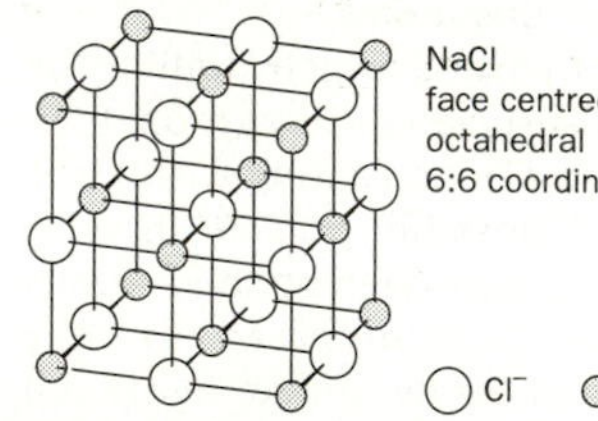

NaCl
face centred cubic
octahedral
6:6 coordination

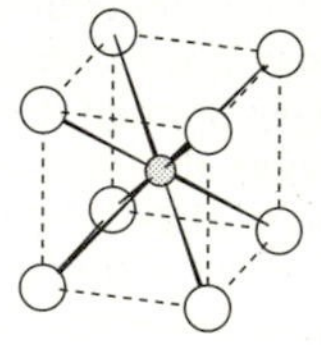

CsCl
body centred cubic
8:8 coordination

lattices (face-centred cubic NaCl or body-centred cubic CsCl) are formed only with those cations of the more electropositive elements, with a low charge to size ratio. Even in calcium chloride, the lattice distortions suggest some covalent character. The scandium salt, $ScCl_3$, forms a layered structure, while aluminum trichloride melts to form a bridged dimer. Titanium(IV), with a very high ratio of charge to size, forms a molecular chloride $TiCl_4$, which is liquid at room temperature (see p. 186). The predominantly ionic chlorides are soluble in water, while the more covalent ones, dominated by a highly polarizing cation, are often hydrolysed.

Chlorine reacts with most non-metals (except the noble gases) to form molecular chlorides; many of these are discussed under the elements concerned. Some, like tetrachloromethane, CCl_4, are very stable; others, such as the six oxides of chlorine, are not. Even its two least exotic oxides, Cl–O–Cl and O–Cl–O (with an unpaired electron) are liable to explode. Dichlorine monoxide, Cl_2O, is, however, manufactured on a large scale since it is the anhydride of hypochlorous acid (see below). Slightly less violent is Cl_2O_7 ($O_3Cl{-}O{-}ClO_3$), which is the anhydride of perchloric acid.

Chlorine forms many interhalogen compounds. The monofluoride is particularly stable, but the higher fluorides ClF_3 and ClF_5, with an expanded octet, are strong fluorinating agents. With bromine, chlorine appears to form only BrCl, but with iodine it forms ICl and a bridged dimer I_2Cl_6, which (unlike Al_2Cl_6) is planar. Many interhalogen compounds act as ionizing solvents dissociating to a polyhalonium cation and a polyhalide anion. Chlorine monofluoride, for example, gives Cl^+(solv) (or Cl_2F^+) and ClF_2^-, and solid salts containing these ions may be made with the aid of suitably large counter ions. Many interhalogen and polyhalogen ions, including the triatomic species Cl_3^+ and Cl_3^-, have been made since 1950.

When chlorine gas is passed into aqueous acid, a little of it disproportionates to form Cl^- and hypochlorous acid, HOCl. This exists only in aqueous solution, but its sodium salt is well known as 'bleaching powder', because in acidic solution it liberates chlorine which oxidizes (and so decolorizes) many organic dyes and stains. It is also a powerful germicide and is manufactured on a large scale. The chlorine atom in HO–Cl has three lone pairs of electrons, each of which may be donated to an oxygen atom, to give chlorous HO–ClO, chloric $HO{-}ClO_2$, and perchloric $HO{-}ClO_3$ acids. The first two, like HOCl, are known only as their salts or in aqueous solution. All three lower acids are thermodynamically unstable to disproportionation; the only stable oxidation states, apart from the element itself, are the chloride ion (–I) and perchloric acid and its

Some molecular metal chlorides

$AlCl_3$ layered crystal sublimes at 180°C to give *dimers* in vapour (see p. 145)

$Cl_2Al(\mu\text{-}Cl)_2AlCl_2$

$TiCl_4$ *monomer*

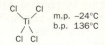

m.p. –24°C
b.p. 136°C

Oxoacids of chlorine

Traditional name	Perchloric	Chloric	Chlorous	Hypochlorous
Formula	$HO\text{–}Cl^{VII}O_3$	$HOCl^{V}O_2$	$HOCl^{III}O$	$HOCl^{I}$
~ p*K* in water (see p. 13)	–10	–3	1.94	7.52

Redox behaviour of chlorine

Simplified Latimer diagrams.[SAL]

Acidic solution $E^{\ominus}/V$

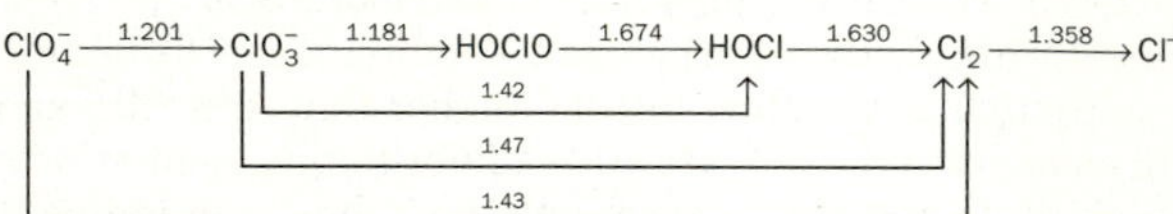

$HClO_2$ is very unstable to disproportionation.

ClO_3^- and HOCl are also thermodynamically unstable with respect to ClO_4^- and Cl_2, but less so.

Alkaline solution $E^{\ominus}/V$

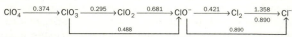

ClO_4^- —0.374→ ClO_3^- —0.295→ ClO_2 —0.681→ ClO^- —0.421→ Cl_2 —1.358→ Cl^-

ClO_3^- —0.488→ ClO^-; ClO^- —0.890→ Cl^-

All intermediate species are unstable relative to ClO_4^- and Cl^-.

anion (+ VII). The acidity of the acids increases with the number of oxygen atoms, since these allow delocalization of the negative charge that is formed on dissociation. Perchloric acid appears to be completely dissociated in water, and its anion binds metal ions less effectively than water does (but although its complexing power is very weak, it is not negligible).

All the oxyoacids of chlorine are oxidizing agents, although chlorous $HClO_2$ disproportionates so readily that it is of little importance. Aqueous solutions of chlorates are used as oxidizing agents in the laboratory; and solid potassium chlorate gives off oxygen when it is heated, and so it is used in fireworks. Aqueous perchloric acid and its salts have little oxidizing power; but anhydrous perchloric acid is an explosively powerful oxidizing agent which has sadly caused many accidents.

If the hydroxyl group in perchloric acid is replaced by a fluorine atom, we get perchloryl fluoride $F-ClO_3$. Although it is not very stable thermodynamically, it has high kinetic stability and has been studied as a possible oxidizer of fuels such as hydrazine, N_2H_4, in rocket propulsion. The analogous compounds $F-ClO$ and $F-ClO_2$ are much less stable, as are the trifluorides F_3ClO and F_3ClO_2 (where one of the oxygen atoms in $F-ClO_2$ or $F-ClO_3$ has been replaced by two fluorine atoms).

Since the chlorine atom can substitute for a hydrogen atom, there are a large number of chlorinated organic compounds. In addition, chlorine compounds make important contributions to synthetic organic chemistry, both as Grignard reagents RMgCl (see p. 135) and as the Lewis acid catalyst $AlCl_3$ (see p. 145).

Although chlorine can and does expand its octet, much of its behaviour simply involves completion of the octet, either by forming the chloride ion, or by sharing an electron from another atom to form one σ bond. So chlorine chemistry lacks the exotic variety of silicon and phosphorus with their extravagant range of oxoanions. Even sulfur, limited to 'two unit' acids but also able to replace O by S, can muster an impressive display compared with the four 'one unit' oxoanions of chlorine. The only bridging group, the $-O-$ in the anhydride Cl_2O_7, breaks down on contact with water. But the oxoacids of chlorine have their own interest, with their pH-dependent tendencies for disproportionation. The main contribution of chlorine to the diversity of inorganic chemistry is, however, in the diverse structures of its binary compounds, both with metal chlorides and with its fellow halogens.

For summary see p. 189.

Summary

Electronic configuration

- $[Ne]3s^2 3p^5$

Element

- σ bonded diatomic gas, yellowish, oxidizing irritant, and very toxic

Occurrence

- As Cl^-, solid deposits, oceans, and salt lakes

Extraction

- By oxidation of Cl^-, electrolytically from melts or solutions, or chemically (small scale)

Chemical behaviour

- Ready electron gain to form stable Cl^-
- Various metal chloride structures depending on electronegativity of metal: ionic, layered, or molecular
- Octet completion by one σ bond
- HCl gaseous, very weak H bonding, completely dissociated in water
- Non-metal chlorides, varied; CCl_4 very stable; oxides explosive
- Range of interhalogen compounds, e.g. Cl–X; many with octet expansion, e.g. ClF_5, I_2Cl_6, Cl_3^+
- Oxoacids, HO–Cl, HOClO, and $HClO_2$, unstable to disproportionation, known only in solution or as anions
- $HO{-}ClO_3$ in water, strong acid with little oxidizing power; explosively oxidizing if anhydrous
- Wide organic chemistry

Uses

- Cl_2 and HCl in industrial and metallurgical processes
- NaCl in food and road clearance
- HOCl as bleach and germicide
- $AlCl_3$ and RMgCl in organic syntheses

Nuclear features

- ^{35}Cl and $^{37}Cl \sim 3:1$

Ar Argon

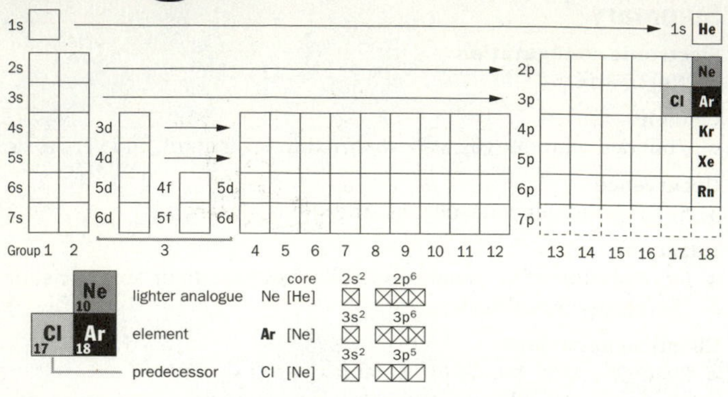

	Predecessor	Element	Lighter analogue
Name	**Chlorine**	**Argon**	**Neon**
Symbol, *Z*	Cl, 17	Ar, 18	Ne, 10
RAM	35.4527	39.948	20.1797
Radius/pm atomic		174	
Melting point/K	172.17	83.78	24.48
Boiling point/K	239.18	87.29	27.10
$\Delta H^{\ominus}_{vap}$**/kJ mol**$^{-1}$	6.41	1.21	0.324
$\Delta H^{\ominus}_{vap}$**/kJ mol**$^{-1}$	20.4033	6.53	1.736
Ionization energy for removal of *j*th electron/kJ mol^{-1} ***j* = 1**	1251.1	1520.4	2080.6
***j* = 2**	2297	2665.2	3952.2
***j* = 3**	3826	3928	6122
***j* = 4**	5158	5770	9370
***j* = 5**	6540	7238	12 177
***j* = 6**	9362	8811	15 238
***j* = 7**	11 020	12 021	19 998

Stable isotopes of argon

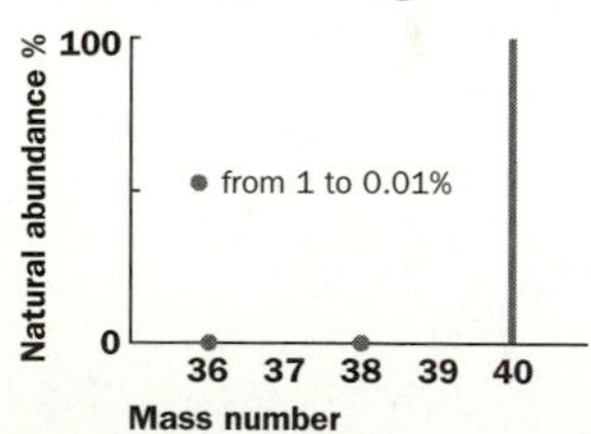

Completion of the $n = 3$ shell gives the 'noble' but far from 'rare' gas, argon, of configuration [Ne]$3s^2p^6$, which accounts for nearly 1% by volume of our atmosphere. Like neon (see p. 119), and for the same reasons, it has no colour, taste, or smell, and it seems to form no compounds except for transient species such as Ar_2^+, produced in discharge tubes.

As argon has a higher mass than neon and a larger and (somewhat) more loosely held electron cloud, it is slightly easier to liquify. It, too, is obtained by fractional distillation of liquid air.

The inertness of argon is widely exploited wherever an unreactive atmosphere is needed, whether for scientific experiments or for preventing the oxidation of hot metals, such as welding material or tungsten light filaments. Argon certainly lives up to its name, perhaps most aptly translated as 'idler'. But although it cannot be coerced into chemical combination, it can be forced into captivity, entrapped in open, hydrogen bonded 'cages'. For example, if β-quinol is crystallized, or water frozen, under a high pressure of argon, some atoms of gas may occupy spaces in the lattice. The resulting clathrate* compounds of limiting composition $Ar(\beta\text{-quinol})_3$ or $Ar_8(H_2O)_{46}$ do not involve any interaction between the argon and its surroundings. The holes in the host crystal can also hold a variety of other species of appropriate size. Neon, however, is not one of them, since it is too small to be retained.

Argon is also found trapped in certain rocks where it is formed when the unstable nucleus of the potassium isotope ${}^{40}_{19}K$ captures an electron and so converts one of its protons into a neutron to become ${}^{40}_{18}Ar$. Since the change has a half-life of 1.25×10^9 years, it is a similar time-scale to geological processes and so measurements of the ratio of the trapped argon to the remaining ${}^{40}_{19}K$ allow us to estimate the age of the rock.

For summary see p. 193.

* The term originates in Oxford, UK, where H.M. ('Tiny') Powell first made such compounds and named them after the Latin word for the cage in which lions were contained before Christians were thrown to them.

Electron capture by K nucleus

$$^{40}_{19}K + ^{0}_{-1}e \xrightarrow[1.25\times10^{9}y]{t_{\frac{1}{2}}} {}^{40}_{18}Ar$$

K nucleus captures an electron, which converts a proton into a neutron and so lowers *Z* by one unit.

lower value of *Z*

Summary

Electronic configuration

- $[Ne]3s^2 3p^6$

Element

- Monatomic gas; difficult to liquefy; not detectable by human senses

Occurrence

- In atmosphere (~ 1%); trapped in rocks

Extraction

- Distillation of liquid air

Chemical behaviour

- No normal compounds known; ionization very difficult; negligible electron affinity
- Transient Ar_2^+ in discharge tubes
- Clathrates of ice or β-quinol (Ar large enough to be trapped)

Uses

- As inert atmosphere
- $^{40}K/^{40}Ar$ dating

Nuclear features

- Mainly ^{40}Ar, traces ^{36}Ar, and ^{38}Ar
- In rocks ^{40}Ar formed from ^{40}K

Part IV

Filling the 4s, 3d and 4p orbitals

The first long period: potassium to krypton

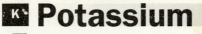

K Potassium

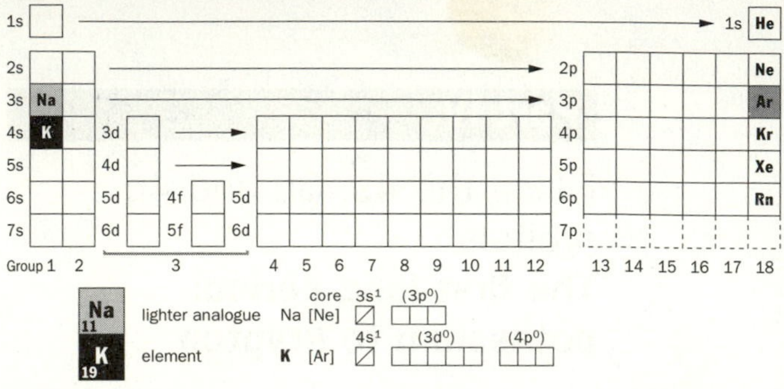

		Element	Lighter analogue
Name		**Potassium**	**Sodium**
Symbol, *Z*		K, 19	Na, 11
RAM		39.0983	22.989768
Radius/pm	**atomic**	227	153.7
	covalent	203	
	ionic, M^+	133	98
Electronegativity (Pauling)		0.82	0.93
Melting point/K		336.8	370.96
Boiling point/K		1047	1156.1
$\Delta H^{\ominus}_{fus}$/kJ mol^{-1}		2.40	2.64
$\Delta H^{\ominus}_{vap}$/kJ mol^{-1}		77.53	89.04
Density (at 293 K)/kg m^{-3}		862	971
Electrical conductivity/Ω^{-1}m^{-1}		1.626×10^7	23.81×10^6
Ionization energy for removal of *j*th electron/kJ mol^{-1}	***j* = 1**	418.8	495.8
	***j* = 2**	3051.4	4562.4
Electron affinity/kJ mol^{-1}		48.4	52.9
Dissociation energy of E_2(g)/kJ mol^{-1}		49.9	73.3
$E^{\ominus}$/V for $M^+(aq) \rightarrow M(s)$		–2.924	–2.713

Stable isotopes of potassium

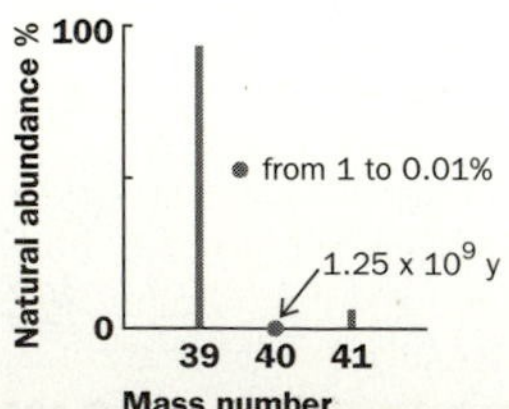

Limiting ionic conductivities AP /(S cm^2 mol^{-1}) in water at 25°C

Li^+	Na^+	K^+	Rb^+
4.01	5.19	7.62	7.92

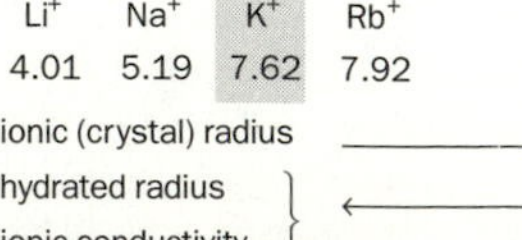

We might wonder if the element following argon, $[Ne]3s^23p^6$, would have one electron in the 3d orbitals; but it does not. The 4s orbitals are, like the 3s and 2s orbitals, able to penetrate appreciably into the filled inner shells, and so a 4s electron has a higher probability of being found much near the nucleus than we might guess from the radius of the 4s subshell. The increased nuclear attraction stabilizes the 4s orbital relative to the much less penetrating 3d orbital, so much so that the 4s orbitals of atoms are of lower energy than the 3d and are filled before them (see p. 198 but also p. 219).

Potassium, then, has the configuration $[Ar]4s^1$ $(3d^0)$, analogous to sodium $[Ne]3s^1$, and its chemistry is almost entirely that of the cation, K^+. The outer electron is, predictably, under even weaker nuclear control for potassium than for sodium. Potassium is easier to atomize and to ionize, and, since its cation is larger than that of sodium, it is even less polarizing.

Potassium is slightly less abundant than sodium in the earth's crust, and much less so in natural waters. It occurs as potassium chloride or, with magnesium, as a double chloride or sulfate. Naturally occurring potassium contains two stable isotopes, together with a trace of ^{40}K, which decays radioactively (partly to argon, see pp. 191–2) but very slowly, with a half-life of 1.25×10^9 years. Potassium metal is made by reducing the chloride with sodium: electrolysis is not feasible because potassium is too soluble in the molten chloride. The chloride and sulfate are widely used in fertilizers, since potassium is essential for plant growth. Its presence in wood is demonstrated by the lilac flames seen in bonfires and around burning logs. Potassium carbonate is used in glass manufacture, while the nitrate and chlorate, which give off oxygen when heated, are components of many fireworks.

Potassium resembles sodium much more closely than sodium resembles lithium, with its small, more polarizing, cation. There are, however, some differences. The larger potassium ion is less attractive to water and so has lower values of both $(-\Delta H^\circ)$ and $(-\Delta S^\circ)$ of hydration (see p. 198); indeed it has a lower *hydrated* radius, which results in higher ionic conductivity. It also forms weaker lattices, and, since this small effect often outweighs the decreased hydration energy (see p. 38), potassium hydroxide and fluoride are more soluble than the corresponding sodium compounds. The reverse is true for salts of larger anions, and potassium perchlorate is only sparingly soluble.

Potassium is even more electropositive than sodium, and more reactive. It decomposes water so exothermically that it ignites both the hydrogen produced and the residual metal, which burns with its characteristic lilac flame. When potassium is heated in excess of

Variation of orbital energies E with increasing atomic number Z [SAL]

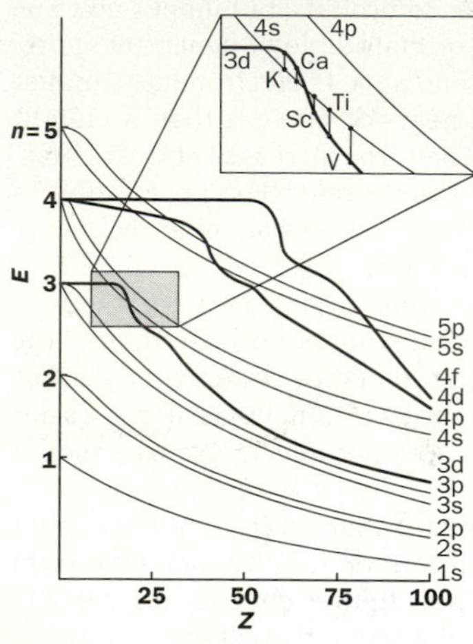

enlargement of shaded part, for elements K ($Z = 19$) to V ($Z = 23$)

As Z increases, the orbitals become more stable. This effect is greater for d orbitals than for s and p (and greater still for f); and it is more marked for cations than for unchanged atoms

Heats and entropies of hydration [D] of M^+ ns^1 ions

	Li^+	Na^+	K^+	Rb^+	Cs^+
$-\Delta H^{\ominus}$ hydn/kJ mol^{-1}	502	406	320	296	264
$-\Delta S^{\ominus}$ hydn/J K^{-1} mol^{-1}	119	89	51	40	37

Calculated lattice enthalpies/kJ mol^{-1}

	F^-	Cl^-	Br^-	I^-
K^+	821	717	689	649
Na^+	926	787	752	705

$K^+O_2^-$ in breathing apparatus

$$2KO_2(s) + CO_2(g) \rightarrow K_2CO_3(s) + \frac{3}{2}O_2(g)$$

Thermal decomposition of KNO_3

$$KNO_3(s) \xrightarrow{\text{heat}} KNO_2(s) + \frac{1}{2}O_2(g)$$

air, it gives the superoxide $K^+ O_2^-$, which is used in breathing apparatus since it absorbs carbon dioxide, liberating oxygen. Potassium, like sodium, forms an ionic hydride $K^+ H^-$, and gives a blue solution in liquid ammonia. The nitrate is decomposed by heat to the nitrite and oxygen. The complexes formed with non-cyclic oxygen donor ligands are weaker than those of sodium, while interaction with crowns and crypts depends on a good fit of the cation within their holes (see p. 128). It may well be that ligands of this type are used by living organisms to control the passage of sodium and potassium ion across cell membranes, and so to manipulate those changes in their relative concentrations that are necessary for all cell functioning, including cells of nerves and muscle.

Summary

Electronic configuration
- $[Ar]4s^1$

Element
- Soft, very electropositive metal, similar to sodium but even more reactive; flame lilac

Occurrence
- As chloride or sulfate (and magnesium) and in wood ash

Extracted
- From KCl by reduction with sodium

Chemical behaviour
- Similar to sodium (but metal burns to KO_2)
- Dominated by K^+, compounds mainly ionic, differ from Na^+ analogues in size-related properties: hydration, lattice energies, crown and crypt complexes

Uses
- KCl and K_2SO_4 as fertilizers
- KNO_3 and $KClO_3$ as oxygen generators in fireworks
- KO_2 in breathing apparatus

Nuclear features
- ^{39}K and ^{41}K stable ~ 12:1 and trace of ^{40}K (decays to ^{40}Ar)

Ca Calcium

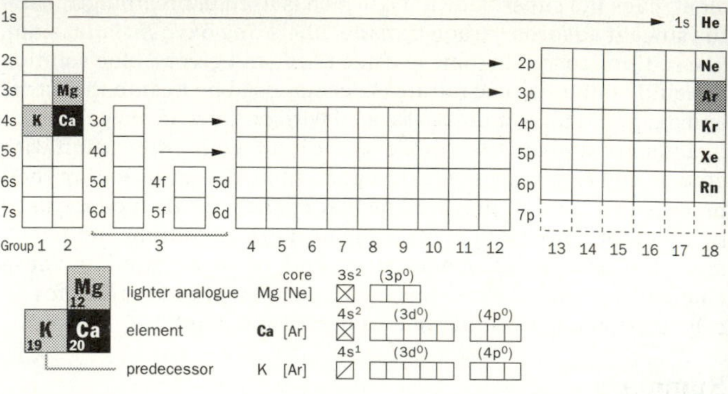

	Predecessor	Element	Lighter analogue
Name	**Potassium**	**Calcium**	**Magnesium**
Symbol, *Z*	K, 19	Ca, 20	Mg, 11
RAM	39.0983	40.078	24.3050
Radius/pm **covalent**	203	174	136
ionic, M^{2+}		106	78
Electronegativity (Pauling)	0.82	1.00	1.31
Melting point/K	336.8	1112	922.0
Boiling point/K	1047	1757	1363
$\Delta H^{\ominus}_{fus}$**/kJ mol**$^{-1}$	2.40	9.33	9.04
$\Delta H^{\ominus}_{vap}$**/kJ mol**$^{-1}$	77.53	149.95	128.7
Density at 293 K)/kg m^{-3}	862	1550	1738
Electrical conductivity/$\Omega^{-1}m^{-1}$	1.626×10^7	2.915×10^7	2.283×10^7
Ionization energy for removal of *j*th electron/kJ mol^{-1} ***j* = 1**	418.8	589.7	737.7
***j* = 2**	3051.4	1145	1450.7
Electron affinity/kj mol^{-1}	48.4	−186	−21
$E^{\ominus}$**/V for** $M^{2+}(aq) \rightarrow M(s)$		−2.84	−2.356

Stable isotopes of calcium

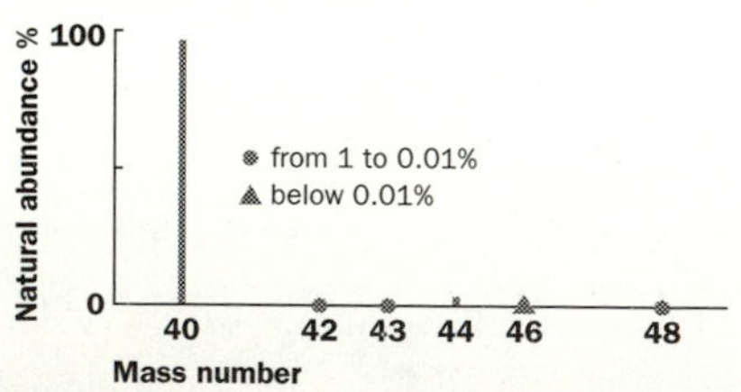

The chemistry of calcium [Ar]$4s^2$ is almost entirely that of the cation Ca^{2+}, as this ion optimizes the difference between the energy needed to atomize and ionize the metal and that evolved on lattice formation or hydration. In this respect calcium behaves more like a doubly charged successor to potassium than a heavier analogue of magnesium, whose cation is formed with more difficulty and is more highly polarizing. The higher charge gives Ca^{2+} salts much higher lattice energies than those of potassium, especially if the anion is small or highly charged. This effect, together with the more unfavourable entropy of hydration, must outweigh the greater hydration enthalpy of Ca^{2+}, since calcium salts are less soluble than the corresponding potassium ones. Although calcium chloride and nitrate dissolve in water, the fluoride and hydroxide are only sparingly soluble; and salts with large multiply charged anions also have low solubilities (e.g. $CaSO_4$) or even very low ones [e.g. $CaCO_3$ and $Ca_3(PO_4)_2$]. Indeed, calcium, the third most abundant metal in nature, is found in rocks containing sulfate, carbonate, and phosphate; and aqueous Ca^{2+} ions are present in so-called 'hard' natural waters. Calcium carbonate is the main component of the shells of molluscs, and calcium phosphate is sufficiently insoluble to be used for vertebrate bones and teeth.

As calcium is so electropositive, the metal, like sodium, is obtained electrolytically from the molten chloride. With two electrons in its conduction band, calcium is tougher than the alkali metals ([core]ns^1), with higher melting and boiling points, and is less violently reactive. The value of $E^\ominus$ is, however, similar to those of sodium and potassium, because the higher energy of formation of Ca^{2+} is balanced by its favourable hydration energy. With liquid ammonia, calcium (like potassium) gives a blue solution from which crystals of the solvated metal may be obtained.

Reactions of calcium with non-metals resemble those of both potassium and magnesium. With oxygen, calcium can form either CaO (like magnesium) or CaO_2 (like potassium). It resembles magnesium and lithium in forming a nitride, Ca_3N_2. The hydride, CaH_2, and the carbide, CaC_2, are both ionic; with the polarizable acetylide ion C_2^{2-}, calcium forms a less stable carbide than does magnesium, but with the less polarizable hydride ion, the reverse is true, as would be expected for a fairly undeformable cation like calcium, with little tendency to form covalent bonds.

The crystal structures of the calcium halides do, however, form a series from the ionic fluoride, through the largely ionic but distorted chloride and bromide, to the iodide, which has a layer structure. Calcium chloride is manufactured on a large scale, for use in refrigeration plants and for snow clearance in cold climates,

Hydration ^{s} of Ca^{2+} ion at 25°C

	Ca^{2+}	Mg^{2+}	Li^+	Na^+	K^+
$-\Delta H^\ominus_{hyd}$/kJ mol^{-1}	1595	1940	528	413	330
$-\Delta S^\ominus_{hyd}$/JK^{-1} mol^{-1}	230	320	140	110	70
$-\Delta G^\ominus_{hyd}$/kJ mol^{-1}	1525	1845	485	379	308

The contribution, $T\Delta S^\ominus$ of the unfavourable entropy change is small (here not more than 8%) compared with the very favourable heat change.

Thermal decomposition of $CaCO_3$

Temp./°C at which p_{CO2} = 1 atm	$CaCO_3$	$MgCO_3$
	840	300

Limewater test

$Ca^{2+}(aq) + 2OH^-(aq) + CO_2(aq) \rightarrow CaCO_3(s)$

sat. aq. $Ca(OH)_2$ 'limewater' (under $Ca^{2+}(aq) + 2OH^-(aq)$) — cloudy precipitate (under $CaCO_3(s)$)

OH^- prevents CO_2 from forming $H_2CO_3(aq)$ or HCO_3^- aq

Chelation of Ca^{2+} by $EDTA^{4-}$

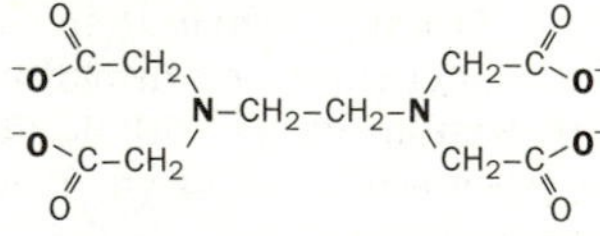

$EDTA^{4-}$ (ethylenediamine tetraacetate) the six donor atoms are in bold

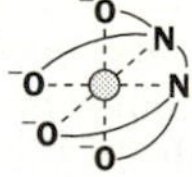

Ca $EDTA^{2-}$ complex
The Ca^{2+} is octahedrally coordinated

○ = Ca^{2+}

For reaction $Ca^{2+}(aq) + EDTA^{4-}(aq) \longrightarrow Ca\,EDTA^{2-}(aq) + xH_2O$

$\Delta H^\ominus = -27$ kJ mol^{-1} favourable but small

exothermic step $Ca^{2+}(aq) + EDTA^{4-}(aq) \longrightarrow Ca\,EDTA^{2-}(aq)$, almost offset by endothermic step $Ca^{2+}(aq) \longrightarrow Ca^{2+}(g) + H_2O$

$\Delta S^\ominus = +113$ J K^{-1} mol^{-1} favourable and large. Water molecules released. Removal of small Ca^{2+} ion and total reduction in charge decreases long-range order

$T\Delta S^\ominus$ at 25°C +33.7 kJ mol^{-1}

$-\Delta G^\ominus$ at 25°C −60.7 kJ mol^{-1} (of which 55.5% is the entropy term) SC

since a mixture of calcium chloride and ice melts at a much lower temperature than the ice and salt (NaCl) mixture used in less severe conditions. The high hydration energy of the Ca^{2+} ion makes the anhydrous chloride a useful drying agent.

By far the most important industrial compound of calcium is the oxide CaO ('quicklime' or more often just 'lime'), obtained by heating the carbonate (limestone). As expected from the Kaputstinskii equation (see p. 36), limestone has a higher decomposition temperature than magnesium carbonate; the smaller the cation, the greater the gain in lattice energy on decreasing the size of the anion. Lime is used mainly for cements, but also for manufacturing steel, magnesium, and glass, and for neutralizing acids in sewage purification, in food technology, and in many other contexts. Calcium oxide reacts rapidly* and exothermically with water to give the sparingly soluble hydroxide $Ca(OH)_2$, which is used as an antacid. The solution, limewater, provides the traditional test for carbon dioxide by turning cloudy as calcium carbonate is precipitated.

Calcium forms fairly weak complexes, usually with ligands that contain several negatively charged oxygen atoms. Solutions of the soluble salts such as the chloride and nitrate show no trace of hydrolysis products of Ca^{2+}(aq). Ligands that form complex anions, like $EDTA^{4-}$ and PO_4^{3-}, are useful for water 'softening', since they remove the free Ca^{2+} ions from solution and so prevent them from forming the insoluble carbonate or stearate that produce 'boiler scale' or soap 'scum'. Calcium ions can also be engulfed by crypts and crowns of a suitable size. The passage of Ca^{2+} (as also of K^+) through biological membranes is probably controlled by complex formation of this type. The role of the Ca^{2+} ion in living systems extends far beyond making structural components such as bones and teeth. Its concentration controls the rates of a wide variety of functions such as muscle contraction (for example heartbeat) and blood clotting; and injection of Ca^{2+} ions into some unfertilized egg cells can stimulate development as effectively as when a similar boost in calcium concentration is triggered by a sperm.

Eutectic ᴬ melting point of $CaCl_2$–H_2O

(i.e. of that mixture with the lowest melting point)

$CaCl_2 - H_2O$	−55°C	$NaCl - H_2O$	−18°C

* Hence the 'quick', which here means lively.

There are, as yet, very few organometallic compounds of calcium; and since calcium has no tendency to form π bonds few would be expected. Some alkyls (CaR_2) and some iodide analogues of Grignard reagents (RCaI) have been made, but they are more reactive than their magnesium counterparts. In the calcium cyclopentadienyl 'sandwich' $Ca(C_5H_5)_2$, there seems to be interaction between the calcium in one molecule and the rings of two adjacent molecules.

Although calcium is an essential element biologically and technologically, its chemistry seems relatively straight forward. The dominant species, Ca^{2+}, being doubly charged but not very small, has only a slight tendency to deviate from totally ionic behaviour.

Summary

Electronic configuration

- $[Ar]4s^2$

Element

- Fairly soft, fairly reactive electropositive metal; flame brick-red

Occurrence

- In rocks with CO_3^{2-} (from mollusc shells), PO_4^{3-}, F^-; in hard water as Ca^{2+}

Extraction

- Electrolysis of molten $CaCl_2$

Chemical behaviour

- Many ionic compounds, e.g. CaC_2, Ca_3N_2, of high lattice energy (but CaI_2 has layer structure); less soluble than K^+ salts
- Complexes moderately weakly with oxygen donors, e.g. $EDTA^{4-}$, crowns, and crypts
- Forms blue solution with liquid NH_3

Uses

- 'Lime', CaO in cements, for making steel and glass, extracting aluminium, purifying sewage, and in food technology
- $CaCl_2$ for refrigerators, snow clearance, and as desiccant
- In biological systems: Ca^{2+} as a regulator; $Ca_3(PO_4)_2$ in bones and teeth; $CaCO_3$ in shells
- 'Limewater', $Ca(OH)_2(aq)$ as an indicator for CO_2

Nuclear features

- Mainly ^{40}Ca, with ~ 2% ^{44}Ca (stable) and traces of four other isotopes

Sc Scandium

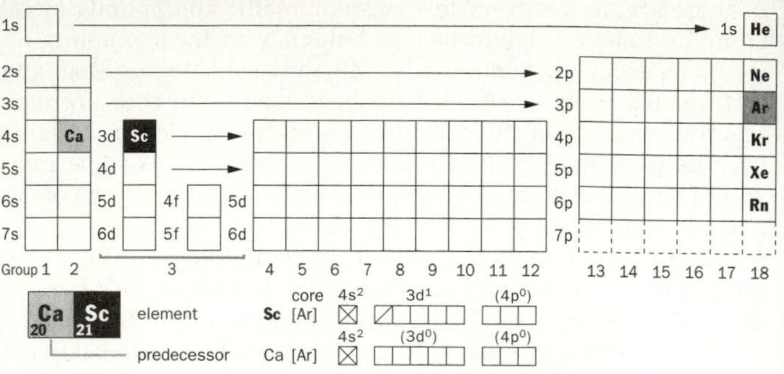

		Predecessor	Element
Name		**Calcium**	**Scandium**
Symbol, *Z*		Ca, 20	Sc, 21
RAM		40.078	44.955910
Radius/pm	**atomic**		160.6
	covalent	174	144
	ionic, M^{3+}		83
Electronegativity (Pauling)		1.00	1.36
Melting point/K		1112	1814
Boiling point/K		1757	3104
$\Delta H^{\ominus}_{fus}$/kJ mol^{-1}		9.33	15.9
$\Delta H^{\ominus}_{vap}$/kJ mol^{-1}		149.95	304.8
Density (at 273 K)/kg m^{-3}		1550	2989
Electrical conductivity/Ω^{-1}m^{-1}		2.915×10^7	1.639×10^6
Ionization energy for removal of *j*th electron/kJ mol^{-1}	***j* = 1**	589.7	631
	***j* = 2**	1145	1235
	***j* = 3**		2389
Electron affinity/kj mol^{-1}		–186	18.1
$E^{\ominus}$/V for $M^{3+}(aq) \rightarrow M(s)$			–2.03

Stable isotopes of scandium

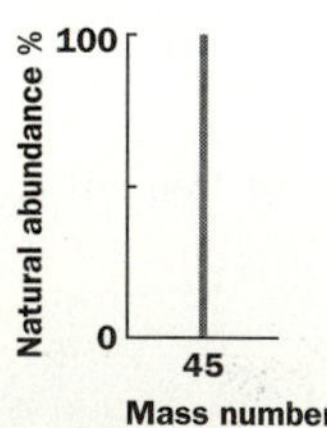

Bipyridyl or 'bipy'

is octahedrally coordinated to Sc^{3+} in $Sc(bipy)_3^{3+}$

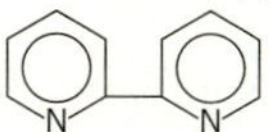

Scandium is a rare, little-used metal which, none the less, occupies a key position among the elements. Discovered in an esoteric silicate ore in Scandinavia, the first (almost) pure sample was made only in 1960. Scandium oxide is present in trace quantities in waste from uranium extraction, and this provides the main source of the element.

After calcium, with its full ns^2 configuration, the next element does not make use of the np orbital, as happened when boron $2s^22p^1$ followed beryllium, or aluminium $3s^23p^1$ followed magnesium. Scandium has three sets of available orbitals, and since the 4s and the 3d are of lower energy than the 4p, the configuration of scandium is $[Ar]4s^23d^1$ (see p. 197). So scandium is our entrée to the vast and exciting chemistry of the d block elements.

The metal, which is obtained by electrolysis of the molten chloride, is moderately electropositive. It is fairly soft, tarnishes to the oxide Sc_2O_3 in air, and reacts with water to form hydrogen, especially when finely divided, or if warmed. Scandium metal conducts electricity far less well than aluminium does, and this *might* suggest that the $3d^1$ electron in scandium is less mobile than the $3p^1$ electron in aluminium. Its chemistry is largely, but not entirely, that of the Sc^{3+} cation, $([Ar]4s^03d^0)^{3+}$. (Although the cation Sc^{2+} plays little part in the reactions of scandium, it is interesting because its configuration, $([Ar]4s^03d^1)^{2+}$, illustrates how the 3d orbital has been stabilized by the charge on the ion.)

The chemical reactivity of scandium metal resembles that of calcium, and so too does the solubility of its salts. Although the nitrate and higher halides of scandium are soluble in water, the fluoride and hydroxide are not, and nor are its salts with highly charged anions. But in water the triply charged cation $Sc^{3+}(aq)$ resembles $Al^{3+}(aq)$, rather than $Ca^{2+}(aq)$, and is hydrolysed to polymeric hydroxo cations. At higher values of pH, there is a gelatinous precipitate of $Sc(OH)_3$, which dissolves in alkali to form, eventually, $Sc(OH)_6^{3-}$. Again like aluminium, scandium forms a six coordinated fluoro complex, ScF_6^{3-}. Double sulfates similar to the 'alums' (see p. 40) are also formed. Scandium combines with several chelating ligands, including nitrogen donors, to give complexes such as $Sc(bipy)_3^{3+}$. Like other metals that have no tendency to form π bonds, scandium seems to form few organometallic compounds, although some alkyls ScR_3 and the cyclopentadienyl complex $Sc(C_5H_5)_3$ have been made.

Scandium, as well as being rare and as yet useless, might also sound boringly predictable, with some similarities to calcium and to the less exotic chemistry of aluminium. But scandium has its surprises, as yet largely unexplained. The hydride is ScH_2 and is a

good electrical conductor. Can it be $Sc^{3+}.2H^{-}$, with a free electron in the conduction band (see p. 425)? Does the compound $CsScCl_3$ really contain Sc^{II}? And what is going on in the six different phases that are formed when scandium metal is dissolved in molten $ScCl_3$? There seems to be good evidence of metal–metal bonding within chains of edge-sharing Sc_6 octahedra; but the detailed structural and electronic properties of the phases have yet to be worked out.

Summary

Electronic configuration

- $[Ar]4s^2 3d^1$

Element

- A fairly soft, fairly reactive metal (tarnishes in air, attacks water)

Occurrence

- Rare: as oxide, with silicon, and in uranium waste

Extraction

- By electrolysis of the molten chloride

Chemical behaviour

- Most compounds contain Sc^{3+}
- Similarities to Ca: reactivity; solubilities of salts
- Similarity to Al: ScF_6^{3-}, polynuclear hydrolysis of Sc^{3+}
- Similarity to La: conducting hydride ScH_2
- Some complexes with nitrogen donors
- $(Sc_6)_x$ clusters in solutions of Sc in molten $ScCl_3$

Uses

- None, as yet

Nuclear features

- Isotopically pure ^{45}Sc

Ti Titanium

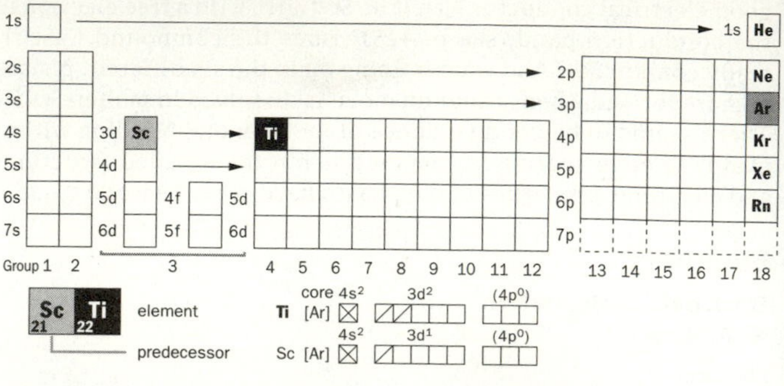

		Predecessor	Element
Name		**Scandium**	**Titanium**
Symbol, *Z*		Sc, 21	Ti, 22
RAM		44.955910	47.88
Radius/pm	**atomic**	160.6	144.8
	covalent	144	132
	ionic, M^{2+}		80
	M^{3+}	83	69
Electronegativity (Pauling)		1.36	1.54
Melting point/K		1814	1933
Boiling point/K		3104	3560
$\Delta H^{\ominus}_{fus}$/kJ mol^{-1}		15.9	20.9
$\Delta H^{\ominus}_{vap}$/kJ mol^{-1}		304.8	428.9
Density /kg m^{-3}		2989 (273 K)	4540 (293 K)
Electrical conductivity/Ω^{-1}m^{-1}		1.639×10^6	2.381×10^6
Ionization energy for removal of *j*th electron/kJ mol^{-1}	***j* = 1**	631	658
	***j* = 2**	1235	1310
	***j* = 3**	2389	2652
	***j* = 4**		4175
Electron affinity/kJ mol^{-1}		18.1	7.6
$E^{\ominus}$/V for TiO^{2+}(aq) → Ti(s) {H^+} = 1			−0.86

Stable isotopes of titanium

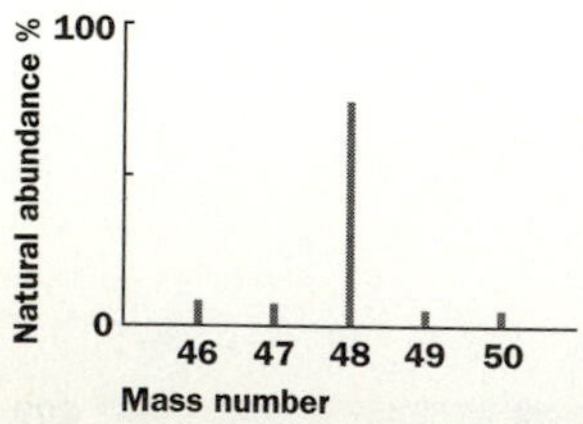

After scandium comes titanium ([Ar]$3d^2 4s^2$); and for the next eight elements successive electrons continue to be added until the 3d orbitals are fully occupied at zinc ([Ar]$3d^{10} 4s^2$). Elements with atoms or ions that have partially filled $(n-1)$ orbitals (and configurations [Core] $(n-1)d^x ns^{1\pm1} np^0$) are known as transition metals, since they come in between those using only their ns orbital and those that have *n*p electrons. There are many ways in which transition metals and their compounds resemble each other, but differ from non-transition metals such as potassium or calcium in the s block, and aluminium in the p block.

Titanium, unlike scandium, is both fairly common and widely used; indeed, it is the second commonest transition element after iron. Those titanium compounds that retain one of the d electrons show typical transition metal behaviour (see p. 281), unlike most scandium compounds, which contain the Sc^{3+} (d^0) ion.

We might expect that, if all the four outer electrons could be removed from titanium, the resulting Ti^{4+} ion would combine strongly with small or highly charged anions such as fluoride or oxide; titanium is found in nature as the dioxide TiO_2 or as a mixed oxide $M^{2+}Ti^{IV}O_3$, where M^{2+} is most commonly iron. Although titanium dioxide is reduced by carbon, this reaction cannot be used to manufacture the metal, since titanium combines with the excess carbon to form a stable carbide. Instead, the oxide is converted to a halide, which is then decomposed to give the metal. Commercially, titanium tetrachloride is treated with magnesium metal under argon, but if a small, very pure sample is needed the volatile tetraiodide is decomposed on a hot wire at low pressure.

Preparation of the metal

Industrial-scale extraction (Kroll):

$$TiO_2(s) + 2Cl_2(g) + 4C(s) \xrightarrow{900^\circ C} TiCl_4(g) + 4CO(g)$$

$$\text{or } 2FeTiO_3(s) + 7Cl_2(g) + 6C(s) \xrightarrow{900^\circ C} 2TiCl_4(g) + 2FeCl_3(s) + 6CO(g)$$

condense

$$\left.\begin{matrix} TiCl_4(g) + 2Mg \\ \text{or } TiCl_4(g) + 4Na \end{matrix}\right\} \xrightarrow[Ar]{1050 \pm 100^\circ C} Ti + \left\{\begin{matrix} 2MgCl_2 \\ 4NaCl \end{matrix}\right.$$

remove salts and excess metal with water or dilute acid.

Laboratory-scale purification (van Arkel–de Boer):

$$Ti(crude) + 2I_2 \rightarrow TiI_4(g) \xrightarrow[\text{low pressure}]{\text{heat}} Ti(s) + 2I_2(g)$$

The rutile form of TiO_2 SM

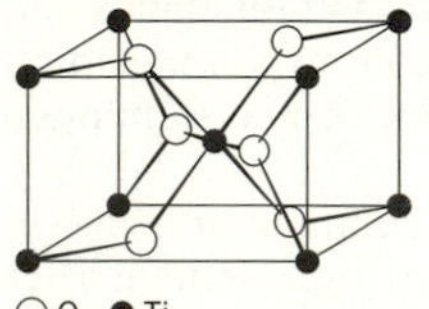

Unit cell, showing central (distorted) TiO_6 octahedron

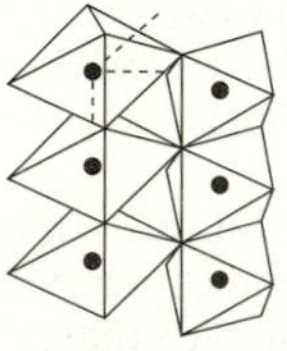

Two columns of TiO_6 built by sharing opposite edges, and joined by sharing verticies

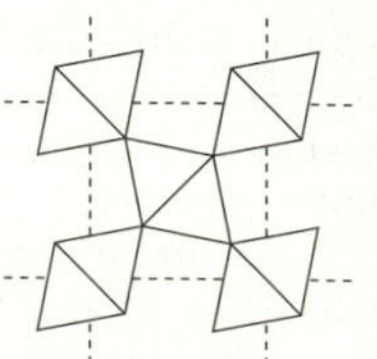

Projection of structure on base of unit cell

Radii/pm of some M^{2+} ions in $MTiO_3$

Ti(IV)	Mg^{2+}	Fe^{2+}	Co^{2+}	Ni^{2+}	Ca^{2+}	Ba^{2+}
60.5	78	82	82	78	106	143
	1 Ti(IV) + 1 of these can fit into the Al_2O_3 carborundum lattice ($r_{Al^{3+}} = 57$)				too large for carborundum: forms perovskite structure	too large for perovskite: expands it

The perovskite structure SM of $CaTiO_3$

● Ti ◎ Ca ○ O

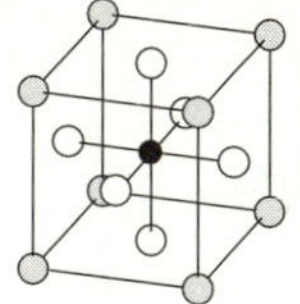

portion showing Ti at centre of TiO_6 octahedron in a cube of Ca

portion showing Ca surrounded by 12 O, in a cube of Ti

* It has even been used as a building-cladding.

** In ferroelectric substances, dipole moments may be aligned in domains, like electron spins in ferromagnetic ones; and since some order may be long term, such materials may, like permanent magnets, have a 'memory'.

Titanium is a moderately strong metal, with higher heats of fusion and atomization than scandium, and it is a somewhat better electrical conductor: the two 3d electrons seem to be making their presence felt. The strength of titanium can be greatly increased by adding small amounts of tin or aluminium and such alloys have the highest ratio of strength to weight of any engineering metal. Titanium is therefore widely used in aircraft, both for the frames and for gas turbine engines.

Titanium is a fairly electropositive metal (although less so than scandium) but (like aluminium) is protected against corrosion and other reactions at room temperature by a passive, very stable, oxide coating*. If the oxide coating is prepared artificially (e.g. by anodic oxidation), its thickness can be controlled to give surface layers thin enough to produce diffraction colours similar to those observed from films of oil on wet roads. Since titanium is so strong and light, the oxide film colours have been used in jewellery, especially ear-rings. At high temperatures, however, titanium metal can take part in reactions, e.g. with carbon and oxygen, and it even burns in nitrogen.

By far the most important titanium compounds are those containing Ti^{IV} and oxygen, which may well involve π donation from the filled oxygen orbitals to the empty d orbitals of the metal, since the small Ti^{4+} ions would be likely to be highly polarizing. The dioxide (which, although written as TiO_2, loses up to nearly 1% of oxygen when it is heated to its melting point) exists in various forms, which all contain TiO_6 octahedra joined in different ways. All forms are chemically stable and, as they have a very high refractive index, finely divided samples scatter light very effectively. Titanium dioxide is widely used as a white pigment in paints and powders.

The mixed oxides $M^{II}TiO_3$ and $M^{II}_2TiO_4$, although loosely called 'titanates', do not contain discrete titanium oxoanions (except for $Ba^{2+}_2[TiO_4^{4-}]$). They consist of a continuous array of oxide ions, with the metal ions occupying the holes in between. The exact structure depends on the size of the M^{2+} ion relative to that of Ti^{IV}. With cations such as Mg^{2+}, Fe^{2+}, Co^{2+}, and Ni^{2+}, which are not grossly larger than Ti^{IV}, the $M^{2+}M^{4+}O_3$ structure is similar to that of the corundum form of $(Al^{3+})_2O_3$; with a larger cation such as Ca^{2+}, the titanium takes up a position as far away from it as possible. In $BaTiO_3$, the large cation forces the structure to expand with the result that, if the crystal is subjected to pressure or an electric field, the titanium ions can move within their cavities. The ability of the small cations to 'rattle' in this way makes the solid piezoelectric and ferroelectric,** and so of value for making electronic devices such as microphones. The mixed oxides $(M^{2+})_2Ti^{IV}O_4$ have struc-

* In some transition metal ions, the magnetic moment of a material depends only on the number (*n*) of unpaired electrons, and the experimental values for Ti^{III} complexes are almost exactly those expected for $n = 1$. But this is largely by chance, since the moment may also be influenced by the orbital momentum of the unpaired electron(s) and by any covalent contribution in the bond. In Ti^{III} complexes both these secondary effects are active but almost cancel each other out.

tures similar to that of the spinels $M^{2+}(M^{3+})_2O_4$ with some cations surrounded by six oxide ions and others by four (see p. 143).

The Ti^{4+} ion is far too polarizing to exist free in solution, but can form basic salts, such as $TiOSO_4.H_2O$. In moderately concentrated aqueous acid, polymeric cations containing –Ti–O–Ti– groups are probably formed. Solutions containing Ti^{IV} complexes combine with hydrogen peroxide, which is thought to chelate to form a three membered ring, giving a bright orange colour which is used in analysis.

Molecular compounds, MX_4, of titanium include $M(OR)_4$ alkoxides (which are used for making non-drip paints) and the volatile halides, MCl_4, MBr_4, MI_4. These compounds can absorb radiation to promote an electron from a predominantly halogen orbital to one associated with the titanium, corresponding to the oxidation of the X^- ion, by the highly charged Ti^{IV}. This charge transfer from X^- to Ti^{IV} occurs most readily for the easily oxidized I^- ion, in the highly coloured TiI_4; Cl^- is more difficult to oxidize and, since $TiCl_4$ absorbs ultraviolet rather than visible radiation, it is colourless. Titanium tetrafluoride is also colourless, but is high melting and of unknown structure. The molecular halides can be hydrolysed to TiO_2. They also act as Lewis acids and are used to make Ziegler–Natta and similar catalysts for the ordered polymerization of olefines (see p. 71). It seems that the catalysts bind the gases strongly enough to keep the molecules in the right orientation, but not so strongly as to prevent them from reacting.

In the compounds we have discussed so far, titanium has lost or shared all four of its outer $3d^24s^2$ electrons. However, it can also form compounds, such as the halides, TiX_3, and the aquo ion, $Ti(H_2O)_6^{3+}$, of electronic structure $[Ar]3d^1$. Here we meet typical transition metal behaviour: two or more oxidation states [for titanium (IV) and (III)] that differ only by one unit, and readily take part in redox equilibria. [The $Ti(H_2O)_6^{3+}$ ion is a moderately strong reducing agent.] The presence of one or more unpaired electrons in many transition metal compounds causes them to be paramagnetic (see p. 104).* Transition metal ions are often coloured, even if only weakly so; $Ti(H_2O)_6^{3+}$ is violet.

The colour of this ion is a direct result of its incompletely filled set of d orbitals. In an isolated gaseous ion of a transition metal, the d orbitals are degenerate, i.e. they are all of the same energy. If the ion is now surrounded by donor groups (such as H_2O), the lone pairs of the Lewis base donors home in on the positively charged Lewis acid acceptor cation, and to some extent repel any d electrons present on its surface. The d orbitals become slightly less stable than in the isolated ion; but some are destabilized more than

Splitting of d orbitals by ligands

Here we shall discuss only six ligands, placed octahedrally (for other arrangements, see pp. 260 and 268)

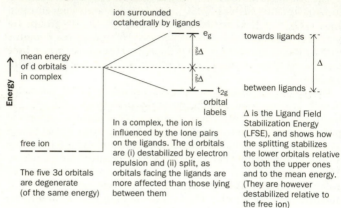

ion surrounded octahedrally by ligands

e_g — towards ligands

$\frac{3}{5}\Delta$

Energy →

mean energy of d orbitals in complex

Δ

$\frac{2}{5}\Delta$

t_{2g} — between ligands

orbital labels

free ion

The five 3d orbitals are degenerate (of the same energy)

In a complex, the ion is influenced by the lone pairs on the ligands. The d orbitals are (i) destabilized by electron repulsion and (ii) split, as orbitals facing the ligands are more affected than those lying between them

Δ is the Ligand Field Stabilization Energy (LFSE), and shows how the splitting stabilizes the lower orbitals relative to both the upper ones and to the mean energy. (They are however destabilized relative to the free ion)

The origin of ligand field splitting [SAL]

Two e_g orbitals are directed towards the ligands L

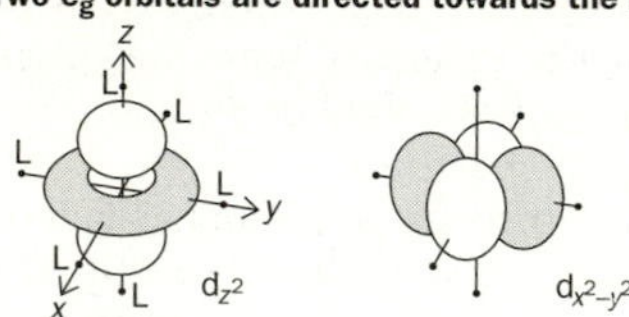

Three t_{2g} orbitals are directed between the ligands

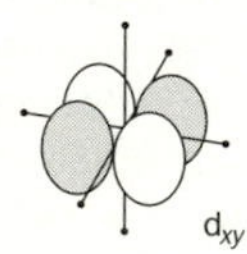

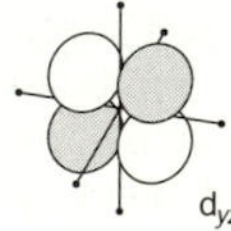

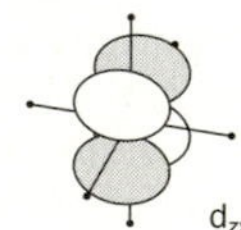

others. The electric field of the ligands will have a greater effect on the orbitals pointing directly towards the donor atoms than on the orbitals that lie between them. So the d orbitals are split into two or more sets, of different energy. The way in which the orbitals are split, and the difference in energy between them, depends on the orientation of the donor groups; and the purely electrostatic 'crystal field' situation outlined above will be complicated by covalent contributions to bonding, which is taken into account in the fuller 'ligand field' treatment of orbital splitting.

The $Ti(H_2O)_6^{3+}$ ion is octahedral, and three of the 3d orbitals point between the ligands, while the other two are less stable because they point towards the ligands. The one 3d electron is of course normally in one of the three more stable orbitals, but it can absorb radiation and be promoted into a higher level. Since the energy gap between the two levels is in the visual region, this absorption alters the composition of any white light that falls on it; and the complex appears coloured. The $Ti(H_2O)_6^{3+}$ ion absorbs light in the green region of the spectrum, and so we see the colour that corresponds to a mixture of the remaining energies (blue to turquoise, yellow to red), which gives violet.* When colours arise from transitions of electrons between sets of orbitals within the same shell, as in this 3d–3d transition, they are usually not very intense; as a general rule of thumb, the more symmetrical the complex, the weaker the colour.

* Since one electron is promoted from one definite level to another, we might expect a sharp spectral line. In fact, we see a moderately broad peak, which is not even symmetrical. The breadth is readily understood because the $Ti(H_2O)_6^{3+}$ is vibrating, and the variation in distance of the water molecules from the metal ion causes changes in the energy gap between the more stable ('t_{2g}') and less stable ('e_g') orbitals and hence in the energy of light absorbed. The ground state of the $Ti(H_2O)_6^{3+}$ ion, with a single electron in the lower, t_{2g}, level, approximates to a regular octahedron; but in the excited state the octahedron is distorted (it is either extended or squashed along one axis) and the pair of e_g orbitals are no longer of the same energy. So *two* transitions can occur, from the t_{2g} level to each of the different higher levels; and the observed 'peak' is really two rather similar peaks, superimposed. The (Jahn–Teller) distortion of an octahedron occurs in the ground state of an atom as well as the excited state whenever the two upper orbitals are unevenly filled; i.e. when they contain a total of one, or three, electrons.

The colour of the Ti^{3+}(aq) ion (simplified)

$Ti(H_2O)_6^{3+}$ is octahedral, d^1

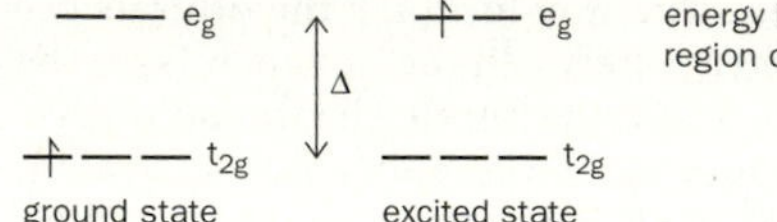

energy is absorbed is in visible (greenish) region of the spectrum: Ti^{3+}(aq) looks violet

The colour of the Ti^{3+}(aq) ion (less simplified)

The scheme above would give one sharp band; the observed peak is neither sharp nor symmetrical. This is because the ion is not a regular octahedron: it is distorted, and it also vibrates

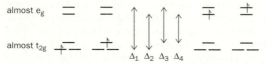

almost e_g

almost t_{2g}

Δ_1 Δ_2 Δ_3 Δ_4

Departure from regularity produces more small splittings, with slightly different values of Δ (here four are shown). So, a mixture of energies is absorbed, and we get a less narrow peak

Metallic conductivity in TiO [SAL]

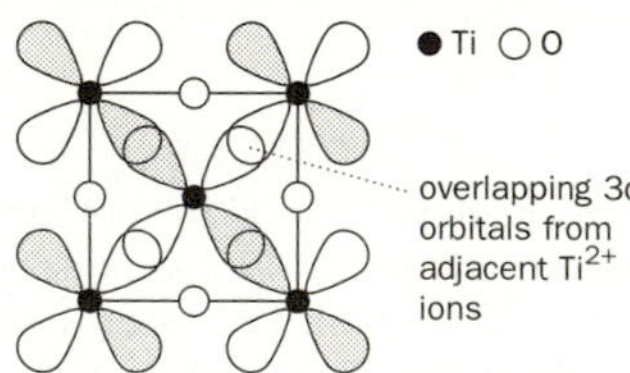

It is often claimed, and rightly, that transition metal ions form a rich variety of complexes; and the Ti^{3+} ion bears this out, usually forming octahedral species. But transition metal ions do not have the monopoly of complex formation. Ions of some non-transitional metals, such as mercury, also show considerable diversity.

Titanium also forms some compounds of formal oxidation state (II). The oxide TiO has a sodium chloride lattice of $Ti^{2+}(3d^2)$ and O^{2-} ions, and is a metallic conductor. This might seem odd behaviour for an ionic compound but can be explained by the octahedral shape both of the t_{2g} orbitals and of the grid of Ti^{2+} ions in the crystal, and the fact that these orbitals are fairly diffuse. So the orbitals of one ion not only point towards those of neighbouring ions, but actually overlap to form a (narrow) band, through which the two $3d^2$ electrons can move as in a metal. The titanium dihalides, TiX_2, are polymers and probably contain metal–metal bonds (similar to those suspected for lower scandium halides, see p. 209). As we should expect, titanium(II) compounds are very readily oxidized.

It might seem odd that, although the 4s orbitals are filled before the 3d in the periodic table, the 4s electrons are ionized from titanium (and from atoms of the other transition metals) more readily than the 3d electrons. But each successive chemical element differs from its prececessor not only in having one more electron in its valence shell but also in having a higher charge on its nucleus. Ionization involves removal of an electron from the atom of a particular element, without changing the nuclear charge. The 4s electrons are more penetrating than the 3d, and so have a greater shielding effect. So if the $4s^2$ electrons are removed, the remaining 3d electrons are stabilized by the increased attraction of the nucleus. Since the 3d electrons do not screen so effectively, there would be less stabilization of remaining outer electrons if they were removed. The electronic configurations of Ti^{3+} and Ti^{2+} are therefore $[Ar]3d^1 4s^0$ and $[Ar]3d^2 4s^0$ (rather than $[Ar]3d^0 4s^1$ and $[Ar]3d^0 4s^2$).

Many transition metal atoms form complexes with carbon monoxide and other small, multiply bonded molecules, by a combination of σ donation from the ligand and π donation to it (see p. 73). It is not surprising that titanium, with so few d electrons to donate, forms no stable carbonyl complex. It does, however, use a mixture of σ and π bonding to form one bizarre compound in which titanium(0) combines with four cyclopentadienyl molecules. It seems that two of the cyclopentadienyl rings are σ bonded, with links from the metal to one carbon atom, on each ring, and that the other two groups are attached by delocalized π bonds, which embrace the metal atom and all five carbon atoms on the ring. The

'Ring-whizzing' in $Ti(C_5H_5)_4$ [cw]

HC=CH
CH
HC=CH

cyclopentadiene

these groups are each π bonded to all five C atoms

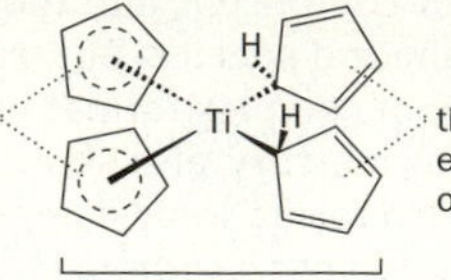

these groups are each σ bonded to one C atom

There is rapid exchange between the type and place of anchorage

amazing part is that NMR spectroscopy (see p. 23) shows that all the twenty hydrogen atoms are in an identical environment, regardless of how the ring is joined to the metal atom. The explanation is that NMR, like any other technique, cannot distinguish between two environments unless they are unchanged over the period needed to measure them (which here may be as much as one hundredth of a second*). Within this brief period, not only are the π bonds interchanging with the σ bonds, but the points of anchorage of the σ bonds are also exchanging in a process graphically called 'ring-whizzing'. So NMR spectroscopy can therefore provide only a smoothed out picture of these events and gives us the *average* environment of each hydrogen atom in this 'fluxional' or rapidly changing molecule.

Titanium is indeed a titanic element, in evidence wherever there is an aeroplane or an area of white paint. It provides us with a colourful introduction to the variable oxidation states, spectra, magnetism, and fluxionality of compounds of transition metals, in species both with partially filled d orbitals and with empty ones; and it gives us a foretaste of some of the complexities of this fascinating series of elements. Although we shall meet broadly similar phenomena in the other eight members of the first transition series, we shall see that the detailed behaviour of each one is intricately related to its own electronic configuration. In order to highlight both the general trends and individual peculiarities, the transition elements will be compared and summarized as a series after the d^9 metal, rather than separately after each member.

For summary, see p. 281–5

* This is longer by a factor of up to five than many exposure times used by amateur photographers.

V Vanadium

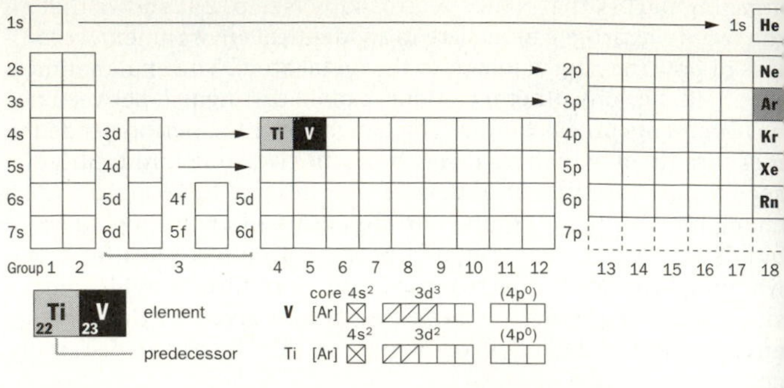

		Predecessor	Element
Name		**Titanium**	**Vanadium**
Symbol, *Z*		Ti, 22	V, 23
RAM		47.88	50.9415
Radius/pm	**atomic**	144.8	132.1
	ionic, M^{2+}	80	72
	M^{3+}	69	65
Electronegativity (Pauling)		1.54	1.63
Melting point/K		1933	2160
Boiling point/K		3560	3650
$\Delta H^{\ominus}_{fus}$/kJ mol^{-1}		20.9	17.6
$\Delta H^{\ominus}_{vap}$/kJ mol^{-1}		428.9	458.6
Density (at 293 K)/kg m^{-3}		4540	6110
Electrical conductivity/Ω^{-1}m^{-1}		2.381×10^6	4.032×10^6
Ionization energy for removal of *j*th electron/kJ mol^{-1}	***j* = 1**	658	650
	***j* = 2**	1310	1414
	***j* = 3**	2652	2828
	***j* = 4**	4175	4507
	***j* = 5**		6294
Electron affinity/kJ mol^{-1}		7.6	50.7

Isotopes of vanadium

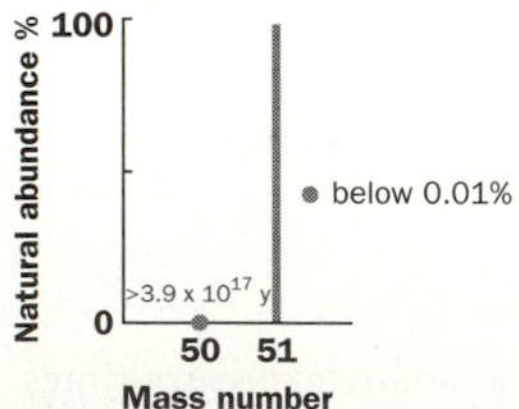

We should expect vanadium ([Ar]$3d^3 4s^2$), with a sparsely populated 3d shell, to be somewhat like titanium; and it does indeed behave in many ways that are typical of transition metals. It forms compounds of several oxidation states, often with unpaired electrons, and in the wide range of colours that accounts for its name: Vanadis was the Scandinavian goddess of beauty. We shall see that each transition metal differs from its predecessor not only because it has one more electron in its d shell, but also because with each added proton the d orbitals become increasingly influenced by the nucleus. So, in the later part of the series, the d electrons are less able to interact with other atoms and are said to be 'sucked into the core'. But this effect is barely noticeable at vanadium, which owes its unusual variety to the balance of the (modestly) generous number of d electrons and their relative freedom to form bonds.

Vanadium has a higher melting point than titanium (it can contribute three 3d electrons to the metallic bonding), but resembles it in forming a tight surface layer of oxide, which protects it from further corrosion. Like titanium, the hot metal combines with oxygen, nitrogen, and carbon to form compounds, in which the non-metal atoms fit into gaps in between the metal ions. Such 'interstitial' compounds may be of irregular structure and are often 'non-stoichiometric' (of very variable composition, which gives no simple ratio of metal to non-metal atoms).

Like titanium, vanadium occurs mainly in combination with oxygen, but it is also found as sulfide ores; the higher nuclear charge of vanadium allows better polarization of the somewhat deformable sulfide ion. Extraction of the metal involves obtaining crude V_2O_5 from the ores. Since the main use of vanadium is in steel hardening, the pure metal is seldom needed and the oxide is often reduced by ferrosilicon (in the presence of lime which removes the resulting silica by converting it to calcium silicate). Purification of the metal is complicated by the formation of interstitial compounds, and small-scale production of the pure metal from vanadium(V) compounds may be achieved by various reductions, including the reaction of vanadium pentachloride with magnesium (compare the production of titanium, p. 211).

The full chemical role played by the 3d electrons of vanadium is illustrated by the fact that its most stable oxidation state is (IV), $3d^1 4s^0$, while (V), $3d^0 4s^0$, is also extremely common and is only very mildly oxidizing. Both states are unusual in that, in acidic aqueous solutions, they exist as oxocations. The vanadium(IV) species, VO^{2+} (formerly the 'vanadyl' ion), is royal blue, and contains a very strong and short V=O bond, with one unpaired electron; the same ion is found in many solids (e.g. VO_2, or $VO^{2+}.O^{2-}$)

Redox behaviour of vanadium

Latimer diagram in acidic solution[SAL]

$E^{\ominus}$/V $\quad VO_2^+ \xrightarrow{1.000} VO^{2+} \xrightarrow{0.337} V^{3+} \xrightarrow{-0.255} V^{2+} \xrightarrow{-1.13} V$

VO_2^+	VO^{2+}	V^{3+}	V^{2+}
orange	royal blue	green	lilac
(mildly oxidizing)			(reducing)

None of these ions disproportionates.

The VO_2^+ ion

```
        O
        ||
 H2O\  V+ /OH2
 H2O/  |  \\O
       OH2
```

In the tetrahydrate, the VO_2^+ ion is bent

Some oxoanions of vanadium(V)

bond [CW] and polyhedral [GE] representations

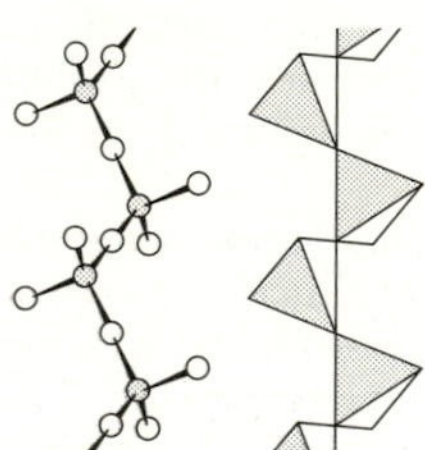

chains of VO_4 tetrahedra, corner linked (as in anhydrous metavanadate, KVO_3)

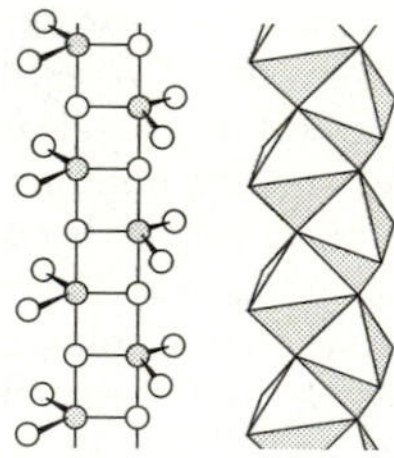

chains of VO_5 bipyramids, edge linked (as in hydrated vanadate, $KVO_3{\cdot}H_2O$)

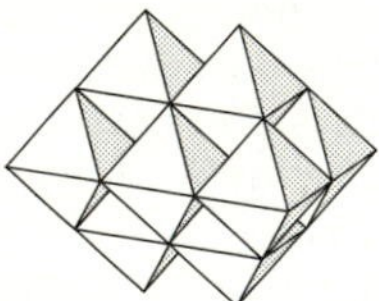

Block of ten VO_6 octahedra, edge linked, in $V_{10}O_{28}^{6-}$

and forms a variety of complexes in solution. The vanadium(V) species, VO_2^+, is bent, allowing effective back donation from the filled pπ orbitals of oxygen to the empty dπ orbitals of vanadium, and its yellow colour is due to this ligand to metal charge transfer (see p. 215). The related oxide V_2O_5 is orange for the same reason. As might be expected, fluorine is the only halogen to form a pentahalide, VX_5, while all except iodine form tetrahalides, VX_4. Both oxidation states form a number of oxohalides (again except with iodine). Vanadium forms a rich and extremely complicated series of compounds with sulfur (ranging from V_3S to VS_4), and with sulfur's heavier analogues selenium and tellurium. But vanadium(V) and (IV) show their richest variety with oxygen, as might be expected for ions of high oxidation state but of modest size.

It is estimated that vanadium(V) is a little too small for very stable octahedral coordination to six oxygen atoms and a little too big for very stable tetrahedral coordination by four of them. The upshot is variety. In V_2O_5, each metal is five coordinate and the resulting (distorted) trigonal pyramids of VO_5 share edges to form zigzag double chains (which will not be illustrated here!). By dissolving the (orange) oxide in concentrated alkali, we can make colourless 'metavanadates' such as KVO_3 which contain simple chains of corner-shared VO_4 tetrahedra. In hydrated metavanadates, however, the vanadium is again five coordinate. If we gradually lower the OH^- concentration of a metavanadate solution, the VO_4^{3-} ions become protonated and condense to give red solutions containing anions such as $V_3O_9^{3-}$ and $V_4O_{12}^{4-}$. In mildly acidic conditions we get orange solutions containing various protonated forms of $V_{10}O_{28}^{6-}$. This improbable looking anion is well established in vanadium minerals and is made up of ten VO_6 octahedra in the form of six edge-shared octahedra forming a central raft, which accommodates two further pairs of octahedra, one pair centrally above it and the other centrally below. The tetraprotonated group $H_4V_{10}O_{28}^{2-}$ is broken down by further acid to the cation VO_2^+. Vanadium(V), like titanium(IV) reacts with hydrogen peroxide to give complexes in which the peroxide ion is joined by both oxygen atoms to the metal ion. The composition of the complex, and its colour, depends on the pH.

Vanadium(IV) oxide, VO_2, like V_2O_5, is amphoteric. Concentrated solutions in alkali contain the spherical polyanion $V_{18}O_{42}^{12-}$, while mildly acidic solutions contain the dimeric cation $(VO)_2(OH)_2^{2+}$, which on further acidification gives VO^{2+}.

Although all the most common vanadium compounds have lost two, or all three, electrons from the 3d shell, compounds of vanadium(III), d^2, and vanadium(II), d^3, can also be prepared. Both

The vanadium(IV) oxoanion [cw], $V_{18}O_{42}^{12-}$

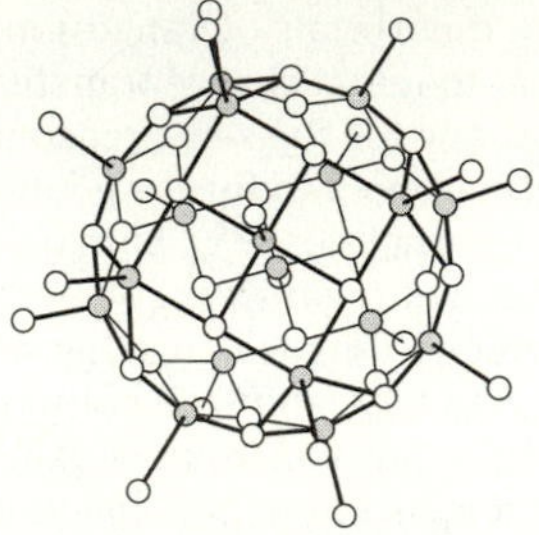

Electron count for $V(CO)_6$

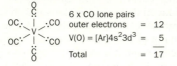

O
C
OC CO
V
OC CO
C
O

6 x CO lone pairs outer electrons	=	12
V(O) = $[Ar]4s^23d^3$	=	5
Total	=	17

V^{3+}(aq) (green) and V^{2+}(aq) (violet) are stable in solution, in that they do not reduce water; nor do they disproportionate. They are, however, oxidized by air. The colours, being a result of d–d transitions, are not very intense (see p. 217). The oxides, of approximate formula V_2O_3 and VO, are non-stoichiometric and basic (rather than amphoteric like V_2O_5 and VO_2). Like TiO (see p. 219), VO is a metallic conductor, which suggests that the 3d orbitals in vanadium are still fairly diffuse. All eight halides, VX_3 and VX_2, are known, and variously coloured.

So far, the chemistry of vanadium, although too rich, complex, and colourful to be predictable, seems none the less to be logically understandable in terms of electronic structure. So has vanadium no surprises for us? To the bioinorganic chemist it poses a number of riddles about the function of various compounds [containing vanadium(V), VO^{2+}, and even V^{4+} and V^{3+}] in a few invertebrates that appeared early in the evolutionary process, and in at least one mushroom. The geologist may speculate on its wide occurrence in petroleums from Venezuela. Many modern inorganic chemists are interested in the behaviour of transition metals in very low oxidation states, such as zero. Vanadium, unlike titanium, does form a hexacarbonyl, $V(CO)_6$, which is stabilized by donation of some of the (high) electron density in the metal atom to the vacant π orbitals of the CO group. In many such compounds, the metal atom is surrounded by 18 electrons, and so has a noble gas configuration of $(n-1)\,d^{10}ns^2np^6$; but this is clearly impossible in a molecule containing one atom of an element such as vanadium [Ar]$3d^34s^2$ that has an odd number of electrons. In $V(CO)_6$, the vanadium has only 17 electrons, and so is unstable; but it readily accepts an electron to complete its tally of 18, and so becomes more stable, as in the acid $HV(CO)_6$ and the anion $V(CO)_6^-$. Vanadium also forms an air-sensitive 17-electron $V(Cp)_2$ sandwich. Still more bizarre is the 17-electron compound of V(O) and dinitrogen $V(N_2)_6$ formed at extremely low temperatures in a matrix of solid noble gas, and the explosively unstable 18-electron compound, $K_3V(CO)_5$, which contains vanadium in its lowest oxidation state (–III) yet known. Vanadium certainly gives us a rich foretaste of the diversity of transition metal chemistry.

For summary, see pp. 281–5.

Cr Chromium

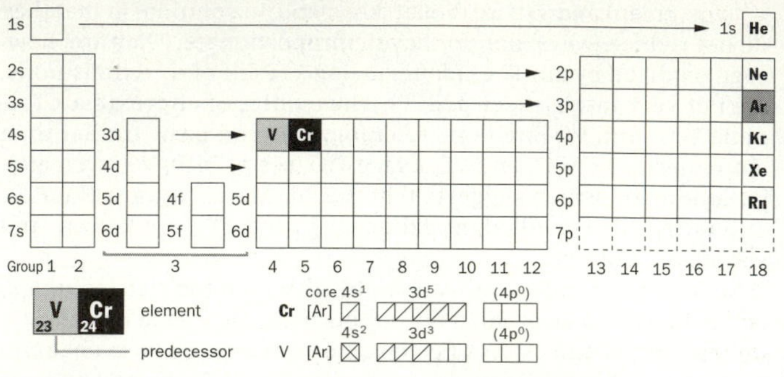

		Predecessor	Element
Name		Vanadium	Chromium
Symbol, Z		V, 23	Cr, 24
RAM		50.9415	51.9961
Radius/pm	atomic	132.1	124.9
	ionic, M^{2+}	72	84
	M^{3+}	65	64
Electronegativity (Pauling)		1.63	1.66
Melting point/K		2160	2130 ± 20
Boiling point/K		3650	2945
$\Delta H^{\ominus}_{fus}$/kJ mol^{-1}		17.6	15.3
$\Delta H^{\ominus}_{vap}$/kJ mol^{-1}		458.6	348.78
Density (at 293 K)/kg m^{-3}		6110	7190
Electrical conductivity/$\Omega^{-1}m^{-1}$		4.032×10^6	7.874×10^6
Ionization energy for removal of jth electron/kJ mol^{-1}	$j = 1$	650	652.7
	$j = 2$	1414	1592
	$j = 3$	2828	2987
	$j = 4$	4507	4740
	$j = 5$	6294	6690
	$j = 6$		8738
Electron affinity/kJ mol^{-1}		50.7	64.3

Stable isotopes of chromium

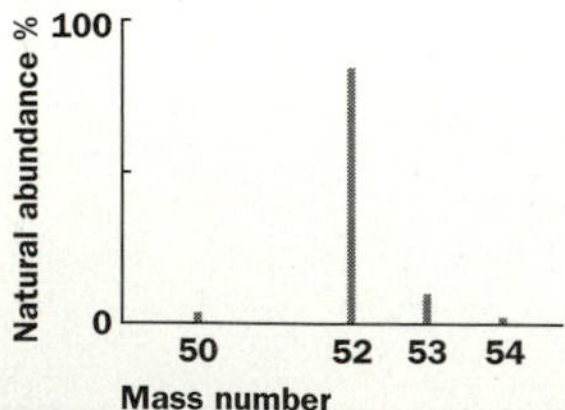

The chromium atom, [Ar](3d + 4s)^6, with one more electron than vanadium, has as many outer electrons as it has valence orbitals. Repulsion between these electrons is least when one electron occupies each orbital, and the most stable configuration of an uncombined chromium atom is $3d^5 4s^1$, unlike the more common arrangement of $3d^n 4s^2$ found in the three previous elements (and in the following ones).

We should expect chromium to show broad similarities to vanadium, and its name (from the Greek for 'colour') suggests that, like vanadium, it has a variety of coloured compounds, based on a range of accessible oxidation states. The higher nuclear charge reduces the size of the atom and the freedom of the 3d electrons. Chromium has lower melting and boiling points than vanadium, and a greater tendency to form compounds in lower oxidation states. The group valency $3d^0 4s^0$ is strongly oxidizing for chromium(VI) but only mildly so for vanadium(V). The (III) state is the most common form of chromium, while vanadium(III) is readily oxidized in air. So it seems that, at chromium, the 3d electrons start their gradual retreat into the core of the atom; but as yet the effect is quite small.

Chromium also provides us with many examples of another feature of transition metal chemistry: that although much of it can be understood in terms of *trends*, many details of behaviour depend on electronic *bookkeeping*. Since chromium (unlike vanadium) has an even number of electrons, we should expect it to form a range of 18-electron compounds with π acceptor ligands like carbon monoxide and unsaturated organic molecules. We shall see that, with an increasing number of d electrons in ions of accessible oxidation states, the splitting of the d orbitals (see p. 217) gives scope for variations that depend on the actual number of d electrons present. Many of the peculiarities of chromium(III) complexes, for example, are a result of the fact that the Cr^{3+} ion is not only small and highly charged, but also has three, rather than two or four, 3d electrons; and chromium(II) complexes have their own peculiarities, some of them specifically owing to the d^4 configuration.

Chromium occurs (with about the same abundance as vanadium or chlorine) as a mixed oxide of iron. If the chromium is needed to make stainless steel, the ore is reduced with ferrosilicon to give an iron alloy (compare with the extraction of vanadium, p. 223). Chromium to be used in non-ferrous alloys is separated from iron by oxidation in molten alkali to Na_2CrO_4 [the (VI) oxidation state is very rare for iron, in which the 3d electrons are under greater restraint; see p. 247]. Reduction with carbon gives Cr_2O_3, which is further reduced to the metal using aluminium or silicon (see p. 230). Chromium is widely used for plating other metals as it

Redox behaviour of chromium

Latimer diagram in acidic solution

$E^\ominus/V$ $\quad Cr_2O_7^{2-} \xrightarrow{1.38} Cr^{3+} \xrightarrow{-0.42} Cr^{2+} \xrightarrow{-0.90} Cr$

oxidizing (Cr₂O₇²⁻) stable in air (Cr³⁺)

(for O_2, H^+, H_2O $E^\ominus = 1.23$ V)

None of these ions disproportionates.

Extraction of chromium metal

$4FeCr_2O_4 + 8Na_2CO_3 + 7O_2 \rightarrow 8Na_2CrO_4 + 2Fe_2O_3 + 8CO_2$

chromite — soluble in water — insoluble in water

H^+, crystallize

$Na_2Cr_2O_7 + C \rightarrow Cr_2O_3 + Na_2CO_3 + CO$

Reduction of Cr_2O_3

$Cr_2O_3 + 2Al \rightarrow 2Cr + Al_2O_3$

or $2Cr_2O_3 + 3Si \rightarrow 4Cr + 3SiO_2$

Chromate and dichromate ions

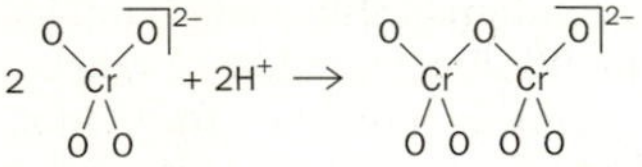

$2\,[CrO_4]^{2-} + 2H^+ \rightarrow [Cr_2O_7]^{2-}$

chromate CrO_4^{2-} (yellow) — dichromate $Cr_2O_7^{2-}$ (orange)

(multiple bonds omitted)

The chromium(V) peroxo complex [CW]

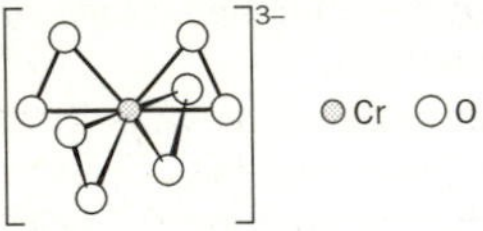

● Cr ○ O

CrO_8^{3-} (red-brown)

resists corrosion at room temperature. At high temperatures, it reacts with most non-metals.

The aqueous redox behaviour of chromium is less tidy than that of vanadium (which behaves in an unusually orderly way, with each decreasing oxidation state merely being less oxidizing than its predecessor). Chromium(VI) is strongly oxidizing; (V) and (IV) have to be coaxed into existence, almost always in the absence of water; (III) is the commonest state; and chromium(II), like vanadium(II), is reducing.

As expected for a strongly oxidizing state, chromium(VI) is almost always found surrounded either by oxygen, as in the red oxide CrO_3, or by the lighter halogens, as in CrF_6, $CrOF_4$, and CrO_2Cl_2; easily oxidized anions such as sulfide or iodide would reduce it to chromium(III). The trioxide consists of chains of corner-sharing CrO_4 tetrahedra, has a fairly low melting point, and is widely used as an oxidizing agent. It dissolves in alkaline solution to give the yellow tetrahedral chromate ion CrO_4^{2-} and in acid to give the orange dichromate ion $Cr_2O_7^{2-}$, in which two such tetrahedra share a corner. Chromium forms no oxocations analogous to VO_2^+ or VO^{2+}. The colours of these oxochromium(VI) compounds, like those of oxovanadium(V), are a result of charge transfer from oxygen to the d^0 metal 'ion'. Although species based on three, and even four, corner-sharing CrO_4 tetrahedra have been made, chromium(VI) oxoanions show little of the variety exhibited by vanadium(V), probably because the central chromium is too small to accommodate six, or even five, surrounding oxygen atoms, and so is limited to a tetrahedral geometry.

Chromium(VI), like vanadium(V) and titanium(IV), reacts with hydrogen peroxide to give coloured, chelated complexes. From acidic solution, $Cr^{VI}O(O-O)_2$ is obtained, but decomposes, through the (V) and (IV) states, to chromium(III). A chromium(V) complex is formed from alkaline solution.

Although chromium(IV) compounds exist only fleetingly in solution, we have probably all used chromium(IV), as the oxide CrO_2, which is the ferromagnetic material present in many audiotapes. Ferromagnetic[A] materials are those that, like the metals iron, cobalt, and nickel, can be magnetized by alignment of (usually d) electrons. If this is to happen, there must be a supply of d electrons of energy around that of the valence orbitals, i.e. they must not be under too firm a control of the nucleus. Chromium dioxide is ferromagnetic because, at chromium, the stabilization of the d electrons by the nucleus is still fairly mild, and so the two d electrons in the Cr^{4+} ion are in orbitals that have the right energy to take part in chemical

The kinetic inertness of Cr^{3+}, d^3

in an octahedral field

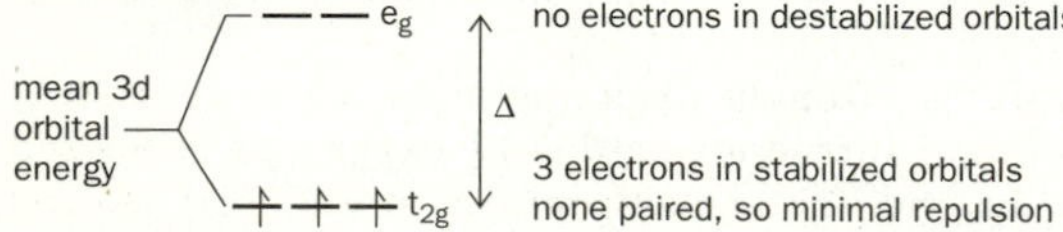

Disruption of so stable an arrangement is improbable, so any change will be slow

Isomers of $CrCl_3 \cdot 6H_2O$

no Cl^- in inner coordination sphere

H_2O, H_2O, H_2O, H_2O, H_2O, H_2O $]^{3+}$ $\cdot\ 3Cl^-$

$Cr(H_2O)_6^{3+} \cdot 3Cl^-$

one Cl^- in inner coordination sphere

H_2O, H_2O, H_2O, H_2O, Cl, H_2O $]^{2+}$ $\cdot\ 2Cl^-$ H_2O

$Cr(H_2O)_5Cl^{2+} \cdot 2Cl^- \cdot H_2O$

○ Cr^{3+}

two Cl^- in inner coordination sphere

H_2O, H_2O, Cl, H_2O, Cl, H_2O $]^{+}$ $\cdot\ Cl^-$ $2H_2O$

cis- (chlorides adjacent)

cis-$Cr(H_2O)_4Cl_2^+ \cdot Cl^- \cdot 2H_2O$

two Cl^- in inner coordination sphere

H_2O, Cl, H_2O, H_2O, Cl, H_2O $]^{+}$ $\cdot\ Cl^-$ $2H_2O$

trans- (chlorides opposite)

trans-$Cr(H_2O)_4Cl_2^+ \cdot Cl^- \cdot 2H_2O$

combination (or to accommodate electrons that can be aligned throughout the sample); see pp. 247–8.

However, (III) is the commonest oxidation state of chromium and its most familiar oxide is Cr_2O_3, which is green because of d–d transitions in the Cr^{3+} ion. This ion forms a large number of octahedral complexes, which have been widely studied because, once formed, they are very reluctant to take part in chemical reactions, even if the change would be thermodynamically favourable. For example, it is very difficult to dissolve the solid trichloride $CrCl_3$ in water, although a solution of Cr^{3+}(aq) and Cl^-(aq) ions is more stable than a mixture of solid $CrCl_3$ and water. This sluggishness is owing to the fact that the Cr^{3+} (d^3) ion is usually found surrounded octahedrally by six other groups, and so three of its d orbitals, like those of the Ti^{3+} ion (see p. 217), become more stable than the other two. In the Cr^{3+} ion, each of the more stable orbitals holds one of the d electrons, and so the ion benefits both from minimal electronic repulsion and from maximal ligand field stabilization energy (LFSE) gained by using the stabilized ('t_{2g}') orbitals rather than the destabilized ('e_g') ones. If any change were to take place in the surrounding ligands, one group would have to leave, or one would have to edge itself in. Whichever way the change occurred, the very stable d^3 octahedron would have to give way to a much less stable intermediate arrangement with either five or seven groups around the chromium. Such a change is improbable, however stable the ultimate product may be, and so it takes place slowly.

Since chromium(III) is kinetically inert, its complexes often exist in different isomeric forms which do not change appreciably from one to the other during the time needed for a laboratory preparation; and so they can be separated. For example, a compound of overall formula $CrCl_3.6H_2O$ may exist in violet, pale green, or dark green forms, depending on whether the central Cr^{3+} ion is surrounded octahedrally by: six water molecules, as in $Cr(H_2O)_6^{3+}.3Cl^-$; five water molecules and one chloride ion, as in $Cr(H_2O)_5Cl^{2+}.2Cl^-.H_2O$; or four water molecules and two chloride ions, as in $Cr(H_2O)_4Cl_2^+.Cl^-.2H_2O$; in each case the remaining water molecules or chloride ions are outside the main coordination sphere. In this last compound, which has two inner chloride ions, these may be next to each other (*cis*) or opposite (*trans*). If the central metal ion were not so inert, the *cis*- and *trans*-isomers, and probably also the inner and outer sphere ones, would be in rapid equilibrium with each other, and the groups would change places so quickly that it would be impossible to separate the different forms.

Chromium(II), d^4, complexes: high spin or low spin?

Two possible configurations in octahedral complex (simplified)

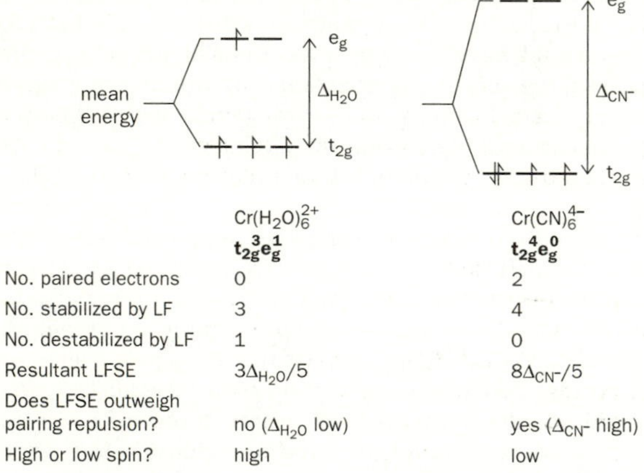

	$Cr(H_2O)_6^{2+}$	$Cr(CN)_6^{4-}$
	$\mathbf{t_{2g}^3e_g^1}$	$\mathbf{t_{2g}^4e_g^0}$
No. paired electrons	0	2
No. stabilized by LF	3	4
No. destabilized by LF	1	0
Resultant LFSE	$3\Delta_{H_2O}/5$	$8\Delta_{CN^-}/5$
Does LFSE outweigh pairing repulsion?	no (Δ_{H_2O} low)	yes (Δ_{CN^-} high)
High or low spin?	high	low

In practice, the high-spin complex may be distorted by the Jahn–Teller effect (see p. 217)

* The Cr^{3+} ion, surrounded by six oxygen atoms is responsible for the green of emerald (2% Cr^{3+} in the beryllium aluminium silicate, beryl) and also for the red of ruby (the compact form of Al_2O_3 in which a few Al^{3+} ions have been replaced by Cr^{3+}, causing lattice distortions that compress the Cr^{3+} ion in one direction and alter the d orbital splitting, and hence the colour).

As the colours of all these complexes are a consequence of d–d transitions it might seem odd that there is such a change when just one water molecule in $Cr(H_2O)_6^{3+}$ is replaced by a chloride ion. The colour of the Cr^{3+} ion is, however, very sensitive to its environment*.

It is claimed that chromium(III) forms thousands of complexes in solution, almost all with six donor atoms. Some contain two or more chromium atoms, bridged by a single –O– or –OH group, or by two –OH groups. Polynuclear hydrolysis products of chromium(III) are, like those of aluminium, used to impregnate both fabrics (to give sites at which dyes can be bound) and hides (to prevent putrefaction and, by providing links between the natural chains, to make the leather supple). Monomeric complexes include those with six simple groups such as halide or cyanide ions, water, ammonia, and amines, together with a vast range of complexes with organic chelating agents.

Since chromium(II) is readily oxidized, it is less common than chromium(III). But Cr^{2+} is an interesting ion because the presence of a fourth d electron introduces possibilities of types of chemical behaviour not possible for the $3d^1$, $3d^2$, and $3d^3$ ions discussed so far.

Most complexes of Cr^{2+} are octahedral, and so the five 3d orbitals are again split in such a way that three are more stable than the other two (see p. 216). With ions such as Cr^{2+}, d^4, each of the lower orbitals contains at least one electron but there are two possible destinations for the fourth electron. It could go into one of the two unoccupied e_g orbitals of higher energy or it could go into one of the partially filled t_{2g} orbitals (where the advantage of the lower energy level would be opposed by repulsion from the electron already in occupation). Which of the two positions gives the more stable ion depends on the energy gap between the higher and lower levels, and this in turn depends on the ligands. Since water molecules produce only a modest splitting, the $Cr(H_2O)_6^{2+}$ has four unpaired electrons, one in the upper level and all with the same spin (and it is therefore known as a 'high spin' complex). The cyanide ion splits the d orbitals much more widely than water does, and in the $Cr(CN)_6^{4-}$ ion all four electrons go into the lower level, to form a 'low spin' complex with two electrons paired and only two unpaired. The gain in stabilization caused by the large splitting more than makes up for the repulsion between the paired electrons. Naturally, the Cr^{2+} ion is slightly smaller in low spin complexes, where the d electrons are stowed between the ligands, than in high spin ones, in which the d electron clouds are more extended. The repulsion between the rather crowded electrons in the d orbitals of low spin complexes is somewhat lowered by par-

The chromium(II) ethanoate dimer [s]

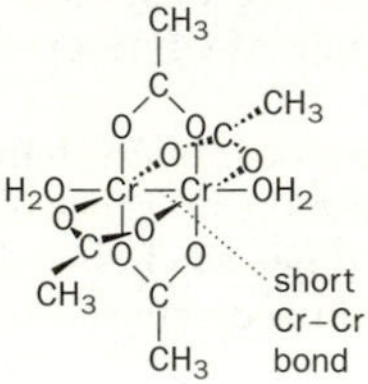

short Cr–Cr bond

The chromium(0) benzene sandwich

electron count

six for each benzene ring	=	12
Cr(O) [Ar]$4s^1 3d^5$	=	6
Total	=	18

rings 'eclipsed' with C atoms aligned
metal interacts with all C atoms

tial transfer of charge from the metal ion to the vacant π orbitals on the ligands. Low spin chromium(II) complexes are therefore intensely coloured as a result of charge transfer, here from metal (in low oxidation state) to the (π acceptor) ligand.

Most complexes of chromium(II) are, however, high spin, with one of the higher d orbitals partially occupied and the other one empty. This arrangement causes further orbital splitting and lowers the symmetry of the octahedron, giving a 'Jahn–Teller' distortion (see p. 217), which usually makes four coplanar bonds a bit shorter and stronger, whilst the two remaining axial bonds become longer and weaker (but the converse may also happen).

A bizarre property of the high spin d^4 chromium(II) ion is a tendency to form dimers in which the four unpaired electrons of one chromium ion are thought to interact with those of an adjacent one to form a short, strong M–M bond, as in $Cr_2(OOCCH_3)_4.2H_2O$; each of the four ethanoate ions is bidentate and acts as an O–C–O bridge between the two metal ions. This compound is bright red (unlike monomeric high spin Cr^{2+}, which is blue), has a magnetic moment only slightly above that for no unpaired electrons, and is much less readily oxidized than the aquo ion.

Chromium(0), with its six outer electrons ($3d^5 4s^1$) seems more likely than vanadium to combine with π acceptors like carbon monoxide and unsaturated organic compounds, both because it has more electrons to donate, and because it has an even number of them. To achieve the stable 18-electron configuration, it would need to accept twelve electrons. With carbon monoxide it forms, predictably, the hexacarbonyl $Cr(CO)_6$ which is quite stable. It also forms a sandwich compound, $Cr(C_6H_6)_2$ between two benzene rings, accepting six electrons from each. The benzene rings appear not to rotate, but to have fixed positions, with the carbon atoms of one ring exactly above those of the other. This disciplined 'eclipsed' structure is in stark contrast to the wild 'ring-whizzing' of the tetracyclopentadienyl compound of titanium (see p. 221); but this is just another illustration of the extent to which chemical behaviour is altered by a small change in the number of d electrons. The length of this chapter compared with that for vanadium, reflects the number of chemical features shown by chromium: ferromagnetism (of CrO_2), the kinetic inertness of d^3 complexes, the possibility of high and low spin complexes in d^4, and, in high spin d^4, Jahn–Teller distortion and the formation of quadruple metal–metal bonds. Chromium is indeed a colourful element, metaphorically as well as literally.

For summary, see p. 281–5.

Mn Manganese

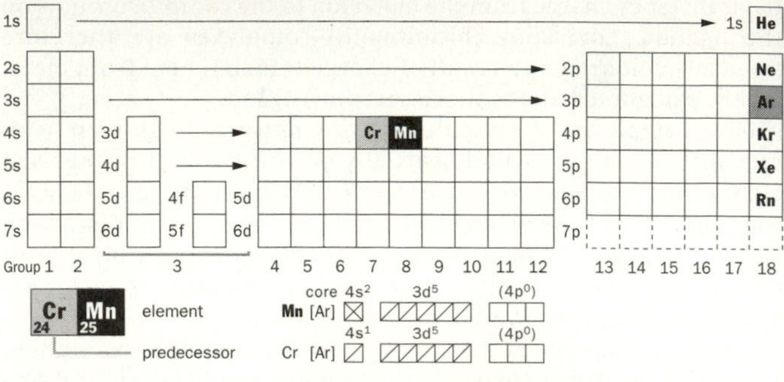

		Predecessor	Element
Name		**Chromium**	**Manganese**
Symbol, *Z*		Cr, 24	Mn, 25
RAM		51.9961	54.93805
Radius/pm	**atomic**	124.9	124
	covalent		117
	ionic, M^{2+}	84	91
	M^{3+}	64	70
	M^{4+}		52
Electronegativity (Pauling)		1.66	1.55
Melting point/K		2130 ± 20	1517
Boiling point/K		2945	2235
$\Delta H^{\ominus}_{fus}$/kJ mol^{-1}		15.3	14.4
$\Delta H^{\ominus}_{vap}$/kJ mol^{-1}		348.78	219.7
Density (at 293 K)/kg m^{-3}		7190	7440
Electrical conductivity/Ω^{-1}m^{-1}		7.874×10^6	5.405×10^5
Ionization energy for removal of *j*th electron/kJ mol^{-1}	***j* = 1**	652.7	717.4
	***j* = 2**	1592	1509.0
	***j* = 3**	2987	3248.4
	***j* = 4**	4740	4940
	***j* = 5**	6690	6990
	***j* = 6**	8738	9200
	***j* = 7**		11 508

Stable isotopes of mangnase

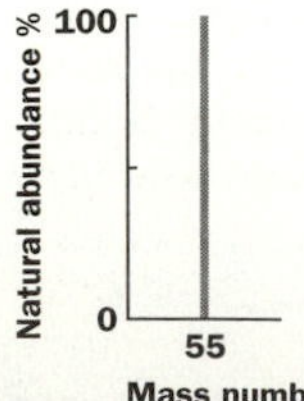

Manganese ([Ar]$3d^5 4s^2$) is the half-way point of the first transition series. In some ways it acts as the end of the first half of the series: it is the last element to lose all its outer electrons to give a readily accessible compound containing a (formally) d^0 ion, as manganese(VII) or permanganate MnO_4^-, and under suitable conditions it can form compounds of all lower oxidation states down to manganese(II), d^5. Like its predecessors, it favours oxygen environments, but can also complex with nitrogen donors. In other respects it behaves as the first member of the second half of the series. Its d electrons are less easily lost than those of earlier elements and it is the first transition metal to form a (II) state in which the M^{2+} ion is non-reducing and the oxide, MO, is antiferromagnetic. As predicted for an element with an odd number of electrons, it forms no monomeric neutral carbonyl. But as well as showing such half-way house behaviour, manganese has its individual peculiarities (for example, the ability of the metal, unlike any of its neighbours, to react with water to give hydrogen).

Predictably, manganese occurs in those minerals in which the metal ion is surrounded by oxygen atoms: in silicates and in oxides, hydrated oxides (in the 'manganese nodules' on ocean beds), and carbonates. The dioxide, MnO_2, is widely converted into manganese steel by mixing it with Fe_2O_3 and reducing it with coke in a blast furnace (see p. 248). The hardest steel, used for excavators, rail crossings, and prison bars, contains 13% manganese. The element is also used in non-ferrous alloys, often as a scavenger to remove sulfur or oxygen by forming MnS or MnO. The pure metal is obtained by reducing the oxide with aluminium or by electrolysing the aqueous sulfate.

At room temperature, manganese metal is hard but brittle, and has a less regular structure than neighbouring elements; there are four different types of site for the metal ions. Its relatively low heat of atomization has been ascribed to a combination of this disorder with the increased retreat of the d electrons from the metal band into the core; and this ease of lattice disruption is a major factor in contributing to the reactivity of the metal with water.

As the nuclear charge increases and the atomic radius decreases across the first transition series, we find fewer interstitial compounds. Manganese does not form any binary hydride, and its carbides are complicated and readily hydrolysed. There are several solid oxides, such as MnO_2, Mn_2O_3, MnO, which contain the metal in a single oxidation state and also Mn_3O_4 [a mixed oxide $Mn^{II}Mn_2^{III}O_4$ which has the 'normal spinel' structure (see p. 142), in which the Mn^{2+} are in tetrahedral sites and the Mn^{3+} ones in octahedral ones, which interact more favourably with the more highly charged ions].

Antiferromagnetism [A] of MnO [SM]

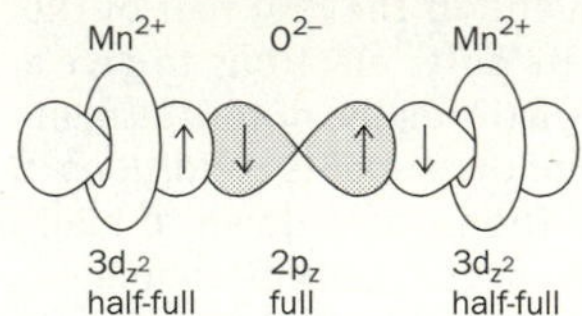

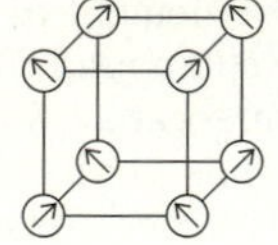

Opposed spins of Mn^{2+} ions in MnO. (The O^{2-} ions at the centre of each face have been omitted)

The opposed spins are aligned by 'super exchange'

Electron configuration of Mn^{2+}, d^5

Two possible configurations in octahedral complex

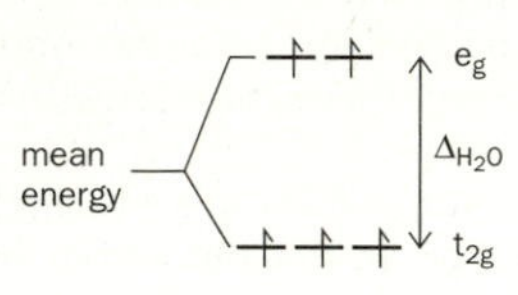

e_g

Δ_{CN^-}

t_{2g}

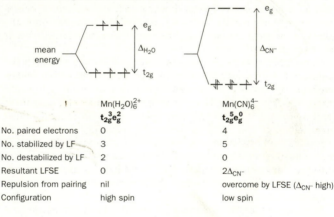

	$Mn(H_2O)_6^{2+}$	$Mn(CN)_6^{4-}$
	$\mathbf{t_{2g}^3e_g^2}$	$\mathbf{t_{2g}^5e_g^0}$
No. paired electrons	0	4
No. stabilized by LF	3	5
No. destabilized by LF	2	0
Resultant LFSE	0	$2\Delta_{CN^-}$
Repulsion from pairing	nil	overcome by LFSE (Δ_{CN^-} high)
Configuration	high spin	low spin

The pale colour of Mn^{2+}(aq)

octahedral high spin d^5

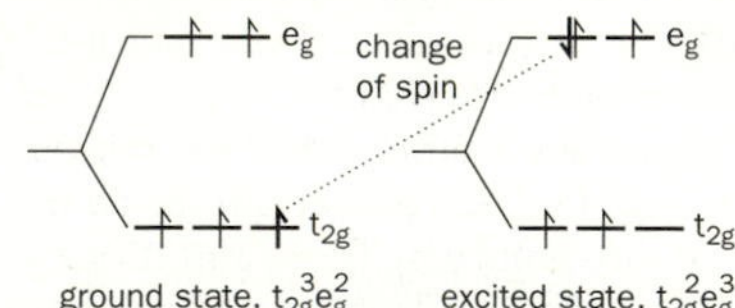

The change of spin is improbable, and only occurs in a very small fraction of the ions.

---

* This green colour is caused by Fe^{2+} (see p. 259), since glass is made from sand and this usually contains traces of iron.

All these oxides are insoluble in water and some are of variable composition. The octahedral environment of the Mn^{3+} in oxides, and in its one halide, MnF_3, is often distorted as a result of the expected Jahn–Teller effect for a high spin d^4 ion (see p. 217).

The oxide MnO, unlike TiO and VO, is not a metallic conductor; since the d orbitals are under the control of a higher nuclear charge for manganese than for the earlier elements, they are not diffuse enough to overlap. But MnO has interesting magnetic behaviour. We should expect that the Mn^{2+} (d^5) ions would be high spin in an oxide, and at room temperature we find that each Mn^{2+} ion does indeed have five unpaired electrons. But below −155°C the ions become alternately aligned, and the bulk magnetism of the sample drops dramatically. This 'antiferromagnetism'[A] is also shown by the oxides, $M^{II}O$, of the transition elements following manganese and by MnO_2 (d^3) and Mn_3O_4 (d^4 and d^5) at still lower temperatures. It is thought to be caused by the influence of the magnetic moment of the metal ion on its neighbouring oxide ions, which in turn cause alignment of their neighbouring metal ions (but in the opposite direction); and this alternating pattern is transmitted by 'super exchange' throughout the sample.

In acidic solution, manganese shows less variety than chromium or vanadium. The most stable state is Mn^{2+}(aq) which, unlike Cr^{2+}(aq) or V^{2+}(aq), is almost non-reducing. The high spin d^5 configuration, with half occupancy of all d orbitals, benefits from minimal electronic repulsion, but of course forfeits any ligand field stabilization; only with those ligands that produce a very large orbital splitting, does Mn^{2+} have a low spin configuration, as in $Mn(CN)_6^{4-}$. The Mn^{2+} ion forms a range of complexes, though without ligand field stabilization they are not very strong. The Mn^{2+}(aq) ion is only very weakly coloured (pink), because any excitation from a lower to a higher d orbital would require the electron to change its spin. Such an improbable change occurs only in a low proportion of the ions, and so the colour is very pale.

All the higher oxidation states of manganese are strongly oxidizing, but (V) exists only transitorily, while (VI) and (III) normally disproportionate; but manganese(III) is stable in concentrated hydrofluoric acid as the MnF_6^{3-} ion, another example of how the fluoride ion stabilizes higher oxidation states. Manganese(IV) is also oxidizing, but is usually encountered as the dioxide MnO_2, which is so insoluble that it takes part in few reactions. It is, however, reduced by concentrated hydrochloric acid, giving chlorine, and by concentrated sulfuric acid, liberating oxygen. Manganese dioxide is used to colour glasses and glazes in the red/brown range, and to neutralize the pale green of many glasses,* in order to

The MnO_4^- ion and Mn_2O_7
(compare CrO_4^{2-}, p. 231)

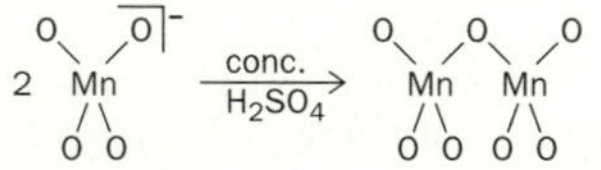

$$2\,MnO_4^- \xrightarrow[H_2SO_4]{conc.} Mn_2O_7$$

permanganate MnO_4^- (intense purple) → dimanganese heptoxide, Mn_2O_7 (dark green oil)

(multiple bonds omitted)

Latimer diagrams for manganese SAL

Acid solution, pH = 0

$$E^\ominus/V\quad MnO_4^- \xrightarrow{0.90} HMnO_4^- \xrightarrow{2.09} MnO_2 \xrightarrow{0.95} Mn^{3+} \xrightarrow{1.5} Mn^{2+} \xrightarrow{-1.18} Mn$$

$MnO_4^- \xrightarrow{1.69} MnO_2$; $MnO_2 \xrightarrow{1.23} Mn^{2+}$

Alkaline solution, pH = 14

$$MnO_4^- \xrightarrow{0.56} MnO_4^{2-} \xrightarrow{0.27} MnO_4^{3-} \xrightarrow{0.93} MnO_2 \xrightarrow{0.146} Mn_2O_3 \xrightarrow{-0.234} Mn(OH)_2 \xrightarrow{-1.56} Mn$$

$MnO_4^- \xrightarrow{0.59} MnO_2$; $MnO_4^{2-} \xrightarrow{0.60} MnO_2$

make the glass very pale grey, or 'colourless'. A metal that forms a dioxide $M^{IV}O_2$ would also be expected to form a tetrafluoride, as does manganese. But unlike chromium, which forms CrF_6 and CrF_5, manganese forms no uncharged fluoride higher than MnF_4, and this again suggests decreasing availability of the d electrons. The compound MnS_2 is not of course $Mn^{IV}(S^{2-})_2$, since the sulfide ion would be oxidised by Mn^{IV}, but contains manganese(II) and the disulfide ion S_2^{2-} (as in iron pyrites, see p. 172).

Manganese is the last member of the first transition series that gives an accessible d^0 compound. The familiar 'permanganate' ion, $Mn^{VII}O_4^-$ owes its intense purple colour to charge transfer from full orbitals mainly associated with the oxygen atoms, to vacant ones derived from the highly positive metal ion. The permanganate ion is strongly oxidizing and is theoretically able to oxidize water to oxygen; but in practice the reaction is very slow. Although its thermodynamic instability makes aqueous permanganate unsuitable as a primary standard, it is widely used in volumetric analysis. It is also used as a germicide, particularly in water purification. With concentrated sulfuric acid, two of the MnO_4^- tetrahedra may be condensed to give the corner-sharing Mn_2O_7, an explosively unstable green oil.

Any decrease in the acidity of a solution may often decrease the oxidizing power of a species, for one of two reasons. Many reductions, such as from MnO_4^- to MnO_2, involve a decrease in the number of oxygen atoms surrounding a metal ion, and since these are removed by H^+ to form water, this type of change is favoured by high acidity. In strongly alkaline solution the tendency for the green manganate(VI), MnO_4^{2-}, to be reduced to MnO_2 is decreased so much that it no longer disproportionates; and even the blue manganate(V) ion, MnO_4^{3-}, can be observed.

Moreover, in alkaline solution, cations such as Mn^{3+} and Mn^{2+} are often precipitated as hydroxides. Those of higher charge will be bound more strongly by hydroxy ions and will form the more insoluble hydroxides, so the alkali makes the higher oxidation state relatively more stable and this lowers its oxidizing power. We find that in alkaline solution, Mn^{III} is no longer highly oxidizing, nor does it disproportionate. A precipitate of $Mn(OH)_2$ is, however, less stable in air than a solution of Mn^{2+}; its surface turns brown by oxidation to the even more insoluble MnO(OH). And because MnO(OH) is so stable, MnO_2 can readily be reduced to it, a change that forms the basis of many alkaline dry batteries. So, although only three oxidation states of manganese, represented by MnO_4^-, MnO_2, and Mn^{2+}(aq), are readily obtained in dilute acid, all its oxidation states, from (VII) to (II), are known in alkaline solutions.

The carbonyl $Mn_2(CO)_{10}$ GE

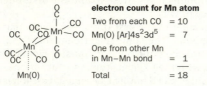

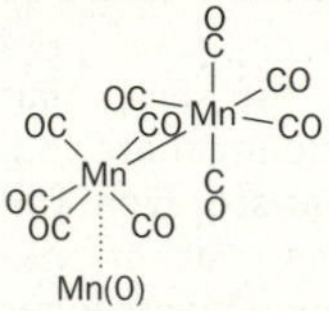

electron count for Mn atom

Two from each CO	= 10
Mn(0) $[Ar]4s^23d^5$	= 7
One from other Mn in Mn–Mn bond	= 1
Total	= 18

the $-Mn(CO)_5$ groups are staggered

As manganese has an odd number of outer electrons, we should not expect it to form a stable monomeric carbonyl, $Mn(CO)_x$. Unlike the previous odd-electron transition metal, vanadium, it forms a dimeric carbonyl $Mn_2(CO)_{10}$ in which two 17-electron $Mn(CO)_5$ groups are joined in a staggered configuration through a Mn–Mn bond. In this way each metal atom has filled all the available orbitals derived from the five 3d, the 4s, and the three 4p atomic orbitals. The dimer can be used to make numerous compounds containing ions such as $Mn(CO)_5^-$, $Mn(CO)_4^{3-}$, and $Mn(CO)_6^+$, all of which also have 18 outer electrons. Manganese, like vanadium, forms an odd-electron sandwich, 'manganocene', with two cyclopentadienyl groups, but it too is very sensitive to air.

So, manganese owes its individuality to being a *transitional* transition metal. It is midway in position and in behaviour between the lower members of the series, with their few 'extrovert' d electrons, rather loosely held by the nucleus, and the later metals, with six or more d electrons, which, under the influence of higher nuclear charge, are partially withdrawn into the core electrons and are less able to participate in chemical reactions. It is predictable that the oxidation state (II) is now the most stable, and that (VII) is attainable, although highly oxidizing. Since states (VI) to (III) are also oxidizing, it is not surprising that these states may disproportionate, particularly in acid solution.

The organometallic chemistry of manganese is as we should expect for an element with enough vacant orbitals and enough outer electrons to act both as a σ acceptor and a π donor, given that it has an odd atomic number. Other features, like the antiferromagnetism of MnO, appear reasonable with hindsight: the high spin d^5 ion would have a high magnetic moment and the Mn^{2+}–O^{2-}–Mn^{2+} distance would be low enough to allow super exchange. If manganese seems to hold no surprises, it is because some of its most interesting chemistry has yet to be unravelled. Manganese is an essential element in plant biochemistry and a component of various enzymes and proteins. It is known to be involved in the production of dioxygen in photosynthesis. Although its ability to complex with both oxygen and nitrogen donors and to change its oxidation state doubtless plays an important role, discussion of how it works are still speculative.

For summary, see p. 281–5.

Fe Iron

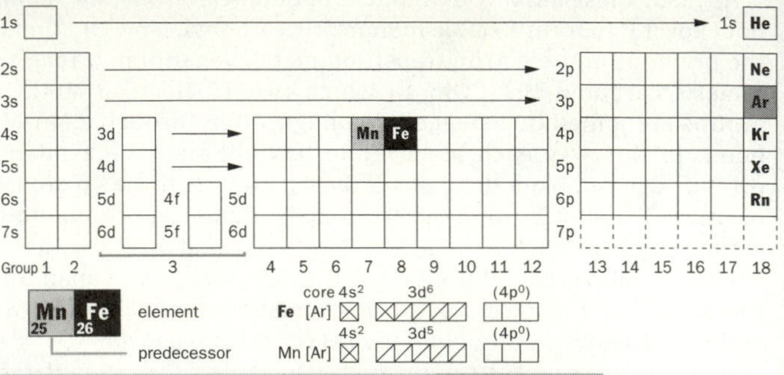

	Predecessor	Element
Name	**Manganese**	**Iron**
Symbol, Z	Mn, 25	Fe, 26
RAM	54.93805	55.847
Radius/pm atomic	124	124.1
covalent	117	116.5
ionic, M^{2+}	91	82
M^{3+}	70	67
Electronegativity (Pauling)	1.55	1.83
Melting point/K	1517	1808
Boiling point/K	2235	3023
$\Delta H^{\ominus}_{fus}$**/kJ mol^{-1}**	14.4	14.9
$\Delta H^{\ominus}_{vap}$**/kJ mol^{-1}**	219.7	351.0
Density (at 293 K)/kg m^{-3}	7440	7874
Electrical conductivity/Ω^{-1}m^{-1}	5.405×10^5	1.030×10^7
Ionization energy for removal of jth electron/kJ mol^{-1} $j = 1$	717.4	759.3
$j = 2$	1509.0	1561
$j = 3$	3248.4	2957
$j = 4$	4940	5290
$j = 5$	6990	7240
$j = 6$	9200	9600
$j = 7$	11 508	12 100
Electron affinity/kJ mol^{-1}		15.7
$E^{\ominus}$**/V for** $M^{3+}(aq) \rightarrow M^{2+}(aq)$	1.5	0.771
$M^{2+}(aq) \rightarrow M^{3+}(aq)$	–1.18	–0.44

Stable isotopes of iron

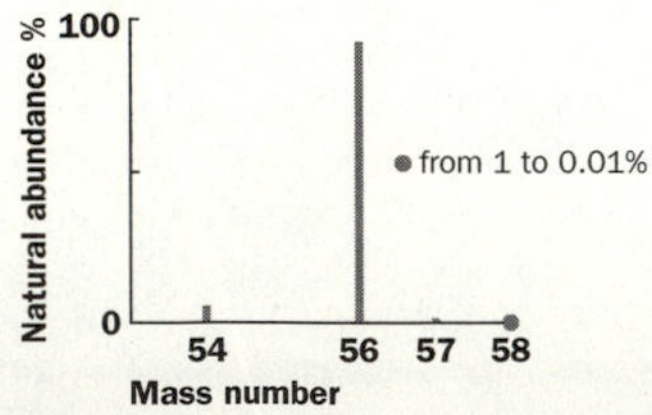

Iron ($[Ar]3d^64s^2$) seems a typical member of the 'later' 3d transition metals, with its non-exotic chemistry being restricted to the metal itself together with oxidation states (II) and (III). The interest of iron (and, to many of us, its romantic appeal) lies in relating its chemical behaviour not only to atomic structure but also to its many sided participation in human affairs. Its biological roles, like its wide technological use, would presumably not have evolved had the element not been so widely available. Not only does the metal itself appear from outer space as meteorites, but it is thought to make up much of the earth's core, possibly because its most common isotope, ^{56}Fe, has, in terms of binding energy per nuclear particle, the most stable atomic nucleus known. Combined in the earth's crust, iron is the second most common metal after aluminum.

It has long been thought that, because of increasing withdrawal of d electrons into the atomic core, manganese is the last member of the 3d series that can formally lose all its outer electrons. However, the Russians have claimed to have prepared Fe^{VIII}, predictably with oxygen, as FeO_4. The red, highly oxidizing $Fe^{VI}O_4^{2-}$ is well established, and a few iron(IV) complexes are known. But iron normally keeps at least five of its d electrons to itself.

Metallic iron has a lower melting point and heat of fusion than titanium, vanadium, or chromium, although its smaller atoms will be expected to hold together more firmly. This effect, too, may arise because only some of the six 3d electrons join the 4s electrons in the delocalized metallic conduction band, while the others remain localized closer to the nucleus. Each atom has as many as possible of its d electrons unpaired and with spins aligned. Indeed, the spins on one atom are aligned with those of neighbouring atoms, to give huge magnetically uniform domains within the solid. When these domains have themselves been aligned by an external magnet, the iron can often retain its magnetism for a long time. Materials such as iron and chromium dioxide (see p. 231), in which spins can be aligned over a long range, are said to be ferromagnetic, as iron was the first substance known to behave in this way, and indeed is still the material from which many large magnets are made. Pure iron is more easily magnetized than steel, but its ferromagnetism is less permanent.

An ion with unpaired electrons is slightly less compact than if the electrons were paired (see p. 253), and metallic iron has a slightly expanded structure, which makes it soft enough to be hot-worked, and open enough to accommodate other atoms, either in between the iron atoms or substituting for them. The most important guest element is carbon, which converts iron into its much tougher alloy steel, although iron can also be hardened by adding

Melting points and heats of fusion

	Ti	V	Cr	Mn	Fe
T_{fus}/K	1933	2160	2130	1517	1808
$\Delta H^{\ominus}_{fus}/kJ\ mol^{-1}$	20.9	17.6	15.3	14.4	14.9
				anomalous solid structure	

The magnetization of iron [SM]

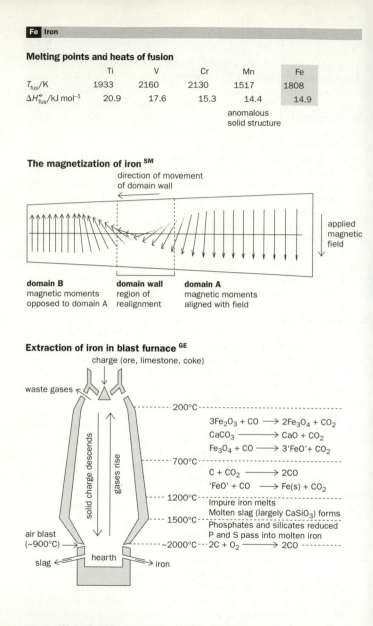

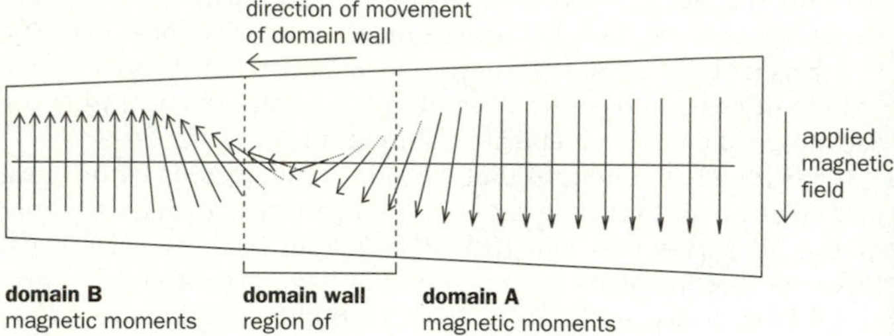

Extraction of iron in blast furnace [GE]

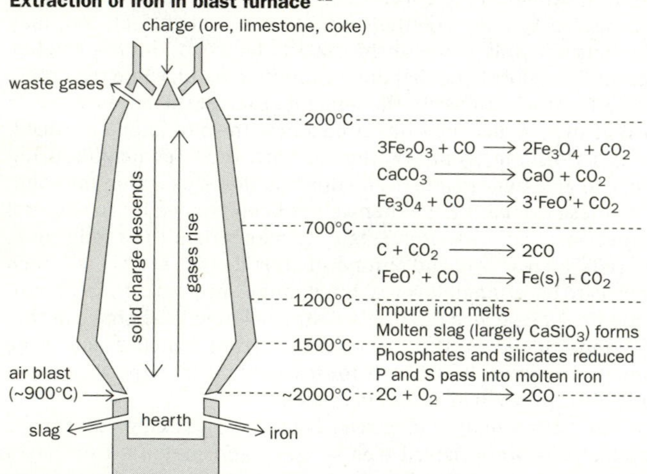

nitrogen or boron. We have seen that the earlier transition elements, such as vanadium, chromium, and manganese, also cause dramatic changes in the mechanical properties of iron; and so do different types of heat treatment such as tempering, quenching, and annealing. The world production of iron and steel (over 700 million tonnes) is 18 times that of all other metals put together, for construction, tools, machinery, blades, firearms, land transport, shipping, vats, and a host of other uses. Composition varies widely. Steel, which is sufficiently strong and ductile for ships' plating, may contain only 0.2% of carbon, compared with 4% in cast iron. A high strength, low alloy steel may be made using only a small amount of a second metal (such as 0.07% of vanadium), whereas prison bars may contain up to 13% of manganese. The kitchen sink may be only 70% iron, together with about 20% chromium, 7% nickel, and 3% molybdenum.

Iron occurs mainly in combination with oxygen, as in the ores Fe_2O_3 (haematite) and its hydrates, Fe_3O_4 (magnetite), and $FeCO_3$ (siderite). Iron pyrites, Fe^{2+} (S_2^{2-}), also occurs widely but is not used as an ore because of environmental problems caused by the sulfur (see p. 175). Reduction of iron ores by carbon in prehistoric times had to await the production of sufficiently high temperatures by the use of closed furnaces and bellows, the forerunner of the blast furnaces used to this day. It is interesting to speculate on how different human history might have been if this reaction had never been performed.

Iron occurs in smaller quantities wherever rocks and the soils derived from them are brown, yellow, orange, or red. These colours are a result charge transfer from surrounding oxygen atoms to Fe^{3+} ions, the hues of the earth pigments (such as raw sienna and burnt umber) varying with the degree of hydration. If Fe^{2+} ions are also present (as in magnetite), the mineral is black, since electrons can be readily transferred from the Fe^{2+} to the Fe^{3+} ions, with light absorbed over much of the visible spectrum. Intense absorption is a feature of charge transfer spectra (see p. 215) and if these involve two oxidation states of one metal, it often occurs over a wide range of energy to give a black compound.

One reaction, which gives an orange-brown product [with oxygen to iron(III) charge transfer], is regrettably familiar and extremely costly: the formation of rust (approximately FeO.OH) wherever unprotected iron is exposed to damp air or to aerated water. Unlike the corrosion products of many metals, which produce a firm, non-porous protective coating, the large flakes of rust are easily detached, exposing more iron to attack and eventual destruction. Rust-like associations of iron and oxygen, wrapped up in

Ligand field effects for Fe^{3+} and Fe^{2+}

High spin octahedral configurations

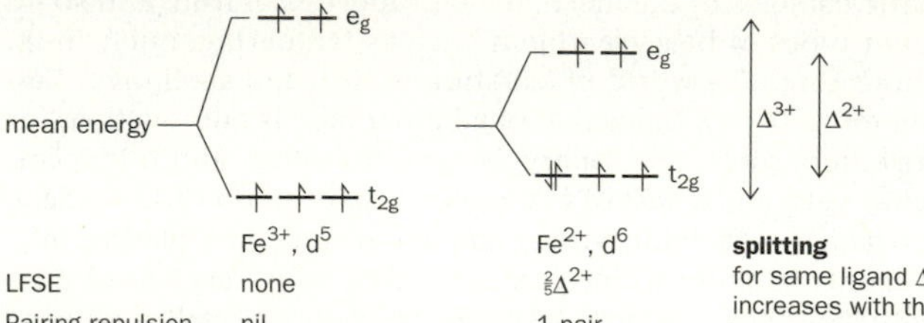

Low spin octahedral configurations

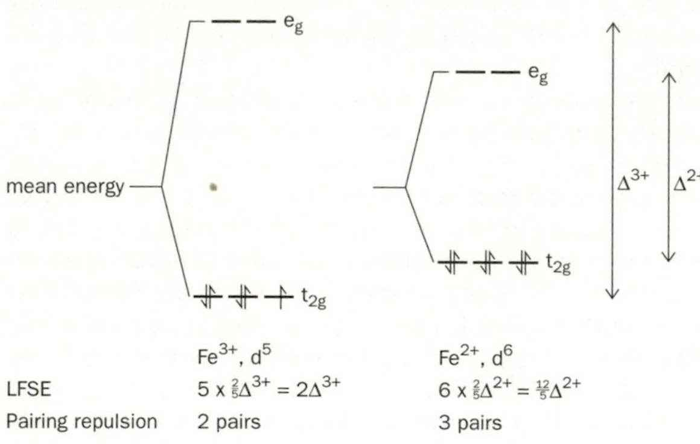

Effect of ligands GE on $E^{\ominus}$

	$E^{\ominus}/V$	
$Fe(bipy)^{3+} + e^- \rightarrow Fe(bipy)^{2+}$	0.96	Fe^{II} stabilized relative to Fe^{III}
$Fe^{3+}(aq) + e^- \rightarrow Fe^{2+}(aq)$	0.77	
$Fe(CN)_6^{3-} + e^- \rightarrow Fe(CN)_6^{4-}$	0.36	Fe^{III} stabilized relative to Fe^{II}
$Fe\ EDTA^- + e^- \rightarrow Fe\ EDTA^{2-}$	−0.12	

For bipy see p. 206; for $EDTA^{4-}$ see p. 202

Stability to aerial oxidation

for $\frac{1}{4}O_2 + H^+ + e^- \rightarrow \frac{1}{2}H_2O$ $E^{\ominus}/V = 1.23$

Since this is more positive than any of the $E^{\ominus}$ values given, all the Fe^{II} complexes are vulnerable to oxidation by air.

phosphate groups and protein molecules, occur in our spleen and liver, and these ‘ferritins’ act as a store of iron, accessible by fast release.

The prevalence of rust and earth colours prompts us to wonder why iron(II) is so readily oxidized to iron(III), while the (II) state of manganese (and of the later 3d elements) is stable to oxidation; and this despite the greater nuclear pull on the 3d electrons of iron than of manganese. The answer, as is usually the case, lies in a very delicate balance of energies, all of which vary from one element to another. The two dominant terms are the energy needed to convert M^{2+} to M^{3+} by removal of an electron, and the energy evolved by the stronger interaction of the M^{3+} ion with neighbouring atoms. This in turn is determined not only by its higher charge and smaller radius, but also by the arrangement of the d electrons in both ions. The most usual environment of six oxygen atoms provides only a low to moderate ligand field and both the hydrated iron ions, Fe^{2+} (d^6) and Fe^{3+} (d^5), are high spin. The high spin d^6 ion has one (ligand field stablized) electron over and above the high spin d^5 structure, which has no ligand field stabilization. But for a doubly charged ion surrounded by oxygen the splitting is small, and this modest stabilization is offset by the electron–electron repulsion (or ‘pairing energy’) within the one full orbital. The d^5 ion, on the other hand, gains not only from its higher charge, but also from the absence of pairing energy. The result is that, while aqueous manganese(II), d^5, is stable in air except under very alkaline conditions [when it is oxidized to manganese(III), d^4, see p. 243], aqueous solutions of iron(II), d^6, are oxidized by air to iron(III), d^5, except in very acidic solutions. Solutions containing Fe^{2+}(aq) are pale green, because of d–d transitions, but as those of iron(III) are always slightly hydrolysed, they are tinged with the colour of rust. At acidities below about pH = 2, the solution contains polymeric cations such as $Fe{-}O{-}Fe^{4+}$, but the hydroxide $Fe(OH)_3$ is precipitated from less acidic solutions. The unhydrolysed violet ion $Fe(H_2O)_6^{3+}$ is, however, found in crystals; the colour [like that of Mn^{2+}(aq)] is very pale, as expected for a high spin d^5 ion, where a d–d transition would require a change in electron spin (see p. 240).

Iron does not restrict its association to oxygen. Most metal ions that combine strongly with OH^- also form stable fluoride complexes, as does Fe^{3+} in FeF_6^{3-}. Small ligands such as fluoride, and highly charged ones such as phosphate and $EDTA^{4-}$ (see p. 202), stabilize iron(III) relative to iron(II). Neither Fe^{3+} nor Fe^{2+} form stable ammonia complexes in aqueous solution, as they combine preferentially with the hydroxy ions present. However, they form complexes with nitrogen donor, π acceptor ligands, such as bipyridyl

Ligand field splitting in tetrahedral complexes [SAL]

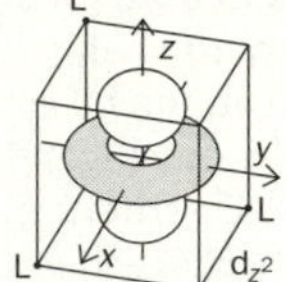

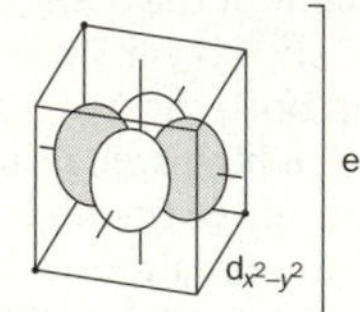

These orbitals are roughly between the ligands

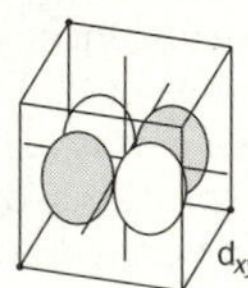

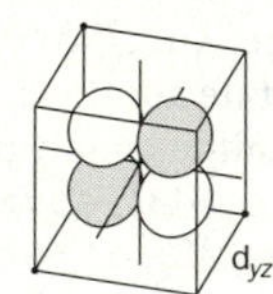

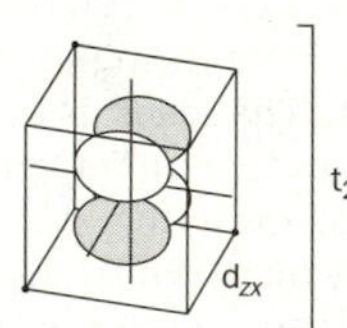

These orbitals are directed more towards the ligand

Comparison with octahedral splitting

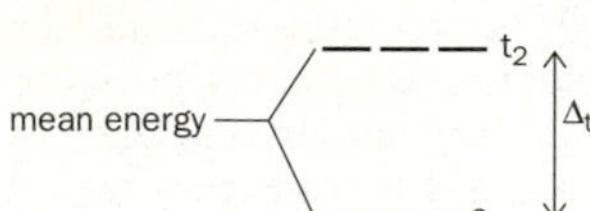

e_g

Δ_o

t_{2g}

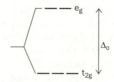

tetrahedral splitting	**octahedral splitting**
3 orbitals destabilized(t_2)	3 orbitals stabilized(t_{2g})
2 orbitals stabilized(e)	2 orbitals destabilized(e_g)
g subscript dropped (it implies centre of symmetry, which is absent in tetrahedron)	
$\Delta_t < \Delta_o$ for same ligand (only four ligands and no orbitals point directly at them)	
complexes always high spin	complexes can be high or low spin

(see p. 143), which may exert a strong enough ligand field to change both iron(II) and iron(III) to low spin configurations; the associated contraction is greater for Fe^{2+} and Fe^{3+} than for other 3d cations, and their radii may decrease by up to 20%. Sometimes the high and low spin forms are so similar in energy that both exist in the same solution. π-Acceptor ligands, by withdrawing charge from the metal ion, stabilize the lower oxidation state. With iron(II) they give complexes that are highly coloured owing to metal to ligand charge transfer; and because these complexes can be readily oxidized to less highly coloured ones containing iron(III) they are used as indicators in redox titrations. Both iron(II) and iron(III) form very stable low spin hexacyanato complexes, $Fe^{II}(CN)_6^{4-}$ and $Fe^{III}(CN)_6^{3-}$.

Since the relative stability of iron(III) and iron(II) are so similar as to be highly sensitive to environment, both oxidation states can be present in the same compound. One example is 'Prussian blue', a family of pigments in which Fe^{2+} and Fe^{3+} ions are bridged by cyanide ions. Although the intense absorption is caused by metal to metal or 'mixed valence' charge transfer (see p. 249), only low energy visible light is absorbed, and the compounds are deep blue, rather than black.

Iron also forms two 'mixed valence' oxides. Although 'FeO' resembles MnO (see p. 241) in being antiferromagnetic at low temperatures, it has a composition around $(Fe^{2+})_{0.85}(Fe^{3+})_{0.10}O$ and the presence of both oxidation states is confirmed by its colour: black. The magnetic ore Fe_3O_4 is a mixed oxide, $Fe^{2+}(Fe^{3+})_2(O^{2-})_4$, that might be expected to adopt the spinel structure of $Mg^{2+}(Al^{3+})_2(O^{2-})_4$ (see p. 143), as do the analagous oxides Mn_3O_4 and Co_3O_4, of the elements on each side of iron. However, for transition metal ions, the greater electrostatic forces within the octahedral sites must be combined with ligand field effects (see p. 217). These are also greater in an octahedral site because the distinction between those orbitals that are repelled to a greater or lesser extent by surrounding ligands is clearer in an octahedral complex than in a tetrahedral one. For high spin species, the larger LFSE provided by an octahedral site favours d^4 and d^6 ions (such as Mn^{3+}, Fe^{2+}, and Co^{3+}) over d^5 ones (such as Mn^{2+} and Fe^{3+}). So, in Mn_3O_4 and Co_3O_4, the effects combine to place the M^{3+} ions in the octahedral sites; but in Fe_3O_4, they are opposed. The ligand field stabilization must, however, be more important because all the Fe^{2+} ions take up octahedral positions, giving the inverse spinel structure of $Fe^{3+}_{tet}(Fe^{2+}Fe^{3+})_{oct}(O^{2-})_4$.

The oxide Fe_3O_4 is a good electrical conductor, as the ions in the octahedral sites are so close together that electrons can move from one to another. The spins of the electrons of the ions in all the octahedral sites are aligned in one direction, while those on the tetra-

Ferrimagnetism of Fe_3O_4

Three 'types' of cation, in equal numbers: all high spin

ions	Fe^{2+} oct(d^6)	Fe^{3+} oct(d^5)	Fe^{3+} tet(d^5)
unpaired electrons per ion	4	5	5
spin alignment	↑	↑	↓

all octahedral ions have aligned spins
all tetrahedral ions spins aligned, and opposed to those of octahedral ones
spins of Fe^{3+} oct cancel those of Fe^{3+} tet
spins of Fe^{2+} oct cause ferrimagnetism

The haem group [CW]

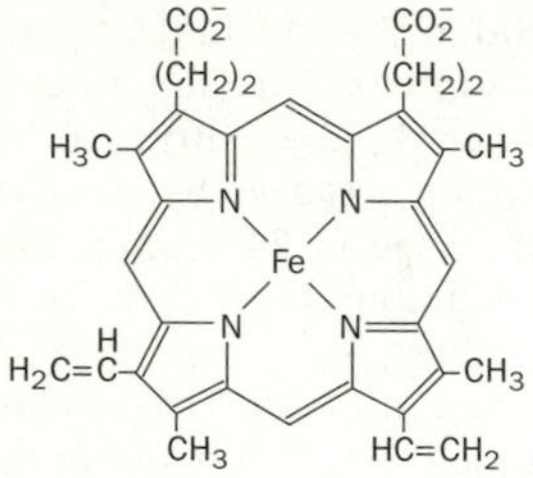

Iron at the centre of a porphyrin ring, joined to 4 N atoms

Ferredoxins [GE]

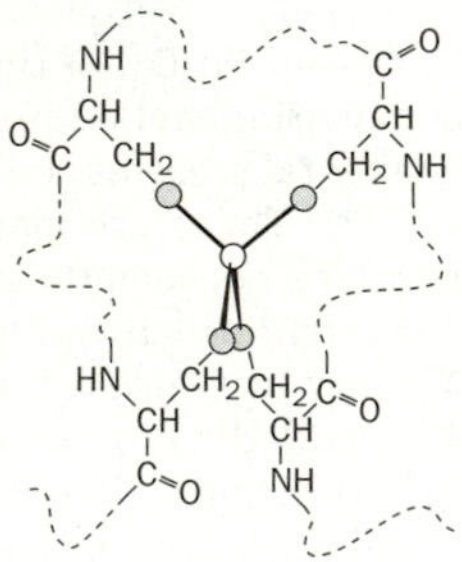

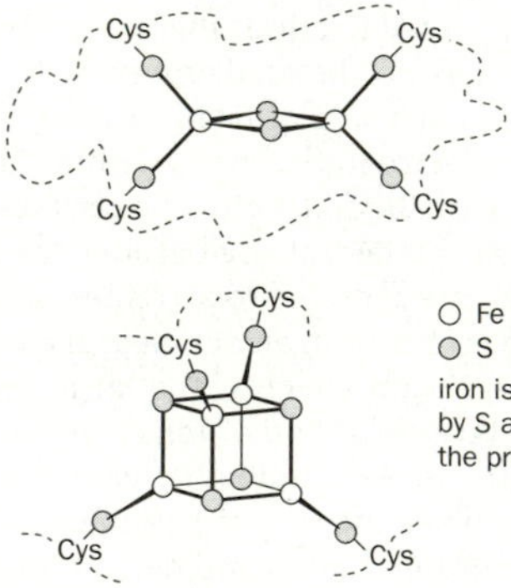

iron is surrounded by S atoms from the protein cysteine

hedral ions are aligned to oppose them. But since there are more unpaired electrons at the octahedral sites, the two effects do not cancel each other completely. This imbalance (known as ferrimagnetism) makes the solid a permanent magnet, the first one known to humans. Once much prized as 'lodestone' for navigation, it is now used for magnetic tapes. Such materials are important industrially as their magnetism is accompanied by low electrical conductivity. Tiny particles of Fe_3O_4 are present in many 'homing' animals such as whales, pigeons, salmon, and bees, and also in some molluscs and anaerobic bacteria. Some have even been found in the membrane covering the human brain.

The extreme sensitivity to environment of the relative stabilities of both the (III) and (II) oxidations states (and also of the low spin and high spin forms of them) is the basis of many delicate physiological processes. The cytochrome enzymes, which contain iron surrounded by a planar ring of four nitrogen atoms in a haem molecule and one or two donor groups of a protein, can promote gradual oxidation or reduction by using simple electron reaction steps.

In both haemoglobin, which transports oxygen in our bloodstream, and in myoglobin, which stores it in red muscles, iron is again associated with a haem ring. On one side of the planar ring is another nitrogen atom, attached to various protein chains. Just protruding from the centre of the ring is an iron ion, which can bind a molecule of dioxygen. When this happens, the iron contracts and sinks into the central cavity formed by the haem nitrogen atoms so that it is almost in the plane of the ring and, in so doing, activates lever-like movements in the chains associated with the fifth (non-haem) nitrogen atom. This finely tuned series of changes shows how sensitively iron, whether as Fe^{2+} or Fe^{3+}, responds to very small changes in its atomic environment.

Other electron transfer enzymes, the ferredoxins, have their iron associated with sulfur. Ferredoxins contain the Fe_4S_4 group, with the atoms at alternate corners of a cube, and this group also occurs in the nitrogen-fixing enzyme nitrogenase (see p. 83).

Although almost all naturally occurring iron is in an oxidation state of either (III) or (II), compounds of iron(0) can be made in the laboratory. Since the iron atom ($3d^6 4s^2$) has eight outer electrons to contribute to its nine valence orbitals, there are five vacant, low lying orbitals available to receive ligand electrons. As expected, iron forms a monomeric carbonyl, $Fe(CO)_5$ (a trigonal bipyramid). It was first made over a century ago, and is the parent of many derivatives, including $Fe_2(CO)_9$ and $Fe_3(CO)_{12}$, in which the iron atoms are joined by metal–metal bonds, sometimes reinforced by

Metal carbonyls [M]

A bridged carbonyl, $Fe_2(CO)_9$

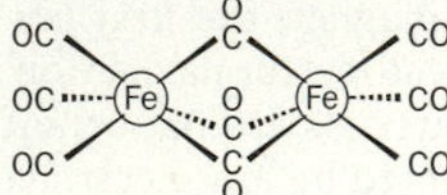

'Ferrocene' $Fe(C_5H_5)_2$ or '$Fe(Cp)_2$'

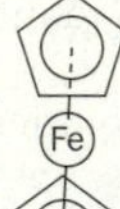

The Fe atom is sandwiched between two cyclopentadienyl, C_5H_5, rings and bound equally to all ten C atoms.

The rings are not exactly one above the other ('eclipsed') but are offset by 9°

bridging carbonyl groups, but in every case associated with 18 electrons (see p. 227).

Since iron (0) needs to gain 10 electrons in order to attain its stable quota of 18, it is no surprise that it also forms a sandwich compound between two cyclopentadienyl groups. Indeed 'ferrocene' was the first of these metal sandwich compounds to be made (by mistake, in the 1930s), and it is much more stable to heat, air, and water than the less-than-18-electron metallocenes of the earlier transition metals (see p. 218).

So iron, with only (III) and (II) as common oxidation states, might have seemed a rather limited transition element, but within these limitations it is indeed impressive. Its d orbitals are neither nearly empty, nor nearly full, and its d electrons are neither totally free, nor mainly withdrawn into the core. The result is a metal that shows strong enough ferromagnetism to make it the main material for permanent or electro-magnets. Its slightly openwork structure allows addition of carbon to form steel, and many different metal atoms to form alloy steels with properties that can be almost tailor-made for a wide range of special jobs, from making railway bridges, to prison bars, to industrial vats. The similar stabilities of iron(II) and iron(III) (achieved by a balance of coulombic attraction, electronic repulsion, and LFSE) give iron the particular redox (and spin state) sensitivity to environment that is exploited in various physiological processes. And for the organometallic chemist, the ability of iron(0) to accept 10 electrons offers scope for two large groups of compounds based on donation of either five pairs, or two quintets, of electrons. But of what significance are these man-made molecules compared with the substances that are the basis not only of our industrial civilization, but also of our very existence?

For summary, see p. 281–5.

Co Cobalt

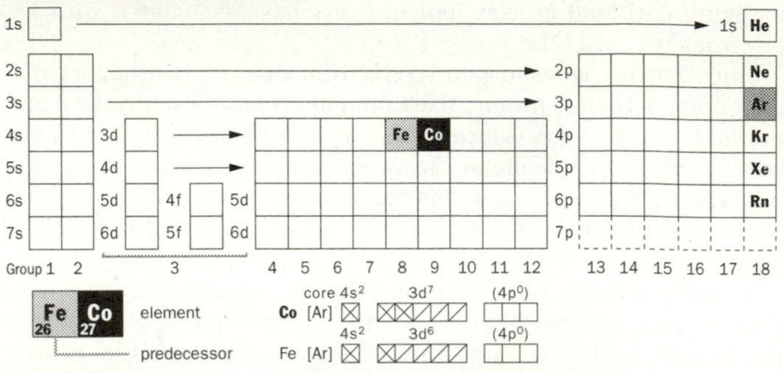

		Predecessor	Element
Name		**Iron**	**Cobalt**
Symbol, *Z*		Fe, 26	Co, 27
RAM		55.847	58.93320
Radius/pm	**atomic**	124.1	125.3
	covalent	116.5	116
	ionic, M^{2+}	82	82
	M^{3+}	67	64
Electronegativity (Pauling)		1.83	1.88
Melting point/K		1808	1768
Boiling point/K		3023	3143
$\Delta H^{\ominus}_{fus}$/kJ mol^{-1}		14.9	15.2
$\Delta H^{\ominus}_{vap}$/kJ mol^{-1}		351.0	382.4
Density (at 293 K)/kg m^{-3}		7874	8900
Electrical conductivity/Ω^{-1}m^{-1}		1.030×10^7	1.602×10^7
Ionization energy for removal of *j*th electron/kJ mol^{-1}	***j* = 1**	759.3	760.0
	***j* = 2**	1561	1646
	***j* = 3**	2957	3232
	***j* = 4**	5290	4950
	***j* = 5**	7240	7670
	***j* = 6**	9600	9840
	***j* = 7**	12 100	12 400
Electron affinity/kJ mol^{-1}		15.7	63.8
$E^{\ominus}$/V for	**$M^{3+}(aq) \rightarrow M^{2+}(s)$**	−0.771	1.416
	$M^{2+}(aq) \rightarrow M(s)$	−0.44	−0.277

Stable isotopes of cobalt

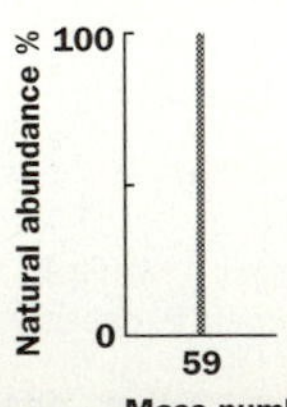

Much of the chemical behaviour of cobalt ([Ar]$3d^7 4s^2$) is predictable, or nearly so, from that of its predecessor, iron. With its higher nuclear charge, the d electrons are increasingly introverted. There is scant evidence for any oxoanion, and, apart from the complex fluoride $K_2Co^{IV}F_6$ almost all readily accessible compounds are in the (III) or (II) oxidation states: cobalt(II) is the only state to combine with chlorine, bromine, or iodine,* or to form a stable aquo complex; and a number of cobalt(I) species such as $Co(bipy)_3^+$ have been made using π acceptor ligands. Much of the detailed chemistry of cobalt, like that of its predecessors, can be explained by the effect of its number of d electrons on either ligand field stabilization (if the d orbitals are not completely filled) or the possibility of attaining an 18-electron configuration.

In metallic cobalt, as in iron, the d orbitals are fairly tightly withdrawn (without being totally localized) and generously (but not completely) filled. Each atom will have its unpaired electrons with the same spin, and will be near enough to its neighbours to have its resultant spin aligned with theirs. So cobalt, like iron, is ferromagnetic. But since the d electrons in cobalt are more withdrawn than in iron, the atoms are less firmly bound together; and, because it loses electrons less readily, cobalt is less reactive.

Cobalt occurs (together with nickel, copper, and lead) mainly with sulfur and arsenic,** rather than with oxygen; with its relatively high nuclear charge and small size it favours more polarizable atoms. Although these ores were used to make blue glazes as long ago as 2600 BC, extraction of the pure metal had to wait over four thousand years. Separation of cobalt from other metals involves oxidizing it (with OCl^-) to cobalt(III), stabilized as the insoluble hydroxide $Co(OH)_3$. Cobalt is still used to make blue pigments, and also to counteract any yellow colour caused by iron(III) in glass [in much the same way as manganese is used to complement the green of iron(II), see p. 241]. More modern uses include catalysis and the manufacture of high temperature alloys for gas turbines, and of alloy steels for magnets. The artificial isotope ^{60}Co is used as a source of γ-rays.

The oxides and sulfides of cobalt are broadly similar to those of iron: in the (II) state, CoO is stable and antiferromagnetic at low

* Although the only cobalt trihalide is the fluoride, iron forms trihalides also with chlorine and bromine.

** Indeed, the name cobalt comes from the German for 'evil spirits', thought to be responsible for the toxicity and metallurgical cussedness of its ores.

Cobalt(II) configurations

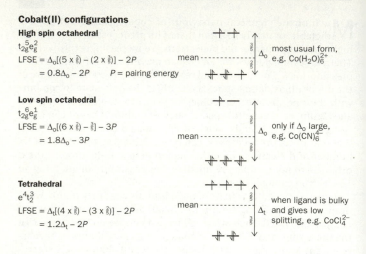

High spin octahedral

$t_{2g}^5 e_g^2$

$\text{LFSE} = \Delta_o[(5 \times \frac{2}{5}) - (2 \times \frac{3}{5})] - 2P$

$= 0.8\Delta_o - 2P$ $\quad P$ = pairing energy

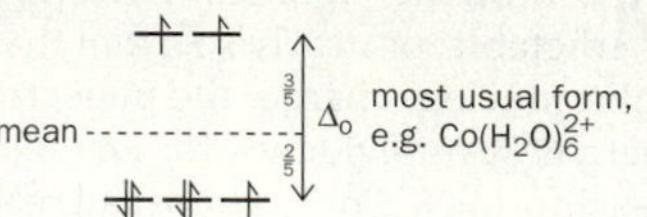

Low spin octahedral

$t_{2g}^6 e_g^1$

$\text{LFSE} = \Delta_o[(6 \times \frac{2}{5}) - \frac{3}{5}] - 3P$

$= 1.8\Delta_o - 3P$

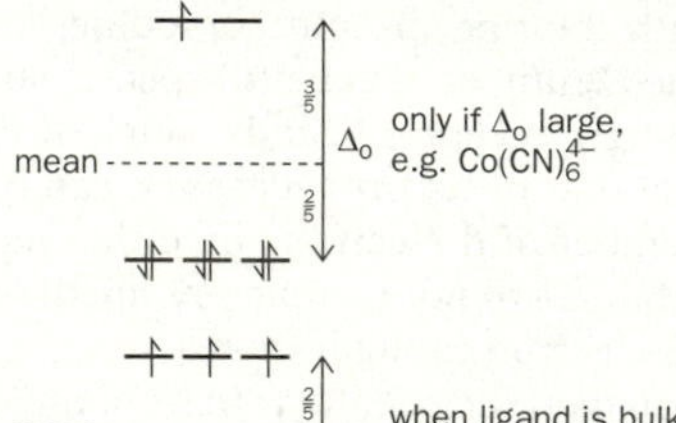

Tetrahedral

$e^4 t_2^3$

$\text{LFSE} = \Delta_t[(4 \times \frac{3}{5}) - (3 \times \frac{2}{5})] - 2P$

$= 1.2\Delta_t - 2P$

mean $\quad \frac{2}{5} \quad \frac{3}{5} \quad \Delta_t$

when ligand is bulky and gives low splitting, e.g. $CoCl_4^{2-}$

Site preference energies for some M^{2+} ions

High spin octahedral, compared with tetrahedral, sites.

Difference in LFSE for the two configurations

Co^{2+}: $(\text{LFSE})_{oct} - (\text{LFSE})_{tet}$

$= (0.8\Delta_o - 2P) - (1.2\Delta_t - 2P)$

$= 0.8\Delta_o - 1.2\Delta_t$

$\sim 0.2\Delta_o$ (because, for many ligands, $\Delta_t \sim 0.5\Delta_o$)

Calculations for Fe^{2+}, Ni^{2+}, and Cu^{2+} give site preference energies of:

Fe^{2+}: $0.4\Delta_o - 0.6\Delta_t \sim 0.1\Delta_o$

Ni^{2+}: $1.2\Delta_o - 0.8\Delta_t \sim 0.8\Delta_o$

Cu^{2+}: $0.6\Delta_o - 0.4\Delta_t \sim 0.4\Delta_o$

temperatures; $Co_{1-x}S$ is non-stoichiometric; and CoS_2 has the pyrites structure $Co^{2+}(S_2^{2-})$. Unlike Fe_3O_4, but like Mn_3O_4 (see pp. 253 and 239), the mixed valence oxide Co_3O_4 has the normal spinel structure with the Co^{3+} low spin d^6 ions in the octahedral sites, where they gain more in stability than would the Co^{2+} ions, both from electrostatic interaction and from large ligand field effects (see p. 251). The Co^{2+} ions are not greatly destabilized in their tetrahedral sites, partly because preference for an octahedral environment increases with Δ_{oct} and hence with charge. Moreover, amongst the common M^{2+} ions, Fe^{2+} and Co^{2+} are less affected by site geometry than are Ni^{2+} and Cu^{2+}.

In aqueous solution, the blue $Co(H_2O)_6^{3+}$ ion (again, low spin d^6) is a strong oxidizing agent, and liberates oxygen from water, while the pink $Co(H_2O)_6^{2+}$ ion is stable. Both colours are a result of d–d transitions in octahedra and are fairly pale. However, the blue, tetrahedral ion $CoCl_4^{2-}$ is more deeply coloured, as a tetrahedron has no centre of symmetry (see p. 217). This colour change of pale pink to mid-blue is familiar as a moisture detector in 'cobalt chloride paper' and as the indicator in silica gel drying agents.

When ammonia is added to aqueous cobalt(II) solutions the complex ion $Co(NH_3)_6^{2+}$ is readily oxidized by air to $Co(NH_3)_6^{3+}$, giving a dramatic example of how changing the ligand can play havoc with the relative stabilities of two oxidation states. Once again we can explain an apparent oddity in terms of the number of d electrons. Ammonia has a stronger ligand field than water, and this produces a larger d orbital splitting, which is more marked for the smaller, more highly charged ion. The most stable configuration of the Co^{3+} (d^6) ion, in any but the weakest of fields, is low spin with all its electrons paired, because the large gain in ligand field stabilization easily offsets the increased repulsion; the only high spin cobalt(III) complex seems to be CoF_6^{3-}. However, the larger Co^{2+} (d^7) ion, with at least one electron in the destabilized orbitals and a much lower splitting, is normally high spin (except in very strong fields like that of cyanide) and it gains little in stability from the replacement of its water molecules by ammonia. Since ammonia stabilizes cobalt(III) relative to Co^{3+}(aq) much more than it stabilizes cobalt(II) relative to Co^{2+}(aq), replacement of the six water molecules by ammonia greatly lessens the tendency of cobalt(III) to be reduced to cobalt(II); and, in practice, $Co^{III}(NH_3)_6^{3+}$ is not an oxidizing agent.

The high LFSE of the low spin d^6 configuration for a triply charged cation gives it not only thermodynamic stability, but also kinetic sluggishness, since the formation of any intermediate could only take place by disrupting this very favourable arrangement. Reactions of low spin cobalt(III) complexes tend to be extremely

Cobalt (III) configuration(s)

low spin octahedral t_{2g}^6

$$LSFE = \Delta_0\left[\left(6 \times \tfrac{2}{5}\right) - \tfrac{3}{5}\right] - 3P$$
$$= 1.8\Delta_0 - 3P$$

most usual form, e.g. $Co(H_2O)_6^{3+}$

high spin octahedral $t_{2g}^5\ e_g^1$

$$LSFE = \Delta_0\left[\left(5 \times \tfrac{2}{5}\right) - \left(2 \times \tfrac{3}{5}\right)\right] - 2P$$
$$= 0.8\ \Delta_0 - 2P$$

occurs only if Δ_o very low, e.g. CoF_6^{3-}

tetrahedral $e^3t_2^3$

$$LSFE = \Delta_t\left[\left(3 \times \tfrac{3}{5}\right) - \left(3 \times \tfrac{2}{5}\right)\right] - P$$
$$= 0.6\ \Delta_t - P$$

very rare

Stabilization of Co^{III} by complex formation[GE]

Low spin Co^{III}		High spin Co^{II}	E°/V	
$Co(H_2O)_6^{3+} + e^-$	→	$Co(H_2O)_6^{2+}$	1.83	Co^{III}(aq) strongly oxidizing
Co $EDTA^- + e^-$	→	Co $EDTA^{2-}$	0.37	Co^{II} complexes oxidized by air
$Co(bipy)_3^{3+} + e^-$	→	$Co(bipy)_3^{2+}$	0.31	
$Co(NH_3)_6^{3+} + e^-$	→	$Co(NH_3)_6^{2+}$	0.11	

An O_2 bridged complex [CW]

$(NH_3)_5-Co^{II}-O_2-Co^{II}-(NH_3)_5$

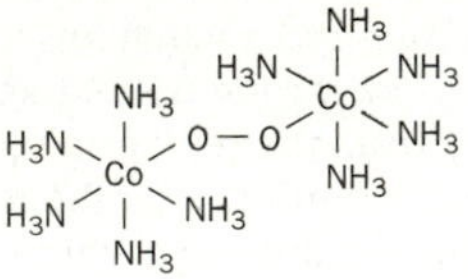

O–O distance/pm	
in complex	131
In O_2	121
In O_2^-	133

slow, and speeds for replacing one ligand by another are often as much as ten times lower even than those for Cr^{3+} (d^3), with its three lower d orbitals half-filled (see p. 233). This kinetic inertness has allowed huge numbers of cobalt(III) complexes to be prepared, frequently in several isomeric forms.

The mechanism of the air oxidation of complexes of cobalt(II) with ammonia and similar ligands seems to involve addition of molecular oxygen to the cobalt ion, probably in the end-on bent position, in a way reminiscent of the binding of dioxygen to the iron(II) in haemoglobin (see p. 255). The electron distribution approximates to that of a complex between cobalt(III) and the superoxide ion O_2^-. The complexes may dimerize through an oxygen–oxygen bridge between the two cobalt ions.

As expected for a metal atom with an odd number of electrons, cobalt does not form a monomeric carbonyl. However, since it has exactly half its complement of 18 outer electrons, we should expect it to combine with four carbon monoxide molecules, provided that it can somehow acquire one extra electron, for example by reduction or by forming a M–M bond; so the existence of $Na^+Co(CO)_4^-$, $HCo(CO)_4$, and $Co_2(CO)_8$ need cause no surprise (see p. 264). In solid $Co_2(CO)_8$ there are two CO bridges (which give a 'bent' cobalt–cobalt bond), but in solution this form is in equilibrium with the unbridged isomer (with a straight metal–metal bond). There is also a carbonyl $Co_4(CO)_{12}$ that contains a central Co_4 tetrahedron, with three of the six metal–metal bonds reinforced by carbonyl bridges. Somewhat surprisingly for a d^7s^2 atom, cobalt combines with cyclopentadienyl to give a 19-electron sandwich that is thermally stable up to 250°C; but, predictably, this is easily oxidized to the 18-electron cation $Co(C_5H_5)_2^+$. The same stable configuration can be achieved in the monocyclopentadienyl dicarbonyl $Co(C_5H_5)(CO)_2$.

So far, cobalt has shown a number of characteristics that may be understood fairly readily from its $3d^74s^2$ configuration: the ferromagnetism and other properties of the metal; the emphasis on oxidation states of (III) and (II); the stability and inertness of the great majority of cobalt(III) d^6 complexes; the small extent to which cobalt(II) favours octahedral to tetrahedral coordination; the possibility of forming cobalt(I) complexes; and the existence of carbonyl and cyclopentadienyl compounds with 18-electron formulae.

However, cobalt has some less predictable behaviour. Its compounds are frequently used as catalysts, often for hydrogenation reactions, which probably involve a cobalt to hydrogen bond of the type formed in $HCo(CO)_4$. A cobalt atom is also the centre-piece of vitamin B_{12}, which is essential to various enzymes that catalyse methyl transfer reactions and rearrangements of organic molecules

Some cobalt carbonyls [GE]

○ Co ● C of CO (O omitted)

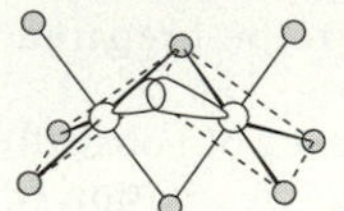

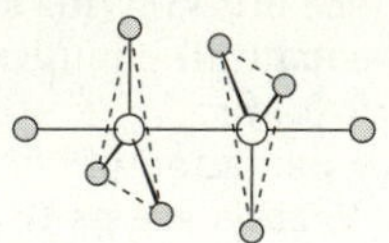

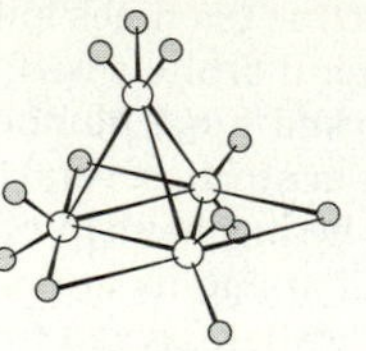

$Co_2(CO)_8$

(a) In solid
'bent' Co–Co bond
2 CO bridges
eclipsed CO groups

(b) In solution
normal Co–Co bond
no CO bridges
staggered CO groups

$Co_4(CO)_{12}$

tetrahedral Co_4 cluster
base: 3 Co–Co bonds
3 CO bridges
base to apex: 3 Co–Co bonds
no CO bridges

in our body. The cobalt is surrounded by five nitrogen atoms in an arrangement much like that of iron in haemoglobin; but in some B_{12} enzymes the sixth site (which in haemoglobin is either empty or occupied by oxygen) is occupied by a side chain with a σ bond from carbon to cobalt. It seems that the activity of the enzyme depends on the breaking and reforming of this bond; and that cobalt, like manganese, may display its most subtle redox chemistry, and least predictable behaviour, within the complex environment of a living system.

For summary, see p. 281–5.

Ni Nickel

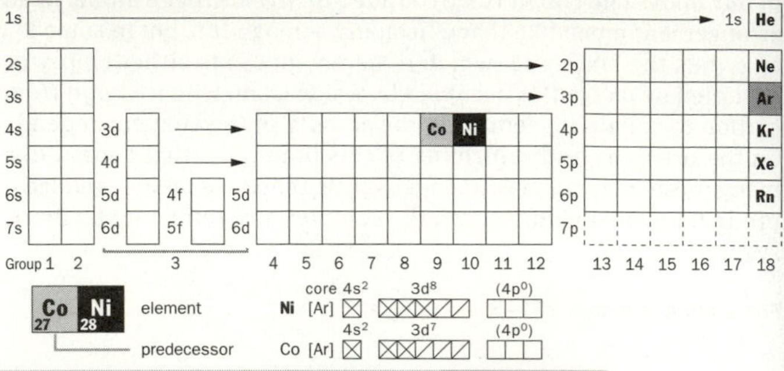

		Predecessor	Element
Name		**Cobalt**	**Nickel**
Symbol, *Z*		Co, 27	Ni, 28
RAM		58.93320	58.69
Radius/pm	**atomic**	125.3	124.6
	covalent	116	115
	ionic, M^{2+}	82	78
	M^{3+}	64	62
Electronegativity (Pauling)		1.88	1.91
Melting point/K		1768	1726
Boiling point/K		3143	3005
$\Delta H^{\ominus}_{fus}$/kJ mol^{-1}		15.2	17.6
$\Delta H^{\ominus}_{vap}$/kJ mol^{-1}		382.4	371.8
Density/kg m^{-3}		8900 (293 K)	8902 (298 K)
Electrical conductivity/Ω^{-1}m^{-1}		1.602×10^7	1.462×10^7
Ionization energy for removal of *j*th electron/kJ mol^{-1}	***j* = 1**	760.0	736.7
	***j* = 2**	1646	1753.0
	***j* = 3**	3232	3393
	***j* = 4**	4950	5300
	***j* = 5**	7670	7280
	***j* = 6**	9840	10 400
	***j* = 7**	12 400	12 800
Electron affinity/kJ mol^{-1}		63.8	156
$E^{\ominus}$/V for $M^{2+}(aq) \rightarrow M(s)$		−0.277	−0.257

Stable isotopes of nickel

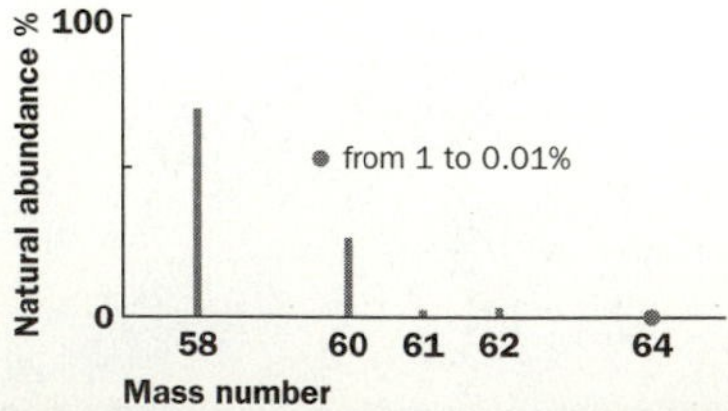

At nickel ([Ar]$3d^84s^2$), the d electrons are even more tightly bound than in cobalt, which it resembles in many ways. The binding energy for nickel is less than that for cobalt, and the metal is resistant to corrosion at room temperature; hence its use for laboratory spatulas. It is ferromagnetic, although less so than cobalt or iron. Apart from the complex fluoride $K_2Ni^{IV}F_6$ and a few compounds containing nickel(III), the non-organometallic chemistry of nickel is dominated by the (II) state.

Nickel occurs with arsenic, sulfur, and oxygen (in both oxides and silicates), and, because its ores did not behave as originally expected, was named after the devil (or, more familiarly, 'Old Nick'). As the first stage in the extraction of nickel, the sulfur and oxygen ores are converted to the oxide, which may be used as such for making steels, many of them for hard materials such as armour plating or (with aluminium and cobalt) for magnets. Nickel metal can be obtained by reducing the oxide with coke, followed by electrolytic purification from an aqueous solution (see p. 268). Alternatively, reduction with 'water gas' (a mixture of carbon monoxide and hydrogen obtained by passing steam over red-hot coke) gives the volatile carbonyl $Ni(CO)_4$ which decomposes to nickel and carbon monoxide on further heating. This 18-electron compound was the first known metal carbonyl, made in 1888. Nickel is also used in non-ferrous alloys, often with copper, as in the corrosion-resistant Monel metal and in cupronickel for 'silver' coins. Alloyed with zinc and copper, it forms a basis for electroplating with silver. Other nickel alloys include invar and nichrome, for which changes of temperature produce extremely small changes of volume and of electrical resistance (respectively), and alloys used as catalysts.

Mond process for nickel extraction (1899)

Steam + red hot coke → $H_2(g)$ + $CO(g)$
'water gas'

'water gas' + oxide ore

$$NiO(s) + H_2(g) \xrightarrow{\text{heat}} \underset{\text{impure}}{Ni(s)} + H_2O(g)$$

'water gas' + impure Ni

$$\underset{\text{impure}}{Ni(s)} + 4CO(g) \xrightarrow{50^\circ C} Ni(CO)_4(g) \xrightarrow{230^\circ C} Ni(s) + 4CO(g)$$

pure (deposited on hot Ni pellets)

The present day process uses the same reactions, but at higher temperatures and pressures.

Different geometries for Ni^{2+}, d^8 GE

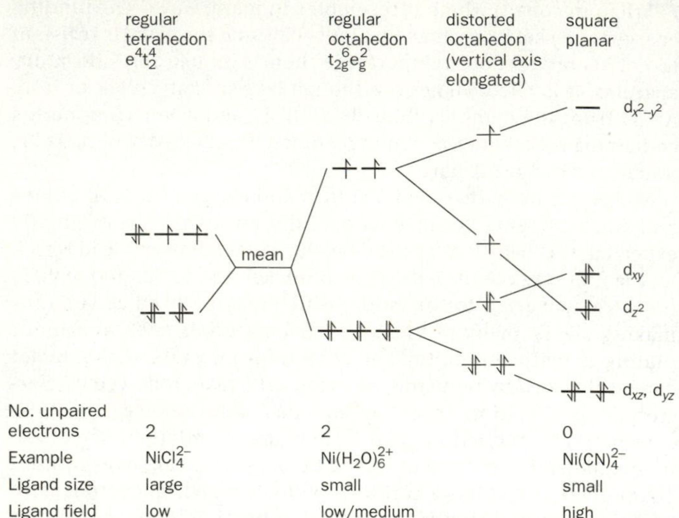

	regular tetrahedon	regular octahedon	square planar
No. unpaired electrons	2	2	0
Example	$NiCl_4^{2-}$	$Ni(H_2O)_6^{2+}$	$Ni(CN)_4^{2-}$
Ligand size	large	small	small
Ligand field	low	low/medium	high

The $(Ni(CN)_2NH_3)_x$ cage GE

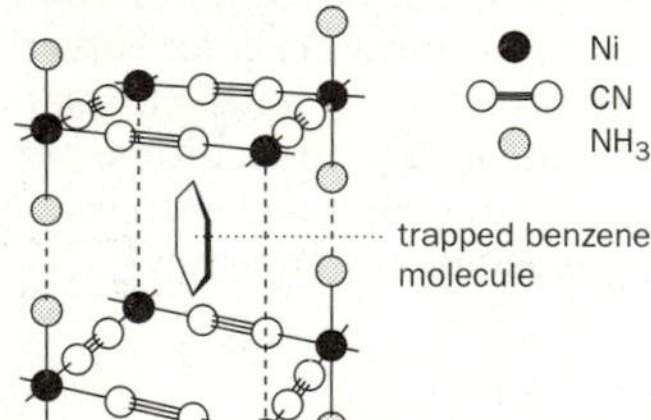

Nickel reacts with many non-metals (but not carbon and nitrogen) when heated and, like many other metals, the finely divided form is pyrophoric at room temperature. Nickel(II) species, such as $Ni(H_2O)_6^{2+}$ and $Ni(OH)_2$ are often green, owing to d–d transitions, but the oxide NiO is black as it is slightly non-stoichiometric and so takes part in mixed valence charge transfer. Like the oxides, MO, of cobalt, iron, and manganese, NiO is antiferromagnetic at low temperatures. The oxide and hydroxide of nickel(II) are purely basic and no oxoanions are known.

Nickel, unlike iron and cobalt, shows no subtle redox chemistry, since its (III) state is too oxidizing to be made readily accessible by use of familiar ligands. Instead, nickel shows how the ligand can affect the delicately balanced energies of different geometrical arrangements *within* the (II) state, in which the Ni^{2+} ion can be surrounded by six ligands octahedrally or by four ligands, either tetrahedrally or in a square. Square complexes are diamagnetic since they contain four pairs of electrons, whilst octahedral and tetrahedral complexes are paramagnetic, with six electrons paired but two unpaired. Bulky ligands such as chloride are best accommodated tetrahedrally, but this arrangement produces little ligand field stabilization. Smaller ligands, such as water, are accommodated octahedrally, unless they produce a large field, in which case a four-coordinate square planar complex such as $Ni(CN)_4^{2-}$ is favoured. The various arrangements may differ little in energy and two configurations may sometimes exist together in equilibrium. The percentage of each configuration will vary with temperature, and, when the two forms are differently coloured, the system will be thermochromic (changing colour with temperature). The equilibrium may involve change from the square planar to the tetrahedral forms of the same complex; or solvation of a square planar species to give an octahedral one. Some chelate complexes, NiL_2, may dimerize or trimerize and so acquire an octahedral arrangement. Nickel(II) may even have two different geometries in the same crystal. In the openwork cage structure of $[Ni(CN)_2NH_3]_x$, all nickel ions are at the centre of a square of cyanide ions; half of the cations have a molecule of ammonia above and below the plane, and so are octahedrally coordinated, while the others remain four coordinate with the axial positions empty. (This compound can absorb molecules of benzene into the cages, to form a clathrate compound.)

Nickel(II) can also be associated with five ligands in either a square pyramid or a trigonal bipyramid. In one complex containing

Five coordinated Ni(II) SAL in $Ni(CN)_5^{3-}$

square pyramid trigonal bipyramid

The two forms can exist together

The cluster carbonyl anions GE $Ni_5(CO)_{12}^{2-}$ and $Ni_6(CO)_{12}^{2-}$

○ Ni ● C of CO (O omitted)

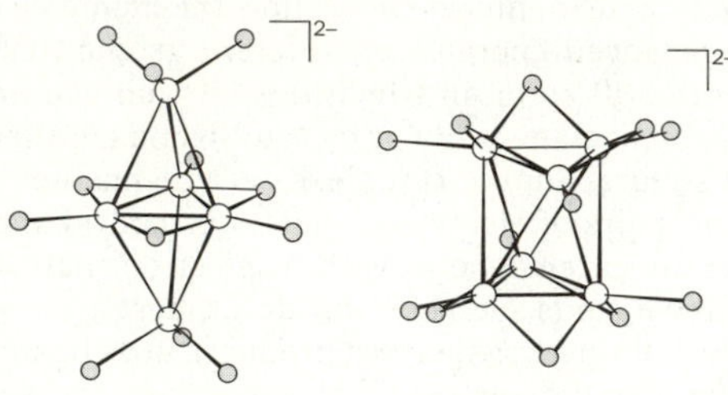

The triple decker sandwich $Ni_2(C_5H_5)_3$ GE

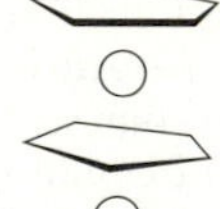

The C_5H_5 rings are neither staggered nor eclipsed.

The Ni atoms are slightly nearer to the outer rings than to the central one

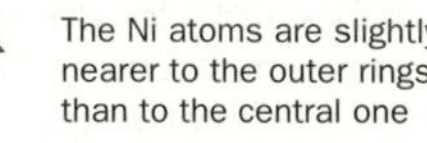

○ Ni

the $Ni(CN)_5^{3-}$ ion, some nickel ions have one geometry and some the other, although both geometries are somewhat distorted. Here, too, there must be very little difference in energy between the two forms.

The interest of nickel(II) stereochemistry is its subtlety; but it is the organometallic chemistry of nickel that is more surprising. Although nickel(0) forms a predictably stable carbonyl, we should expect that, with its 10 outer electrons, it has not got enough unoccupied outer orbitals to form a stable sandwich with cyclopentadienyl. However, it does. The 20-electron $Ni(C_5H_5)_2$ has two electrons in antibonding orbitals, but it is easily oxidized to the 19-electron cation $Ni(C_5H_5)_2^+$. Other interesting organonickel compounds are complex carbonyls based on Ni_5 and Ni_6 clusters; a triple-decker cyclopentadienyl $C_5H_5{-}Ni{-}C_5H_5{-}Ni{-}C_5H_5$ sandwich; and complexes with alkynes that act as catalysts for their polymerization.

For summary, see p. 281–5.

Cu Copper

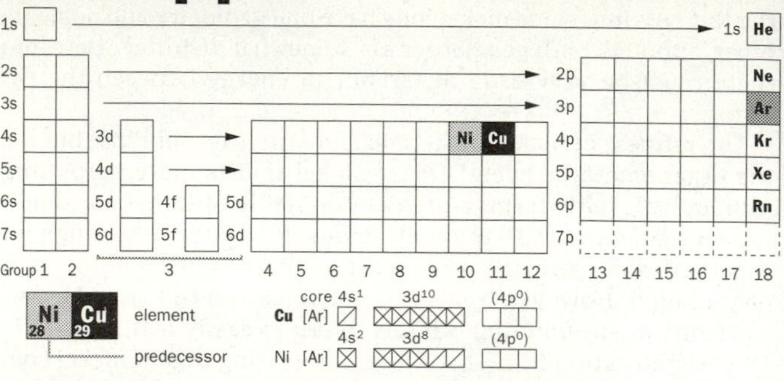

		Predecessor	Element
Name		Nickel	Copper
Symbol, Z		Ni, 28	Cu, 29
RAM		58.69	63.546
Radius/pm	atomic	124.6	127.8
	covalent	115	117
	ionic, M^+	78	96
	M^{2+}	62	72
Electronegativity (Pauling)		1.91	1.90
Melting point/K		1726	1356.6
Boiling point/K		3005	2840
$\Delta H^{\ominus}_{fus}$/kJ mol^{-1}		17.6	13.0
$\Delta H^{\ominus}_{vap}$/kJ mol^{-1}		371.8	304.6
Density/kg m^{-3}		8902 (298 K)	8960 (293 K)
Electrical conductivity/Ω^{-1}m^{-1}		1.462×10^7	5.977×10^7
Ionization energy for removal of jth electron/kJ mol^{-1}	$j = 1$	736.7	745.4
	$j = 2$	1753.0	1958
	$j = 3$	3393	3554
	$j = 4$	5300	5326
	$j = 5$	7280	7709
Electron affinity/kJ mol^{-1}		156	118.5
$E^{\ominus}$/V for	$M^{2+}(aq) \rightarrow M^{+}(aq)$		0.159
	$M^{+}(aq) \rightarrow M(s)$		0.520
	$M^{2+}(aq) \rightarrow M(s)$	−0.257	0.34

Stable isotopes of copper

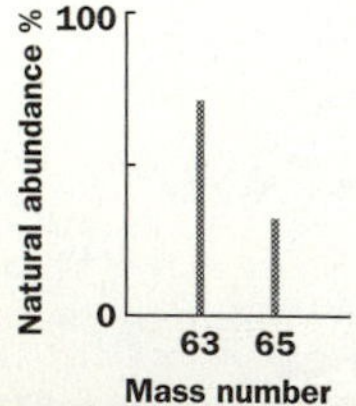

Copper is the first metal that humans extracted from an ore, probably 5500 years ago, by reduction with charcoal. It also occurs free in nature, and had already been in use for over a thousand years. Although the history of copper suggests that the metal is rather unreactive, it forms a wide range of compounds, and its complexes, both of copper(II) and of copper(I), are amongst the strongest known for their oxidation states.

Copper ($[Ar](3d + 4s)^{11}$), with one more electron than nickel and a higher nuclear charge, has a smaller atom and even more tightly bound d electrons. Its configuration, both in the isolated atom and in the metal, is $3d^{10}4s^1$ (rather than $3d^94s^2$). With only one electron in the sp band, copper is a soft metal, but it can form harder alloys with metals that contribute more outer electrons. Bronze (copper and tin) has been made for about five thousand years, and brass (copper and zinc) for over three thousand years. Metallic copper looks orange-pink because the electrons in the filled d band can be promoted to the sp conduction band by absorbing light in the blue-green region of the spectrum. If zinc is added, there are more electrons in the conduction band, the d to sp energy gap is a little wider, and so higher energy, violet-blue light, is absorbed, giving brass its yellow colour. Copper has a high electrical conductivity and is used in electric cables, as well as for plumbing. Its alloys have a range of uses, ranging from coinage (see p. 267) to bronze sculptures and brass ironmongery.*

Copper, with its high nuclear charge and small size, is found in combination with both oxygen and sulfur. Copper is the only member of the 3d series that, in nature, occurs in the (I) state (as Cu_2O or Cu_2S), as well as in the (II) state [as the basic carbonate $Cu_2CO_3(OH)_2$]. It is extracted from its main ore, $CuFeS_2$, by a complicated process that involves roasting in air (to convert the iron and some of the copper to their oxides), removal of the FeO with silica (to form silicate slag), and reaction of the copper(I) oxide with the residual copper(I) sulfide to give the metal and sulfur dioxide. The pure copper, which is often needed as an electrical conductor, may be obtained by electrolysis of aqueous copper sulfate, using the impure metal as anode.

Copper metal is fairly unreactive, partly on account of its high nuclear charge, small size, and consequently high ionization potential. It is stable in pure dry air at room temperatures and does

---

* No wonder scientific nomenclature is confusing when even in everyday language plumbers use *copper* pipes and ironmongers sell *brass* fittings.

Extraction of copper from mixed Fe, Cu, S ore

Heat ore with air and silica

$$2FeS(s) + 3O_2(g) \rightarrow 2SO_2(g) + FeO \xrightarrow{SiO_2} \text{iron silicate slag}$$

$$2Cu_2S(s) + 3O_2(g) \rightarrow 2Cu_2O(s) + 2SO_2(g)$$

$$\xrightarrow{Cu_2S(s)} 6Cu + SO_2(g)$$

impure

Purify electrolytically

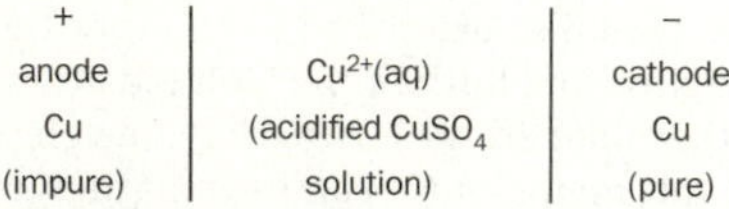

When voltage applied, Cu^{2+} ions removed from anode and pure Cu deposited on cathode.

Action of nitric acid on copper

Various reactions, including

$$Cu(s) + 4HNO_3(aq) \rightarrow 2NO_2(g) + Cu(NO_3)_2(aq) + 2H_2O$$

brown

Jahn–Teller distortion in Cu^{2+}, d^9

(exaggerated)

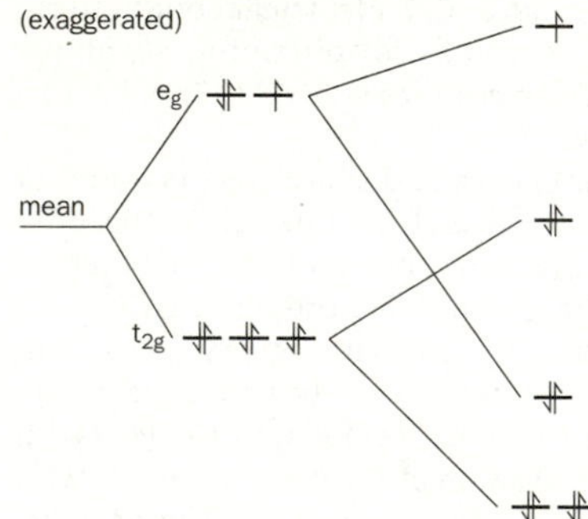

The distorted octahedron is more stable than the regular one, particularly with high-field ligands, since the stabilized former e_g orbital holds two electrons, while the destabilized one hold only one

regular octahedon

octahedon with one axis elongated (or shortened)

not liberate hydrogen from dilute non-oxidizing acids. However, in damp air it forms a green patina of basic salts such as the carbonate (and, near the sea, the chloride) and, if red-hot, it reacts with oxygen to form copper(I) oxide. Copper dissolves in those acids where it can reduce the anion: for example, in nitric acid to form NO_2.

It is no surprise that copper, with its $d^{10}s^1$ configuration, forms a number of stable compounds in the (I) oxidation state; and since there can be no d–d transitions in a d^{10} ion, these are colourless. But the d^{10} shell in copper is not inviolate, particularly since the Cu^{2+} (d^9) ion can benefit from strong complex formation owing not only to its small size and high nuclear charge, but also to favourable stabilization by high field ligands. The high LFSE is enhanced by the Jahn–Teller effect because the uneven occupancy (one pair and one single) of the two upper d orbitals in a d^9 ion produces a stabilizing distortion similar to that shown by d^4 ions such as Cr^{2+} and Mn^{3+} (see pp. 236 and 241).

Copper can also exist in higher oxidation states, but these are rare. Predictably, they can be prepared as the complex fluorides $Cu^{IV}F_6^{2-}$(d^7) and $Cu^{III}F_6^{3-}$ (high spin d^8) and as some low spin square-planar complexes of copper(III), with ligand field stabilization that is even more favourable than for nickel(II) (d^8) (see p. 269), because of the higher charge.

The aqueous chemistry of copper is restricted to copper(II) and Cu^{2+}(aq) is its only stable aquo ion. It is mildly hydrolysed in near-neutral solution, forming the dimer $Cu_2(OH)_2^{2+}$. Addition of alkali gives a precipitate of the hydroxide which is somewhat soluble in excess, probably forming the $Cu(OH)_4^{2-}$ anion. Ammonia also precipitates the hydroxide, which then redissolves to give various deep blue ammine complexes $Cu(NH_3)_x^{2+}$, where the value of x depends on the concentration of free ammonia: it is usually between 1 and 4, but can rise to 5 in saturated aqueous ammonia and even to 6 in pure, liquid ammonia. The familiar blue crystals of copper sulfate pentahydrate contain the planar $Cu(H_2O)_4^{2+}$ ion, with each copper ion bound more distantly to two oxygen atoms from sulfate groups in a typically Jahn–Teller-distorted octahedron (see p. 217). The fifth water molecule is accommodated between the copper and the sulfate ions and can be readily driven off by heat. Copper sulfate is widely used as a fungicide and in electroplating. Copper(II) nitrate is unusual in that the anhydrous salt sublimes to form planar molecules of $Cu(NO_3)_2$ where the copper is chelated by oxygen donors. We have seen that the Cu^{2+} ion forms stronger complexes than other doubly charged metal ions, and, because part of this enhanced stability is a result or ligand field effects, it is not surprising that nitrogen donor ligands,

The $Cu(NO_3)_2$ chelate in the gas phase [GE]

```
    O       O
O–N   Cu     N–O
    O       O
```

The dimeric hydrate of copper(II) ethanoate [GE]

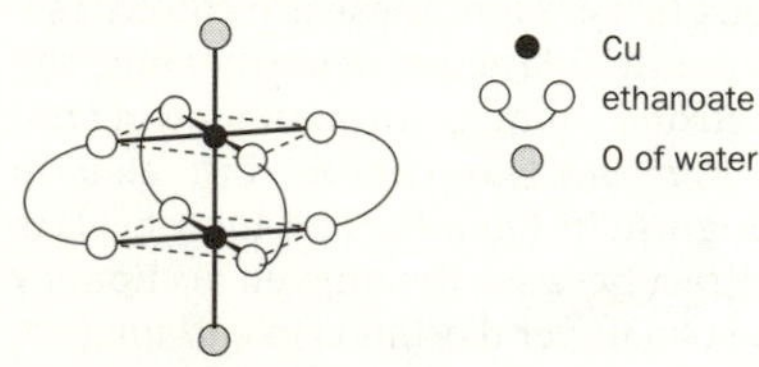

Reaction of Cu^{2+}(aq) with I^-(aq)

$$2Cu^{2+}(aq) + 4I^-(aq) \rightarrow 2CuI + I_2(s)$$

CuI: white precipitate

$I_2(s) \xrightarrow{I^-(aq)} I_3^-(aq)$ brown

This reaction is used in the volumetric determination of Cu^{2+}; the I_3^- ion is titrated with $Na_2S_2O_3$(aq), which it oxidizes:

$$I_3^- + 2S_2O_3^{2-} \rightarrow S_4O_6^{2-} + 3I^-$$

which produce relatively high fields, bind copper(II) even more strongly than do lower field oxygen donors. Copper(II) complexes are almost always blue or green, owing to d–d transitions, and since the structures are usually distorted the colours are often somewhat more intense than those caused by d–d transitions in more symmetrical arrangements such as $Ni(H_2O)_6^{2+}$.

Copper(II), like chromium(II) (see p. 237), forms ethanoate dimers, $Cu_2(OOCCH_3)_4(H_2O)_2$. The four ethanoate groups bridge the two copper ions, forming two $Cu(-O-)_4$ squares, one on top of the other, with a water molecule attached to the outer face of each square. So each copper ion is in a distorted octahedral environment, but, although the electron spins are opposed, it is difficult to say that there is a Cu–Cu bond, rather than antiferromagnetic coupling (see p. 241) between pairs of cations. There is a similar dimer containing four N donor bridging ligands. Copper(II) can also form tetrameric clusters, usually based on a Cu_4 tetrahedron, again with bridging ligands.

With halide ions, copper(II) shows a rich diversity (except with iodide, which it oxidizes to I_2). Many of the CuX_2 molecules, and CuX_3^- and CuX_4^{2-} ions, are based on square-planar copper or on distorted octahedra. But the shape of the $CuCl_4^{2-}$ ion depends on which cation is present; in the green ammonium salt it seems to be a square (or an extremely distorted octahedron), while in the orange salts of larger cations such as Cs^+, it is very nearly tetrahedral. In CuF_2 and K_2CuF_4 the copper is octahedrally surrounded by six fluoride ions, but while the difluoride has the usual distorted structure, with two elongated bonds, the complex ion has two short bonds and four longer ones. The formula of the black sulfide CuS is deceptively simple. It is a mixed valence compound (hence its colour), containing both S^{2-} and S_2^{2-} ions, and is better represented as $Cu^I_2Cu^{II}(S_2)S$.

Copper(I) is unstable in aqueous solution, and disproportionates to Cu^{2+}(aq) and the metal, the change being favoured by the high hydration energy of Cu^{2+} compared with Cu^+. The Cu^+ ion is, however, stabilized by solvents such as CH_3CN, which solvate it strongly; and by groups with which it forms insoluble solids, such as CuI and CuCN, or stable complexes, such as $Cu(NH_3)_2^+$ and $CuCl_2^-$ (which is a complicated spiral chain).

Copper(I), like copper(II), forms a number of polymeric species, but in even richer variety. Tetramers are particularly favoured. The metal atoms are always connected by a bridging group; indeed, we should expect no metal–metal bonding between d^{10} ions.

'1–2–3'

The 'high temperature' superconductor $YBa_2Cu_3O_{7-x}$ S

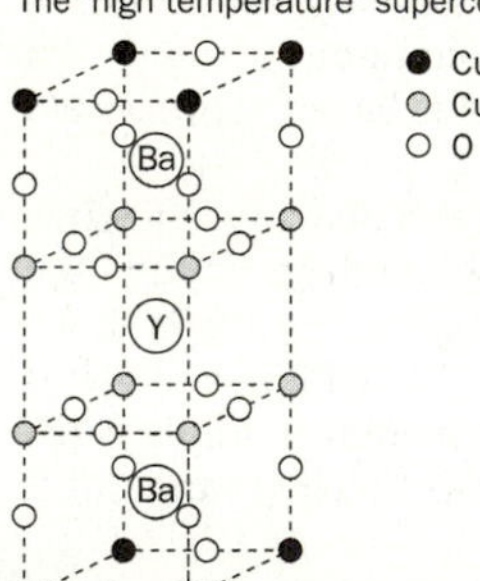

Copper(I) also forms a variety of simple organometallic compounds: the explosively unstable ionic acetylide Cu_2C_2; σ bonded aryls and alkyls, CuR; and, despite the withdrawn nature of its d electrons, complexes containing π bonded ethene and ethyne. The chloride combines with CO to form the 18-electron compound $Cu(CO)_3Cl$, but no other copper carbonyl compound has been made at ordinary temperatures, perhaps because it has fewer vacant orbitals than the previous odd-electron element, cobalt.

We might guess that a readily available element such as copper, with two oxidation states that are accessible (given suitable ligands), differ by one unit, and can tolerate a variety of geometries, would be exploited by biological systems in electron transfer catalysts. Copper is indeed widely used by microorganisms, plants, and animals as a component of many electron transfer enzymes. A dimeric dioxygen-bridged copper protein, haemocyanine, is used as an oxygen carrier in some invertebrates. Like many other functional proteins treated under bioinorganic chemistry, their mode of action, although certainly very complicated, is incompletely understood. The unravelling of these processes, together with a deeper insight into reasons for the varied stereochemistry of copper(II), and for the formation of so many copper dimers and cluster compounds, should stimulate research for many years. Another area of very intense investigation is that of mixed oxide ceramic superconductors which copper forms with many other cations. These function at relatively high temperatures (up to 125 K for $Tl_2Ba_2CaCu_3O_{10}$). The oxide $YBa_2Cu_3O_{7-x}$ (nicknamed '1–2–3') appears to contain copper in oxidation states (I), (II), and (III), but again we know little about how it functions. Even less easy to understand are preliminary reports that a spinning disc of 1–2–3 causes a decrease in the force of gravity (but as yet only by 2%).

Clearly, there is much more to the modern chemistry of this ancient material than we could predict merely from its electronic configuration.

For summary, see p. 281–5.

The first (3d) transition series: trends and summaries Ti-Cu

Here we compare the elements from titanium ($3d^2 4s^2$) to copper ($3d^{10} 4s^1$) to pinpoint and tabulate both similarities and regular trends. Scandium ($3d^1 4s^2$) and zinc ($3d^{10} 4s^2$) are not included as their behaviour is dominated by $3d^0$ or $3d^{10}$ ions without partially filled d shells.

As nuclear charge increases, radii decrease, and the d electrons become less available for bonding, in both the metal and its compounds. Since the atoms can lose a varying number of d electrons as well as the 4s ones, the elements show a number of oxidation states. Indeed, elements up to and including manganese (and perhaps even iron) may lose *all* their d electrons. When some d electrons are retained, they often (but not always) occupy separate orbitals if possible; and unpaired electrons produce interesting magnetic effects. Except in isolated atoms, the d orbitals are not all of the same energy; and, as electrons may be promoted from one orbital to another by absorbing visible light, transition metal compounds are often coloured. Compounds with no d electrons, such as $KMnO_4$ may also be coloured, much more intensely, by transfer of electrons from the electron-rich non-metal to the highly oxidized metal atom. We also find deep charge transfer colours when an electron-rich metal ion is bound by a π acceptor ligand, as in $Fe^{II}(bipy)_3^{2+}$, or when a substance such as Fe_3O_4 contains the same metal in different oxidation states.

Transition metal ions form a wide variety of complexes: elements early in the series favour oxygen donors, while later ones bind more strongly to nitrogen or sulfur atoms. The redox properties, and sometimes also the magnetic behaviour, can be changed grossly by complex formation. The metals often form non-stoichiometric compounds with small atoms of non-metals, such as boron, carbon, and nitrogen, which interact with electrons in the conduction band; and similar bonding may account for the catalytic properties of many transition metals. Metal atoms with some full and some empty 3d orbitals often react with carbon monoxide by combination of σ acceptance and π donation; simple metal carbonyls, $M(CO)_n$, are most likely to survive if they contain 18 outer electrons, and so are formed mainly by metals with an even number of d electrons. Interesting compounds, often 'sandwiches', are formed with other π donors such as cyclopentadienyl and benzene. And, of course, like all other elements, each transition metal has its individual peculiarities.

Comparative summaries of elements

Element	Ti	V	Cr	Mn
Outer configuration	$3d^2 4s^2$	$3d^3 4s^2$	$3d^5 4s^1$	$3d^5 4s^2$
Found combined with (and in association with)	O	O,S	O (and Fe)	O
Extraction of metal	$TiO_2 \rightarrow TiX_4$ $TiCl_4 \xrightarrow{Mg} Ti$ $TiI_4 \xrightarrow{heat} Ti$	$V_2O_5 \rightarrow VCl_5 \xrightarrow{Mg} V$	Separate from Fe as $Cr^{VI} \xrightarrow{C} Cr_2O_3$ $\xrightarrow{Al\ or\ Si}$ Cr	$MnO \xrightarrow{Al} Mn$
Freedom of d electrons	Little control by nucleus	Little control by nucleus	Some retreat into core	Increased retreat into core
Main oxidation states, redox stability, e.g.	**IV** stable TiO_2 (white) TiX_4 (no Ti^{4+}) **III** reducing Ti^{3+} (aq) (violet), TiX_3	**V** mildly oxidizing VO_2^+ (orange), VO_3^- (yellow) **IV** stable VO^{2+} (blue), VO_2 **III** reducing V^{3+} (green) **II** strongly reducing	**VI** oxidizing $Cr_2O_7^{2-}$ (orange) CrO_4^{2-} (yellow) **III** stable Cr^{3+} (aq) (green) + complexes **II** reducing Cr^{2+} (aq) (blue)	**VII** strongly oxidizing MnO_4^- (purple) **IV** oxidizing MnO_2 (blac **II** stable Mn^{2+} (aq) (v. pale pink
[Less common oxidation states]			[IV, V]	[VI, V and III: often dispropo tionate]
Physical properties of metal	High strength, low density	High melting	Shiny	Disordered structure with low ΔH of atomization Liberates hydrogen from water
Reaction with damp air	Corrosion-resistant oxide film (diffraction colours)	Corrosion-resistant oxide coating	Corrosion-resistant	Oxidized: burns if finely divided
Uses of metal	Aircraft manufacture (and jewellery)	Steel hardening	Plating and stainless steel	Steel hardening and for non-ferrous alloys

Ti to Cu: $[Ar]3d^y4s^2$ or $[Ar]3d^{y+1}4s^1$

Fe	Co	Ni	Cu
$3d^64s^2$	$3d^74s^2$	$3d^84s^2$	$3d^{10}4s^1$
O	S, As (and Ni, Cu, Pb)	O, S, As (and Co)	O, S (and native)
Oxides + C → Fe	Separate as Co $(OH)_3$ C → Co	NiO + C → Ni purify via $Ni(CO)_4$	$Cu_2O + Cu_2S \rightarrow Cu + SO_2$ purified by electrolysis
Appreciably withdrawn into core	Tightly withdrawn into core	Largely withdrawn into core	Almost totally withdrawn
III v. mildly oxidizing Fe^{3+} (aq) (lilac, in crystals) Fe^{3+} hydrolysed in solution (yellow) **II** v. mildly reducing Fe^{2+} (aq) (green)	**III** Co^{3+} (aq) (blue) oxidizing Co^{III} complexes, stable **II** Co^{2+} (aq) (pink) stable Co^{II} complexes, reducing	**II** stable Ni^{2+} (aq) (green)	**II** (usually) stable Cu^{2+} (aq) (turquoise) CuO (black) **I** [Cu^+ (aq) disproportionates] Stable solids CuI (white) Cu_2O (red)
[VI, in $BaFeO_4$] [VIII in FeO_4 ?]	[IV, with F^-, I, with π acceptors]	[IV, with F^-; III]	[IV, III]
Brittle if pure, but hardened by carbon (→ steel) and by other metals, ferromagnetic	Similar to iron but less reactive and less ferromagnetic	Softer than iron, ferromagnetic (less so than Co)	Soft, pink metal, high electrical conductivity
RUSTS	Stable	Stable	Forms basic carbonate, chlorides, etc. (green patina)
On vast scale for construction and for magnets	High temperature alloys (e.g. in turbines), magnets, catalysts	Hard steels, magnets, non-ferrous alloys (coins and scientific instruments), catalysis, and spatulas	Electric cables, water pipes, alloys (brass, bronze, coinage alloys)

Element	Ti	V	Cr	Mn
Main complexes with CO, C_5H_5, and other organic ligands	No carbonyls Ring-whizzing $Ti(C_5H_5)_4$	$V(CO)_6$ oxidizing $\rightarrow V(CO)_6^-$ $V(C_5H_5)_2$ unstable	$Cr(CO)_6$ $Cr(C_6H_6)_2$ sandwich	$Mn_2(CO)_{10}$ $Mn(CO)_5^-$ $Mn(CO)_6^+$ $Mn(C_5H_5)_2$ very sensitive to air
Points of special chemical interest	• Ti^{3+} (aq) simplest, ion with d–d transition • TiO metallic conductor • Peroxo complexes	• VO metallic conductor • Peroxo complexes	• Highest fluorides: CrF_6, CrF_5 • CrO_2, ferromagnetic • Cr^{III} (d^3), complexes often inert; many isomers • Cr^{II}(d^4), usually high spin (Jahn–Teller distortion) • $Cr(CN)_6^{4-}$ low spin • Cr^{II} carboxylates: dimers with short Cr–Cr • Peroxo complexes	• Highest fluoride: MnF_4 • MnS_2 is $Mn^{2+}S_2^{2-}$ (pyrites structure) • MnO, Mn_3O_4, and MnO_2, antiferro-magnetic* • Mn_3O_4 normal spinel
Uses of compounds	• TiO_2 white pigment and filler • $BaTiO_3$ piezoelectric • $Ti(OR)_4$ in non-drip paints • TiX_4 catalysts in ethylene polymerization		• Chromate pigments and primers • $K_2Cr_2O_7$ oxidizing agent	• $KMnO_4$ widely used as oxidizing agent
Biological functions		Biological function(s) in early forms of life		Component of plant enzymes
Nuclear features				

* At low temperatures.

Fe	Co	Ni	Cu
$Fe(CO)_5$ and polymeric carbonyls	$Co_2(CO)_8$, $HCo(CO)_4$,	$Ni(CO)_4$ stable	$Cu^1(CO)_3Cl$
$Fe(C_5H_5)_2$ fairly stable	$Co(C_5H_5)_2$ easily oxidized to $Co(C_5H_5)_2^+$	$Ni(C_5H_5)_2$ easily oxidized	Explosive acetylide: $(Cu^+)_2C_2^{2-}$
• $Fe^{III/II}$ redox v. sensitive to ligands • Fe^{III} and Fe^{II} spin states v. sensitive to ligands • Many mixed valence compounds, e.g. Fe_3O_4 (ferromagnetic), Prussian blue • FeS_2 is $Fe^{2+}S_2^{2-}$ (pyrites) • FeO antiferromagnetic* • Fe_3O_4 inverse spinel	• $Co^{III/II}$ redox v. sensitive to ligands • $Co^{III}d^6$ usually low spin and inert: many isomers • $Co^{II}d^7$ geometry sensitive to ligands (oct. or tetr.) • CoO antiferro-magnetic* • Co_3O_4 normal spinel	• Ni^{II} d^8 geometry sensitive to ligands (oct., tetr., or sq.pl.) • NiO antiferro-magnetic*	• Cu^{II} complexes strongest of 3d series. Strong preference for N donors • Cu^{II} carboxylates dimers • $Cu(NO_3)_2$ sublimes • Cu^{II} d^9 Jahn–Teller distorted with varying geometry
• Fe_2O_3 as pigments and fine abrasive	• Blue pigment in ceramics Co^{II} • Tetr. → oct. change blue → pink moisture detector • ^{60}Co as γ-ray source		• Mixed (e.g. with Ba and Y) as 'high' temperature super_ conductors
In oxygen carriers (haemoglobin) and many biological redox systems	In Vitamin B_{12}		In wide variety of proteins
^{56}Fe very stable nucleus	^{60}Co source of γ-rays		

* At low temperatures.

Zn Zinc

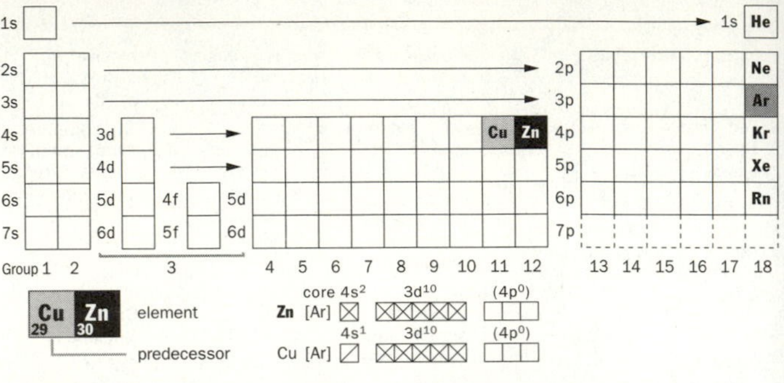

		Predecessor	Element
Name		Copper	Zinc
Symbol, Z		Cu, 29	Zn, 30
RAM		63.546	65.39
Radius/pm	atomic	127.8	133.2
	covalent	117	125
	ionic, M^{2+}	72	83
Electronegativity (Pauling)		1.90	1.65
Melting point/K		1356.6	692.73
Boiling point/K		2840	1180
$\Delta H^{\ominus}_{fus}$/kJ mol^{-1}		13.0	6.67
$\Delta H^{\ominus}_{vap}$/kJ mol^{-1}		304.6	115.3
Density (at 293 K)/kg m^{-3}		8960	7133
Electrical conductivity/Ω^{-1}m^{-1}		5.977×10^7	1.690×10^7
Ionization energy for removal of jth electron/kJ mol^{-1}	$j = 1$	745.4	906.4
	$j = 2$	1958	1733.3
Electron affinity/kJ mol^{-1}		118.5	9
$E^{\ominus}$/V for $M^{2+}(aq) \rightarrow M(s)$		0.159	−0.763

Stable isotopes of zinc

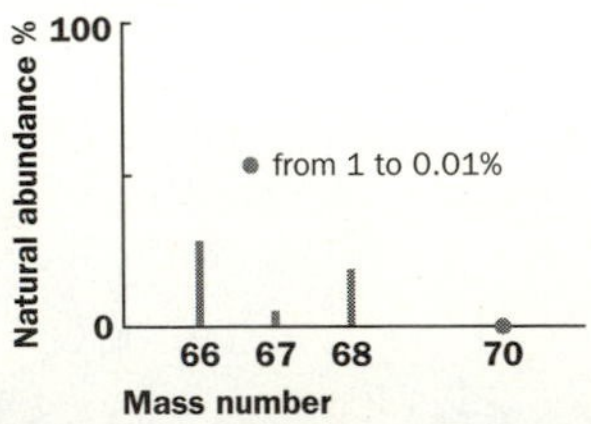

At zinc, [Ar]$3d^{10}4s^2$, the 3d shell is full and, as part of the inner core of electrons, is unbreachable by chemical reagents, so the reactions of zinc involve only the $n = 4$ shell, which supplies two electrons, together with orbitals derived from its 4s, 4p, and perhaps also 4d atomic orbitals. The 3d electrons are too withdrawn to take part in metallic bonding and so metallic zinc is soft, not particularly dense, low melting, and relatively volatile. Almost all zinc compounds exhibit the (II) oxidation state. No higher states have been encountered, nor has zinc any tendency to donate 3d electrons to π acceptors such as carbon monoxide, ethene, or ethyne. The full d shell removes the possibility of d–d transition and most zinc compounds are white. With no ligand field stabilization, species containing four-coordinated zinc are usually tetrahedral, since this arrangement minimizes steric repulsion. However, zinc also forms six-coordinate octahedral groups with some small ligands such as water and ammonia.

The Zn^{2+} ion, with its small size and high nuclear charge, is much more polarizing than Ca^{2+}, but not so much so as to favour large, deformable donor atoms over smaller, less polarizable ones. While the Ca^{2+} ion is usually found in the company of fluorine or oxygen, the Zn^{2+} ion shows little preference between oxygen and sulfur, or between fluorine and chlorine; and although the main zinc ore is the sulfide, the element is also found combined with oxygen atoms in carbonates, silicates, and phosphates.

The metal may be obtained by roasting the sulfide in air and reducing the oxide with hot coke in an enclosed inert atmosphere (to prevent the escape of the zinc vapour, or its reoxidation). Alternatively, the oxide may be converted to the aqueous sulfate, and the zinc obtained by electrolysis. Although zinc tarnishes to the oxide in damp air, the metal is widely used as the protective antirust coating on 'galvanized' iron, and also as a component of brass and diecast alloys, and as the negative pole of many dry batteries.

Zinc oxide also has a number of important uses: to speed the vulcanization of rubber, as a pigment, as a soothing agent for wounds, and in the manufacture of 'custom built' magnetic oxides containing iron. (As the zinc ousts iron (III) from the tetrahedral sites, it changes the magnetic properties of the material.) Zinc oxide turns yellow if heated, because it loses a small amount of oxygen, to give the 'defect' solid $Zn_{1+x}O$, which has a slight excess of electrons. These absorb energy in the UV and visible region, producing the colour, which fades as the oxide reabsorbs oxygen on cooling.

Zinc sulfide can exist in two forms, both consisting of two interpenetrating close-packed lattices of zinc ions and sulfide ions. In

The two forms of zinc sulfide [GE]

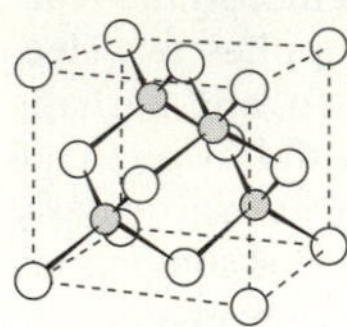

(a) zinc blende

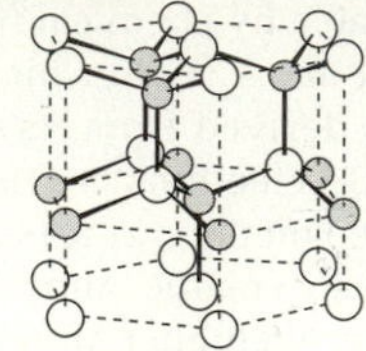

(b) wurtzite

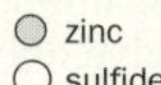

zinc blende the lattices are cubic and in wurtzite they are hexagonal, but in both structures each ion is surrounded tetrahedrally by the other element. The optical properties of zinc sulfide, unlike those of the oxide, are exploited commercially. The sulfide absorbs energy from electron beams, or even from high energy UV radiation, uses some of it to increase its lattice vibration, and re-emits the rest as visible light. Rutherford used a zinc sulfide screen to track α-particles in his famous experiments that led to the discovery of the atomic nucleus; and phosphors of this type, many based on zinc sulfide, together with traces of other metal ions, are used today in TV tubes and fluorescent lighting.

Hot zinc also combines with phosphorus and the halogens, but not with hydrogen, carbon, or nitrogen. The fluoride, ZnF_2, is ionic, but the other halides have some covalent characteristics. Zinc dissolves in dilute hydrochloric acid to give hydrogen, and in nitric acid to give oxides of nitrogen. The water molecules can be displaced from the aquo ion, $Zn(H_2O)_6^{2+}$, by a variety of ligands, often with nitrogen or oxygen donor atoms, and often to give tetrahedral species, with the orbitals derived from the $4sp^3$ levels completely filled. Because of the small size and high nuclear charge of the Zn^{2+}ion, its complexing ability would be expected to be stronger than that of any other doubly charged $3d^{0-10}$ ion, were it not for ligand field stabilization of the many ions with only partially filled d orbitals.

When aqueous alkali is added to a solution of a zinc salt, the hydroxide, $Zn(OH)_2$, is precipitated but, like $Al(OH)_3$, it redissolves readily in excess alkali; the end-product is the zincate ion, written as $Zn(OH)_4^{2-}$ or ZnO_2^{2-}(aq). Many zinc salts such as the hydrated sulfate $ZnSO_4{\cdot}7H_2O$ are isomorphous with the corresponding magnesium compound; and as Zn^{2+} has the same charge and almost the same radius as Mg^{2+}, it is no surprise that the two salts have the same structure and that the cations are interchangeable. Zinc forms a basic ethanoate which is isomorphous with the beryllium derivative, but the zinc compound, with its larger cation, hydrolyses much more rapidly, doubtless because zinc can add another two ligands (compare with silicon tetrachloride, p. 151).

Zinc forms a strong cyanide complex, $Zn(CN)_4^{2-}$, which must owe its stability to σ overlap; the 3d electrons are too tightly held for back-donation to π acceptor ligainds, and, predictably, zinc forms no carbonyls. There are, however, some four-electron (sp) σ bonded dialkyls and diaryls, ZnR_2 and also haloalkyl compounds, RZnX, which were widely used in organic syntheses before Grignard made the analogous, but more effective, magnesium compounds, RMgX

Methyl (C_5H_5) zinc SAL

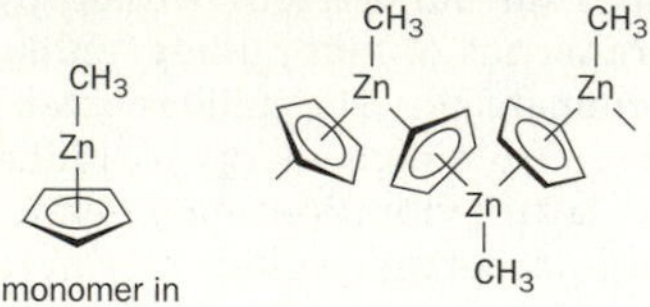

CH_3
Zn

monomer in gas phase

zigzag chain in solid

Summary

Electronic configuration

- ([Ar]$3d^{10}$)$4s^2$

Element

- Non-transition, fairly soft, volatile and low-melting, tarnishes in damp air

Occurrence

- As sulfide, and with oxygen (with carbonate, silicate, phosphate)

Extraction

- Roast sulfide ore to oxide and reduce electrolytically or with coke in inert atmosphere

Chemical behaviour

- d electrons inactive
  - no oxidation state above (II),
  - no π complexes
- Fairly electropositive
  - gives H_2, will dilute HCl
  - salts similar to those of Mg^{2+},
  - basic ethanoate (like Be)
- Small polarizing cation
  - fairly strong complexes, often tetrahedral

(see p. 135). The eight-electron (sp^3) σ bonded methyl (cyclopentadienyl) zinc seems unsurprising as the gaseous monomer; but in the solid it polymerizes to a zigzag chain somewhat reminiscent of the triple-decker sandwich formed by nickel (see p. 271).

Despite this last bizarre aberration, it might seem that zinc compounds are not only colourless but also dull. However, we could not survive without them, as there are several zinc enzymes in almost every cell of our body, and the element accounts for about two grams of our mass. These enzymes, unlike others incorporating metals of the 3d series, cannot be directly involved in electron transfer, since in aqueous solution zinc is restricted to the (II) state. We know that one zinc enzyme is involved in breaking peptide links during digestion, and another in the transport of carbon dioxide (as the hydrogen carbonate ion HCO_3^-) from the muscles to the lungs. However, as with so many biochemical processes, the enzyme mechanisms are not yet known. Zinc is also deeply involved with proteins that bind to DNA and so help to control the expression of coded information.

  - Zn^{2+}(aq.) octahedral
  - $Zn(OH)_2$ amphoteric
  - viable sulfides (two forms ZnS)
- With oxygen gives defect oxide, $Zn_{1+x}O$
- With halogens gives ionic ZnF_2; other halides partly covalent
- No hydride, carbide, or nitride
- σ bonded ZnR_2 and RZnX

Uses
- Metal
  - for 'galvanizing' iron with rust-resistant coating
  - in alloys (e.g. brass)
  - in dry batteries
- Oxide
  - in vulcanization of rubber
  - as pigment
  - to replace Fe in magnetic materials
  - to sooth wounds
- Enzymes
  - many (non-redox)

Nuclear features
- Various stable isotopes

Ga Gallium

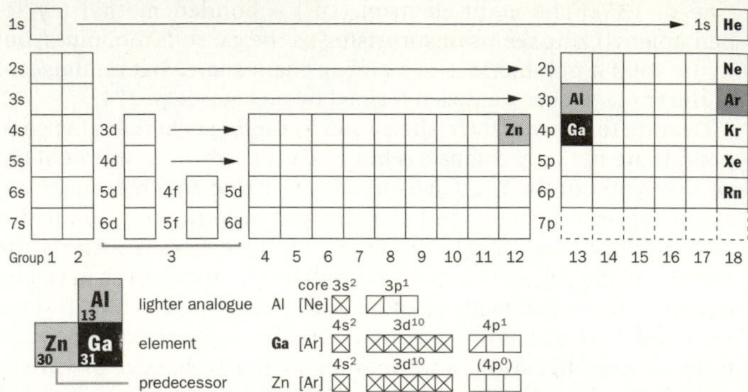

		Predecessor	Element	Lighter analogue
Name		**Zinc**	**Gallium**	**Aluminium**
Symbol, *Z*		Zn, 30	Ga, 31	Al, 13
RAM		65.39	69.723	26.98154
Radius/pm	**atomic**	133.2	122.1	143.1
	covalent	125	125	125
	ionic, M^+		113	
	M^{3+}		62	57
Electronegativity (Pauling)		1.65	1.81	1.61
Melting point/K		692.73	302.93	933.52
Boiling point/K		1180	2676	2740
ΔH°_{fus}/kJ mol^{-1}		6.67	5.59	10.67
ΔH°_{vap}/kJ mol^{-1}		115.3	256.1	293.72
Density (at 293 K)/kg m^{-3}		7133	5907	2698
Electrical conductivity/Ω^{-1}m^{-1}		1.690×10^7	3.704×10^6	37.67×10^6
Ionization energy for removal of *j*th electron/kJ mol^{-1}	***j* = 1**	906.4	578.8	577.4
	***j* = 2**	1733.3	1979	1816.6
	***j* = 3**		2963	2744.6
Electron affinity/kJ mol^{-1}		9	~ 30	44
E°/V for $M^{3+}(aq) \rightarrow M(s)$			–0.53	

Stable isotopes of gallium

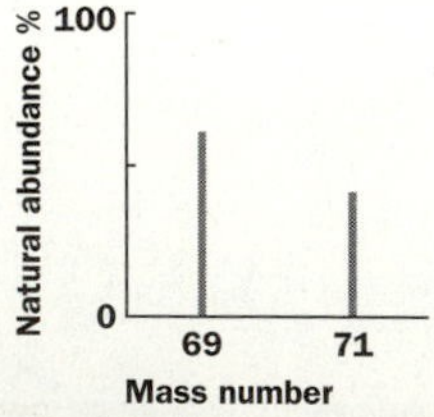

Many of the properties of gallium are predictable from its electronic configuration, ([Ar]$3d^{10}$)$4s^24p^1$, and from the known properties of other elements such as aluminium, [Ne]$3s^23p^1$, and zinc, ([Ar]$3d^{10}$)$4s^2$. Indeed, in 1871 Mendeleev predicted that a new element ('eka-aluminium') with certain specified properties would be discovered, probably spectroscopically; and four years later he was proved right (see p. 296).

Gallium, like aluminium, is a metal. It can lose its three outer electrons to the valence band in the solid and it liberates hydrogen from dilute acids to give the $Ga(H_2O)_6^{3+}$ ion. Again like aluminium, it dissolves in aqueous alkalis to give an anion, GaO_2^-(aq). But unlike aluminium, its 'core' contains the $3d^{10}$ electrons, which are poor shielders. So, despite the larger primary quantum number, the outer $4s^24p^1$ electrons of gallium are under much stronger nuclear control than are the $3s^23p^1$ valence electrons of aluminium and this has a marked effect on its properties. The ionization energies of gallium are higher than those of aluminium, and the atom is smaller, despite its much higher mass. Gallium has a lower electrical conductivity than aluminium, and in its reactions, too, it is 'less metallic'; it has a lower (theoretical) tendency to liberate hydrogen from acids, and a greater tendency for the oxide M_2O_3 to dissolve in alkalis. In the solid metal each gallium atom has one neighbour much closer than the others, in a structure approaching that of iodine with its diatomic molecules. Gallium has a very low melting point and, like ice, contracts on melting. As it has the longest liquid range of any known substance, it is used in thermometers.

Gallium, unlike aluminium, occurs in small amounts in sulfide ores, but also combined with oxygen, usually as a very minor impurity in aluminium ores. It is obtained from bauxite (see p. 139) and, as it is less electropositive than aluminium, it can be separated from it during the electrolytic extraction of aluminium.

Gallium hydride, GaH_3, unlike the solid AlH_3, is a liquid which decomposes at room temperature. It acts as a Lewis acid and combines with donors such as $N(CH_3)_3$ to give complexes in which the gallium is usually tetrahedrally coordinated.

Some halides are reminiscent of those of aluminium: the high melting 'ionic' trifluoride contains six-coordinate gallium, but the trichloride is a four-coordinate dimer, Ga_2Cl_6, even (unlike aluminium) in the solid, which melts at 77.8°C. Like aluminium, gallium forms (unimportant) monohalides; but it also has two interesting, formally gallium (II), chloro species which have no aluminium analogues. The stable 'dichloride' (which is diamagnetic and so cannot contain the odd-electron ion Ga^{2+}([Ar]$4s^1$)) is actually

Some band gaps *E* in diamond–like solids [SAL]

Solid	E/kJ mol^{-1} at 25°C
Diamond	527.8
SiC	289.4
Si	108.1
Ge	63.7
GaAs	137.0
InAs	34.7

GaAs as a semi-conductor [SM]

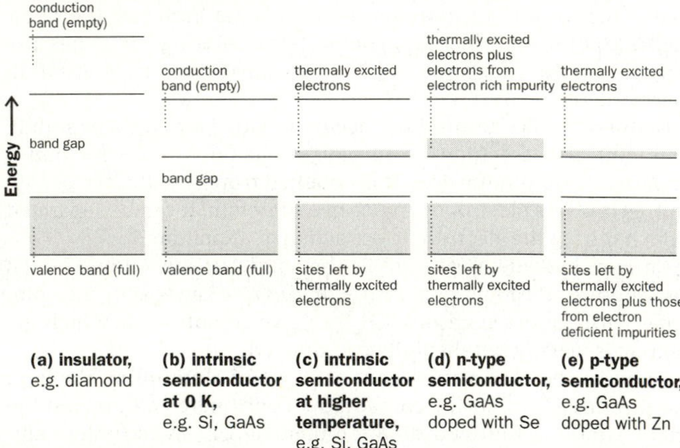

(a) insulator, e.g. diamond

(b) intrinsic semiconductor at 0 K, e.g. Si, GaAs

(c) intrinsic semiconductor at higher temperature, e.g. Si, GaAs

(d) n-type semiconductor, e.g. GaAs doped with Se

(e) p-type semiconductor, e.g. GaAs doped with Zn

a mixed valence compound with the structure $Ga^+[Ga^{III}Cl_4^-]$, in which one of the gallium ions, $Ga^+[Ar]3d^{10}4s^2$), has kept its $4s^2$ pair of electrons. The ion $Ga_2Cl_6^{2-}$ is $[Cl_3Ga^{II}-Ga^{II}Cl_3]^{2-}$ in which the gallium has a formal oxidation state of (II) by virtue of its metal–metal bond.

Gallium forms a variety of compounds, such as Ga_2S_3, with sulfur (and also with selenium and tellurium); but much more important are the so-called III–V compounds formed by gallium with the group 15 elements nitrogen, phosphorus, arsenic, and antimony. Since gallium has three outer electrons and the other element ([core]ns^2np^3) has five, these 1:1 compounds have the same structure as diamond or silicon. However, the energy gap between the top of the full valence band and the bottom of the empty conduction band is much smaller than in diamond. Except at absolute zero, some electrons have enough thermal energy to enter the conduction band, and so leave a hole in the valence band. So the electrons have some mobility and the compounds are semiconductors.

Conductivity can be increased by 'doping' with very small amounts of other elements. If, say, some arsenic atoms in gallium arsenide are replaced by selenium, which has one more electron, there will be a higher concentration of *n*egative charge; this enables the material to carry more current and makes it an '*n*-type' semiconductor. If, on the other hand, some gallium atoms are replaced by zinc, with one less electron, there will be a deficiency of negative charge and current can be carried by movement of electrons into the *p*ositive sites that are formed, making the material into a '*p*-type' semiconductor. Junctions between *p*-type, *n*-type, and undoped semiconductors such as gallium arsenide are used in a number of electronic devices. Electrons can be promoted across the band gap by visible light, and this increase in conductivity is the basis of a photoelectric cell. Alternatively, electrons may be excited by an applied voltage, and then fall back to the valence band, emitting the visible light familiar from the light emitting diodes (or LEDs) of video clocks and cash registers. Or excited electrons may be stimulated to lose their energy in a series of coherent pulses, to give the infrared laser used to read computer disks.

Gallium seems to play no part in living processes and as yet its organometallic chemistry has not been extensively explored. But not even Mendeleev could have predicted its significance for the electronics industry.

For summary, see p. 297.

Mendeleev's predictions for 'eka–aluminium'[GE]

Mendeleev's predictions (1871) for eka–aluminium, M	Observed properties (1977) of gallium (discovered 1875)
Atomic weight ~ 68	Atomic weight 69.72
Density/g cm^{-3} 5.9	Density/g cm^{-3} 5.904
M.P. low	M.P./°C 29.78
Non-volatile	Vapour pressure 10^{-3} mmHg at 100°C
Valence 3	Valence 3
M will probably be discovered by spectroscopic analysis	Ga was discovered by means of the spectroscope
M will have an oxide of formula M_2O_3, *d* 5.5 g cm^{-3}, soluble in acids to give MX_3	Ga has an oxide Ga_2O_3, *d* 5.88 g cm^{-1}, soluble in acids to give salts of the type GaX_3
M should dissolve slowly in acids and alkalis and be stable in air	Ga metal dissolves slowly in acids and alkalis and is stable in air
$M(OH)_3$ should dissolve in both acids and alkalis	$Ga(OH)_3$ dissolves in both acids and alkalis
M salts will tend to form basic salts; the sulfate should form alums; M_2S_3 should be precipitated by H_2S or $(NH_4)_2S$; anhydrous MCl_3 should be more volatile than $ZnCl_2$	Ga salts readily hydrolyse and form basic salts; alums are known; Ga_2S_3 can be precipitated under special conditions by H_2S or $(NH_4)_2S$; anhydrous $GaCl_3$ is more volatile than $ZnCl_2$

Summary

Electronic configuration

- ([Ar]$3d^{10}$)$4s^2 4p^1$

Element

- Low-melting metal, nearly dimeric in solid; long liquid range; ionization harder than for Al

Occurrence

- In Al ores (with oxygen); and in sulfide ores

Extraction

- By-product of electrolytic extraction of aluminium

Chemical behaviour

- Similar to aluminium, but electrons more tightly held
- Metal liberates H_2 from dilute acids and dissolves in aqueous alkali; Ga_2O_3 amphoteric
- (III) state dominant [with (I) and (III) in complex chlorides]
- GaF_3ionic, Ga_2Cl_6 low-melting solid, GaH_3 unstable liquid (Lewis acid)
- GaAs (and other group 15 analogues) semiconductors

Uses

- Metal in thermometers
- GaAs, etc. in electronic devices (photocells, LEDs, lasers)

Nuclear features

- ^{69}Ga:$^{70}Ga \sim 3:2$

Ge Germanium

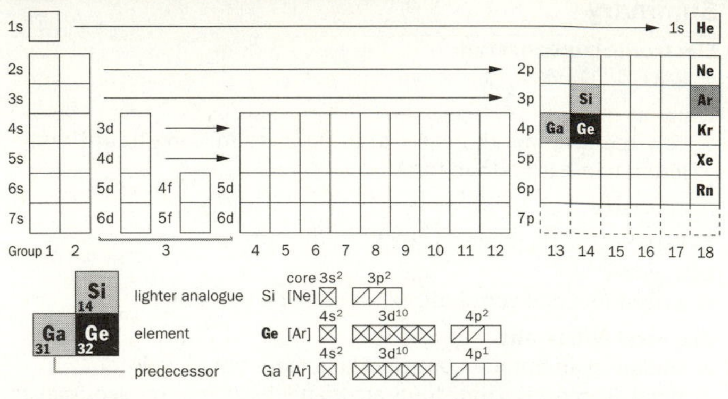

		Predecessor	Element	Lighter analogue
Name		**Gallium**	**Germanium**	**Silicon**
Symbol, *Z*		Ga, 31	Ge, 32	Si, 14
RAM		69.723	72.61	28.0855
Radius/pm	**atomic**	122.1	122.5	117
	covalent	125	122	117
	ionic, M^{2+}		90	
Electronegativity (Pauling)		1.81	2.01	1.90
Melting point/K		302.93	1210.6	1683
Boiling point/K		2676	3103	2628
$\Delta H^{\ominus}_{fus}$/kJ mol^{-1}		5.59	34.7	39.6
$\Delta H^{\ominus}_{vap}$/kJ mol^{-1}		256.1	334.3	383.3
Density (at 293 K)/kg m^{-3}		5907	5323	2329
Electrical conductivity/Ω^{-1}m^{-1}		3.704×10^6	2.174	
Energy band of gap in solid/kJ mol^{-1}		0	64.2	106.8
Ionization energy for removal of *j*th electron/kJ mol^{-1}	***j* = 1**	578.8	762.1	786.5
	***j* = 2**	1979	1537	1577.1
	***j* = 3**	2963	3302	3231.4
	***j* = 4**		4410	4355.5
Electron affinity/kJ mol^{-1}		~ 30	116	133.6
Bond energy/kJ mol^{-1}	**E–E**		163	226
	E–H		288	326
	E–O		363	452
	E–F		340	582
	E–Cl		163	391

Germanium is widely distributed in the earth's crust, but in such low concentration that it was discovered only about a century ago. From its outer electronic structure of ([Ar]$3d^{10}$)$4s^2 4p^2$ we might expect it to be a more metallic version of silicon [Ne]$3s^2 3p^2$ were it not for the fact that gallium seems *less* metallic than aluminium. Germanium does indeed resemble silicon in many ways: with as many valence electrons as low lying vacant orbitals, it too has the diamond structure, but it is softer than silicon and has a much smaller band gap. A semiconductor, it was the basis of early transistor technology; but nowadays its main use is for making infrared optical instruments, as it is transparent to radiation in this range. The energy needed to remove the outer s and p electrons is slightly greater for germanium than for silicon (as it was for gallium compared with aluminium); but germanium loses its p^2 electrons more easily and its s^2 electrons less easily than does silicon. Germanium is larger than silicon (as we should expect, were it not for the unusual shrinkage from aluminium to gallium).

Germanium is more reactive than silicon, and is attacked by both concentrated acids and fused alkalis. The red-hot element reacts with air, with sulfur or hydrogen sulfide, and with the halogens to give binary compounds in the (IV) state, and with hydrogen chloride to give a mixture of $GeCl_4$ and $HGeCl_3$. The bond strengths to germanium are slightly lower than those in similar silicon compounds, as expected from its larger radius. The tetrahalides, apart from the orange iodide (m.p. 146°C), are liquids and the volatility of $GeCl_4$ is exploited in the extraction of germanium from zinc flue dust; the two elements cannot be separated via their oxides or oxoanions since both GeO_2 and ZnO are amphoteric. The recondensed germanium tetrachloride is hydrolysed to the dioxide, which is then reduced to germanium with hydrogen.

Similarities to silicon include GeO_2, which can have many of the different structures adopted by silicon dioxide, and the germinates, which resemble the silicates both in structural variety and in chemical behaviour. Germanium, like silicon, forms hydrides of formula Ge_nH_{2n+2}; they are gaseous up to $n = 5$ and liquid for $6 \leqslant n \leqslant 9$.

Although germanium is only slightly larger than silicon, it is too large for any π overlap, and no multiply bonded germanium compounds are known. Moreover, since larger atoms usually form weaker single bonds because of less effective σ overlap, it may be less advantageous for such an atom to be in its highest oxidation state: the energy needed to remove or promote the valence electrons may be insufficiently repaid by the strengths of the bonds formed. So, germanium has a much greater tendency than silicon to form compounds using only its p^2 electrons, just as the M(I) state is

Stable isotopes of germanium

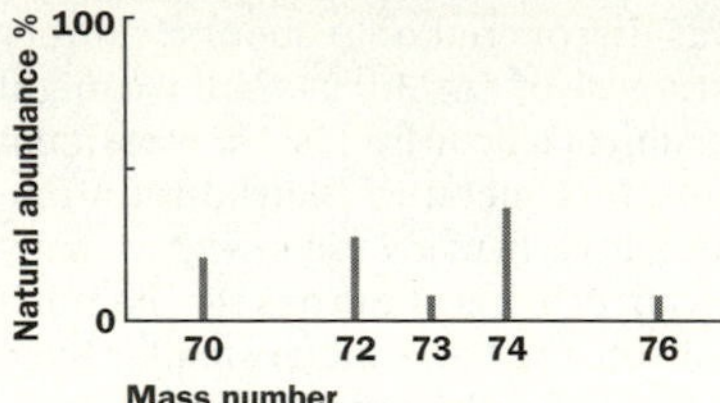

better defined for gallium (see p. 293) than for aluminium. The monoxide, GeO, and the dihalides, GeX_2, are well established, but they are much less important than the germanium(IV) compounds, and the dihalides react readily with the halogens to form the tetrahalides. Germanium, unlike silicon, can form salts with (a few) oxoanions. These are unstable and unimportant, except to show glimmerings of incipient metallic behaviour. Another, more diverting, way in which germanium differs from silicon is its ability to form cluster anions (often of nine atoms), which can be obtained by adding an uncharged chelating agent, together with sodium or potassium ions; $Na_4(en)_5Ge_9$, for example, can be made with ethylenediramine (en) in liquid ammonia. A recent cryptate* species $[K(crypt)^+]_6.Ge_9^{4-}.Ge_9^{2-}$, contains two such anions with different geometries according to their number of valence electrons, which of course depends on their charges. So germanium is not merely the predictable heavy analogue of silicon that it might have seemed.

Summary

Electronic configuration

- ([Ar]$3d^{10}$)$4s^2 4p^2$

Element

- Semiconductor, with diamond structure; ionization of s electrons harder than for Si (but p electrons easier)

Occurrence

- Very sparse, often with zinc

Extraction

- From zinc flue dusts; $GeCl_4$ volatized and hydrolysed to GeO_2 which is reduced with H_2

Chemical behaviour

- Like Si, but more electropositive
- (IV) state dominant but some (II) (GeO, GeX_2)
- Reacts with concentrated acid and fused alkalis; red-hot with air, sulfur, H_2S, halogens, and HCl
- Hydrides, Ge_nH_{2n+2}, gaseous to $n = 5$, liquid to $n = 9$
- No π bonded compounds
- Clusters Ge_9^{2-} and Ge_9^{4-}

Uses

- For transistors (formerly); infrared optics

Nuclear features

- Five stable isotopes: none dominant

* For cryptate see p. 326.

As Arsenic

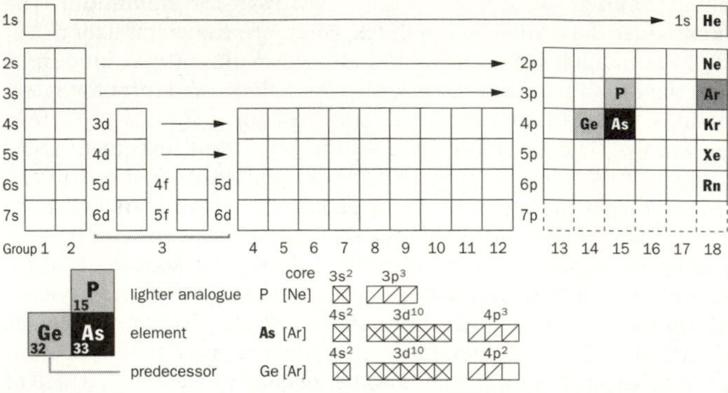

		Predecessor	Element	Lighter analogue	
Name		**Germanium**	**Arsenic**	**Phosphorus**	
Symbol, *Z*		Ge, 32	As, 33	P, 15	
RAM		72.61	74.9216	30.973762	
Radius/pm	**atomic**	122.5	125	93 (white); 115 (red)	
	covalent	122	121	110	
	ionic, M^{3+}		69	(44)	
Electronegativity (Pauling)		2.01	2.18	2.19	
Melting point/K		1210.6		317.3	
Boiling point/K		3103	889 (sublimes)	553	
$\Delta H^{\ominus}_{fus}$/kJ mol^{-1}		34.7	27.7	2.51	
$\Delta H^{\ominus}_{vap}$/kJ mol^{-1}		334.3	31.9	51.9	P_4
Density (at 293 K)/kg m^{-3}		5323	5780 (α)	1820	
			4700 (β)		
Electrical conductivity/Ω^{-1}m^{-1}		2.174	2×10^{-2} (α)	10^{-9}	
Ionization energy for removal of *j*th electron/kJ mol^{-1}	***j* = 1**	762.1	947.0	1011.7	
	***j* = 2**	1537	1798	1903.2	
	***j* = 3**	3302	2735	2912	
	***j* = 4**	4410	4837	4956	
	***j* = 5**		6042	6273	
Electron affinity/kJ mol^{-1}		116	78	72.0	
Bond energy/kJ mol^{-1}	**E–E**	163	348	209	
	E–H	288	~ 245	328	
	E–C		200	264	
	E–O	363	477	407	
	E–F	340	464	490	
	E–Cl	163	293	319	

Arsenic, ([Ar]$3d^{10}$)$4s^24p^3$, like phosphorus, [Ne]$3s^23p^3$, has more electrons than low lying vacant orbitals; so we should expect it neither to be a metal, like gallium, nor to form a single bonded three-dimensional lattice, like germanium. The three p electrons could be accommodated in the three free valence orbitals by forming three σ bonds, leading, as in phosphorus, to tetrahedral tetrameric molecules, as in the vapour phase As_4, or to a layer structure of puckered sheets of hexagons, which is the usual form* of the solid. The energy gap between occupied and unoccupied bands decreases as n increases, but in arsenic it is still too high for metallic conductivity. Since the solid, although brittle, has a metallic sheen, arsenic is classed as a metalloid, but in many ways it behaves as a non-metal and, as expected, particularly resembles phosphorus.

Arsenic shows none of the anomalous effects shown by germanium, $4s^24p^2$, and even more by gallium, $4s^24p^1$, as compared with their $n = 3$ analogues. It is larger than phosphorus, and slightly more electropositive, with all five of its outer electrons easier to ionize. Bond strengths are usually weaker than in the corresponding phosphorus compounds, except in compounds with oxygen or in the elements themselves.

Arsenic occurs mainly as sulfide ores, including mixed sulfides with metals. The element can be obtained by smelting the ore in the absence of air. In the presence of air, it tarnishes and, if heated, the As_4 vapour burns to As_4O_6 which can be converted into As_2O_3 (either crystalline or glassy). With pure oxygen, As_4O_{10} is also formed. The structures of the arsenic oxides are similar to those of phosphorus.

Arsenic dissolves in oxidizing acids and in fused alkali; the usual products are in the (III) state (as in H_3AsO_3, As_4O_6, or Na_3AsO_3), but with concentrated nitric acid $H_3As^VO_4$ is formed. The (V) state of arsenic, unlike the highest ('group') oxidation states of phosphorus, germanium, or gallium, is strongly oxidizing; and the oxide As_4O_{10}, which is hydrolysed by acid to $H_3As^VO_4$ is used as an oxidizing agent in quantitative analysis. Arsenic(III), unlike phosphorus(III), never disproportionates.

Arsenic forms binary compounds with many other elements. Metal arsenides are very varied and often non-stoichiometric; those formed with ns^2p^1 elements, such as gallium, are 'III–V' semiconductors (see p. 295). Some arsenides, such as Na_3As and NiAs, are based on alternate hexagonal arrangements of the metals and minor (or gross) absences of the metal ion cause deviation from

* A yellow solid is formed by condensing the vapour, but its structure is unknown because it is destroyed by X-rays.

Stable isotopes of arsenic

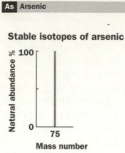

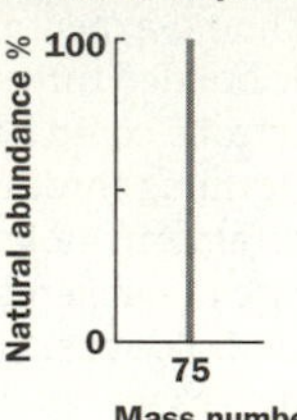

The alloy-like structure of NiAs [SM]

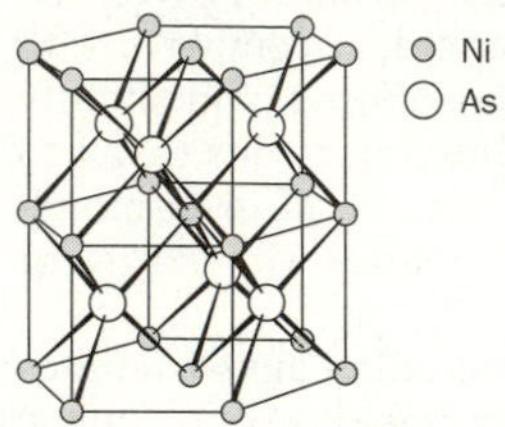

The chain anion $(AsO_2^-)_n$ [GE]

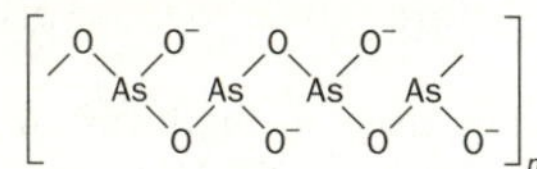

Some sulfides of arsenic [GE]

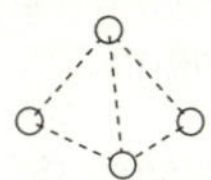
As_4 vapour

○ As

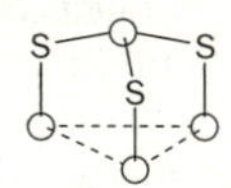

α- (and β-) As_4S_3

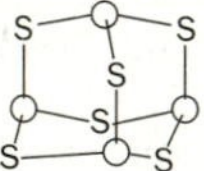

As_4S_6

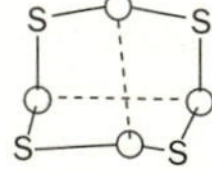

Realgar α-As_4S_4 (also β-As_4S_4). see below for alternative viewpoint

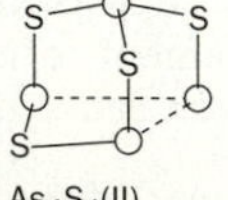

As_4S_4(II)

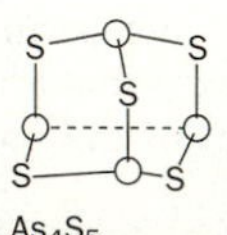

As_4S_5

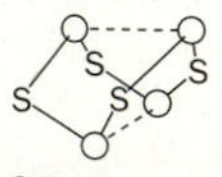

β-As_4S_4

stoichiometry. Other arsenides seem to contain $(As)_n$ units. Cobalt arsenide, $CoAs_2$, contains $CoAs_6$ octahedra with their corners forming (almost) planar As_4 squares, while LiAs, which has metallic conductivity and lustre, contains infinite spirals of arsenic. Tin arsenides, SnAs and Sn_4As_3, are low temperature superconductors.

The one stable gaseous hydride, 'arsine' AsH_3, is formed by direct reaction; unlike phosphorus, elemental arsenic does not disproportionate. Arsine decomposes at 250°C to give a shiny deposit of arsenic, which was used as a forensic test for its presence. Diarsine, $H_2As{-}AsH_2$, readily decomposes but forms more stable organo derivatives.

As might be guessed, arsenic forms all four trihalides, AsX_3, which are readily hydrolysed by water. The three lighter halides are molecular but the triiodide has a layer structure. Arsenic can also form the pentafluoride which is a very strong fluorinating agent, though the anion AsF_6^- is more stable. Arsenic is too small to accommodate five larger halogen atoms (except in the transient $AsCl_5$), although the $As^VCl_4^+$ ion is known. There is also an iodide, As_2I_4, that contains an As–As bond and disproportionates on heating.

The aqueous solution chemistry of arsenic is very sketchy. There is no definite evidence of the cation As^{3+} (nor, predictably, of As^{5+}). 'Arsenious acid' is thought to be $As(OH)_3$ (unlike 'phosphorous acid' $HPO(OH)_2$, see p. 166). Although polymeric chain anions, $(AsO_2^-)_n$, exist in crystals, the salts of heavy metals are insoluble in water, while those of sodium and potassium dissolve to give simpler, hydrolysed species such as $H_2AsO_3^-$. The solid $H_3AsO_4{\cdot}\frac{1}{2}H_2O$ has been prepared, but arsenic, in either its (III) or its (V) state, lacks the variety of oxoacids or condensed oxoanions that is displayed by phosphorus.

Since arsenic is usually found in sulfide minerals, it is no surprise that a rich series of sulfides can be made in the laboratory. Of the stoichiometrically obvious compounds, As_2S_3 has a layer structure and dissolves in aqueous sulfide solution, but the structure of As_2S_5 is unknown. There is also a series of sulfides $As_4S_{3,4,5\text{ or }6}$ that resemble the simpler sulfides of phosphorus and are based on an As_4 tetrahedron, with three or more As–S–As bridging sulfur atoms, but no terminal ones. The affinity of arsenic for sulfur is to be expected for a moderately heavy 'near-metal' which can deform the somewhat polarizable outer electrons of sulfur and this may account for its toxicity, since it could then interfere with the functioning of sulfur-containing enzymes. Although trace quantities of arsenic are necessary for human life, the element has been used for many centuries in pesticides and fungicides and as an instrument of murder. Arsenic compounds were amongst the first drugs used

Some cluster anions of arsenic M

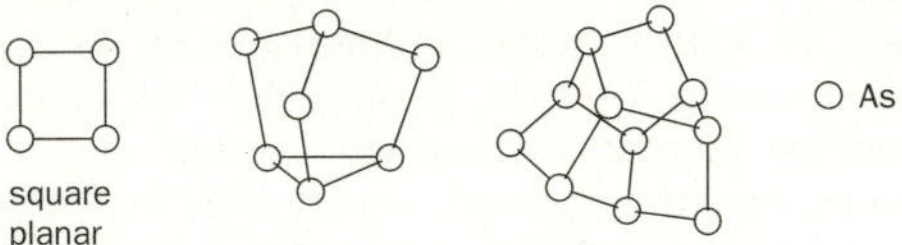

Summary

Electronic configuration
- ([Ar]3d^{10})4s^{2}4p^3

Element
- Brittle, shiny metalloid; sheets of puckered hexagons
- As_4 tetrahedra in vapour
- Easier to ionize than P

Occurrence
- In sulfide ores, often with metal sulfides

Extraction
- Heat ore in absence of air

Chemical behaviour
- Like P but larger and more electropositive, with most bonds weaker
- Dominant state (III), e.g. As_4O_6, H_3AsO_3, trihalides
- Hydrides
  - AsH_3 (decomposes on heating)
  - As_2H_4 (unstable but forms organic derivatives)

to treat syphilis, and are still used against amoebic dysentery and sleeping sickness. But these uses have declined, and arsenic is now used mainly in lead alloys (it seems to 'improve the sphericity' of lead shot) and to make gallium arsenide for a variety of electronic and optical devices (see p. 295).

We have seen that, in addition to As_4 in arsenic vapour, As–As bonds are present in some unstable compounds such as $I_2As-AsI_2$ and $H_2As-AsH_2$ but, unlike its predecessors germanium and gallium, arsenic does not seem to form longer chains. However, like germanium, it can form clusters. A side or a triangular face of As_4 is retained in many organoarsenic compounds as an As_2 or As_3 unit, and the sulfide, As_4S_3, may be represented as $As_3.S_3As$, with three As–S–As bridges. Since arsenic ($4s^24p^3$) has one fewer outer electron than sulfur ($3s^23p^4$), it is no surprise that each of the sulfur atoms can be replaced by a bridging As^- anion, to give the cluster As_7^{3-}. Compounds containing this ion, and other cluster anions such as As_4^{2-} (square) and As_{11}^{3-}, may be prepared with very large cations such as $K(crypt)^+$, where the metal ion is protected from reduction by its all-embracing ligand. These cluster anions are analogous to the Zintl ions of phosphorus (see p. 161). Arsenic may indeed form the basis of a diverting detective novel,* and still act as a killer: preferably restricted to unwelcome microorganisms. But as one of the middle-weight elements poised between metals and non-metals, it also has a lot of interest to offer the research chemist.

- Many metal arsenides, varied and non-stiochometric, e.g. GaAs (semiconductor), NiAs (hexagonal structure)
- As^V very strongly oxidizing, e.g. As_4O_{10}, AsF_5
- Aqueous chemistry uncertain, much less variety in oxoanions than P
- Sulfides
  - wide variety (similar to P)
  - As–S interaction may account for its toxicity
- Anion clusters, e.g. As_4^{2-}, As_7^{3-}, and As_{11}^{3-} (like Ge)

Uses

- Manufacture of GaAs
- In alloys
- In drugs, fungicides, and pesticides

Nuclear features

- Isotopically pure ^{75}As

* Sayers, D., *Strong Poison*

Se Selenium

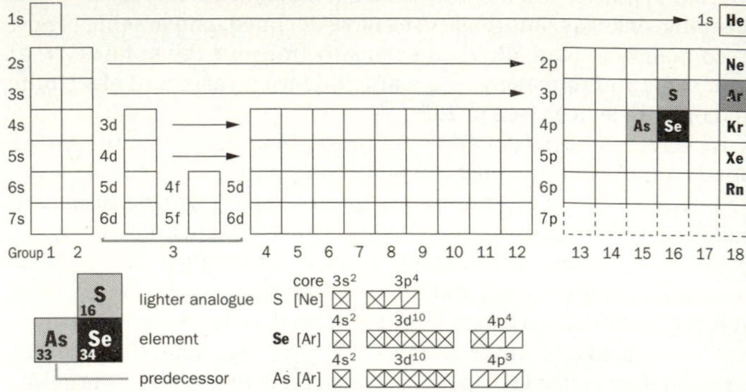

		Predecessor	Element	Lighter analogue
Name		**Arsenic**	**Selenium**	**Sulfur**
Symbol, *Z*		As, 33	Se, 34	S, 16
RAM		74.9216	78.96	32.066
Radius/pm	**atomic**	125	215.2	104
	covalent	121	117	104
	ionic, X^{2-}		191	184
Electronegativity (Pauling)		2.18	2.55	2.58
Melting point/K		889 (sublimes)	490	386.0 (α)
Boiling point/K			958.1	717.824
$\Delta H^{\ominus}_{fus}$/kJ mol^{-1}		27.7	5.1	1.23
$\Delta H^{\ominus}_{vap}$/kJ mol^{-1}		31.9	26.32	9.62
Density (at 293 K)/ kg m^{-3}		5780 (α)		
		4700 (β)	4790 (grey)	2070 (α)
Electrical conductivity/Ω^{-1}m^{-1}		2×10^{-2} (α)	100	5×10^{-16}
Energy band gap in solid/kJ mol^{-1}			178	
Ionization energy for removal of *j*th electron/kJ mol^{-1}	***j* = 1**	947.0	940.9	999.6
	***j* = 2**	1798	2044	2251
	***j* = 3**	2735	2974	3361
	***j* = 4**	4837	4144	4564
	***j* = 5**	6042	6590	7013
	***j* = 6**		7880	8495
Electron affinity/kJ mol^{-1}		78	195	200.4
Bond energy/kJ mol^{-1}	**E–E**	348	330	226
	E–H	~ 245	305	347
	E–C	200	245	272
	E–O	477	343	265
	E–F	464	285	328
	E–Cl	293	245	255

As the number of p electrons in the series $([Ar]3d^{10})4s^24p^x$ increases, the effect of the 3d contraction becomes less noticeable and so it becomes easier to predict the behaviour of a 4p element from that of its 3p analogue. So the relationship between selenium $([Ar]3d^{10})4s^24p^4$ and sulfur $[Ne]3s^23p^4$ should show none of the anomalies we found between, say, gallium and aluminium (see p. 293). As we should expect, selenium is larger than sulfur, with lower ionization energies, lower electronegativity, and lower bond energies. Like sulfur, it shows the oxidation states (VI), (IV), (II), (0), and (–II), but, as expected, selenium(VI) is more strongly oxidizing than sulfur(VI). Selenium also forms rings or chains by catenation; and the band gaps in extended structures are smaller than in sulfur.

It is no surprise that selenium occurs in sulfide ores and is recovered from the left overs of sulfide processing, such as slime from copper refining and sludge from sulfuric acid manufacture. Metal selenides are roasted in air with sodium carbonate, and the resulting sodium selenite is treated with sulfur trioxide to give sulfuric acid and selenium.

There are at least six forms of the element, of which three are red and contain Se_8 rings similar to S_8, and two (one ordered and one amorphous) contain spiral chains of –Se–Se–Se– atoms. The ordered spiral form is thermodynamically the more stable, with some delocalized interaction between the chains; it is a grey photo-

One route for extracting selenium

$$Cu_2Se + Na_2CO_3 + 2CO_2 \rightarrow Na_2SeO_3 + 2CuO + CO_2$$

from Cu anode 'slime'

$$Na_2SeO_3 \xrightarrow{H_2SO_4(aq)} H_2SeO_3(aq) \xrightarrow{SO_3 + H_2O} Se(s) + H_2SO_4(aq)$$

Purification:

$$\text{Se (+ As, S and Te)} + H_2 \longrightarrow H_2Se \text{ (+ } H_2S\text{, but no hydrides of As or Te)}$$

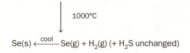

$$\downarrow 1000°C$$

$$Se(s) \xleftarrow{\text{cool}} Se(g) + H_2(g) \text{ (+ } H_2S \text{ unchanged)}$$

Elemental Se GE

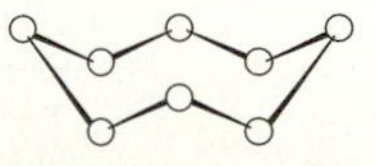

Se_8 ring

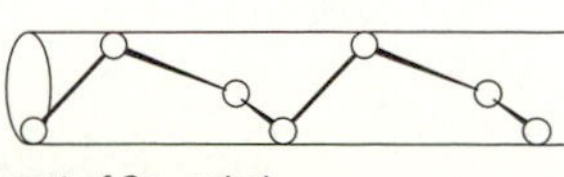

part of Se_∞ spiral

Some Se_x^{2+} cluster cations [GE]

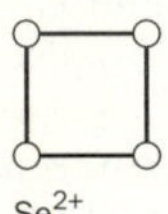

Se_4^{2+}

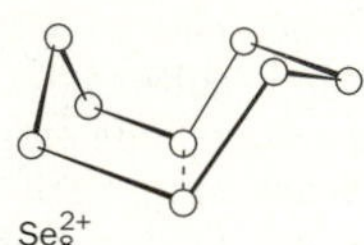

Se_8^{2+}

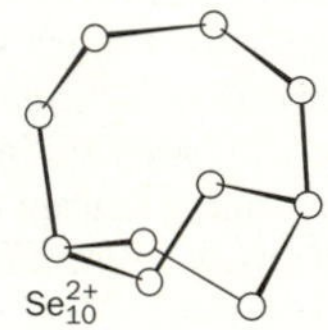

Se_{10}^{2+}

The $(SeO_2)_\infty$ chain [GE]

```
 O      O      O
/  \   /  \   /  \   /
    Se     Se     Se
    ||     ||     ||
    O      O      O
```

The $(SeO_3)_4$ ring [GE]

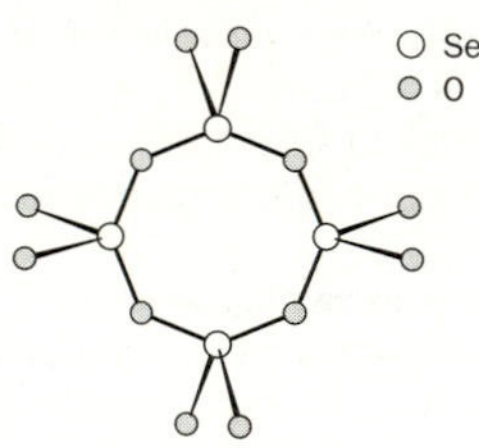

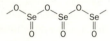

Latimer diagrams $E^{\ominus}/V$

Acidic solution pH = 0: $SeO_4^{2-} \xrightarrow{1.15} H_2SeO_3 \xrightarrow{0.74} Se \xrightarrow{-0.11} H_2Se$

Basic solution pH = 14: $SeO_4^{2-} \xrightarrow{0.03} SeO_3 \xrightarrow{-0.36} Se \xrightarrow{-0.67} Se^{2-}$

conductor widely used in xerography. The amorphous chain form is red. Commercial selenium is black and vitreous, and is thought to contain large rings (*very* large rings, with up to one thousand members). As well as forming the photosensor in every office copying machine, selenium is also used in other photoelectric devices such as photocells and solar panels, in alloys, and in pharmaceuticals. A red form is used as a glass decolorant or, in larger quantities, as a colorant; and very deep red glass is produced by cadmium selenide, together with some (yellow) cadmium sulfide.

Selenium behaves like sulfur in a number of ways. It forms cation clusters; Se_4^{2+} is square, while Se_8^{2+} and Se_{10}^{2+} are bridged cyclic structures. There are binary compounds with many other elements. The selenides of highly electropositive metals are salt-like, but those of other metals are often non-stoichiometric and semiconducting, with alloy-type structures. Hydrogen selenide is a gas at room temperature; indeed, its melting and boiling points are lower than those of hydrogen sulfide, presumably because there is even less hydrogen bonding. It dissolves in water (to give a solution more acidic than hydrogen sulfide) and burns in air to give the dioxide; but (unlike sulfur dioxide) selenium dioxide is a solid chain polymer of $-O-Se(=O)-$ units. Selenium dioxide dissolves in water to give selenious acid, $H_3Se^{IV}O_3$, which (unlike sulfurous acid) neither disproportionates nor is readily oxidized. Indeed, it is used as a mild oxidizing agent in organic chemistry. Selenium trioxide is also solid, a cyclic molecule containing four $-O-Se(=O)_2-$ units. It is the anhydride of selenic acid, H_2SeO_4, which has the distinction of being such a strongly oxidizing acid that it will dissolve gold and palladium. The oxoacids of selenium are much less diverse than those of sulfur, but condensation of oxoselenium(IV) anions gives $[O_2Se-O-SeO_2]^-$, while oxoselenium(VI) acid dissolves the trioxide to give $H_2Se_2O_7$ and $H_4Se_3O_{11}$.

Predictably, selenium forms a variety of halides: fluorides up to SeF_6, and chlorides and bromides up to SeX_4, but no iodides. There are also species with more than one selenium atom, such as Se_2X_2 and Se_4X_{16}, and oxohalides such as $SeOX_2$. Like arsenic, but unlike sulfur, selenium forms anionic halide complexes such as $SeBr_6^{2-}$, which excites chemists because, despite having 14 (rather than 12) outer electrons, it is a regular octahedron. This suggests that the $4s^2$ electrons of the selenium have no influence on the structure, but behave as part of the electron 'core'.

Selenium nitride, Se_4N_4, has the same structure as S_4N_4 but is much less stable; indeed, it explodes on shock. Selenium also forms a number of sulfides, SeS_7, Se_2S_6, Se_4S_4, merely by replacing one or more sulfur atoms in the S_8 ring. It is not surprising that selenium

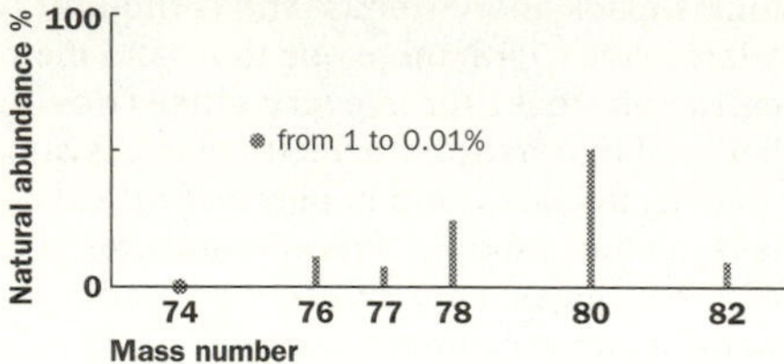

Summary

Electronic configuration

- ([Ar]$3d^{10}$)$4s^2 4p^4$

Element

- Six solid forms: Se_8 rings (three red allotropes), ordered chains (grey photoconductor), disordered chains (red), megacyclic vitreous (black)
- Larger atom than S, with electrons more ionizable
- Band gaps smaller

Occurrence

- In sulfide ores (and in waste from copper refining and sulfuric acid manufacture)

Extraction

- Roast selenides with air and Na_2CO_3 to give Na_2SeO_3; add SO_3 to get H_2SO_4 and Se

Chemical behaviour

- Same oxidation states as S (VI, IV, II, 0, –II)

can also replace sulfur in many organic compounds, giving rise to the rich field of organoselenium chemistry. Selenium also occurs in a number of proteins, and even in enzymes in some mammals. Traces of it are essential to human metabolism, acting in part as an antioxidant and maybe in other processes as yet unknown. In appreciable concentrations, it is toxic. Even were it not a health hazard, it is best avoided for social reasons: if ingested it lingers long in the body, causing foul breath and sweat.

- Forms mixed $(S + Se)_8$ rings and cation clusters, Se_4^{2+}, Se_8^{2+}, Se_{10}^{2+} (like S)
- Many selenides: salt-like with very electropositive metals, alloy-type with others
- H_2Se has lower m.p. and b.p. than H_2S and is more acidic
- SeO_2 solid chain polymer with water gives H_2SeO_3, mildly oxidizing, does not disproportionate
- SeO_3 solid cyclic tetramer, with water gives H_2SeO_4, very strongly oxidizing
- Few condensed oxoanions, up to three Se atoms
- Range of halides and oxohalides; $SeBr_6^{2-}$ octohedral
- Se_4N_4 explosive

Uses

- As photosensor; and in alloys, in pharmaceuticals, and in glass manufacture

Nuclear features

- Roughly half ^{80}Se, with five other stable isotopes

Br Bromine

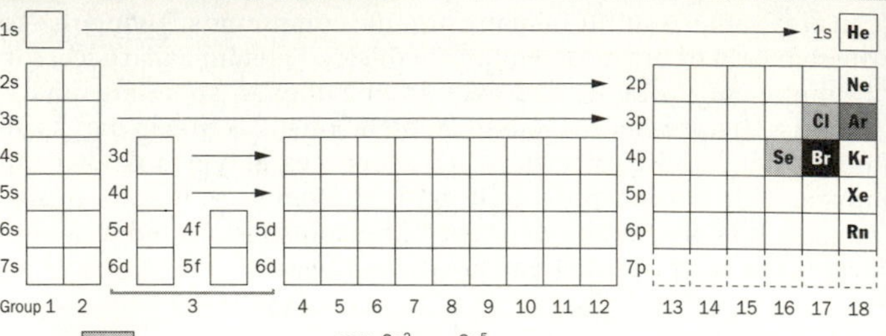

		core			
Cl 17	lighter analogue	Cl [Ne]	$3s^2$ ⊠	$3p^5$ ⊠⊠◸	
Br 35	element	**Br** [Ar]	$4s^2$ ⊠	$3d^{10}$ ⊠⊠⊠⊠⊠	$4p^5$ ⊠⊠◸
Se 34	predecessor	Se [Ar]	$4s^2$ ⊠	$3d^{10}$ ⊠⊠⊠⊠⊠	$4p^4$ ⊠◸◸

		Predecessor	Element	Lighter analogue
Name		**Selenium**	**Bromine**	**Chlorine**
Symbol, *Z*		Se, 34	Br, 35	Cl, 17
RAM		78.96	79.904	35.4527
Radius/pm	**covalent**	117	114.2	99
	ionic, X^-		196	181
Electronegativity (Pauling)		2.55	2.96	3.16
Melting point/K		490	265.9	172.17
Boiling point/K		958.1	331.93	239.18
$\Delta H^{\ominus}_{fus}$**/kJ mol^{-1}**		5.1	10.8	6.41
$\Delta H^{\ominus}_{vap}$**/kJ mol^{-1}**		26.32	30.0	20.4033
Ionization energy for removal of *j*th electron/kJ mol^{-1}	***j* = 1**	940.9	1139.9	1251.1
	***j* = 2**	2044	2104	2297
	***j* = 3**	2974	3500	3826
	***j* = 4**	4144	4560	5158
	***j* = 5**	6590	5760	6540
	***j* = 6**	7880	8550	9362
	***j* = 7**		9940	11 020
Electron affinity/kJ mol^{-1}		195	324.7	349.0
Bond energy/kJ mol^{-1}	**E–E**	330	193	242
	E–H	305	366	431
	E–C	245	285	325
	E–O	343	234	206
	E–F	285	285	257

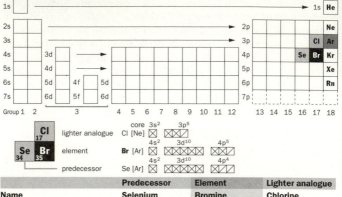

Lattice energies [D] of some ionic halides

U_{expt}/kJ mol^{-1}	Cl^-	Br^-
Li^+	–852	–815
Na^+	–786	–752
K^+	–717	–681
Cs^+	–675	–654

The hydration of halide ions [D]

$X^-(g) + H_2O \rightarrow X^-(aq)$

	Cl^-	Br^-
$\Delta H^{\ominus}$/kJ mol^{-1}	–378	–348
$\Delta S^{\ominus}$/JK^{-1} mol^{-1}	–96	–80

We have seen that, as the number of electrons in the 4p shell increases, the chemistry of an element becomes all the more predictable from that of its 3p analogue. Since bromine, ([Ar]$3d^{10}$)$4s^2 4p^5$, does indeed behave in many ways like a heavier version of chlorine, [Ne]$3s^2 3p^5$, we shall discuss mainly the differences between them.

At room temperature, bromine is a volatile, red, pungent liquid, rather than a gas: the increase of 18 in atomic number between chlorine and bromine has been accompanied by a large increase not only in mass, but also in the number of electrons involved in van der Waals interaction. Dibromine is less reactive than dichlorine, even though its greater size and less effective overlap causes a weaker interatomic bond. As we saw when comparing difluorine with dichlorine (see p. 183), we must think about other energy changes as well. The electron affinity of the bromine atom is lower than that of chlorine and so are the strengths of the bonds formed with other elements, whether they are ionic or covalent. And, on balance, these last factors outweigh the relative ease of dissociation of dibromine. Bromine is not such a strong oxidizing agent as chlorine; the element was discovered when chlorine was added to brine, which was rich in bromide ions, and it is still prepared in this way. But the ratio of bromine to chlorine in sea water is only about 1:300.

Hydrogen bromide, like hydrogen chloride, is gaseous and its aqueous solution is an even stronger acid. In crystals that are predominantly ionic, the larger bromide ion gives rise to a weaker lattice than the chloride, but it also has less negative values of heat and entropy of hydration. Moreover, we should expect bromide to be more readily deformed than chloride by highly polarizing metal ions such as silver(I) and mercury(II). As these factors counteract each other, it is difficult to predict the relative solubilities of a metal chloride and bromide, or the relative stabilities of the halogeno complexes such as $HgCl_4^{2-}$ and $HgBr_4^{2-}$. However, experience shows that cations such as K^+ and Ca^{2+}, with lower polarizing power, form bromides that are more soluble than the chlorides; but for highly polarizing cations, the bromides are less soluble than the chlorides and also form more stable anionic complexes.* So, although sodium bromide is more soluble than sodium chloride, silver bromide is even less soluble than silver chloride. Both of

* Some chemists find it helpful to classify cations as 'class a' or 'hard' if they favour chloride, and 'class b' or 'soft' if they are so highly polarizing as to favour bromide.

Stable isotopes of bromine

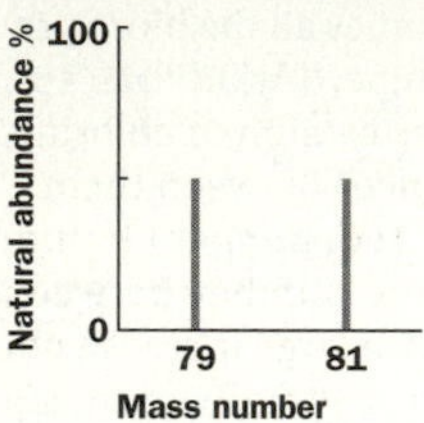

Solubilities of chlorides and bromides in water [D]

For the reaction MX(s) → MX(aq)

$\Delta G^{\ominus}$/kJ mol^{-1} at 25°C

NaCl	−9	AgCl	+56
NaBr	−19	AgBr	+70
bromide more soluble		bromide less soluble	

Stabilities of some halide complexes [SC]

For $Ag^{+}(aq) + 2X^{-}(aq) \rightarrow AgX_2^{-}(aq)$

log equilibrium constants: Cl^- 2.3 Br^- 7.2

(in 5M $NaClO_4$ at 25°C)

For $Hg^{2+}(aq) + 4X^{-}(aq) \rightarrow HgX_4^{2-}(aq)$

log equilibrium constants: Cl^- 15 Br^- 21

(in 0.5M $NaClO_4$ at 25°C)

Summary

Electronic configuration

- ([Ar]$3d^{10}$)$4s^24p^5$

Element

- Red, volatile, pungent liquid; diatomic; larger radius than chlorine and lower electron affinity

Occurrence

- Low concentrations in sea water

Extraction

- Oxidation of bromide ions with chlorine

Chemical behaviour

- Similar to chlorine, but less reactive; Br_2 weaker oxidizing agent

these silver salts dissolve somewhat in excess halide ions, but $AgBr_2^-$ is more stable than $AgCl_2^-$.

The simple metal bromides provide two of the best known uses of bromine: the photosensitive, insoluble silver salt forms the basis of photography (both colour and black and white), while the soluble potassium salt was formerly used as a sedative. But the main bulk use of bromine is to make ethylene dibromide for a number of unrelated purposes: as a lead scavenger in petrol, a pesticide (particularly against worms), a fire retardant, and as a drilling fluid.

Although oxides of bromine are even less stable than those of chlorine, its oxoacids are very similar to the chlorine oxoacids, except that there appears to be no BrO_2^-. The acid HOBr disproportionates in acidic solution, while the element itself disproportionates in alkaline solution. Bromine(VII), in the form of the perbromate ion or as perbromic acid, is a much stronger oxidizing agent than chlorine(VII), just as the highest states of the two preceding elements, selenium(VI) and arsenic(V), are unexpectedly strong oxidizing agents for reasons that are not yet clear. Perbromate compounds are, however, kinetically rather sluggish, and so are not as reactive as their thermodynamic oxidizing ability would suggest. Bromine(VII) created a great stir when it was first prepared at the end of the 1960s, not least because papers had been published explaining why it was theoretically impossible to make it. It was first obtained in trace quantities by radioactive emission of an electron from the artificially prepared *nucleus* of ^{83}Se in $^{83}SeO_4^{2-}$, giving $^{83}BrO_4^-$, but soon afterwards it was made more conventionally by very aggressive oxidation of bromate [bromine(V)] by adding xenon difluoride or fluorine, or by electrolysis. Although bromine(VII) no longer causes surprise, let it (together with our next element) remain a warning against 'theoretical impossibility'.

- Redox behaviour like chlorine, but no BrO_2^-, and BrO_4^- more oxidizing (but sluggish)
- Br^- forms stronger complexes than Cl^- with highly polarizing metal ions, and vice versa

Uses
- In manufacture of $C_2H_4Br_2$ for various uses
- AgBr in photography
- KBr (formerly) as a sedative

Nuclear features
- ^{79}Br and $^{81}Br \sim 1:1$
- BrO_4^- first observed as $^{83}BrO_4^-$ from β-decay of $^{83}SeO_4^{2-}$

Kr Krypton

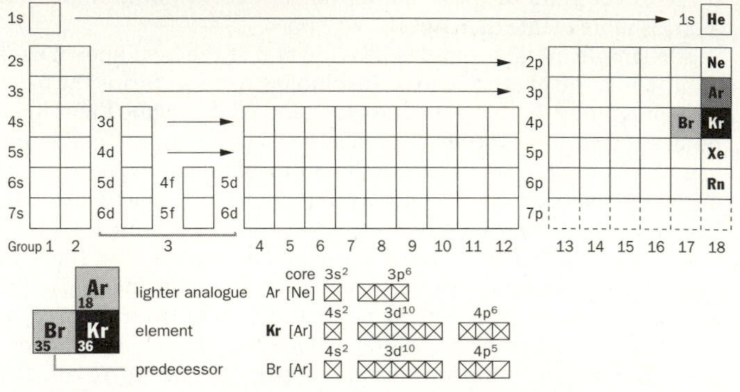

		Predecessor	Element	Lighter analogue
Name		**Bromine**	**Krypton**	**Argon**
Symbol, *Z*		Br, 35	Kr, 36	Ar, 18
RAM		79.904	83.80	39.948
Radius/pm	**covalent**	114.2	189	174
Melting point/K		265.9	116.6	83.78
Boiling point/K		331.93	120.85	87.29
ΔH°_{fus}/kJ mol^{-1}		10.8	1.64	1.21
ΔH°_{vap}/kJ mol^{-1}		30.0	9.05	6.53
Ionization energy for removal of *j*th electron/kJ mol^{-1}	***j* = 1**	1139.9	1350.7	1520.4
	***j* = 2**	2104	2350	2665.2
	***j* = 3**	3500	3565	3928
	***j* = 4**	4560	5070	5770
	***j* = 5**	5760	6240	7238
	***j* = 6**	8550	7570	8811
	***j* = 7**	9940	10 710	12 021
Bond energy/kJ mol^{-1}	**E–F**	285	50	

Synthesis of KrF_2

$Kr(g) + F_2(g) \rightarrow KrF_2(s)$ $\quad \Delta H^{\circ}/\text{kJ mol}^{-1} = +63$

Reaction is endothermic (enthalpically unfavourable).

Reaction results in increased order: 2 moles of gas consumed (entropically unfavourable).

Irradiation supplies required energy.

Low temperature discourages decomposition.

At krypton, ([Ar]$3d^{10}$)$4s^2 4p^6$, the 4p shell is complete and so it is no surprise that krypton, like argon [Ne]$3s^2 3p^6$, is a monatomic gas. As it is heavier than argon, it has a higher boiling point; and it too is obtained by fractional distillation of liquid air (which contains about one thousandth of a per cent of krypton). Like neon and argon (although much more expensive) it is used in strip lighting and for lasers. The wavelength, λ, of the orange-red line in the spectrum of ^{86}Kr is used to define the standard of length.*

Krypton is larger than argon and it too can be trapped into ice and be contained as a clathrate inside the hydrogen bonded but rather openwork cage of β-quinol. With its ns^2np^6 configuration, krypton has a very high ionization energy (although lower than that of argon), and it might seem very unlikely that this extremely stable arrangement could be cajoled to change. But, remembering bromine(VII), we must treat 'improbability of reaction' with caution. As the value of n increases, less energy is needed to promote an electron from the highest occupied np orbital to the lowest vacant one, and the energy might be provided by bond formation with a very electronegative element which prevents accumulation of too much negative charge on the krypton.

The most electronegative element is, of course, fluorine; and if a mixture of krypton and fluorine at −184° C is grossly energized by irradiation with high energy protons or electrons, or by an electric discharge, there appears a volatile white solid, containing linear KrF_2 molecules. As this product is thermodynamically unstable and decomposes at higher temperatures, it may not seem impressive, particularly as it is the only uncharged krypton compound yet known. But we must remember that krypton is, so far, the lightest member of the former 'inert' gases that can breach its super-stable ns^2p^6 configuration. The exact nature of the disruption is still under discussion, but it is thought that four electrons (two from krypton and one from each fluorine) are accommodated in a bonding and a non-bonding orbital to give a three-centre bond (cf. xenon, p. 411).

When krypton difluoride reacts with some molecular halides, cations such as KrF^+ and $Kr_2F_3^{3+}$ are formed. Arsenic pentafluoride, for example, is a strong fluoride acceptor and the anion formed, AsF_6^-, gives a (relatively) stable lattice with the large krypton fluoride cations. Not only are ions of a similar size geometrically compatible, but the larger the anion, the less marked is the tendency (predicted by the Kaputstinskii equation, see p. 36) for

* 1 m = 1650763.73 λ.

Stable isotopes of krypton

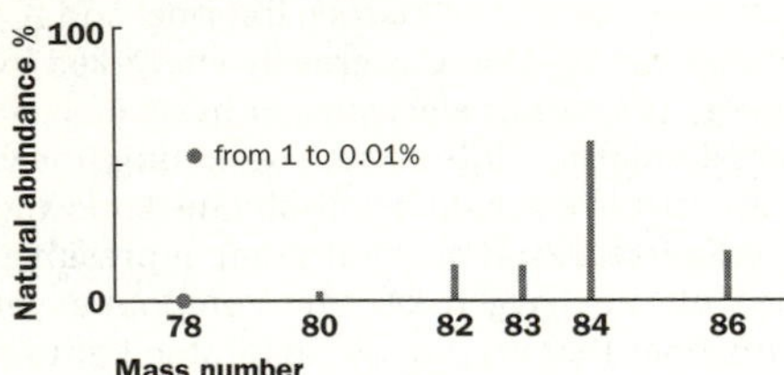

decomposition of the cation to produce a more stable lattice. Even so, the krypton fluoride salts, like the molecular difluoride, are very powerful fluorinating agents. But, with the promotion of one 4p electron, the ns^2p^6 gases have been stripped of their 'inert' status and are now merely 'noble' and no more inviolate than that ancient noble metal gold, which can be dissolved in an acid [selenium(VI), p. 311] without even adding a complexing anion.

Summary

Electronic configuration

- ([Ar]$3d^{10}$)$4s^2 4p^6$

Element

- Monatomic gas, higher b.p. than Ar, and lower ionization energy

Occurrence

- In air (about 0.001%)

Extraction

- Fractional distillation

Chemical behaviour

- Extremely unreactive
- Inclusion into ice; clathrate with β-quinol
- Aggressive fluorination at very low temperature gives KrF_2 (unstable); KrF^+ + $K_2F_3^{3+}$ stabilized by large anions

Uses

- Lasers and strip lighting

Nuclear features

- Six stable isotopes; ^{84}Kr dominant (57%)

Part V

Filling the 5s, 4d, and 5p orbitals

The second long period; rubidium to xenon

Rb Rubidium

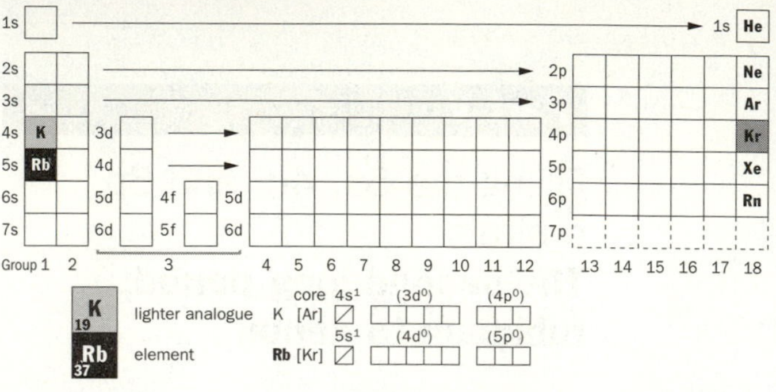

	Element	Lighter analogue
Name	**Rubidium**	**Potassium**
Symbol, *Z*	Rb, 37	K, 19
RAM	85.8478	39.0983
Radius/pm atomic	247.5	227
ionic, M^+	149	133
Electronegativity (Pauling)	0.82	0.82
Melting point/K	312.2	336.8
Boiling point/K	961	1047
$\Delta H^{\ominus}_{fus}$/kJ mol^{-1}	2.20	2.40
$\Delta H^{\ominus}_{vap}$/kJ mol^{-1}	69.2	77.53
Density (at 293K)/kg m^{-3}	1532	862
Electrical conductivity/Ω^{-1}m^{-1}	8.0×10^6	1.626×10^7
Ionization energy for removal of *j*th electron/kJ mol^{-1} $j = 1$	403.0	418.8
$j = 2$	2632	3051.4
Electron affinity/kJ mol^{-1}	46.9	48.4
Dissociation energy of E_2(g)/kJ mol^{-1}	47.3	49.9
$E^{\ominus}$/V for $M^+(aq) \rightarrow M(s)$	−2.924	−2.924

Stable and long–lived isotopes of rubidium

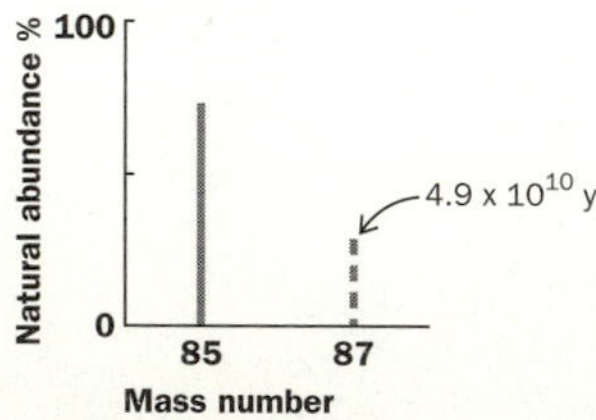

The element following krypton starts a new valency shell with $n = 5$. We should expect rubidium ($[Kr]5s^1$) to be very like potassium ($[Ar]4s^1$); and indeed it resembles potassium even more than potassium resembles sodium ($[Ne]3s^1$). The increase (of 18) in nuclear charge between potassium and rubidium, rather poorly shielded by the ten 3d electrons, counterbalances the increased radius of the outer s orbital, with the result that the nuclear pull on the $5s^1$ electron of rubidium is rather similar to that on the $4s^1$ electron of potassium. The ionization energy of rubidium is only slightly lower than that of potassium, and its atomic and ionic radii are only slightly higher. It is not surprising that rubidium occurs in some of the same minerals as potassium, widely but sparingly distributed.

Rubidium is obtained (as a by-product of lithium extraction) from lepidotite, a lithium potassium aluminosilicate, which also contains rubidium, fluoride, and hydroxyl ions. The element was first detected by the red line in its spectrum (hence its name). The metal can be prepared by reducing the molten chloride with calcium, but, as it is even more reactive than potassium, it has to be handled under argon. It is soft, white and low melting, with heats of fusion and evaporation very similar to those of potassium.

It seems that rubidium differs markedly from potassium in only one respect: the formation of metallic oxides. Both metals burn in a free supply of air to give the peroxides $Rb^+O_2^-$ and $K^+O_2^-$, but partial oxidation at low temperatures gives, for rubidium only, two oxides, Rb_6O and Rb_9O_2, which are metallic conductors. Both oxides are based on a regular octahedron of six metal atoms around a central oxygen, with rubidium atoms much closer together in the oxide than in the pure metal. It is pleasing that we find such a variation between two elements that are otherwise so similar; and salutary to

The lower oxides of rubidium

$$2\ Rb_6O \xrightarrow{-7.3°C} Rb_9O_2 + 3\ Rb$$

The structure of Rb_9O_2[GE]

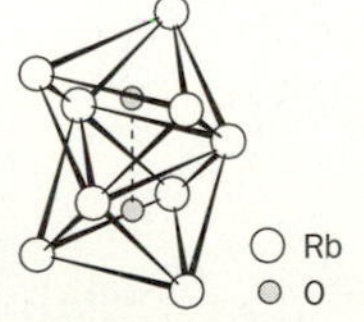

2 Rb_6O octahedra share a face

Rb–Rb in Rb_6O unit	352 pm
Rb–Rb in metal	485 pm

'Rb_6O' is better written as $(Rb_9O_2)Rb_3$ since it contains the same Rb_9O_2 groups, each accompanied by three Rb atoms

Lattice enthalpies/kJ mol^{-1} of halides [D]

	MF	MCl	MBr	MI
M = K	–821	–717	–689	–649
Rb	–789	–695	–668	–632

Some cryptate complexes of Rb

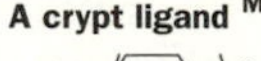

A crypt ligand [M]

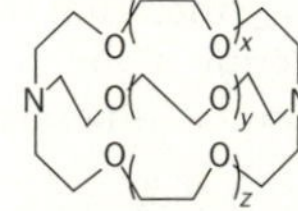

Such species are named according to the number of O atoms in each strand.

The crypt 2,2,2 has values of $x = y = z = 1$

The cation [Rb crypt 2,2,2]$^+$

(in the monohyrdate thiocyanate complex)

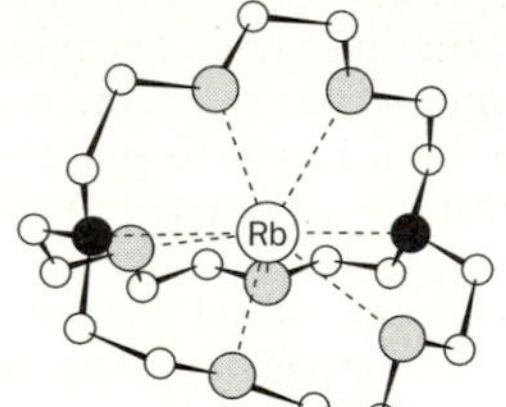

(From M.R. Truter *Chem. Br.*, 1971, 203)

Stability constants, *K* of some crypt complexes [CW]

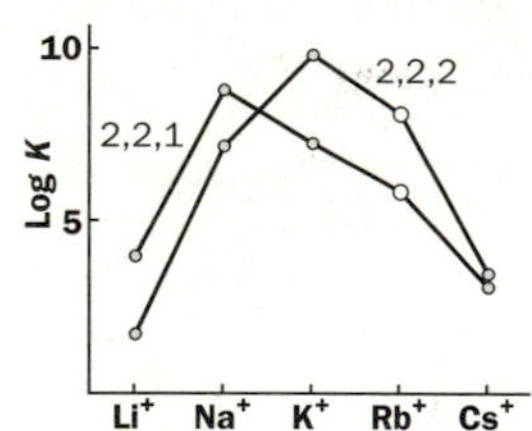

K is the equilibrium constant for:

M^+ (solv) + crypt (solv) = Mcrypt$^+$ (solv)

in 95% MeOH–H_2O

Rb^+ forms a stronger complex with crypt 2,2,2 (preferred cation K^+) than with crypt 2,2,1 (preferred cation Na^+)

think how such small differences in size and energy can give rise to apparent chemical eccentricity.

The few other differences between the behaviour of rubidium and potassium are doubtless a result of size. The slightly larger rubidium forms slightly weaker dimers in the vapour and slightly weaker ionic lattices. Its preferred crowns and crypts have a slightly larger cavity than those of potassium. The larger rubidium ion is somewhat less heavily hydrated than the potassium ion and so has a *smaller* hydrated radius, which results in a higher ionic mobility and a greater tendency to be held by some cation exchange resins; but these are small variations in degree rather than in kind.

Summary

Electronic configuration
- $[Kr]5s^1$

Element
- Soft, white, low melting, very reactive metal; red flame

Occurrence
- With potassium

Extraction
- (From a lithium potassium aluminosilicate, as by-product of Li manufacture): reduce molten RbCl with Ca under Ar

Chemical behaviour
- Very like potassium, but more reactive and ion slightly larger (with lower hydrated radius)
- Slightly weaker lattices, hydration, M–M bonds in vapour
- Larger crown and crypt cavities
- Burns in excess air to $Rb^+O_2^-$; but (unlike K) in restricted air to Rb_6O and Rb_9O_2 metallic conductors

Uses
- None (cost over 300 times that of K and 6000 times that of Na)

Nuclear features
- Isotopes ^{85}Rb:$^{87}Rb \sim 3.5:1$

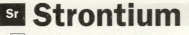

Sr Strontium

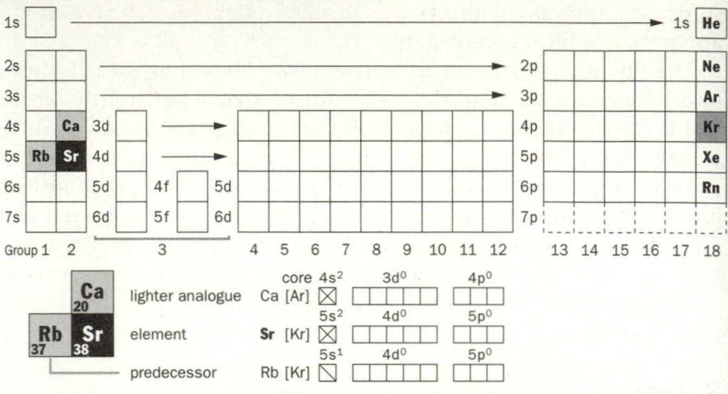

		Predecessor	Element	Lighter analogue
Name		**Rubidium**	**Strontium**	**Calcium**
Symbol, *Z*		Rb, 37	Sr, 38	Ca, 20
RAM		85.8478	87.62	40.078
Radius/pm	**atomic**	247.5	215.1	
	covalent		192	174
	ionic, M^{2+}		127	106
Electronegativity (Pauling)		0.82	0.95	1.00
Melting point/K		312.2	1042	1112
Boiling point/K		961	1657	1757
$\Delta H^{\ominus}_{fus}$/kJ mol^{-1}		2.20	9.16	9.33
$\Delta H^{\ominus}_{vap}$/kJ mol^{-1}		69.2	138.91	149.95
Density (at 293 K)/kg m^{-3}		1532	2540	1550
Electrical conductivity/Ω^{-1}m^{-1}		8.0×10^6	4.348×10^6	2.915×10^7
Ionization energy for removal of *j*th electron/kJ mol^{-1}	***j* = 1**	403.0	549.5	589.7
	***j* = 2**	2632	1064.2	1145
Electron affinity/kJ mol^{-1}		46.9	−146	−186
$E^{\ominus}$/V for $M^{2+}(aq) \rightarrow M(s)$			−1.085	−2.84

Stable isotopes of strontium

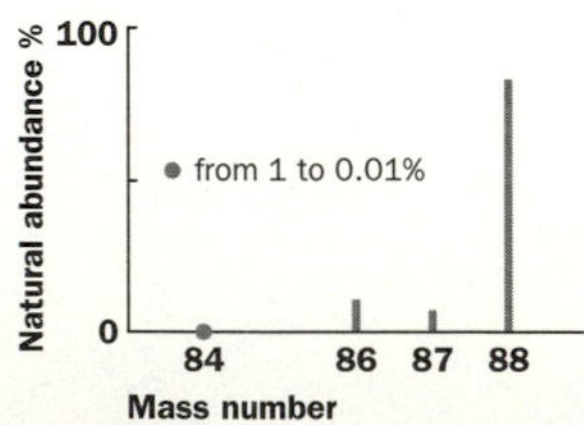

Strontium, with the electronic configuration [Kr]$5s^2$, behaves almost exactly as we should expect, and resembles calcium ([Ar]$4s^2$) more closely than calcium resembles magnesium ([Ne]$3s^2$). Strontium is predictably slightly larger than calcium and slightly easier to ionize; and the differences between the two elements, although small, are greater than those between potassium and rubidium.

Like calcium, strontium occurs fairly commonly, as the sparingly soluble carbonate or sulfate. It can be obtained by reducing the oxide with aluminium or by electrolysis of the fused chloride. The metal is like a denser version of calcium, but with lower melting and boiling points, and almost identical heats of fusion and vaporization. Even the flame colours are fairly similar, although distinguished in the text books as 'bright red' for strontium and 'brick red' for calcium. Strontium, like calcium, reacts with both oxygen and nitrogen, and since it is less volatile, it is used as a 'getter' to scavenge the remaining traces of these gases from vessels that have been (almost) evacuated.

The larger ionic radius of strontium has predictable consequences. As we should expect (see p. 203), strontium carbonate decomposes at a higher temperature than calcium carbonate. Although ligands that complex with unhydrated ions favour the smaller calcium cation, those such as nitrate and sulfate, which pair up with hydrated cations, form complexes of similar (low) stability with strontium and calcium, as the two cations have almost the same *hydrated* radius.

The few differences in kind between strontium and calcium are not very dramatic but reflect the ability of the slightly larger strontium ion to have seven nearest atoms (as in the iodide and hydroxide), whereas the analogous calcium compounds have only six.

Although the chemistry of strontium may not seem very interesting, it is of vital concern to anyone who has the misfortune to ingest any radioactive strontium from nuclear waste, tests, or disasters. As strontium is so like calcium, it can replace it in human teeth and bone and is secreted in milk; and for the same reason, it is difficult to persuade the body to part with the strontium without also removing the essential calcium as well. One of the strontium isotopes, strontium-90, has a half-life of 29 years. So if some were imbibed by a one-year-old child, at least 10% of it would still be left, giving out electrons that might do untold genetic and other damage, over a 90-year lifespan.

For summary see p. 331.

Stability constants, *K* of some 18-crown-6 complexes [M]

(in water at 25°C)

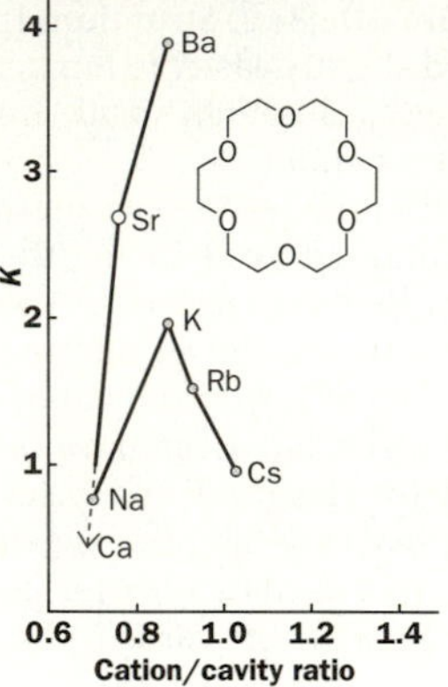

The less favourable 'fit' for Sr^{2+} compared with K^+ is more than counterbalanced by increased charge. [A high charge/radius ratio produces a more unfavourable $\Delta H^{\ominus}$ of desolvation, somewhat offset by a more favourable $\Delta S^{\ominus}$. For Ca^{2+}, this reinforces the bad 'fit', to give a very low stability constant.]

Summary

Electronic configuration

- [Kr]$5s^2$

Element

- Fairly soft metal, very like calcium; red flame

Occurrence

- As carbonate or sulfate (sparingly soluble)

Extraction

- Reduction of SrO with Al, electrolysis of fused $SrCl_2$

Chemical behaviour

- Very similar to calcium, ion slightly larger, and lower hydrated radius
- Combines with both O_2 and N_2 in air
- Can have up to seven nearest neighbours
- Can replace Ca in bones, teeth, milk

Nuclear features

- Several natural isotopes; ^{89}Sr dominant ~ 82%
- ^{90}Sr from radioactive fall-out can get lodged in human body

Y Yttrium

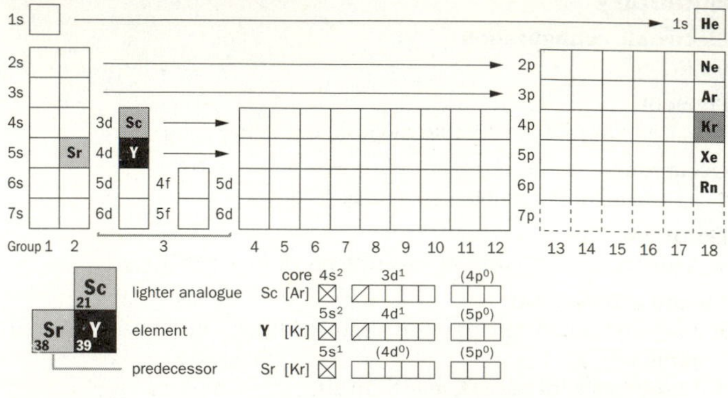

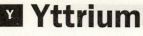

	Predecessor	Element	Lighter analogue
Name	**Strontium**	**Yttrium**	**Scandium**
Symbol, *Z*	Sr, 38	Y, 39	Sc, 21
RAM	87.62	88.90585	44.955910
Radius/pm **atomic**	215.1	181	160.6
covalent	192	162	144
ionic, M^{3+}		106	83
Electronegativity (Pauling)	0.95	1.22	1.36
Melting point/K	1042	1795	1814
Boiling point/K	1657	3611	3104
$\Delta H^{\ominus}_{fus}$**/kJ mol^{-1}**	9.16	17.2	15.9
$\Delta H^{\ominus}_{vap}$**/kJ mol^{-1}**	138.91	393.3	304.8
Density/kg m^{-3}	2540 (293 K)	4469 (293 K)	2989 (273K)
Electrical conductivity/Ω^{-1}m^{-1}	4.348×10^6	1.754×10^6	1.639×10^6
Ionization energy for removal of *j*th electron/kJ mol^{-1} *j* = 1	549.5	616	631
***j* = 2**	1064.2	1181	1235
***j* = 3**		1980	2389
Electron affinity/kJ mol^{-1}	–146	29.6	18.1
$E^{\ominus}$**/V for $M^{3+}(aq) \rightarrow M(s)$**		–2.37	–2.03

Stable isotopes of yttrium

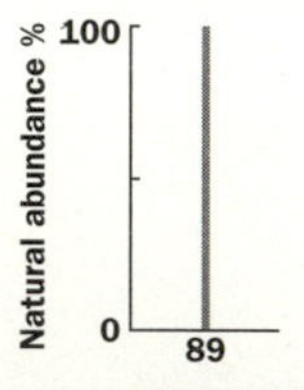

It may seem ludicrous to have four chemical elements called after the same place, particularly if its name is as prosaic as Ytterby, which in Swedish means merely 'outer village'. Yttrium ($[Kr]4d^1 5s^2$) is the lightest of the quartet and, like scandium ($[Ar]3d^1 4s^2$), exists mainly as the triply charged ion. The Y^{3+} ion is, as we shall see (p. 426), almost the same size as the M^{3+} ions of three of the heavier lanthanides, terbium (p. 444), erbium (p. 480), and ytterbium (p. 449) so it is not surprising that they occur together. In the late eighteenth century mineral deposits near Ytterby yielded an oxide residue that was named yttria, but it was later found to contain the four different metals: hence the four related names. Separation of such similar cations is extremely difficult and was first carried out by many successive fractional crystallizations. Nowadays, small differences in the hydrated radii allow them to be separated by multistage cation exchange. The pure metal can be made by reducing the fused chloride with calcium.

Yttrium is, of course, the first element in the second transition series in which the 4d shell is being filled. But, like scandium, it nearly always loses all three outer electrons and so has a predominant oxidation state of (III) which, being d^0, is diamagnetic and uncoloured. From the comparison of rubidium with potassium, and of strontium with calcium, we should predict, rightly, that yttrium would be a fairly soft metal, which is a little more electropositive than scandium but less so than strontium; but it seems less reactive as it acquires a protective coating in air.

Yttrium behaves very like scandium: it, too, forms some lower oxidation state species when the metal is dissolved in the molten chloride (see p. 209) and it has some unstable trialkyls and triaryl compounds. Differences may seem trivial. In the complex anion $Y(NO_3)_5^{2-}$, all nitrate ions chelate yttrium, giving it 10 donor atoms, while the smaller scandium ion can only accommodate nine.

Although yttrium has received little attention in the past, it is now acquiring a much higher profile because of applications of its compounds in various branches of solid-state technology, such as the red phosphors in TV screens. Useful mixed oxides abound: $Y_3Fe_5O_{12}$ as microwave filters in radar; $Y_3Al_5O_{12}$ for artificial diamonds and as a matrix for lasers (known as 'YAG': yttrium aluminium garnet); and $YBa_2Cu_3O_{7-8}$ (known as '1–2–3') as a 'high' temperature superconductor (see p. 279). There may well be many more uses to come.

For summary see page 335.

Summary

Electronic configuration

- [Kr]$4d^1 5s^2$

Element

- Fairly soft, electropositive metal, with protective oxide coating

Occurrence

- With Tb, Er, and Yb

Extraction

- Separate from Ln^{3+} ions by cation exchange
- Metal obtained from molten YF_3 and Ca

Chemical behaviour

- resembles scandium (and heavier lanthanides)
- Main species is $Y^{3+}(d^0)$
- Metal dissolves in molten YCl_3 (like Sc) to give lower oxidation state
- Can accept 10 donor atoms

Uses

- In red phosphors for TV
- In mixed oxides for various solid uses, e.g. artificial diamonds and high temperature superconductors

Nuclear features

- Isotopically pure ^{89}Y

Zr Zirconium

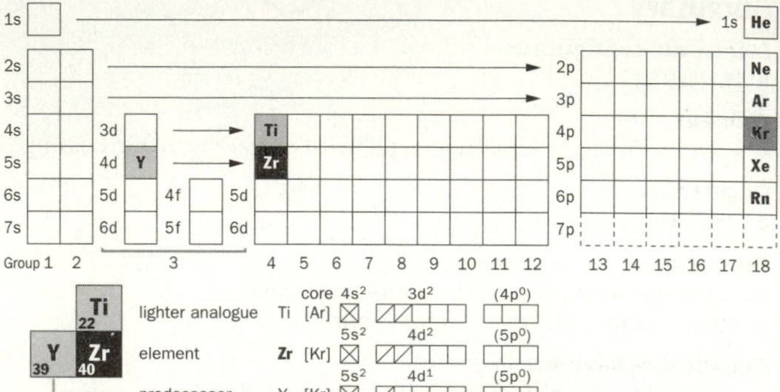

		Predecessor	Element	Lighter analogue
Name		Yttrium	Zirconium	Titanium
Symbol, Z		Y, 39	Zr, 40	Ti, 22
RAM		88.90585	91.224	47.88
Radius/pm	atomic	181	160	144.8
	covalent	162	145	132
	ionic, M^{3+}		109	
	M^{4+}		87	80
Electronegativity (Pauling)		1.22	1.33	1.54
Melting point/K		1795	2125	1933
Boiling point/K		3611	4650	3560
$\Delta H^{\ominus}_{fus}$/kJ mol^{-1}		17.2	23.0	20.9
$\Delta H^{\ominus}_{vap}$/kJ mol^{-1}		393.3	581.6	428.9
Density (at 293 K)/kg m^{-3}		4469	6506	4540
Electrical conductivity/Ω^{-1}m^{-1}		1.754×10^6	2.375×10^6	2.381×10^6
Ionization energy for removal of jth electron/kJ mol^{-1}	$j = 1$	616	660	658
	$j = 2$	1181	1267	1310
	$j = 3$	1980	2218	2652
	$j = 4$		3313	4175
Electron affinity/kJ mol^{-1}		29.6	41.1	7.6

Stable and long-lived isotopes of zirconium

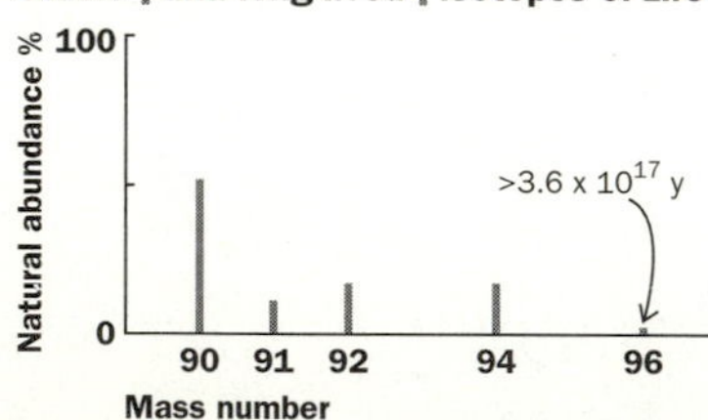

Since yttrium is so like scandium, we should expect zirconium ($[Kr]4d^2 5s^2$) to be very similar to titanium ($[Ar]3d^2 4s^2$). Like titanium, it occurs in combination with oxygen, as the gemstone zircon, $ZrSiO_4$ (which contains discrete SiO_4^{4-} ions, see p. 153), and as the oxide ZrO_2 It, too, can be extracted as the volatile chloride, which is reduced to the metal by magnesium. Zirconium can be purified by heating the crude metal with iodine and then decomposing the volatile tetraiodide at a much higher temperature. As zirconium is heavier than titanium and has twice as many 4d electrons as yttrium, it has a higher melting and boiling point than either, and is an appreciably better metallic conductor than yttrium. Zirconium, with its strong affinity for oxygen, forms a very stable oxide layer which protects it from further corrosion. It is used in chemical plants and as a component of stainless steels. Natural zirconium contains macro-quantities of four different isotopes, none of which has much tendency to capture neutrons and this makes the corrosion-resistant metal useful for cladding uranium dioxide rods in water-cooled nuclear reactors (but the zirconium must be pure, see p. 461). It is also used, alloyed with niobium, for making superconducting magnets.

The chemical differences between zirconium and titanium, although not dramatic, are greater than those between yttrium and scandium. In the dioxides zirconium has seven neighbouring oxygen atoms, while the smaller titanium has only six. Zirconium dioxide has a high melting point and expands little on heating and so is used in furnace linings (some calcium or magnesium oxide must be added to prevent the phase change, which, in the pure material, occurs at 1100°C and would lead to cracking after repeated thermal recycling). A fibrous form of zirconium dioxide is used for filters and insulators. The disulfide, ZrS_2, is, like titanium disulfide, a semiconductor with a layer structure.

As we shall find with other members of the second (and third) transition series, simple hydrated cations are fairly scarce. Although it is possible to make hydrates of simple salts such as $Zr(NO_3)_4 \cdot 5H_2O$ and $Zr(SO_4)_2 \cdot 4H_2O$ acidic solutions must be used to

Seven-coordinated Zr CW

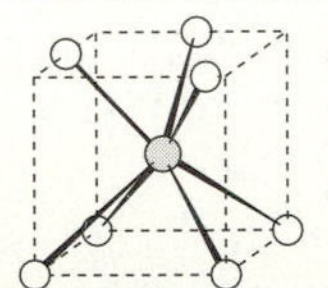

ZrO_2 (baddeleyite form)

Chains in solid $ZrCl_4$ CW

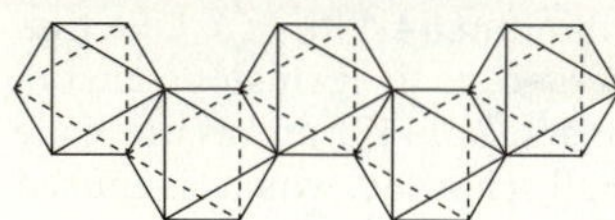
zigzags of edge-shared $ZrCl_6$ octahedra

$Zr(BH_4)_4$ GE

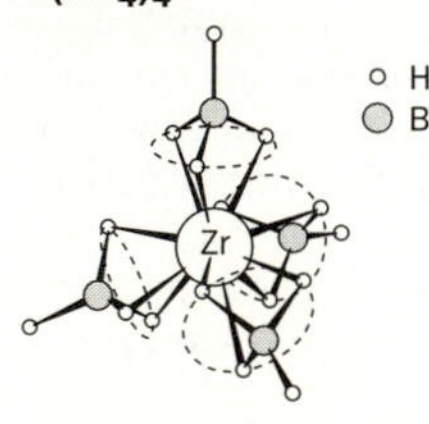

prevent hydrolysis. There is no evidence of oxocations analogous to TiO^{2+}, and the hydrolysed solutions probably contain –Zr–O–Zr– chains. Nor do we know of any monomeric $ZrO_x^{(4-2x)-}$ oxoanions.

Zirconium also shows higher coordination numbers than titanium in its halogen compounds. Although the gaseous tetrahalides are ZrX_4 molecules, eight fluoride ions or six chloride ions can fit round the metal in the solids. The larger zirconium ion, with its outer energy levels closer together, forms compounds of more varied geometry than that shown by titanium.

It is usually found that lower oxidation states are less stable in the 4d (and 5d) transition series than for the 3d elements. Aqueous titanium(III) ([Ar]$3d^1$) is indeed a reducing agent (see p. 215); but aqueous zirconium(III) cannot be made since it reduces water. The coloured trihalides, 'ZrX_3', have been made, but with difficulty, and appear to be dimers with two halide bridges and a low magnetic moment, which suggests some type of interaction between the two zirconium atoms. But the magnetic properties (and spectra) of the 4d and 5d elements are much more complicated than those of the 3d elements and so should be interpreted only with great caution. Zirconium, like scandium and yttrium, can also be persuaded, with even more difficulty, into forming halides of oxidation states (II) and (I).

Zirconium, like titanium, can combine with four cyclopentadienyl groups; but (instead of having two attached by a σ bond, and two π bonded to all five carbon atoms) the zirconium compound contains *one* σ bond and *three* five-point π bonds. There is also a diverting compound, which may be written as $Zr^{4+}(BH_4^-)_4$, in which three hydrogen atoms from each BH_4^- group seem to be attached to the zirconium. But it is the more prosaic chemistry of zirconium that gives us a foretaste of the behaviour of the 4d and 5d elements: the scarcity of simple aqueous or oxocations and of lower oxidation states; manifestations of high coordination numbers and geometrical variety; and the difficulty of interpreting magnetic moments or spectra.

For summary, see pp. 371–5.

Nb Niobium

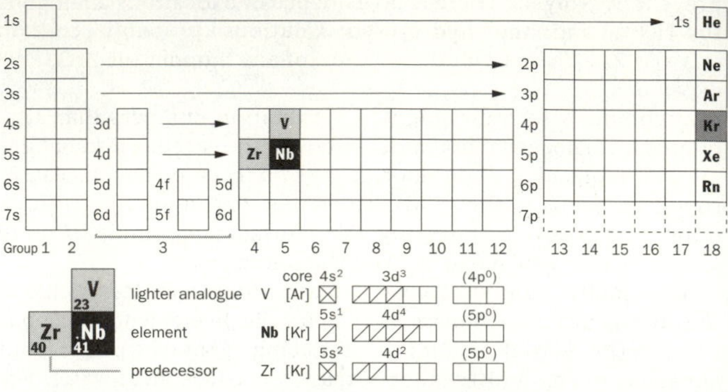

		Predecessor	Element	Lighter analogue
Name		**Zirconium**	**Niobium**	**Vanadium**
Symbol, Z		Zr, 40	Nb, 41	V, 23
RAM		91.224	92.90638	50.9415
Radius/pm	**atomic**	160	142.9	132.1
	covalent	145	134	
	ionic, M^{4+}	87	74	
Electronegativity (Pauling)		1.33	1.64	1.63
Melting point/K		2125	2741	2160
Boiling point/K		4650	5015	3650
$\Delta H^{\ominus}_{fus}$/kJ mol^{-1}		23.0	27.2	17.6
$\Delta H^{\ominus}_{vap}$/kJ mol^{-1}		581.6	696.6	458.6
Density (at 293 K)/kg m^{-3}		6506	8570	6110
Electrical conductivity/Ω^{-1}m^{-1}		2.375×10^6	8×10^6	4.032×10^6
Ionization energy for removal of jth electron/kJ mol^{-1}	**$j = 1$**	660	664	650
	$j = 2$	1267	1382	1414
	$j = 3$	2218	2416	2828
	$j = 4$	3313	3695	4507
	$j = 5$		4877	6294
Electron affinity/kJ mol^{-1}		41.1	86.2	50.7

Stable isotopes of niobium

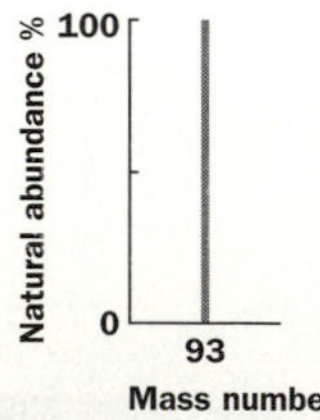

Niobium was formerly called columbium but is probably equally unfamiliar under either name. Although we might have expected it to have three 4d electrons outside a filled 5s orbital, the electronic structure of the free atom is actually $[Kr]4d^4 5s^1$, indicating both the small difference in energy between the 5s and the 4d orbitals, and the greater relative stability of the 4d electrons with increasing atomic number. This effect continues until the 4d orbital is full (see also palladium, p. 363); but since compound formation involves either loss or promotion of the outer electrons, the precise configuration of an isolated atom has little effect on the overall behaviour of the element.

Like its predecessor zirconium ($[Kr]4d^2 5s^2$), niobium is found as a mixed oxide; in MNb_2O_6, the M represents a mixture of iron and manganese and some of the niobium is almost always replaced by its $5d^3$ analogue, tantalum, which resembles it extremely closely (see p. 463). Originally prepared by reduction of its halides, the metal is now obtained from its oxide, using carbon or sodium. Since niobium is harder and higher melting than zirconium, it seems that its three d electrons are actively contributing to metallic bonding. Niobium is used in alloys: with iron for stainless steels; and with zirconium for superconducting magnets. Although protected by an oxide layer from corrosion in air, it dissolves in hot concentrated acids and (slowly) in fused alkalis.

Since zirconium $4d^2$ differs markedly from titanium $3d^2$ (see p. 211), it is no surprise that niobium $(4d + 5s)^6$ behaves very differently from its lighter analogue, vanadium $(3d + 4s)^6$. The predominant oxidation state of niobium is (V), with the (IV) state limited mainly to halides. Lower formal oxidation states are either niobium cluster compounds, or complexes of strong π acceptors, which drain negative charge away from the metal atom. Niobium appears to form no aquo- or oxo-cations and its (ill-defined) basic sulfates and nitrates probably contain hydrolysed polymeric niobium(V) groups.

Much of niobium chemistry is based on a few very simple geometrical units, which are combined in so many ways as to lead to bewildering chaos, or sumptuous splendour, depending on one's viewpoint. Although niobium can be surrounded by up to eight groups in a variety of arrangements, an octahedron of six oxygen or halogen atoms around a central niobium is far more common. The octahedra are then joined up, often with some distortion, by sharing edges or corners, or by almost any combination of these. The oxide Nb_2O_5 can be made in a large number of different forms ('polymorphs'), and is resistant to all acids except hydrofluoric acid. Niobium pentahalides are based on similar octahedra. In the fluoride tetramer, four octahedra share corners so that the niobium atoms form a square; but the chloride and bromide are dimers with

Nb(V) halides, $(NbX_5)_x$ GE

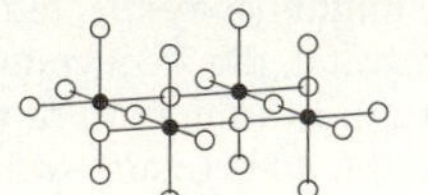

fluoride tetramer
$(NbF_5)_4$

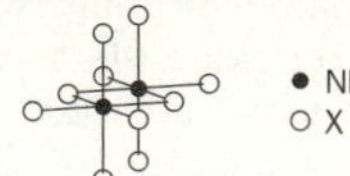

chloride and bromide dimers
$(NbCl_5)_2$ and $(NbBr_5)_2$

The anion GE $[Nb^{V}_6O_{19}]^{8-}$

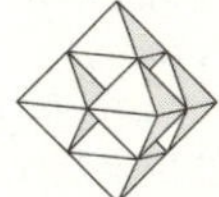

A group of six edge-shared NbO_6 octahedra (one obscured)

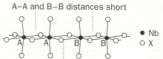

Nb(IV) halides, $(NbX_4)_\infty$ GE

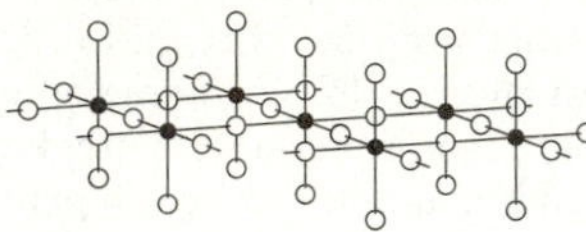

fluoride sheets
$(NbF_4)_\infty$

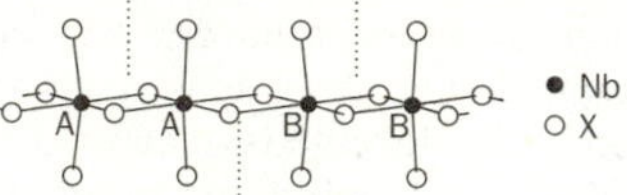

chloride and bromide chains
$(NbCl_4)_\infty$ and $(NbBr_4)_\infty$

The Nb_6 cluster in NbO GE

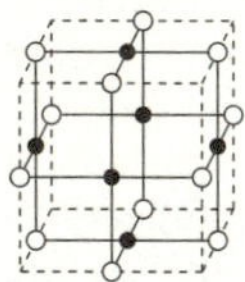

NbO showing planar coordination of Nb (and O) and vacancies at the cube corners (Nb) and centre (O)

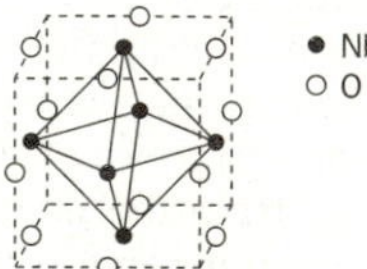

NbO showing octahedral Nb_6 cluster (joined by corner-sharing to neighbouring unit cells)

The Nb_6 cluster in $[Nb_6X_{12}]^{n+}$ ions CW

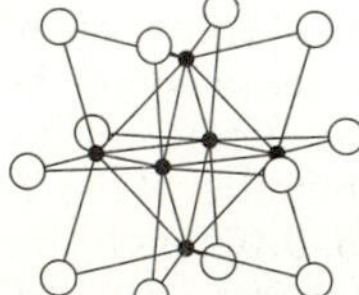

• Nb
○ X

The halide ions bridge each of the 12 edges of the Nb_6 octahedron

two edge-sharing octahedra. The anion $Nb_6O_{19}^{8-}$, which occurs in mildly alkaline solution contains, six NbO_6 octahedra arranged like a smaller version of the analogous vanadium(V) anion, $V_{10}O_{28}^{4-}$ (see p. 224).

The niobium(IV) halides are also based on NbX_6 octahedra, again corner shared for the smaller fluoride, to give sheets. The chloride, bromide, and iodide are edge-shared chains, but the niobium atoms no longer sit centrally inside their octahedra; instead, they are displaced towards one edge in pairs so that the chain appears to be built around Nb–Nb pairs. Since these compounds are diamagnetic, there is probably some interaction between the two niobium atoms.

Reduction of the pentoxide gives an oxide, NbO, which is a metallic conductor. Its structure is like that of NaCl, but with vacancies that are ordered in such a way as to leave octahedral groups of six niobium atoms. This Nb_6 cluster is also found in a number of complex halides and some oxides, which contain Nb_6X_{12} ions of various charge. The X atoms lie outside the 12 edges of the central Nb_6 octahedron, and the Nb_6X_{12} ions are so stable that they do not disintegrate even when the substance dissolves in water.

Occasionally, niobium has fewer than six nearest neighbours. Separate $Nb^{V}O_4^{3-}$ ions exist in niobates of some triply charged cations, such as those of scandium and the lanthanoid metals. But the compound $LiNbO_3$ is a continuous mixed oxide lattice and does not contain NbO_3^- ions. It is ferro- and and piezo-electric and is used in electronic devices.

Niobium forms a layered sulfide, NbS_2, which can accommodate other species, such as pyridine, between sheets of atoms, much as graphite does (see p. 69). Like vanadium, niobium forms a stable 18-electron hexacarbonyl anion, $Nb(CO)_6^-$, but with cyclopentadienyl it forms $Nb(C_5H_5)_4$ rather than a sandwich. The binding, however, resembles that in the titanium compound (two rings σ bonded and two five-atom π-bonded rings) rather than in $Zr(C_5H_5)_4$. However, niobium does not yet have a rich range of stable organic derivatives. Its importance for this book is more as an example of a rather low profile but typical member of the 4d series, exhibiting mainly the two highest possible oxidation states, and having no simple cation chemistry. Many of its compounds are built up of octahedra, which are often NbX_6 groups (where X is oxygen or a halogen) but are sometimes Nb_6 clusters; and great variety is achieved by joining up the octahedra in different ways. We shall find that behaviour of this type is quite common amongst the 4d (and 5d) elements.

For summary, see pp. 371–5.

Mo Molybdenum

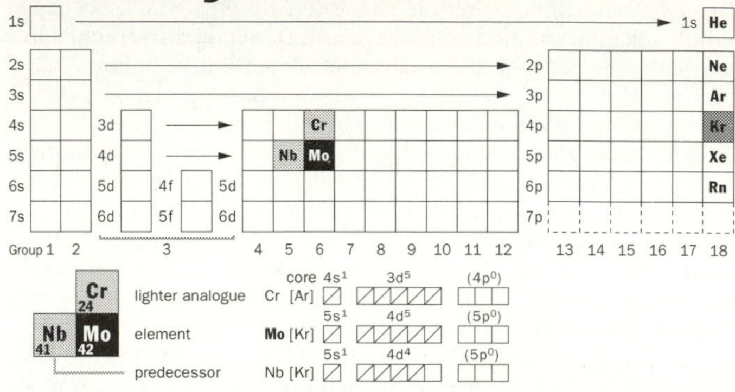

		Predecessor	Element	Lighter analogue
Name		**Niobium**	**Molybdenum**	**Chromium**
Symbol, *Z*		Nb, 41	Mo, 42	Cr, 24
RAM		92.90638	95.94	51.9961
Radius/pm	**atomic**	142.9	136.2	124.9
	covalent	134	129	
	ionic, M^{3+}		92	84
Electronegativity (Pauling)		1.64	2.2	1.66
Melting point/K		2741	2890	2130 ± 20
Boiling point/K		5015	4885	2945
$\Delta H^{\ominus}_{fus}$/kJ mol^{-1}		27.2	27.6	15.3
$\Delta H^{\ominus}_{vap}$/kJ mol^{-1}		696.6	594.1	348.78
Density (at 293 K)/kg m^{-3}		8570	10220	7190
Ionization energy for removal of *j*th electron/kJ mol^{-1}	***j* = 1**	664	685	652.7
	***j* = 2**	1382	1558	1592
	***j* = 3**	2416	2621	2987
	***j* = 4**	3695	4480	4740
	***j* = 5**	4877	5900	6690
	***j* = 6**		6500	8738
Electron affinity/kJ mol^{-1}		86.2	72.0	64.3

Stable isotopes of molybdenum

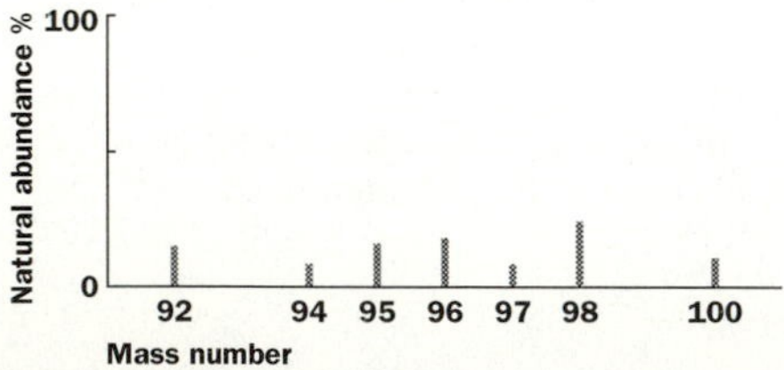

Although the name of this element sounds awkward, we shall see later how it arose. From the electronic configuration of molybdenum ($[Kr]4d^5 5s^1$) we should expect it to be a hard, corrosion resistant, high melting metal [though perhaps with slightly weaker bonds than niobium, if the 4d electrons (like the 3d electrons, see p. 281) are becoming more under the control of the nucleus as the d orbitals are filled]. We should predict that, like chromium ($[Ar]3d^5 4s^1$), it could show oxidation states up to (VI), but with greater emphasis towards the higher states, except for some low oxidation state species with metal–metal bonding. As molybdenum is larger than chromium, we should also guess that its oxospecies, like those of niobium (and vanadium), are built up from MoO_6 octahedra rather than from tetrahedra; and great variety, as in the niobium oxoanions, would be no surprise.

Molybdenum has in fact a higher melting point than niobium, although a lower heat of atomization. At room temperature it is attacked only by fluorine. Molybdenum(VI) occurs combined with oxygen as $CaMoO_4$, in which it is often partially replaced by tungsten(VI) (see p. 465). However, its most usual ore and most important compound is $Mo^{IV}S_2$ (see p. 347). This is converted, via the acidic oxide MoO_3, to ammonium molybdate, which can be reduced to the powdered metal with hydrogen. However, because molybdenum has such a high melting point, it cannot be cast but has to be formed into shape by compressing the powder at very high temperature under hydrogen. It is used for electrodes, as a catalyst, and for removing sulfur from petrol.

The white trioxide MoO_3 (which can also be obtained by passing air over red hot molybdenum) is made up of distorted layers of MoO_6 octahedra, sharing corners. If it is heated, it loses oxygen, turns various shades of blue and violet, and eventually becomes MoO_2. This, too, is made up of layers of MoO_6 octahedra, but now sharing edges. Between the two extremes are a large number of different structures in which there are both edge- and corner-sharing, octahedra and (since they are metallic conductors) also electron delocalization. The dioxide itself is a metallic conductor and its distorted structure gives rise to pairs of adjacent Mo–Mo atoms reminiscent of the metal–metal pairs in niobium(IV) halides (see p. 343).

The trioxide dissolves in aqueous alkali to give the simple anion MoO_4^{2-}. When the solution is gradually acidified, the anions condense quite rapidly to give a motley collection of *iso*polymolybdates,*

* Iso: the same, because the anions contain only one type of metal atom.

Some isopolymolybdates (idealized) GE

Discrete ions

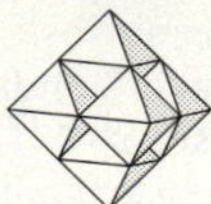

$[Mo_6O_{19}]^{2-}$
(the sixth octahedron is obscured)

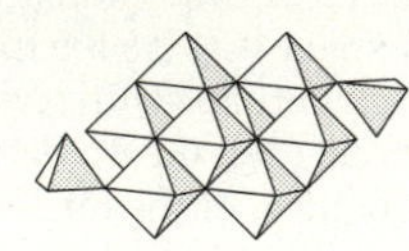

$[Mo_{10}O_{34}]^{8-}$

Chain

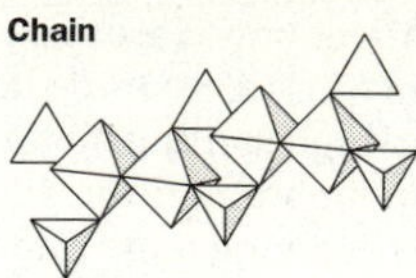

the polyanion $[Mo_2O_7^{2-}]_n$

including $Mo_7O_{24}^{6-}$, $Mo_8O_{26}^{4-}$, and $Mo_{36}O_{112}^{8-}$ (which is the largest oxo-anion known so far); but there are certainly some others, and probably many more. These anions, too, are based on MoO_6 groups joined in various ways, but given extra variety by having tetrahedral MoO_4 groups, either around the edges or as bridging groups. However, there is scope for yet greater variety. Some of the corner-joined structures contain tetrahedral cavities which can be occupied by ions of other elements to give *hetero*polymolybdates,* such as the phosphomolybdates $PMo_{12}O_{40}^{3-}$ and $P_2Mo_{18}O_{62}^{6-}$. Other fairly small atoms, such as arsenic(V), silicon(IV), and titanium(IV), can replace the phosphorus. There are also some other polymolybdates that have octahedral cavities and so can accommodate larger atoms, such as iodine and those metal ions such as cobalt(III), $3d^6$, that favour octahedral sites (see p. 261–2).

If MoO_3 or heteropolymolybdates are reduced in the presence of water, some of the molybdenum becomes Mo^V and intense blue colours are produced. There is a series of highly coloured mixed oxides containing a mixture of molybdenum(VI) and (V), together with sodium or potassium ions. Of formula $M^I_xMoO_3$ (where *x* runs from about 0.3 to nearly 1), they have metallic conductivity and lustre. They are less well known than the similar 'bronzes' formed by tungsten (see p. 465), perhaps because the (V) state disproportionates more readily for molybdenum, which, being slightly smaller, is subject to less repulsion when in the edge-sharing sites required for the (IV) oxidation state.

Molybdenum is indeed in the (IV) state in its best known compound, MoS_2, which is made up of MoS_6 units; but these are trigonal prisms rather than octahedra. They are linked in sheets to form a semiconducting layer structure, which resembles graphite, and is used as a lubricant and a catalyst. Molybdenum sulfide, too, forms 'intercalation' compounds by slipping other molecules such as pyridine between the layers. Like graphitic oxide (see p. 69) and lead, the mineral MoS_2 is so soft that it can make black marks on pale materials like paper and its forebears; and it was called molybdite, from the Greek word for lead, for the same reason that we call the graphite innards of pencils 'leads' to this day. Lower, less important, sulfides are also known.

Molybdenum forms halides in states oxidation (VI) to (II). As expected, the (VI) state occurs only with fluorine, and MoF_6 is molecular and a strong oxidizing agent. There are also a number of

* Hetero: different, because some molybdenum atoms are replaced by those of other elements.

A Mo(III) chloro complex [GE]

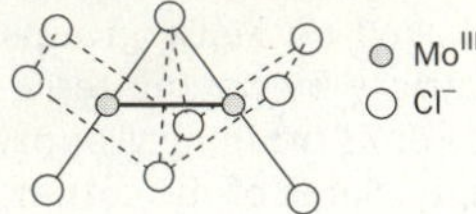

$[Mo_2Cl_9]^{3-}$ showing Mo–Mo bond

Some Mo(II) chloro complexes [GE]

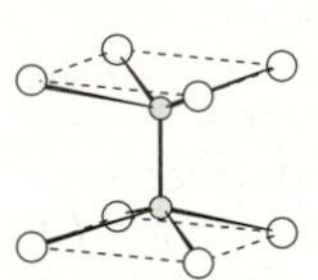

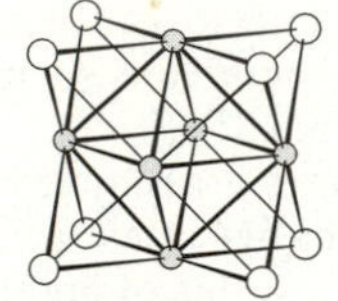

Mo^II
Cl^−

$[Mo_2Cl_8]^{4-}$
Mo–Mo bond
no bridging Cl

$[Mo_6Cl_8]^{4+}$
Mo_6 octahedral cluster
one Cl^- ion over each of its eight faces

The Mo(II) ethanoate dimer [GE]

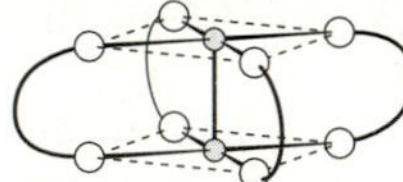

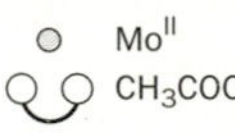

strongly oxidizing oxohalides. The halides of molybdenum(V) (with fluorine and chlorine) and (IV) (with fluorine, chlorine, and bromine) are, like those of niobium, based on octahedra. The four trihalides have ionic lattices similar to those of chromium(III) compounds. Octahedral anions such as $MoCl_6^{3-}$ and $Mo_2Cl_9^{3-}$ are also known, and the low magnetic moment of the dimer suggests strong interaction between the metal atoms. Dihalides are formed with only chlorine, bromine, and iodine: fluorine is too oxidizing. They contain the familiar M_6 octahedron with a halide ion over each of its eight faces, giving the cluster cation $M_6X_8^{4+}$, which is associated with four more halide ions, to give Mo_6X_{12} or 'MoX_2'.

Association between metal ions is also common in many of the so-called 'aquo' ions of the lower oxidation states of molybdenum. The only example of a non-hydrolysed mononuclear species appears to be $Mo(H_2O)_6^{3+}$, which can be hydrolysed to polynuclear cations. The higher oxydation states are represented by $Mo_3^{IV}O_4^{4+}$(aq) and $Mo_2^{V}O_4^{2+}$(aq). The (II) state, although unhydrolysed, is the dimeric and quadruply bonded Mo_2^{4+}.aq (similar to the quadruply bonded $Mo_2^{II}Cl_8^{4-}$ ion). Strong Mo–Mo bonding also occurs in the Mo^{II} carboxylate dimers, which are similar to those of chromium(II) (see p. 237) but without the two axial water molecules.

Although the hexacarbonyl, $Mo(CO)_6$, is unstable, molybdenum, like chromium, forms a stable 18-electron sandwich compound with two benzene molecules. But, despite all the variety it exhibits in its octahedral stacking and metal clusters, perhaps the most interesting aspect of molybdenum is its enzyme action. Without molybdenum, albeit in very small quantities, plants would not be able to reduce atmospheric nitrogen to ammonia, and the manufacture of plant, and hence of animal, protein would not be possible. Molybdenum is also essential for other biological reductions: of sulfate and of nitrate. Presumably, it achieves these changes because the energy gaps between its own oxidation states are not too great. But there is much work to be done on how molybdenum carries out its catalytic roles.

For summary, see pp. 371–5.

Tc Technetium

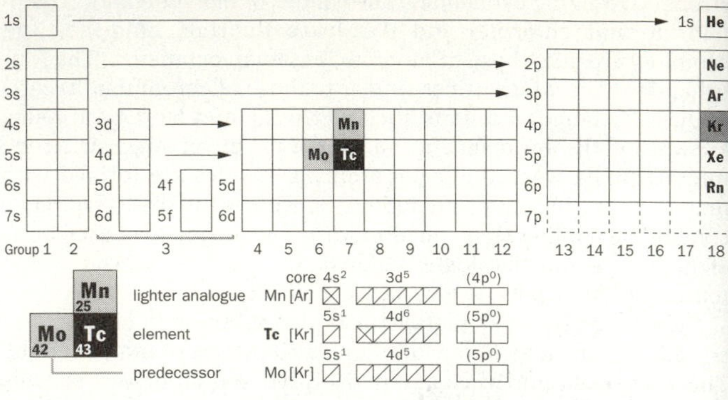

		Predecessor	Element	Lighter analogue
Name		**Molybdenum**	**Technetium**	**Manganese**
Symbol, *Z*		Mo, 42	Tc, 43	Mn, 25
RAM		95.94	98.9062	54.93805
Radius/pm	**atomic**	136.2	135.8	124
	ionic, M^{2+}	92	95	91
	M^{4+}		72	52
Electronegativity (Pauling)		2.2	1.9	1.55
Melting point/K		2890	2445	1517
Boiling point/K		4885	5150	2335
ΔH°_{fus}/kJ mol^{-1}		27.6	23.81	14.4
ΔH°_{vap}/kJ mol^{-1}		594.1	585.22	219.7
Electrical conductivity/Ω^{-1}m^{-1}			4.4×10^{6}	5.405×10^{5}
Ionization energy for removal of *j*th electron/kJ mol^{-1}	***j* = 1**	685	702	717.4
	***j* = 2**	1558	1472	1509.0
	***j* = 3**	2621	2850	3248.4
Electron affinity/kJ mol^{-1}		72.0	96	

Key isotopes of technetium

Mass number	97	98	99	99m
Half-life	2.6×10^{6} y	4.2×10^{6} y	2.13×10^{5} y	6.01 h

Technetium is the lightest of those elements that do not occur in any appreciable amounts on earth; traces, however, are formed by decay of naturally occurring uranium. It can be made artificially by bombarding molybdenum nuclei ($Z = 42$) with neutrons, which are absorbed to give unstable molybdenum nuclei that then decay, emitting electrons, to become technetium nuclei ($Z = 43$); and it was discovered in this way in 1937. Nowadays, however, it is obtained from fission products of uranium.

All its isotopes are radioactive and, although they decay emitting only modest energy, some are very long-lived, and so great care must be taken in their handling and disposal. As technetium is so scarce, it is also very expensive, and so it is not surprising that it has been studied less than its neighbouring elements; but enough has been done to back up, or to refute, our predictions about its behaviour. Its electronic configuration is $[Kr]4d^6 5s^1$, so we should expect similarities to its predecessor, molybdenum, but with indications that the d electrons are slightly more withdrawn into the core.

Technetium is obtained (like molybdenum, usually as a powder) from reduction of either ammonium technate, NH_4TcO_4, or of the sulfide, Tc_2S_5, with hydrogen. It is indeed a hard metal, resistant to corrosion, but with a slightly *lower* enthalpy of atomization than molybdenum; however, the decrease is smaller than that from chromium to manganese. We should also expect technetium $\{[Kr](4d + 5s)^7\}$ to behave like manganese $\{[Ar](3d + 4s)^7\}$, but with a shift in favour of the higher oxidation states. Few, if any, cations would be predicted, but metal–metal bonds would cause no surprise.

The metal is more reactive than molybdenum. With oxygen it forms a volatile solid, Tc_2O_7, which contains two corner-shared TcO_4 tetrahedra, and is much more stable than the explosive, oily, manganese analogue. As a 4d element, technetium is predictably more stable in high oxidation states than its 3d counterpart. Technetium reacts with fluorine to give TcF_6, and with sulfur to give TcS_2, a layer compound like MoS_2. It seems to form no cations and dissolves in acids only if they are oxidizing; it does not react even with hydrofluoric acid. But it does dissolve in bromine water to give pertechnic acid, $HTc^{VII}O_4$. The pertechnate ion, unlike the permanganate ion, MnO_4^-, is not a particularly strong oxidizing agent; and since oxygen to metal charge transfer needs much more energy than in the permanganate ion, the transition to technetiun is in the UV, and so the pertechnate ion is colourless. The pertechnate ion reacts with hydrogen sulfide to give Tc_2S_7 (unlike the permanganate ion, which is reduced to MnS). Technetium(II) compounds, on the

other hand, are much less stable than those of manganese(II) and many disproportionate to technetium(IV) and the metal. Technetium(IV) oxide, TcO_2, has a structure similar to that of MoO_2, and also seems to contain pairs of metal ions (see p. 345).

The 18-electron dimeric carbonyl $Tc_2(CO)_{10}$ has the same structure as the manganese analogue (see p. 245), but is less stable; and technetium does not seem to form a cyclopentadienyl sandwich. Like other transition metals, technetium does, of course, form a wide variety of complexes; but here we shall mention only one, the hydrido anion TcH_9^{2-}, which is isomorphous* with ReH_9^{2-} and so must have six hydrogen atoms arranged around the technetium in a trigonal prism, with another hydrogen atom above each of the three rectangular faces (see p. 469).

We might expect that a scarce, artificial, radioactive element would be of little practical use, but one of its several unstable nuclei is important for medical diagnoses. There are two technetium isotopes of mass number 99: one, ^{99m}Tc, is metastable and decays to the other, ^{99}Tc, by emitting energy as γ-rays. If this nucleus can be incorporated in a compound that is concentrated in a certain part of the body, such as the heart, a γ-ray scan of that area may show up irregularities. In addition, with greater use of nuclear reactors, the world supply of technetium is increasing, so our lightest but oldest artificial element may be put to other uses in the future.

For summary, see pp. 371–5.

* Isomorphous substances have crystals of identical geometry.

Ru Ruthenium

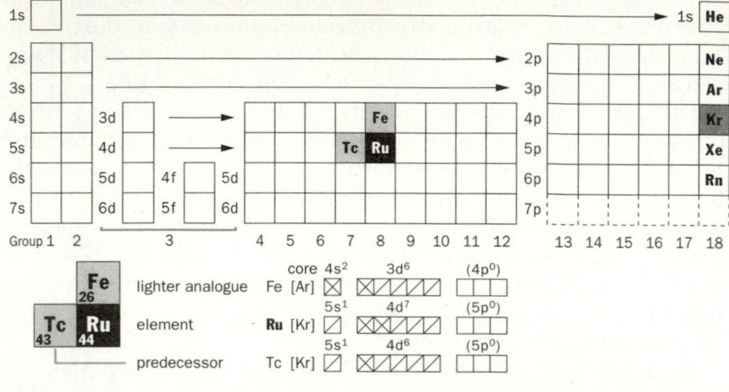

		Predecessor	Element	Lighter analogue
Name		**Technetium**	**Ruthenium**	**Iron**
Symbol, *Z*		Tc, 43	Ru, 44	Fe, 26
RAM		98.9062	101.07	55.847
Radius/pm	**atomic**	135.8	134	124.1
	covalent		124	116.5
	ionic, M^{3+}		77	67
	M^{4+}	72	65	
Electronegativity (Pauling)		1.9	2.2	1.83
Melting point/K		2445	2583	1808
Boiling point/K		5150	4173	3023
$\Delta H^{\ominus}_{fus}$/kJ mol^{-1}		23.81	23.7	14.9
$\Delta H^{\ominus}_{vap}$/kJ mol^{-1}		585.22	567.8	351.0
Density (at 293 K)/kg m^{-3}			12370	7874
Electrical conductivity/Ω^{-1}m^{-1}		4.4×10^6	1.316×10^7	1.030×10^7
Ionization energy for removal of *j*th electron/kJ mol^{-1}	***j* = 1**	702	711	759.3
	***j* = 2**	1472	1617	1561
	***j* = 3**	2850	2747	2957
Electron affinity/kJ mol^{-1}		96	101	

Stable isotopes of ruthenium

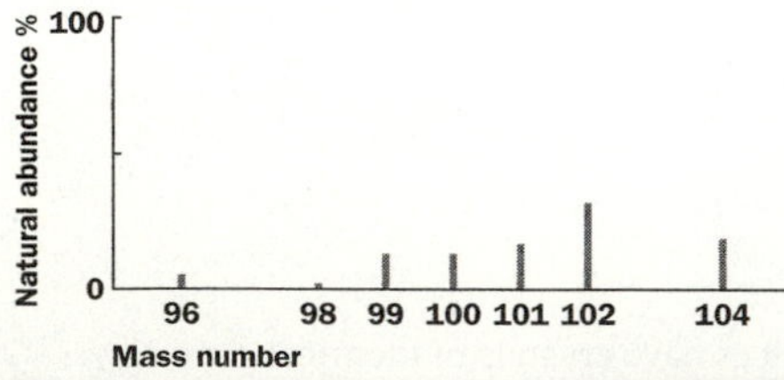

Ruthenium is extremely rare, being only slightly more abundant than natural technetium. It occurs as the metal, together with platinum and other rare metals, all with a total of from eight to ten electrons in their (4d + 5s) or (5d + 6s) outer shells, and is obtained from the waste products of platinum manufacture.

Following technetium ([Kr]$4d^6 5s^1$), the free ruthenium atom has an outer electronic configuration of $4d^7 5s^1$, indicating both the small difference in energy between the 5s and 4d levels, and the increasing stabilization of the 4d electrons as this shell becomes more occupied. So it is no surprise that the heat of atomization and the transition temperatures are somewhat lower for ruthenium than for technetium; but they are still high enough to require powder metallurgy (see p. 345). The metal is resistant to air and to acids but dissolves in fused alkali. Ruthenium, although so rare, has uses, nevertheless: to harden platinum, and as a catalyst.

Ruthenium, like iron, ([Ar]$3d^6 4s^2$) has eight outer electrons; but, predictably, its favoured oxidation states are higher than those of iron. It can even use all eight of its outer electrons to form the volatile, toxic oxide $Ru^{VIII}O_4$; and the anion $Ru^{VI}O_4^{2-}$ can be oxidized to $Ru^{VII}O_4^-$ (which, however, oxidizes water). There is a linear cation $Ru^{VI}O_2^{2+}$. With fluorine, which, like oxygen, is always a spur to high oxidation states, ruthenium forms RuF_6. Unlike iron, ruthenium forms no oxide lower than RuO_2 (which differs from its technetium analogue in showing no sign of metal atom pairing). The 'disulfide', RuS_2, is, however, a three-dimensional pyrites, $Ru^{II}(S_2^{2-})$, like its iron counterpart, rather than a layer sulfide like $Mo^{IV}(S^{2-})_2$ and $Tc^{IV}(S^{2-})_2$; but there are few other simple compounds containing ruthenium(II) except for the chloride, bromide, and iodide.

The more usual oxidation states of ruthenium are (III) d^5 and (IV) d^4, which form a variety of halides. Many are based on RuX_6 octahedra, and all of those are low spin, with as many electrons as possible shared in the three more stable 4d orbitals (see p. 235). (Halide ions produce only a mild ligand field, but as any field has much more effect on the larger 4d and 5d metal ions than on their 3d counterparts, low spin complexes are the norm in the second half of the second and third transition series.) The ruthenium(V) halides are cyclic tetramers, similar to the niobium(V) ones (see p. 343), with four distorted corner-sharing RuX_6 octahedra joined in a ring. Anions, $Ru^{IV}X_6^{2-}$, are common, and can readily be reduced to ruthenium(III) or, by using hydrogen, to the metal.

Ruthenium, like iron, forms an 18-electron monomeric carbonyl, $Ru(CO)_5$; but $Ru_3(CO)_{12}$ differs from $Fe_3(CO)_{12}$ in having no bridging carbonyl groups. Ruthenium also forms numerous cyclopentadienyl compounds.

The distorted octahedra of $(RuF_5)_4$ GE

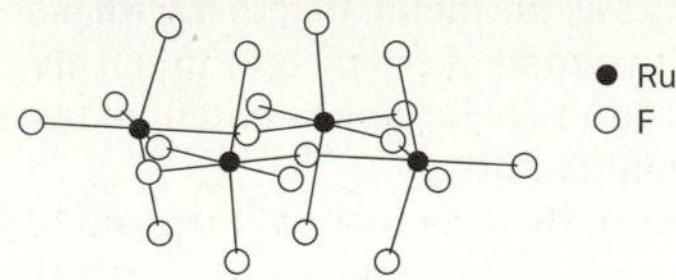

The Ru_3 cluster in $Ru_3(CO)_{12}$ GE

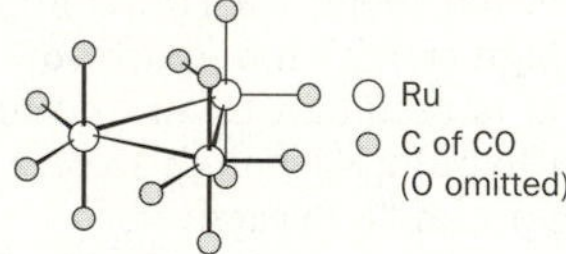

no bridging CO groups

Some Ru^{III}/Ru^{II} complexes (CW)

Complex	*Ligands*	*Charge*	*Ru atoms*
$[(NH_3)_5RuN\bigcirc NRu(NH_3)_5]^{5+}$	Symmetrical	Delocalized	Identical
$[(bipy)_2ClRuN\bigcirc NRuCl(bipy)_2]^{3+}$	Symmetrical	Probably localized	Probably one Ru^{III} and one Ru^{II}
$[(NH_3)_5Ru^{III}N\bigcirc NRu^{II}Cl(bipy)_2]^{4+}$	Unsymmetric	Localized	Non-identical: π accepting bipy stabilizes Ru^{II}

So far, the behaviour of ruthenium is unsurprising, and predictably resembles both that of technetium (but favouring lower oxidation states) and of iron (but favouring higher ones); but some of its complexes are far more interesting. Although $Ru^{2+}(aq)$ is extremely readily oxidized to $Ru^{3+}(aq)$ (a cation with no stable technetium counterpart), lower states can of course be stabilized by complex formation with π acceptor ligands to attract some of the high electron density away from the metal. So, although the species $Ru(NH_3)_6^{2+}$ is very easily oxidized, the aromatic $Ru(bipy)_3^{2+}$ is much more stable. This complex, and many similar ones, can transfer an electron completely from the metal to the ligand if excited by light, and in this condition they are much more reactive. It is hoped that such complexes may one day be used for converting solar energy into chemical energy, perhaps to catalyse the photochemical decomposition of water in order to provide dihydrogen for use as a fuel.

Other complexes that have attracted a lot of attention are those formed when one of the ammonia molecules in $Ru(NH_3)_6^{2+}$ is replaced by another ligand such as dinitrogen; the first dinitrogen complex to be made was $Ru(NH_3)_5N_2^{2+}$ and this triggered off much work aimed at the fixation of atmospheric nitrogen (see p. 83). Ruthenium(II), like iron(II), can add a molecule of nitric oxide, NO: only one, but the reaction is so favourable that it seems to happen whenever ruthenium(II) meets nitric acid or the nitrite ion. This causes problems in nuclear reactors, where ^{106}Ru is a major waste product of uranium and plutonium fission. Uranium and plutonium can be separated from the other fission products because the two heavy metals can be extracted into an organic solvent from nitric acid, while the nitrates of most lighter metals stay in the aqueous solution. Unfortunately, because the $RuNO^{2+}$ complex is also appreciably soluble in the organic solvent, multistage extraction must be used.

Another family of complexes consists of two $-Ru(NH_3)_5$ or similar groups bridged together, usually by two connected nitrogen atoms such as $-N-N-$, or, more often, two nitrogen atoms *para*- to each other in an aromatic system such as pyrazine. If one of the ruthenium atoms is formally in a lower oxidation state than the other, the excess charge may be delocalized over the two atoms; or it may not; it depends on the nature of the bridge, and perhaps also on the other ligands present. Work on such complexes helps us to understand how electrons are transferred during redox reactions. Despite ruthenium's tendency to favour higher oxidation states for its simpler compounds, its lower oxidation state complexes seem to be much more fun.

For summary, see pp. 371–5.

Rh Rhodium

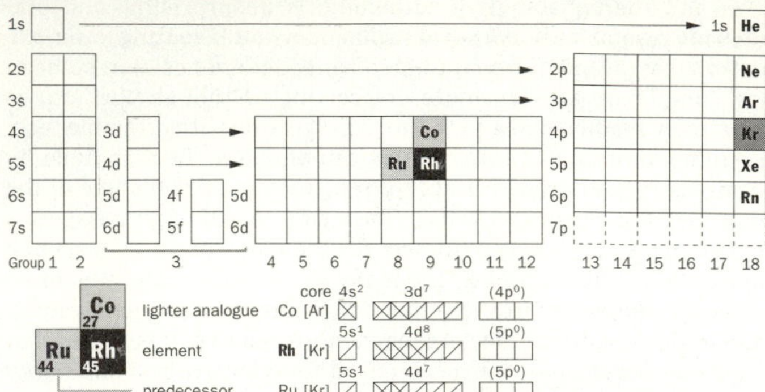

		Predecessor	Element	Lighter analogue
Name		Ruthenium	Rhodium	Cobalt
Symbol, Z		Ru, 44	Rh, 45	Co, 27
RAM		101.07	102.90550	58.93320
Radius/pm	atomic	134	134.5	125.3
	covalent	124	125	116
	ionic, M^{2+}		86	82
	M^{3+}	77	75	64
	M^{4+}	65	67	
Electronegativity (Pauling)		2.2	2.3	1.88
Melting point/K		2583	2239	1768
Boiling point/K		4173	4000	3143
$\Delta H^{\ominus}_{fus}$/kJ mol^{-1}		23.7	21.55	15.2
$\Delta H^{\ominus}_{vap}$/kJ mol^{-1}		567.8	495.4	382.4
Density (at 293 K)/kg m^{-3}		12 370	12 410	890
Electrical conductivity/Ω^{-1}m^{-1}		1.316×10^7	2.217×10^7	1.602×10^7
Ionization energy for removal of jth electron/kJ mol^{-1}	$j = 1$	711	720	760.0
	$j = 2$	1617	1744	1646
	$j = 3$	2747	2997	3232
Electron affinity/kJ mol^{-1}		101	109.7	63.8
$E^{\ominus}$/V for $M^{3+}(aq) \rightarrow M(s)$			0.76	−0.277

Stable isotopes of rhodium

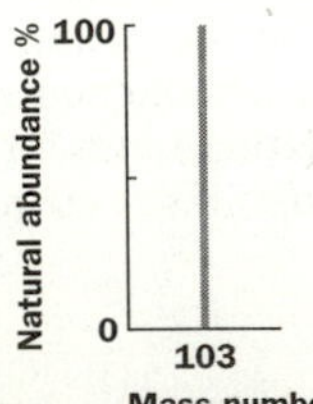

Rhodium ([Kr]$4d^85s^1$), like ruthenium ([Kr]$4d^75s^1$), is a very rare, corrosion-resistant metal, which is extracted as a sponge or powder from platinum residues. It, too, is used as a catalyst; for example, to lessen pollution from car exhausts. It is attacked by oxygen and halogens only if it is red hot, and is unaffected by any acid, even highly oxidizing ones. Very forceful methods are needed to get rhodium into solution: the metal is heated in a sealed tube with concentrated hydrochloric acid and sodium chlorate (using stringent safety precautions!).

As expected, withdrawal of d electrons into the core is more marked for rhodium than for ruthenium; the metallic bonding is weaker and the higher oxidation states of rhodium(VI) and (V) are rare, predictably occurring mainly with fluoride.

The hexafluoride, RhF_6, and the tetrameric cyclic pentafluoride, $(RhF_5)_4$, resemble their ruthenium counterparts. The commonest oxidation state is rhodium(III) ($4d^6$) which resembles cobalt(III) ($3d^6$) in forming a huge range of octahedral complexes that are diamagnetic, thermodynamically stable, and kinetically inert because of their low spin configuration of three pairs of electrons filling the three low energy t_{2g} orbitals (see p. 263). Solutions of rhodium(III) are often rose-pink, hence its name, from the Greek for 'rose'. As with cobalt(III) complexes, many different isomers can be made. There is a simple aquocation, $Rh(H_2O)_6^{3+}$, but, unlike ruthenium, no oxoanion; rhodium and later elements in the series have too many d electrons to act as π acceptors of electrons from oxygen. The (III) state provides the most important oxide, Rh_2O_3, and all four halides, MX_3. Even with low field ions such as halides, the anionic complexes MX_6^{3-} are low spin and inert (except for the iodide). Rhodium(III) is a somewhat more polarizing, and so 'softer', ion than cobalt(III) and has a greater tendency to form complexes with phosphorus or sulfur donors, rather than with those that bind through nitrogen or oxygen. It also forms a variety of hydride complexes, such as $Rh^{III}H(NH_3)_5^{2+}$, which is stable in air even though it contains no π acceptor ligands to drain off excess charge.

Rhodium(II) compounds are not very common, and often have metal–metal bonds, as in the aquo ion $(Rh_2)^{4+}$ (aq) and the carboxylate dimers, which resemble those of chromium(II), copper(II), and molybdenum(II) (see pp. 237, 277, and 349).

A rhodium complex that has received a lot of attention is the (almost) square-planar $Rh^ICl(PPh_3)_3$.* Known as Wilkinson's catalyst, it can catalyse homogeneous hydrogenation by undergoing a

* Ph = phenyl, C_6H_5.

Two Rh carbonyls [GE]

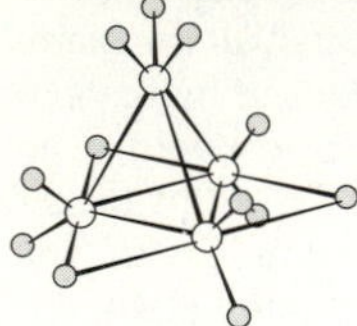

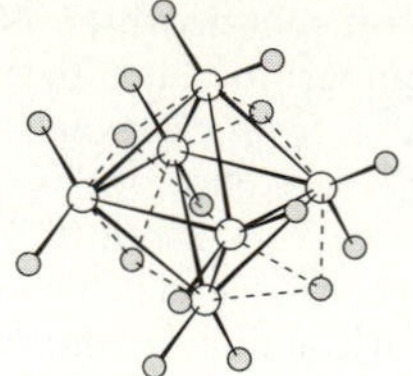

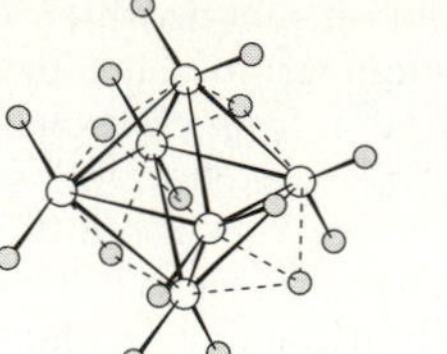

$Rh_4(CO)_{12}$

Rh_4 tetrahedron

bridging CO groups on 3 Rh–Rh bonds at base

2 terminal CO on each basal Rh

3 terminal CO on Rh at apex

$Rh_6(CO)_{16}$

Rh_6 octahedron

4 opposite faces have a CO group above them

2 terminal CO on each Rh

series of reactions in which the coordination number of the rhodium can change up to 5 and 6, and down to 3, before returning to 4.

We should not expect rhodium, with an odd number of electrons, to form a monomeric carbonyl; and its most stable compound with carbon monoxide is not the dimeric analogue of $Co_2(CO)_8$, but the tetramer, $Rh_4(CO)_{12}$, which contains the tetrahedral Rh_4 cluster. Another carbonyl, $Rh_6(CO)_6$, is based on the Rh_6 octahedron. Both have bridging carbonyl groups like those that occur in the iron and cobalt carbonyls (but not in the ruthenium carbonyl $Ru_3(CO)_{12}$, see p. 355). Rhodium also forms a number of bizarre compounds containing up to 22 rhodium atoms, 35 carbonyls, and one or more atoms of another element (such as hydrogen, phosphorus, or sulfur). Sometimes these form the filling of a triple decker crown ether sandwich. Fortunately, such complications are outside the scope of this book, except to indicate the scope of diversity now open to synthetic organometallic chemists.

For summary, see pp. 371–5.

Pd Palladium

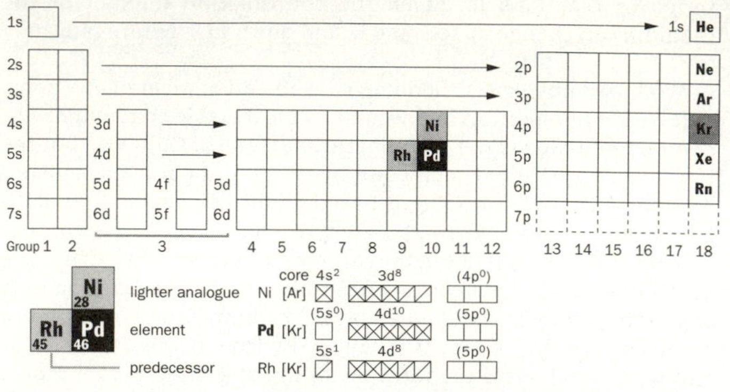

		Predecessor	Element	Lighter analogue
Name		**Rhodium**	**Palladium**	**Nickel**
Symbol, *Z*		Rh, 45	Pd, 46	Ni, 28
RAM		102.90550	106.42	58.69
Radius/pm	**atomic**	134.5	137.6	124.6
	covalent	125	128	115
	ionic, M^{2+}	86	86	78
	M^{4+}	67	64	62
Electronegativity (Pauling)		2.3	2.2	1.91
Melting point/K		2239	1825	1726
Boiling point/K		4000	3413	3005
$\Delta H^{\ominus}_{fus}$/kJ mol^{-1}		21.55	17.2	17.6
$\Delta H^{\ominus}_{vap}$/kJ mol^{-1}		495.4	393.3	371.8
Density/kg m^{-3}		12 410 (293 K)	12 020 (293 K)	8902 (298 K)
Electrical conductivity/Ω^{-1}m^{-1}		2.217×10^7	9.259×10^6	1.462×10^7
Ionization energy for removal of *j*th electron/kJ mol^{-1}	***j* = 1**	720	805	736.7
	***j* = 2**	1744	1875	1753.0
	***j* = 3**	2997	3177	3393
Electron affinity/kJ mol^{-1}		109.7	53.7	156
$E^{\ominus}$/V for $M^{2+}(aq) \rightarrow M(s)$			0.915	–0257

Stable isotopes of palladium

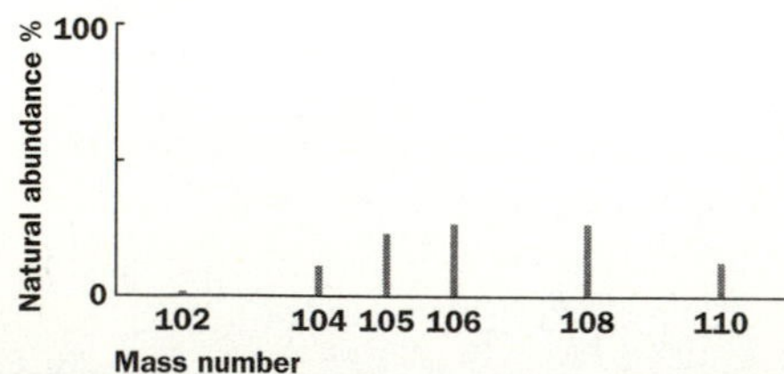

Palladium is another rare metal that occurs with platinum and is extracted from its residues. It, too, is resistant to corrosion by air non-oxidizing acids, and is used mainly as a catalyst (particularly for addition or removal of hydrogen). Its metallic bonding is appreciably weaker than that of rhodium. The electronic structure of its isolated atom is $[Kr]4d^{10}5s^0$ (rather than $[Kr]4d^n5s^1$, like most of the 4d transition elements), doubtless reflecting the increasing stability of the 4d orbitals and the similar energies of the 4d and 5s shells.

Palladium is more reactive than rhodium and dissolves slowly in oxidizing acids as well as in fused alkali. It reacts at red heat not only with oxygen and fluorine, but also with chlorine. But its most surprising characteristic is its ability to absorb up to 935 times its own volume of hydrogen. At a composition of about $PdH_{0.5}$, the material becomes a semiconductor but does not lose its metallic ductility. The hydrogen (unlike any other gas, except deuterium) seems to be mobile within the lattice, which makes palladium useful in gas separation. In other ways palladium behaves much as we should expect, favouring an oxidation state of (II), $4d^8$ (as opposed to rhodium, which favours (III), $4d^6$). Unlike nickel(II), $3d^8$, palladium(II), with its greater sensitivity to the ligand field, forms mainly diamagnetic square-planar complexes, including even the aquo ion $Pd(H_2O)_4^{2+}$, the paramagnetic fluoride. PdF_2 is an exception. The usual oxide of palladium is PdO, and all four palladium dihalides are known. There are two forms of the dichloride, in both of which the palladium ion is surrounded by a square of chloride ions. One is a chain linked by pairs of chlorine bridges; but the other contains Pd_6Cl_{12} units, similar to those formed by niobium (see p. 342).

Palladium(II) forms many complex ions ML_4^{2+} with uncharged ligands such as ammonia and amines, some of which may be replaced by halide ions, often to form ML_2X_2 molecules. The square $4d^8$ configuration is so stable that any disruption to form reaction intermediates is unfavourable, and so the complexes react sluggishly, and different isomers can often be made (see p. 233). As with many other transition metal ions, the ethanoate is not a monomer; but palladium(II), unusually, forms a trimer, held together only by carboxylate bridges with no reinforcement by Pd–Pd bonds. Palladium(II) can be classified as a class 'b' metal ion (see p. 315), as it combines preferentially with ligands that contain heavier donor atoms such as sulfur, phosphorus, and arsenic. Many of the complexes of palladium(II) resemble those of platinum(II), which have been more fully studied (see p. 483).

The Pd ethanoate trimer GE

Bonding of Pd to an allyl group GE

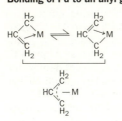

Other oxidation states of palladium are much less important. The 'trifluoride' is not PdF_3 but $Pd^{2+}[Pd^{IV}F_6]^{2-}$, in which fluorine, as usual, brings out the highest possible oxidation state. The palladium(IV) is in a low spin octahedral environment. [The ligand field splitting is greater for the 4d metal ion than for those in the 3d series, and in both cases increases with the charge on the ion; so even the very weak field provided by fluoride provides enough stabilization for $Pd^{4+}(4d^6)$ to overcome the greater electronic repulsion in the spin paired state.]

Oxidation states below (II) are also rare. Palladium (unlike nickel) forms no stable simple carbonyl. It can, however, form addition compounds with ethene and its analogues; and it can combine with allyl radicals such as $CH_2=CH-CH_2-$, forming both σ bonds and three-point attachment π bonds. Nickel and platinum behave in the same way, but palladium shows a particular preference for such ligands.

For summary, see pp. 371–5.

Ag Silver

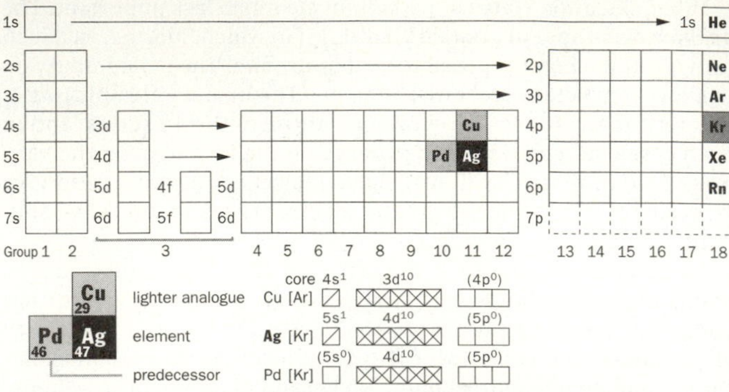

		Predecessor	Element	Lighter analogue
Name		**Palladium**	**Silver**	**Copper**
Symbol, *Z*		Pd, 46	Ag, 47	Cu, 29
RAM		106.42	107.8682	63.546
Radius/pm	**atomic**	137.6	144.4	127.8
	covalent	128	134	117
	ionic, M^+		113	96
	M^{2+}	86	89	72
Electronegativity (Pauling)		2.2	1.9	1.90
Melting point/K		1825	1235.08	1356.6
Boiling point/K		3413	2485	2840
$\Delta H^{\ominus}_{fus}$/kJ mol^{-1}		17.2	11.3	13.0
$\Delta H^{\ominus}_{vap}$/kJ mol^{-1}		393.3	255.1	304.6
Density (at 293 K)/kg m^{-3}		12 020	10 500	8960
Electrical conductivity/Ω^{-1}m^{-1}		9.259×10^6	6.289×10^7	5.977×10^7
Ionization energy for removal of *j*th electron/kJ mol^{-1}	***j* = 1**	805	731	745.4
	***j* = 2**	1875	2073	1958
	***j* = 3**	3177	3361	3554
Electron affinity/kJ mol^{-1}		53.7	125.7	118.5
$E^{\ominus}$/V for	**$M^{2+}(aq) \rightarrow M(s)$**		1.98	0.159
	$M^{+}(aq) \rightarrow M(s)$		0.80	0.520

Stable isotopes of silver

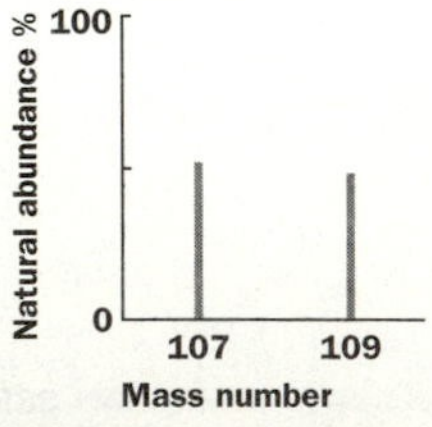

Unlike the other 4d transition metals, silver has been known since ancient times. It is less rare than its immediate predecessors and it, too, can occur uncombined (or as 'electrum', a natural alloy with gold). As it is stable in damp air, it was used for coinage and for jewellery and small artefacts.

Silver ([Kr]$4d^{10}5s^1$) differs in many ways from ruthenium, rhodium, and palladium. The d electrons are less involved in bonding and the metal is much softer and lower melting. It has high thermal and electrical conductivities. Silver is also much less resistant to chemical attack, which usually leads to oxidation state (I). It tarnishes in sulfur-ridden atmospheres to black Ag_2S, and often occurs combined with sulfur. The main source of silver is the residues of processes for extracting other metals (such as copper, zinc, and lead) from their sulfide ores. The metal is still used for jewellery and silver ware, and also for the backs of mirrors. Silver salts form the basis of photography (in colour as well as in black and white) and are also used in batteries.

Silver dissolves not only in hot concentrated sulfuric acid, but also in cold nitric acid to form the $Ag^+(aq)$ ion (which has the d^{10} configuration and so is dilute, colourless, and diamagnetic). As we can guess from its ready reaction with sulfur, silver(I) is a strongly polarizing, 'class b' metal, which favours donor groups such as iodide and cyanide. However, silver(I) also forms complexes with nitrogen donors. The familiar cation $Ag(NH_3)_2^+$, which causes solid AgCl to dissolve in aqueous ammonia, demonstrates the preference of silver(I) for coordination with only two groups to form a linear complex, and it is unusual in that, contrary to statistical probability, the second ammonia molecule is bound more strongly than the first. Ligands with two donor atoms often combine with the Ag^+ ion

Stepwise stability constants of silver ammines

A stepwise stability constant, K_n, is the equilibrium constant for the formation of the nth complex from the previous one.

$ML_{n-1} + L \rightleftharpoons ML_n \quad K_n$

and an overall stability constant, β_n, is that for its formation from a metal ion and n ligands

$M + nL \rightleftharpoons ML_n \quad \beta_n$

(charges are omitted and solvation of all species is assumed)

For Ag^+–NH_3 complexes at 30°C in 2M NH_4NO_3

$\log K_1 = 3.20 \qquad \log K_2 = 3.83$

[More typical behaviour is shown by Cu^{2+}–NH_3 complexes under the same conditions:

$\log K_1 = 4.15 \qquad \log K_2 = 3.50 \qquad \log K_3 = 2.89 \qquad \log K_4 = 2.13$]

Chains containing (almost) linearly bound Ag^{I}

–Ag–C–N–Ag–C–N–

$(AgCN)_\infty$

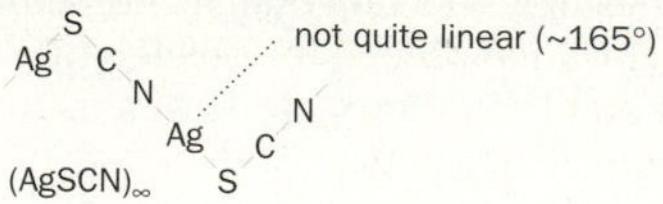

$(AgSCN)_\infty$

Ag–S–C–N···Ag–S–C–N (not quite linear (~165°))

Solubilities of Ag^+ [and Na^+] halides [D]

Data/(kJ mol^{-1}) for reactions

$$MX(s) \xrightarrow{aq} M^+(aq) + X^-(aq)$$

	ΔG°			ΔH°		$T\Delta S^\circ$	
AgF	–15	soluble	decreasing solubility ↓	–22	stronger (partly covalent) lattice ↓; weaker hydration of X^- ↓	–7	decreased ordering of solvent by X^- ↓
AgCl	56	insoluble		66		10	
AgBr	70	insoluble		85		15	
AgI	91	insoluble		111		20	
					both factors decrease solubility		increases solubility but outweighed by ΔH°
NaF	3	sparingly soluble	increasing solubility ↓	1	weaker (mainly ionic) lattice ↓; weaker hydration of X^- ↓	–2[a]	decreased ordering of solvent by X^- ↓
NaCl	–9	soluble		4		13[b]	
NaBr	–19	soluble		–1		18[c]	
NaI	–32	soluble		–9		23[c]	
					factors opposed and similar magnitude		

[a] $T\Delta S^\circ$ reinforces ΔH° to give positive ΔG°

[b] $T\Delta S^\circ$ overrides small ΔH° to give negative ΔG°

[c] $T\Delta S^\circ$ reinforces ΔH° to give negative ΔG°

(as with other d^{10} ions such as Cu^+ and Hg^{2+}) to give bridged polynuclear complexes rather than chelate ones, and so allow the metal ion to keep its linear environment.

Silver forms all four monohalides. The fluoride alone is soluble in water. The ionic structures of the heavier (and more polarizable) halides, particularly of the iodide, are stabilized by covalent overlap, while the smaller fluoride ion has the greater tendency to be hydrated. As usual, however, the various energy terms are rather finely balanced. The insoluble halides are all somewhat soluble in concentrated solutions of halide ions, forming linear AgX_2^- complexes. The sensitivity of the silver halides to light is caused by photoreduction of the metal ions by the halide, to give clusters of silver atoms together with halogen molecules (which escape by diffusion). Photography makes use of the bromide or iodide, and the minute amount of photoreduction that occurs when light falls on the film is then amplified by 'developing' it. Naturally, the remaining photosensitive halide must be removed before the film is brought into the light, and this is accomplished by 'fixing' the film with sodium thiosulfate,* $Na_2S_2O_3$, and washing out the $Ag(S_2O_3)_2^{3-}$ ions that are formed.

Silver(I) cyanide and thiocyanate are also insoluble in water but form AgX_2^- ions with excess of ligand. Unlike the halides, they have a chain structure. The Ag^+ ion also forms complexes with a number of simple π acceptors such as derivatives of ethene and ethyne.

The main oxide of silver is Ag_2O which, when strongly oxidized, gives 'AgO'. As this higher oxide is diamagnetic, it cannot contain silver (II) d^9, but must be a mixed oxide $Ag^IAg^{III}O_2$. Silver does, in fact, form a few compounds in the (II) state. In some, the cation is protected from reduction by bulky ligands, as in $Ag(py)_4^{2+}$ in combination with a non-reducing anion; and in others, such as AgF_2 and $Ba^{2+}AgF_4^{2-}$, fluorine can again stabilize a higher oxidation state. Indeed there is also a fluoro silver(III) anion, AgF_6^{3-}. But these are oddities, in which silver is being forced into transition metal behaviour. More typically, it forms the $4d^{10}$ ion, Ag^+, with no unpaired electrons, no ligand field stabilization, and no colour: with no claim, in fact, to be considered a transition metal ion.

For summary, see pp. 371–5.

* Known to photographers as 'hypo'.

The second (4d) transition series: trends and summaries Zr–Ag

The metals in which the 4d shell is being filled show typical transition metal behaviour in that they have compounds of different oxidation states, often coloured and sometimes paramagnetic; they may bond with either or both σ donors and π acceptors; and many of their compounds are non-stoichiometric. However, the members of the second transition series differ from their 3d analogues in a number of important ways. The metals are denser, but also harder and more resistant to corrosion; although, in the latter part of the series, the metals become softer as the d electrons are increasingly withdrawn into the core. Several elements in the middle of the series have melting points so high that they must be worked by powder metallurgy. The later metals, such as palladium and silver, are also more reactive. The 4d metals form fewer simple cations and aquo ions, and may exist in aqueous solution as polymerized hydroxo complexes. The earlier metals often form a wide range of oxoanions.

Compared with the 3d metals, the 4d series favours higher oxidation states: the (II) state is often strongly reducing, while the highest states, such as niobium(V), molybdenum(VI), and technetium(VII), are more weakly oxidizing. As with the 3d elements, the higher oxidation states of the 4d metals are less common towards the end of the series. Those that exist as cations form complexes with 'softer', more deformable donor atoms, such as sulfur and phosphorus, in preference to oxygen and nitrogen. Complexes are more often low spin and, if paramagnetic, their magnetic moments are more difficult to interpret (since the spin and orbital moments of any unpaired electron interact more strongly). Spectra are also more complicated in the 4d series. Although some higher carbonyls of the 3d metals [and the ethanoates of chromium(II) and copper(II)] contain metal–metal bonds, these are much more common in the 4d elements; compounds of the metal in a low formal oxidation state commonly contain metal clusters stabilized by π acceptor ligands.

As 4d atoms are larger than their 3d counterparts, more donor atoms can be fitted round them. The octahedral groups favoured by the 4d series give much more scope for geometrical variety than the tetrahedra favoured by some 3d metals (compare, for example, chromium and molybdenum) and this accounts for the much greater diversity of oxoanions and halides. The disulfides of the earlier metals contain M^{IV} and have graphitic layer structures, which can often accommodate guest molecules; but those of the later elements contain M^{II} and have the pyrites structure, $M^{2+}S_2^{2-}$. The table that follows gives a very brief comparative summary of the elements zirconium to silver.

Comparative summaries of elements

Element	Zr	Nb	Mo	Tc
Outer configuration	$4d^25s^2$	$4d^45s^1$	$4d^55s^1$	$4d^65s^1$
[3d analogue]	[Ti, $3d^24s^2$]	[V, $3d^34s^2$]	[Cr, $3d^54s^1$]	[Mn, $3d^54s^2$]
Freedom of 4d electrons	Little nuclear control	Little nuclear control	Slight withdrawal	Increased withdrawal
Main oxidation states, redox stability, e.g.	**IV** ZrO_2 high m.p (CN 7—unlike TiO_2) ZrS_2 layer (like TiS_2); simple salt hydrates; polymerized hydrolysis products. No monomeric oxocations or anions.	**V** Nb_2O_5 and NbO_3^-, huge variety of oxoanions from NbO_6 groups, no aquo or oxocations; $(NbX_5)_4$ cyclic **IV** mainly halides (containing Nb_2 and NbX_6), NbS_2 graphitic layer (intercalates)	**VI** MoO_3 not oxidizing [MoF_6 strongly oxidizing] **V** halides (like Nb) **IV** halides (like Nb); MoO_2 contains Mo_2 pairs (like Nb); MoS_2 graphitic (like Nb) **III** ionic halides (like Cr), simple Mo^{3+}(aq)	**VII** TcO_4^- less oxidizing th[illegible] MnO_4^-; Tc_2O[illegible] more stable than Mn_2O_7 Tc_2S_7; TcF_6 **IV** TcO_2 has Tc[illegible] pairs (like Mo); TcS_2 graphitic (like Mo)
[Less common oxidation states]	[III reduces water 'ZrX_3'] [prob. dimeric II] [I unstable halides]	[II rare, oxides and halides (contain Nb_6)]	[II only in complexes, usually Mo–Mo bonding]	[II dispropor-tionates]
Found combined with (and in association with)	O	O, (Fe, Mn, Ta)	O (MoO_3) and S (MoS_2)	No natural isotope
Extraction of metal	Reduction of volatile chloride by Mg (like Ti)	Reduction of Nb_2O_5 by C (formerly from halides)	Reduction of MoO_3 with H_2 (worked by powder metallurgy)	Made by neutron bombardment of U

Zr to Ag: [Kr]$(4d + 5s)^{y+2}$

Ru	Rh	Pd	Ag
$4d^7 5s^1$	$4d^8 5s^1$	$4d^{10} 5s^0$	$4d^{10} 5s^1$
[Fe, $3d^6 4s^2$]	[Co, $3d^7 4s^2$]	[Ni, $3d^8 4s^2$]	[Cu, $3d^9 4s^2$]
Appreciable withdrawal	Higher withdrawal	Largely withdrawn	Almost totally withdrawn
IV RuO_2 (no Ru_2 pairs, unlike Fe, Mo, etc.) **III** halides (often from RuX_6 units) low spin d^5; no oxide	**III** $Rh(H_2O)^{3+}_6$ Rh_2O_3, RhX_3; many complexes (inert); favours soft ligands; (no oxoanions)	**II** favours soft ligands, low spin, square, diamag., fairly inert	**I** Ag^+(aq), favours soft ligand, linear, CN = 2
[VIII RuO_4 strongly oxidizing] [VII RuO_2^- oxidizes water] [VI RuO_4^{2-}, RuF_6] [II RuS_2 (pyrites $Ru^{2+}S_2^{2-}$ unlike graphitic $M^{IV}(S^{2-})_2$ in Zr–Tc)]	[VI RhF_6] [V $(RhF_5)_4$ cyclic (like Nb, Mo)] [II Rh_2^{4+} and carboxylate dimers (like Mo)] [I Rh_2^{4+}, Wilkinson's catalyst RhCl $(Ph_3)_3$]	[IV in IV/II mixed valence fluorides]	[II with F^- (AgF_4^{2-} + $Ag(py)_2^{2+}$] [III with F^- and O(AgF_6^{3-}, $Ag^I Ag^{III} O_2$)]
Native (v. rare, with Pt group)	Native (v. rare, with Pt metals)	Native (rare, with Pt group)	Native, and with S (Cu, Zu, Pb)
Reduction of $RuCl_6^{3-}$ by H_2. Worked by powder metallurgy	Reduction of $RhCl_6^{3-}$ with H_2. Worked by powder metallurgy	Heat $Pd(NH_3)_2Cl_2$	From residues of other processes

Element	Zr	Nb	Mo	Tc
Physical properties of metal	High melting metal, fair metallic conductor	Hard, high melting metal	Hard, v. high melting metal	Hard
Reaction with damp air	Corrosion-resistant oxide layer	Protective oxide layer (but attacked by conc. acids and fused alkali)	V. corrosion resistant	Corrosion resistant (less so than Mo)
Uses of metal	• Cladding in nuclear reactors, • in chemical plants • stainless steel • super-conducting magnets (with Nb)	• Alloys, • stainless steel • super-conducting magnets (with Zr)	• Electrodes, • catalyst • sulfur scavenger	
Main complexes with CO, C_5H_5, and other organic ligands	$Zr(C_5H_5)_4$ $1 \times \sigma, 3 \times \pi$	$Nb(CO)_6$ (like V) $Nb(C_5H_5)_4$ [bonded like $Ti(C_5H_5)_4$]	$Mo(CO)_6$ unstable Mo (benzene)$_2$ sandwich (like Cr)	
Points of special chemical interest	• Resembles Ti (but less than Y does Sc) CN 7 in fibrous ZrO_2	• $LiNbO_3$ piezoelectric	• VI many iso- and hetero-polyanions based on MoO_6 and MoO_4 groups • VI/V, mixed oxide bronzes • II, $Mo_6X_8^{4-}$ (Mo_6 cluster, like Nb). • Mo_2^{4+}(aq)$Mo_2Cl_8^{4-}$ (quad-ruple bond), • carboxylate-bridged dimers (like Cr)	• CN 9 in TcH_9^{2-}
Uses of compounds	ZrO_2 in filters, insulators, furnace linings		MoS_2 as lubricant	• γ-emitting isotope in medical diagnoses
Biological function			Enzyme for reductions, e.g. N fixation	

Ru	Rh	Pd	Ag
Hard and corrosion resistant (less so than Tc)	Hard and extremely corrosion resistant	Hard (less so than Rh)	Fairly soft, low melting metal, good electrical conductor
Stable (but sol. in fused alkali)	None	Fairly unreactive (less so than Rh). Soluble in oxid. acids	Soluble in oxid. acids, Stable in air unless S present
• Catalyst • hardens Pt	• Catalyst (cars)	• Catalyst • H_2 absorbent	• Jewellery, • coins (formerly) • small artefacts • mirrors
$Ru(CO)_5$ $Ru_3(CO)_{12}$ (no CO bridges) several C_5H_5 derivatives	$Rh_4(CO)_{12}$ and $Rh_6(CO)_6$ cont. metal clusters (unlike Co) and CO bridges (unlike Ru) Many complex CO derivatives	No stable carbonyl, ethene or allyl derivatives	No CO or C_5H_5 complexes Ag^+ combines with C_2H_4, C_2H_2 and derivatives
• Ru^{II} π acceptor complexes, e.g. with N_2, NO		• Metal absorbs large volumes H_2 • Ethanoate trimeric (no Pd–Pd bonds),	AgX photoreducible
Ru^{II} π complexes potentially for solar energy and N_2 fixation			AgX (photography), Ag_2O in batteries

Cd Cadmium

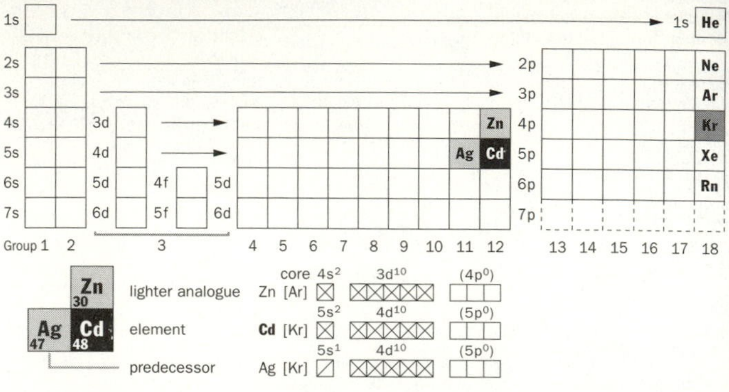

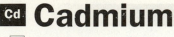

		Predecessor	Element	Lighter analogue
Name		**Silver**	**Cadmium**	**Zinc**
Symbol, *Z*		Ag, 47	Cd, 48	Zn, 30
RAM		107.8682	112.411	65.39
Radius/pm	**atomic**	144.4	148.9	133.2
	covalent	134	141	125
	ionic, M^+	113	114	
	M^{2+}	89	103	83
Electronegativity (Pauling)		1.9	1.7	1.65
Melting point/K		1235.08	594.1	692.73
Boiling point/K		2485	1038	1180
$\Delta H^{\ominus}_{fus}$/kJ mol^{-1}		11.3	6.11	6.67
$\Delta H^{\ominus}_{vap}$/kJ mol^{-1}		255.1	99.87	115.3
Density (at 293 K)/kg m^{-3}		10 500	8650	7133
Electrical conductivity/Ω^{-1}m^{-1}		6.289×10^7	1.464×10^7	1.690×10^7
Ionization energy for removal of *j*th electron/kJ mol^{-1}	***j* = 1**	731	867.6	906.4
	***j* = 2**	2073	1631	1733.3
Electron affinity/kJ mol^{-1}		125.7	−26	9
$E^{\ominus}$/V for $M^{2+}(aq) \rightarrow M(s)$			−0.402	−0.763

Stable isotopes of cadmium

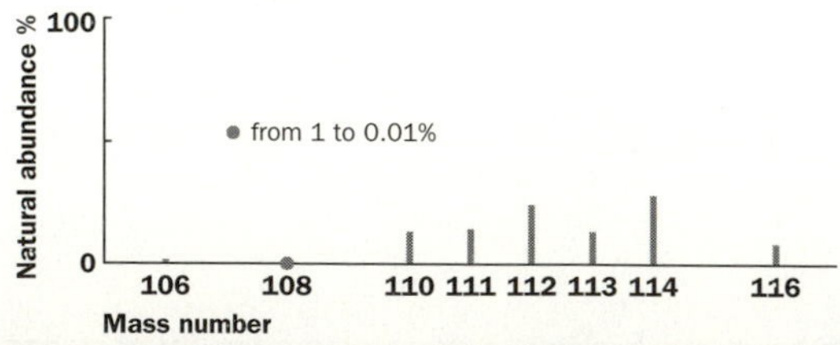

At cadmium ([Kr]$4d^{10}5s^2$) both the 4d and the 5s shells are complete. The 4d shell has been totally absorbed into the inner 'core' and no longer plays any part in bonding. The 4d 'transition series' is over. The metallic bonding is much weaker even than in silver (see p. 367), as only the two 5s electrons are available. Cadmium metal is quite volatile at high temperatures and can be separated from zinc by fractional distillation. Moreover, the third ionization potential is dramatically higher for cadmium than for silver, $4d^{10}5s^1$, in which the d electrons are more accessible.

The behaviour of cadmium is governed largely by the 5s and 5p orbitals. Since the two 5s electrons are fairly easily removed or promoted, cadmium almost always adopts the (II) oxidation state. Despite its krypton core, the radius of the Cd^{2+} ion, ([Kr]$4d^{10}5s^0$, is very similar to that of the calcium Ca^{2+}ion, ([Ar]$3d^0$)$4s^0$, because the d electrons are poor screeners and so only partly counteract the increased charge on the nucleus. However, since its higher effective nuclear charge makes the Cd^{2+} ion much more polarizing than the Ca^{2+}ion, the two cations behave very differently. Predictably, cadmium is more like zinc ([Ar]$3d^{10}$)$4s^2$ than calcium, even though zinc is smaller than either and less polarizing than cadmium. Compared with zinc (see p. 287), cadmium shows more 'class b' character, forming bonds with more covalent contribution and complexing more favourably with heavier donor groups such as sulfur.

Cadmium occurs naturally as the sulfide, but is also often present in zinc sulfide or carbonate ores, from which it is extracted. It can be used in much the same way as zinc: for coatings, alloys, and dry batteries. But, unlike zinc, cadmium is very toxic, possibly because it can replace zinc in enzymes and because it is bound to sulfur atoms in thio proteins.

Cadmium metal, like zinc, tarnishes in air and if hot combines with oxygen, sulfur, phosphorus, and the halogens, but not with hydrogen, carbon, or nitrogen. It too dissolves in dilute acids, giving off hydrogen, although the reaction is slightly less favourable than for zinc. Cadmium oxide is more basic than zinc oxide and dissolves in only concentrated aqueous alkali. The halides of the two metals are also similar: the fluorides are high melting 3D ionic structures, only sparingly soluble in water, while the other halides are very soluble, layer structures with lower melting points, and probably with appreciable covalent overlap. Cadmium forms a very stable tetrahedral tetraiodo anion, CdI_4^{2-}, and similar, but somewhat less stable, complexes with chloride and bromide.

Cadmium sulfide, unlike its zinc analogue, is coloured and is used as a pigment; it may be either red or yellow, depending on trace replacement of cadmium by zinc or mercury, and of sulfur by

The structure of CdI_2 GE

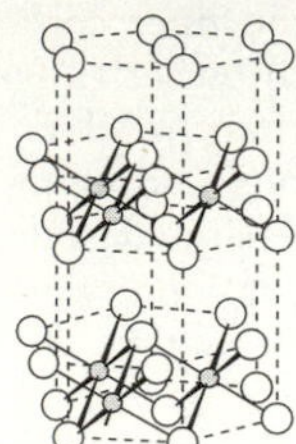

Showing CdI_6 octahedra, and hexagonal close packing of iodide ions.

[$CdCl_2$ has a similar layer structure, but with cubic close packed anions]

Summary

Electronic configuration

- ([Kr]$4d^{10}$)$5s^2$

Element

- Metal, fairly volatile, tarnishes in air and dissolves in dilute acids

Occurrence

- As sulfide, and with zinc sulfide and carbonate

Extraction

- With zinc, by reduction of the oxide with carbon; and separated from it by fractional distillation

Chemical behaviour

- 4d electrons totally withdrawn

selenium. Like zinc sulfide, cadmium sulfide and selenide are phosphors; cadmium telluride, like gallium arsenide, is a semiconductor. The greater polarizing power of cadmium, compared with zinc, is illustrated by their two tetrathiocyanato anions. Cadmium binds to the sulfur to form $Cd(SCN)_4^{2-}$, while zinc binds to the nítrogen to give $Zn(NCS)_4^{2-}$.

Cadmium does not take part in π bonding. The full $4d^{10}$ shell cannot accept electrons; and since it is now part of the 'core', it cannot donate them. So cadmium forms no compounds with carbon monoxide, and those that it forms with cyclopentadienyl groups are σ bonded and ring-whizzing (see p. 221). Cadmium also forms aryls and alkyls but these decompose in air or water to give cadmium oxide.

So far, cadmium seems to form compounds only of oxidation state (II). Has it no variety to offer us? In fact it has. The dimeric cadmium(I) cation can be made, but only under somewhat strained conditions: if cadmium metal is added to a melt of cadmium chloride and $NaAlCl_4$, the ion Cd_2^{2+} can be detected spectroscopically in the resulting $Cd_2^{-}(AlCl_4)_2$. But although this is an interesting forerunner of the familiar dimeric mercury(I) ion (see p. 497), it has no significance for the (rather prosaic) everyday chemistry of cadmium.

- Resembles zinc
- Dominant state (II) ($5s^0$) but Cd^{2+} more polarizing ('class b'), favouring 'soft' donors
- Tetrahedral halogeno anions
- No π complexes, but σ bonded organometallics
- Toxic (probably displaces zinc from enzymes)
- Cadmium(I) in melts as Cd_2^{2+}

Uses

- Metal used in coatings, alloys, and dry batteries
- Sulfide and selenide as phosphors
- Telluride as semiconductor

Nuclear features

- Eight stable isotopes; major component ^{114}Cd, 28%

In Indium

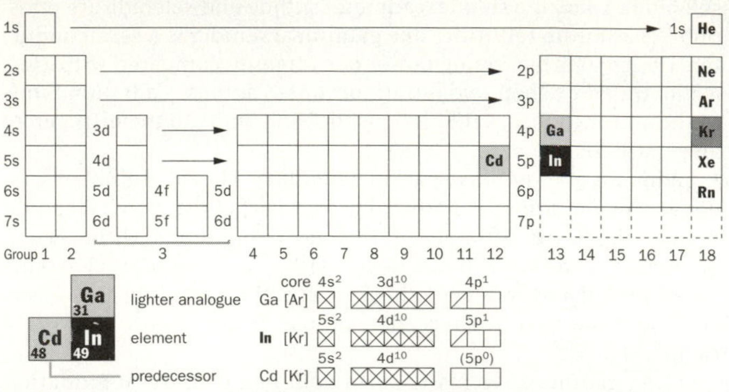

		Predecessor	Element	Lighter analogue
Name		**Cadmium**	**Indium**	**Gallium**
Symbol, *Z*		Cd, 48	In, 49	Ga, 31
RAM		112.411	114.82	69.723
Radius/pm	**atomic**	148.9	162.6	122.1
	covalent	141	150	125
	ionic, M^+		132	113
	M^{3+}		92	62
Electronegativity (Pauling)		1.7	1.8	1.81
Melting point/K		594.1	429.32	302.93
Boiling point/K		1038	2353	2676
ΔH°_{fus}/kJ mol^{-1}		6.11	3.27	5.59
ΔH°_{vap}/kJ mol^{-1}		99.87	226.4	256.1
Density/kg m^{-3}		8650 (293 K)	7310 (298 K)	5907 (293K)
Electrical conductivity/Ω^{-1}m^{-1}		1.464×10^7	1.195×10^7	3.704×10^6
Ionization energy for removal of *j*th electron/kJ mol^{-1}	***j* = 1**	867.6	558.3	578.8
	***j* = 2**	1631	1820.6	1979
	***j* = 3**	3616	2704	2963
Electron affinity/kJ mol^{-1}		–26		~ 30
E°/V for	**$M^{3+}(aq) \rightarrow M^+(aq)$**		–0.444	
	$M^+(aq) \rightarrow M(s)$		–0.338	
	$M^{3+}(aq) \rightarrow M(s)$		–0.409	–0.53

Stable and long-lived isotopes of indium

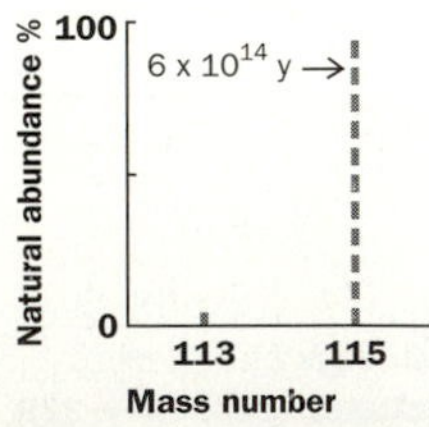

Indium was first detected from the indigo line in its spectrum and was named accordingly. With the configuration ([Kr]$4d^{10}$)$5s^2 5p^1$, it behaves much as we should expect. It is a soft, fairly low melting metal, but as it has an additional electron in the sp band, its metallic bonding is stronger than that of its predecessor cadmium, ([Kr]$4d^{10}$)$5s^2$. It, too, occurs combined with sulfur, often together with zinc, which is no surprise since both metals have full d shells and the two cations, In^{3+} and Zn^{2+}, are of similar size. Indium occurs in sulfide ores of zinc and lead and is obtained by roasting the flue dusts from the extraction of those metals. It can be purified by electrolysis. The metal is used in low melting alloys for solder and safety devices, such as sprinklers, and also in 'III–V' semiconductors (see p. 295): the arsenide and antimonide are used for photoconductors and transistors at low temperatures, while the phosphide is used for high temperature transistors. The nucleus of its major isotope, ^{115}In, has a high neutron capture cross-section, which makes indium useful for control rods in nuclear reactors.

As the outer electronic configuration of indium is the same as that of gallium, ([Ar]$3d^{10}$)$4s^2 4p^1$, we expect similar reactions, but perhaps with weaker covalent bonds formed by the larger indium; for example, indium appears to form no bond to hydrogen. Indium is both more basic than gallium and less electropositive: more basic because neither the metal, nor any oxide or hydroxide, reacts with aqueous alkali to form oxoanions, and less electropositive because, although indium dissolves in dilute acid to form $In^{3+}(aq)$, the reaction is less thermodynamically favourable than for gallium. As with gallium, the predominant oxidation state of indium is (III). The halides of the two elements are very similar: indium trifluoride is non-volatile with a 3D structure, while the other indium trihalides are volatile dimers. (However, again like gallium, but unlike aluminium, the trialkyls, InR_3, are monomers.) The 'dihalides' of indium, too, are mixed valence compounds, $In^{I}[In^{III}X_4]$, containing indium(I), which retains its two 5s electrons; and the tetrahedral ion InX_4^- can be extracted as $H^+InX_4^-$ into either. However, as indium is larger than gallium, it can accommodate further ligands; $InCl_5^{2-}$ has a square pyramidal structure, and the octahedral $InCl_6^{3-}$ has also been made.

Indium also forms monohalides and, as expected, the stability increases from the (unstable) fluoride to the iodide. Aquo cations of lower oxidation states can be made by electrolysis, but $In^+(aq)$ reduces water to dihydrogen.

Since indium is found in sulfide ores, it is no surprise that it combines also with sulfur's heavier analogues, selenium and tellurium. Its compounds with all three elements show considerable diversity.

A mixed valence indium selenide, In_4Se_3 or $In^+[In_3^{III}]^V[Se^{2-}]_3$ GE

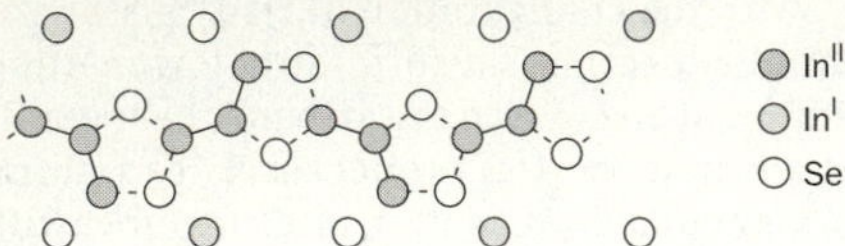

Since compounds that have an apparent formula of 'InS' are diamagnetic, they cannot contain In^{2+} (which would have an unpaired electron) but must, like the 'dihalides', be mixed valence compounds containing indium (I) and (III). Unlike most indium compounds, but like cadmium sulfide, many of these substances are coloured red, yellow, or even black. Some contain $(In^{III})_2$ pairs [analogous to the $(Cd^{II})_2$ pairs in melts] or $(In^{III})_3$ bent chains.

Despite the diversity of its sulfides and their analogues, indium seems a rather low-key element, at least compared with its more colourful 4d predecessors. It has, however, the curious ability to protest if manhandled, a talent shared only with tin. It is claimed that when it is bent it emits a high-pitched cry; but it is unlikely that many readers, any more than the author, have had the opportunity to test this for themselves.

Summary

Electronic configuration

- ([Kr]$4d^{10}$)$5s^2 5p^1$

Element

- Fairly soft, low-melting metal, but stronger than cadmium; blue flame

Occurrence

- In sulfide ores of zinc and lead

Extraction

- By roasting flue dusts from extraction of zinc and lead; purified by electrolysis

Chemical behaviour

- Similar to gallium, but forms weaker covalent bonds
- Dominant state (III)
- Dissolves in acids to give In^{3+}(aq) (but trialkyls monomers); complex halides $In^{I}In^{III}X_4$ and anions to $InCl_6^{3-}$
- InF_3 ionic; other trihalides In_2X_6
- Lower states: $In^{III}-In^{III}$ and In^+ (strongly reducing); monohalides: InI most stable

Uses

- Metal in low melting alloys
- III–IV semiconductors with As and Sb
- Control rods in nuclear reactors

Nuclear features

- Mainly ^{115}In (high neutron capture) ~ 4% ^{113}In

Sn Tin

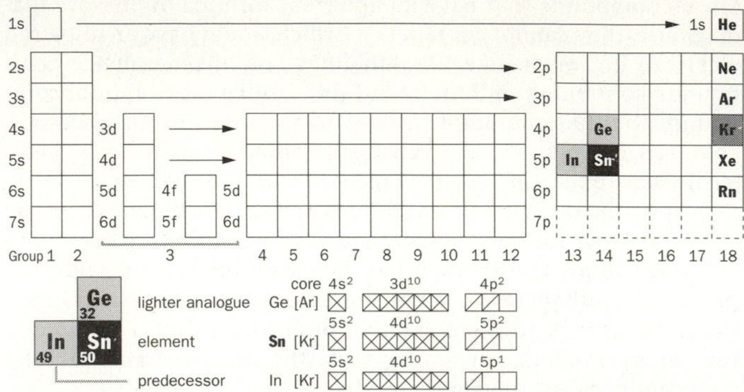

	Predecessor	Element	Lighter analogue
Name	**Indium**	**Tin**	**Germanium**
Symbol, *Z*	In, 49	Sn, 50	Ge, 32
RAM	114.82	118.710	72.61
Radius/pm atomic	162.6	140.5	122.5
covalent	150	140	122
ionic, M^{2+}		93	90
M^{3+}		74	
Electronegativity (Pauling)	1.8	2.0	2.01
Melting point/K	429.32	505.118	1210.6
Boiling point/K	2353	2543	3103
$\Delta H^{\ominus}_{fus}$/kJ mol^{-1}	3.27	7.20	34.7
$\Delta H^{\ominus}_{vap}$/kJ mol^{-1}	226.4	290.4	334.3
Density/kg m^{-3}	7310 (298 K)	5750 (α) (293 K)	5323 (293 K)
		7310 (β)	
Electrical conductivity/Ω^{-1}m^{-1}	1.195×10^7	9.091×10^6	2.174
Energy of band gap in solid/kJ mol^{-1}	0	0	64.2
Ionization energy for removal of *j*th electron/kJ mol^{-1} *j* = 1	558.3	708.6	762.1
***j* = 2**	1820.6	1411.8	1537
***j* = 3**	2704	2943.0	3302
***j* = 4**		3930.2	4410
Electron affinity/kJ mol^{-1}		116	116
Bond energy/kJ mol^{-1} E–E		195	163
E–H		< 314	288
E–C		225	
E–O		557 Sn^{II}	363
E–F		322	340
E–Cl		315	163
$E^{\ominus}$/V for $M^{4+}(aq) \rightarrow M^{2+}(aq)$		0.15	
$M^{2+}(aq) \rightarrow M(s)$		–0.137	

Tin, ([Kr]$4d^{10}$)$5s^2 5p^2$, like its predecessor, indium, is a white metal, which is said to 'cry' if you bend it. As for indium and for tin's lighter analogue, germanium, ([Ar]$3d^{10}$)$4s^2 4p^2$, most tin compounds are formed by using or losing all their (s + p) electrons, although there are also compounds in which they retain their ns^2 electrons, and these are more numerous for tin(II) than for germanium(II) or indium(I). Indeed, tin has a much higher profile than indium, chemically, industrially, and historically. It also has a larger number of stable isotopes than any other element (see p. 386).

Tin occurs mainly as the easily reducible dioxide SnO_2, which, about 3500 BC was added to copper ores to give an alloy, bronze, that was hard enough to be cast into tools, statues, and, of course, weapons. The production of pure tin is more difficult, since the ore is usually contaminated with iron which has to be removed by selective oxidation. The metal is widely used for plating iron food storage containers ('tins') as it is resistant to both damp air and food juices. Molten tin, also resistant to air, is used as a 'float' in the manufacture of sheet glass. Many alloys, in addition to bronze, contain tin: solder, type metal, pewter, Babbitt's metal (used for heavy bearings), and the niobium alloy Nb_3Sn, which is used for superconducting magnets.

Although germanium is a semiconductor with the diamond structure, tin, with its less tightly bound outer electrons, is a metal: at least at room temperature. After long exposure to low temperatures, however, tin expands grossly and becomes brittle because of a change to the diamond structure. In the low temperature 'grey' form, the Sn–Sn distance is smaller than in 'white' metallic tin, but in the lower density grey form each atom has only four close neighbours, while in the metal it has six. This change in struc-

Tin allotropes[GE]

	β	13.2°C* ⇌	α
Characteristics	tetragonal white metallic		diamond grey non-metallic
Density/g cm^{-3}	7.265		5.769
Nearest neighbours at *d*/pm	*d* 4 at 302 2 at 318 4 at 377 8 at 441		*d* 4 at 280 12 at 459

*In practice, white tin changes to grey only after long exposure to temperatures well below the transition point.

Stable isotopes of tin

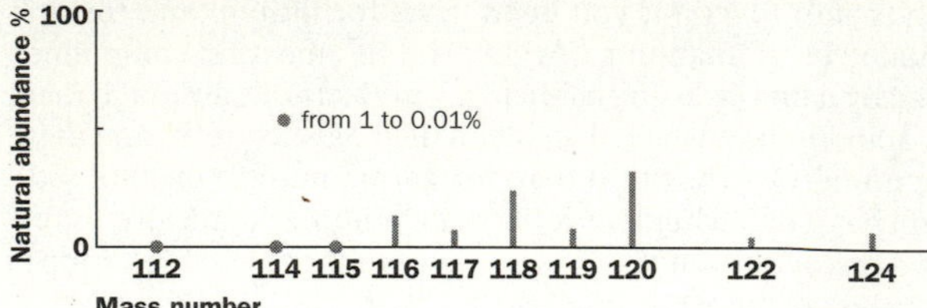

Tin difluoride 'SnF_2' GE

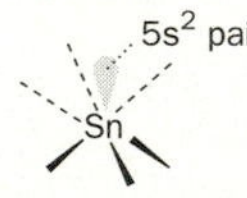

The compound is a cyclic tetramer, $[SnF(F)]_4$, with half the fluorides much nearer to the tin than the others

----- longer bonds (not all the same)
▬ shorter bonds (not all the same)

Some Sn^{II} fluoro ions GE

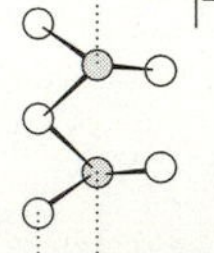

$Sn_2F_5^-$ consists of two slightly skew tetrahedra, SnF_3, with Sn at the apex of each. Corner-linked through one F, one Sn points up, and one down

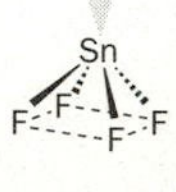

A slightly skew square pyramid SnF_4 is the basis of **$Sn_3F_{10}^{4-}$**, with three corner-linked pyramids and of **$(SnF_3^-)_\infty$** in which the pyramids are corner-linked to form a chain

ture has been responsible for loss of petrol through crumbling of the solder on the fuel cans in polar expeditions; but the story of Napoleon's army losing all its (pewter) buttons during the retreat from Moscow may, unfortunately, be only legend.

Metallic tin resists not only food juices but also dilute hydrochloric and sulfuric acids. It is, however, attacked by steam and dilute nitric acid to form Sn^{2+}(aq), and even in concentrated oxidizing acids it exists as tin(II). But it dissolves in hot concentrated alkalis to give $Sn(OH)_6^{2-}$ in which the (IV) state is stabilized by the hydroxy ligands; and with all four halogens tin forms the *tetra*halide. So it seems that for tin the energies of the (II) and (IV) states ($5s^25p^0$ and $5s^05p^0$) are fairly similar, whereas for indium the higher state is by far the more important.* Like indium, tin forms a variety of compounds with sulfur and selenium, its oxidation state depending on the proportions used; but with tellurium, tin gives only SnTe. Unlike indium, it can form hydrides, but only SnH_4 and Sn_2H_6, and these are much less stable than those of germanium or silicon.

Tin(II) compounds often have complex structures in which the metal atom sits at the apex of the pyramid; the pair of $5s^2$ non-binding electrons is not spherically symmetrical as it is in the free Sn^{2+} ion, but is occupying a directional orbital (much like the nitrogen lone pair in ammonia, see p. 9). This s orbital is sometimes described as 'distorted' or 'stereochemically active' or 'acquiring some p character'. It enables tin(II) to act as a donor; and since tin(II) can also accept electrons into unfilled 5p or 5d orbitals, it can act as both donor and acceptor simultaneously, as in $F_3B{\leftarrow}SnCl_2{\leftarrow}N(CH_3)_3$.

The structural complexity of tin(II) can be illustrated by the dihalides. The difluoride** consists of interconnected rings of $Sn_4F_4(F_4)$ tetramers in which each tin sits on top of a slightly skew triangular pyramid. Tin can also form a variety of complex ions in which the metal atom is the apex of a distorted square pyramid. The other halides are equally complicated in different ways (although the chloride, too, favours pyramids).

In contrast, the tin(IV) halides are almost predictable. The chloride, bromide, and iodide are volatile, with isolated tetrahedral molecules, which can add two halide ions to form octahedral SnX_6^{2-}

* For a more detailed discussion, see p. 503.

** Used in fluoride toothpaste (see p. 111) since tin(II) is non-toxic to mammals.

SnO GE

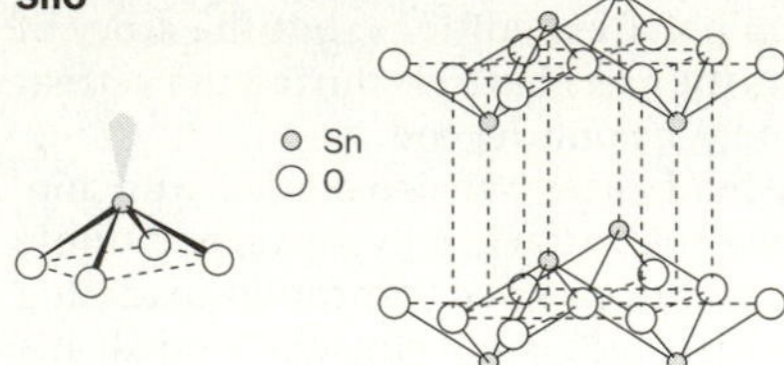

Layer structure of SnO, with Sn atoms at apex of pyramid pointing alternately up and down

Some hydrolysed Sn^{II} species GE

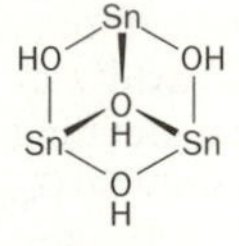

$Sn_3(OH)_4^{2+}$

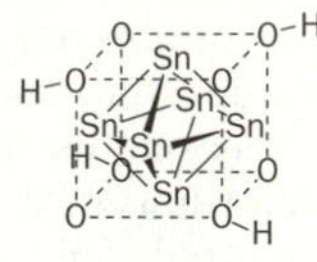

$(H^+)_4Sn_6O_8^{4-}$

SnO_2 GE

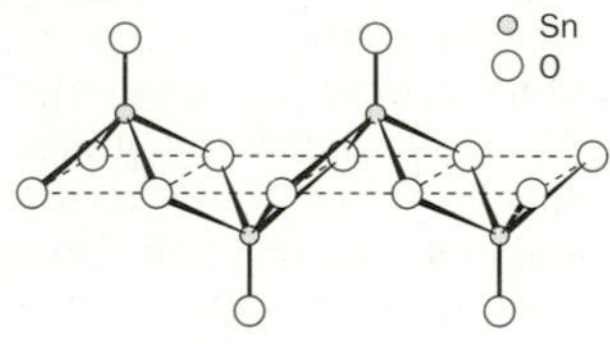

TheSnO_5 square pyramids are edge-linked, alternately up and down (compare SnO, above)

Some Sn cluster anions GE

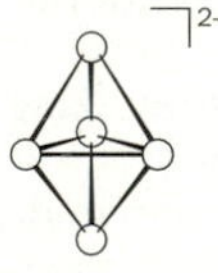

Sn_5^{2-} is a trigonal bipyramid

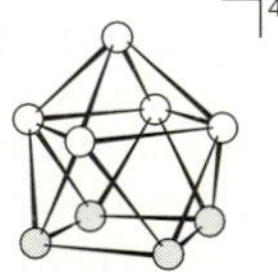

Sn_9^{4-} is a 'distorted monocapped square antiprism', i.e. a square pyramid placed diagonally over a second, smaller square (shaded)

A mixed Sn metal carbonyl complex GE

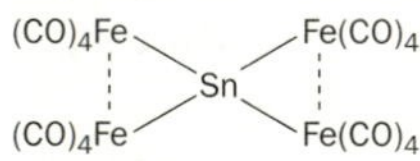

$Sn(Fe(CO)_4)_4$

complex anions. The fluoride is not quite the extended 3D ionic structure we might expect, but a layer solid that sublimes at 705°C and is made up of edge-shared SnF_6 octahedra.

There are also oxides in both oxidation states. Tin(II) oxide, SnO, can exist in several forms, the commonest one consisting of layers of square pyramidal SnO_4 units, pointing alternately up and down. It is amphoteric and dissolves in alkali to give the pyramidal $Sn(OH)_3^-$ ion. In less alkaline solutions there are condensed oxocations, such as $Sn_3(OH)_4^{2+}$, while the hydrous oxide, $3SnO.H_2O$ contains $Sn_6O_8^{4-}$ clusters, based on Sn_6 octahedra and reminiscent of $Mo_6Cl_8^{4+}$ (see p. 349). Tin(IV) oxide, SnO_2, is also amphoteric and dissolves in alkali to form $Sn(OH)_6^{2-}$. When $K_2Sn(OH)_6$ is heated, it gives a SnO_3^{2-} chain, containing tin atoms in a square pyramid that shares opposite basal faces with two adjacent groups. Tin(IV) oxide and its derivatives are used in glass making, as tougheners and opacifiers.

Tin(II) and tin(IV) both form a range of salts of oxoacids, some of which are polymerized; and some 'basic' salts containing the cation $Sn_3(OH)_4^{2+}$. If $Sn^{2+}(aq)$ is treated with a non-oxidizing but complexing ligand, such as a carboxylate, a tin(II) salt will probably be formed, whereas oxidizing but weakly complexing ligands, such as nitrate, give the tin(IV) salt, giving us yet another illustration of the delicate balance between the two oxidation states of tin. The tin(II) carboxylates are widely used as homogeneous catalysts in polymer manufacture.

An atom of tin can also join to a second metal atom. Tin–tin bonds are present in clusters with hydroxy ions (see above), and in the polyanions, Sn_5^{2-} and Sn_9^{4-}, which are formed by reduction of the metal in liquid ammonia. There are also compounds in which tin is linked to an atom of some other metal which forms part of a carbonyl complex, as in $Sn[(Fe(CO)_4]_4$ and $SnCl_2[Co(CO)_4]_2$.

Tin(IV), but not tin(II), forms a variety of organometallic compounds, mostly alkyls or aryls, SnR_4, or their halides, R_2SnX_2. These are widely used as silicone tougheners and as PVC stabilizers and coating agents. As they are toxic to fungi, insects, mites, and marine worms, but harmless to mammals, they are valuable fungicides and pesticides. Tin(II) can be lured into forming a few organic derivatives, but only if it is protected by bulky groups, as in SnL_2, where L is cyclopentadienyl or $-CH[Si(CH_3)_3]_2$.

Tin's claim to a chemically high profile certainly does not rest only on its distinguished historical past.

For summary, see p. 391.

Summary

Electronic configuration

- ([Kr]$4d^{10}$)$5s^2 5p^2$

Element

- 'White'metal at normal and high temperatures; resistant to air and acid fruit juice
- 'Grey' diamond structure at low temperature

Occurrence

- mainly as SnO_2, together with iron

Extraction

- reduction of SnO_2 with carbon, but removal of iron complicated

Chemical behaviour

- Forms (II) and (IV) states, depending on environment; with steam and dilute HNO_3 gives Sn^{2+} (aq); with hot concentrated alkali gives $Sn^{IV}(OH)_6^{2-}$; with halogens gives SnX_4; with S and Se depends on proportions; with Te gives SnTe
- Hydrides SnH_4 and Sn_2H_6 only; unstable
- Sn^{II} lone pair ($5s^2$) stereochemically active (Lewis base); 5d orbitals empty (Lewis acid)
- Geometry complex, e.g. SnX_2, SnO (several forms, and amphoteric with condensed oxocations; hydrous oxide contains $Sn_6O_8^{4+}$)
- Sn^{IV} simpler, but SnF_4 has layer structure, and SnO_3^{2-} is a chain
- Bonds to metals: Sn–Sn bonds in polymers Sn_5^{2-} and Sn_9^{4-} and in hydroxoanions; and Sn–M bonds to metals in carbonyls
- Rich, mainly Sn^{IV}, organometallic chemistry: alkyls, aryls, and halides

Uses

- Metal: for tins and liquid as glass float
- alloys
  - bronze, brass, pewter, solder, type metal, Babbitt's metal
  - Nb_3Sn superconductivity magnet
- Sn^{II} carboxylates: polymerization catalysts
- SnO_2: glass toughener and opacifier
- Sn(IV) organometallics: in plastics industry and as fungicides and pesticides

Nuclear features

- 10 stable isotopes; most abundant $^{120}Sn \sim 32\%$

Sb Antimony

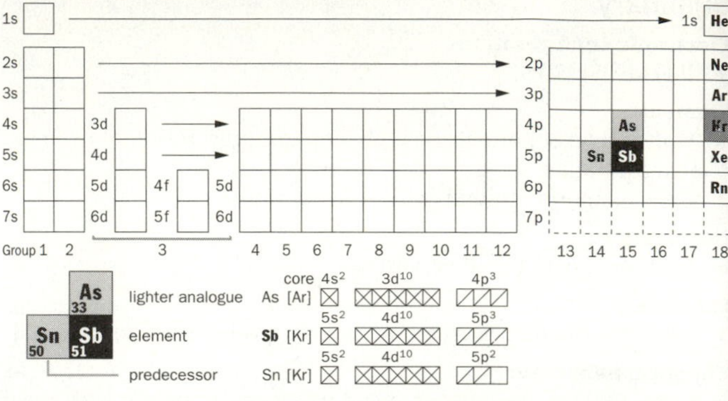

		Predecessor	Element	Lighter analogue
Name		**Tin**	**Antimony**	**Arsenic**
Symbol, Z		Sn, 50	Sb, 51	As, 33
RAM		118.710	121.75	74.9216
Radius/pm	**atomic**	140.5	182	125
	covalent	140	141	121
	ionic, M^{3+}		89	69
Electronegativity (Pauling)		2.0	2.05	2.18
Melting point/K		505.118	903.89	
Boiling point/K		2543	1908	889 (sublimes)
$\Delta H^{\ominus}_{fus}$/kJ mol^{-1}		7.20	20.9	27.7
$\Delta H^{\ominus}_{vap}$/kJ mol^{-1}		290.4	67.91	31.9
Density (at 293 K)/kg m^{-3}		5750 (α)	6691	5780 (α)
		7310 (β)		4700 (β)
Electrical conductivity/Ω^{-1}m^{-1}		9.091×10^6	2.564×10^6	2×10^{-2} (α)
Ionization energy for removal of *j*th electron/kJ mol^{-1}	***j* = 1**	708.6	883.7	947.0
	***j* = 2**	1411.8	1794	1798
	***j* = 3**	2943.0	2443	2735
	***j* = 4**	3930.2	4260	4837
	***j* = 5**		5400	6042
Electron affinity/kJ mol^{-1}		116	101	78
Bond energy/kJ mol^{-1}	**E–E**	195	299	348
	E–H	<314	257	~ 245
	E–C	225	215	200
	E–O	557 Sn^{II}	314	477
	E–F	322	389	464
	E–Cl	315	313	293

Antimony, ([Kr]$4d^{10}$)$5s^25p^3$, has an even longer history than either its predecessor, tin ([Kr]$4d^{10}$)$5s^25p^2$, or its lighter analogue, arsenic ([Ar]$3d^{10}$)$4s^24p^3$; its compounds were known in early biblical times and the element itself was extracted at least six thousand years ago. As it has five electrons in its four outer (s + p) orbitals, we should expect less delocalization than in tin (which itself has a non-metallic allotrope) but since the orbitals for $n = 5$ are closer together in energy than those for $n = 4$, we can guess that although antimony will resemble arsenic it will be more metallic. We should expect antimony to form compounds in both oxidation states (III) and (V) by retention or loss of its $5s^2$ electrons, and probably also to form clusters.

Antimony occurs mainly as the sulfide Sb_2S_3, from which it can be obtained either by heating the ore with iron or by roasting it in air and then reducing the oxide with coke. Its major use is in alloys: with lead to improve lead storage batteries, and for solders, bearings, ammunition, and so forth. Recently, it has become important in the electronics industry in 'III–V semiconductors' (see p. 295).

There are several allotropes of antimony. The most important (α) form is black, shiny, brittle, and fairly low melting (630°C). It is composed of puckered hexagonal layers, like those in arsenic or black phosphorus (see p. 161), but with the layers closer together relative to the ring size. Although this makes the structure more 'metal-like', the electrical conductivity is still rather low. Like arsenic, antimony forms a rich variety of alloys, which are often non-stoichiometric. Some contain $(Sb)_x$ spirals and others may also contain sulfur atoms.

Antimony is stable in damp air at room temperature, but is more reactive when hot (although less so than arsenic). It can be oxidized to Sb_2O_n (where $n = 3$, 4, or 5). With chlorine and heavier halogens it gives the trihalides SbX_3, and with hydrogen the very poisonous hydride stibnine SbH_3, which, like antimony's other hydride, Sb_2H_4, is even more unstable than the hydrides of tin or arsenic. Although antimony does not react with dilute acids, it is attacked by concentrated oxidizing ones; it may form antimony(V) (nitric acid gives Sb_2O_5 and aqua regia, see p. 489, gives $SbCl_5$), but with hot concentrated sulfuric acid it gives only antimony(III) sulfate.

The trihalides of antimony and their complex anions are, like those of tin, often based on pyramidal structures. The fluoride SbF_3 is less stable than arsenic trifluoride and its packing approaches a 3D array. The other three antimony trihalides are more stable than the arsenic ones: the chloride and bromide are molecular, while the higher melting, red iodide consists of layers of linked SbI_6 octahedra.

Stable isotopes of antimony

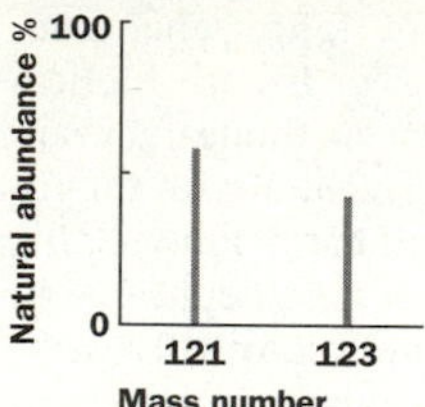

Some fluoro anions of Sb^{III} GE

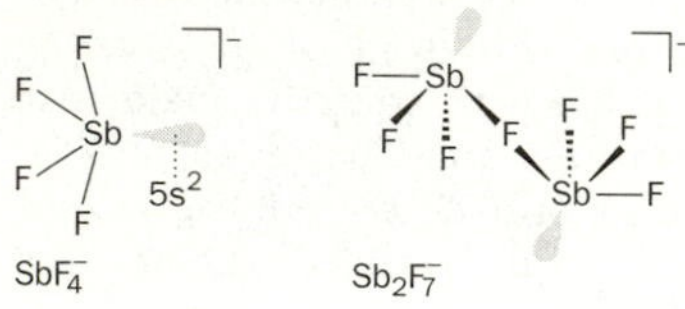

Solid $(SbF_5)_4$ GE

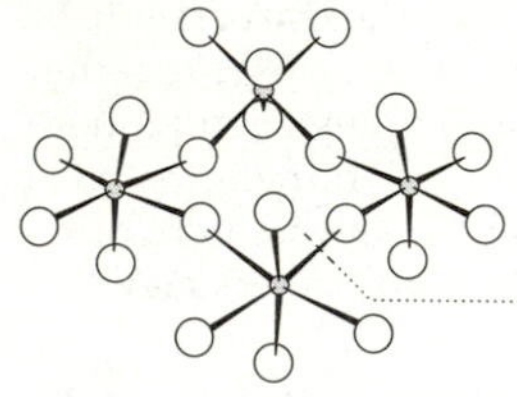

When the solid tetramer melts one of the SbF bridges breaks, and the $(SbF_5)_4$ chains polymerize

The anion cluster Sb_7^{3-}

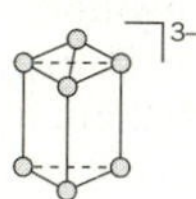

There are only two antimony pentahalides (naturally, the fluoride and chloride) and these, although very reactive, are predictably more stable than those of arsenic; both are low melting. The solid fluoride contains tetramers of *cis*-F-linked SbF_6 octahedra, but melts to a chain polymer; the chloride molecule is a trigonal bipyramid. Since all antimony halides, both (III) and (V), are strong Lewis acids, it is not surprising that there are also many and varied halide complexes. We shall pick only a few examples: the salt $KSbF_4$ contains the cyclic, F-linked tetramer $(SbF_4^-)_4$, reminiscent of tin(II) difluoride; the anion $Sb_2F_7^-$ is $F_3Sb{-}F{-}SbF_3$; and there is a black, mixed valence ammonium salt containing the two anions $Sb^{III}Br_6^{3-}$ and $Sb^{V}Br_6^-$.

Antimony shows less variety with oxygen and sulfur. Its main oxide is Sb_4O_6 which has the same structure as P_4O_6 in the vapour and in one solid form; its other form (which is used as a flame retardant) consists of O-linked sheets of SbO_3 pyramids, similar to one form of As_2O_3. There are a few antimony(III) salts, such as $NaSbO_2$ and $NaSb_3O_5.H_2O$, but no well-characterized acid. Nor is there any acid or well-defined oxide of antimony(V), although the mixed valence oxide, $Sb^{III}Sb^{V}O_4$, is known. There is only one sulfide, predictably Sb_2S_3, which, like As_2S_3, dissolves in aqueous sodium sulfide.

Antimony, like phosphorus, tin, and arsenic, can form polyanions such as Sb_7^{3-} in combination with large cations such as $Na(crypt)^+$ and can, again like tin, join to other metal atoms with attached carbonyl groups. Multiple antimony–antimony bonds can also be made in organometallic complexes: one even contains $Sb{\equiv}Sb$ in a coordination site often occupied by $N{\equiv}N$. Antimony(V) gives fairly stable R_5Sb alkyls and aryls, which can form the anions R_6Sb^-. But the collector's prize oddity of antimony chemistry must surely be its splendidly complex selenide, $Ba_4Sb_4^{III}Se_{11}$, which contains four different types of anion (see p. 396).

For summary, see p. 397.

The four types of anion in $Ba_4Sb_4^{III}Se_{11}$ GE

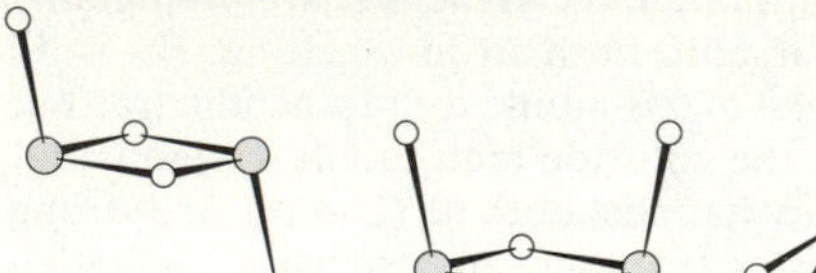

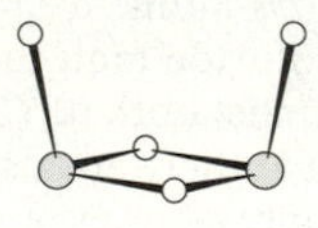

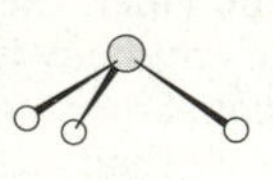

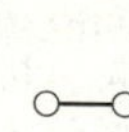

trans-$[Sb_2Se_4]^{2-}$ *cis*-$[Sb_2Se_4]^{2-}$ $[SbSe_3]^{3-}$ Se_2^{2-}

Summary

Electronic configuration

- ([Kr]$4d^{10}$)$5s^2 5p^3$

Element

- Metalloid; commonest form is black, shiny, fairly low-melting; poor electrical conductor
- Stable in damp air; reactive when hot; forms alloys

Occurrence

- As sulfide

Extraction

- Roast sulfide in air to oxide and reduce with coke

Chemical behaviour

- More metallic than arsenic, less so than tin
- Less reactive than arsenic; (V) state less oxidizing; (III) and (V) of similar stability, like tin(II) and (IV)
- Main oxide Sb_4O_6 (various forms); a few oxoanions but no acid; also mixed oxide $Sb^{III}Sb^{V}O_4$ and ill-defined Sb_2O_5
- Only sulfide: Sb_2S_3
- Wide variety of halides: (all SbX_3, and SbF_5 and $SBCl_5$) and complex halides in (III) and (V) states, and mixed
- Sb–Sb bonds in anion clusters and Sb–M bonds to metal carbonyls (like tin)
- Organoantimony, $RSb^{III}X_2$ and $R_2Sb^{III}X$ halides; R_5Sb^{V} aryls and alkyls (less stable than organoarsenic compounds)

Uses

- In alloys, often with lead
- In III–V semiconductors

Nuclear features

- ^{121}Sb:^{123}Sb ~ 5.7:4.3

Te Tellurium

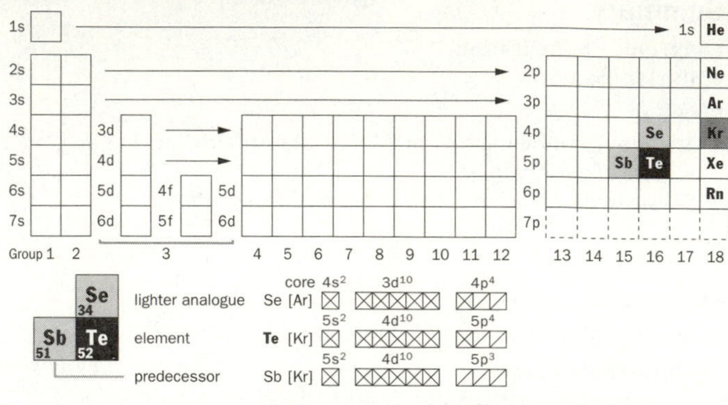

		Predecessor	Element	Lighter analogue
Name		**Antimony**	**Tellurium**	**Selenium**
Symbol, *Z*		Sb, 51	Te, 52	Se, 34
RAM		121.75	127.60	78.96
Radius/pm	**atomic**	182	143.2	215.2
	covalent	141	137	117
	ionic, M^{4+}		97	
	X^{2-}		211	191
Electronegativity (Pauling)		2.05	2.1	2.55
Melting point/K		903.89	722.7	490
Boiling point/K		1908	1263.0	958.1
$\Delta H^\ominus_{fus}$/kJ mol^{-1}		20.9	13.5	5.1
$\Delta H^\ominus_{vap}$/kJ mol^{-1}		67.91	50.63	26.32
Density (at 293 K)/kg m^{-3}		6691	6240	4790 (grey)
Electrical conductivity/Ω^{-1}m^{-1}		2.564×10^6	2.293×10^2	100
Energy of band gap in solid/kJ mol^{-1}			32.2	178
Ionization energy for removal of *j*th electron/kJ mol^{-1}	***j* = 1**	833.7	869.2	940.9
	***j* = 2**	1794	1795	2044
	***j* = 3**	2443	2698	2974
	***j* = 4**	4260	3610	4144
	***j* = 5**	5400	5668	6590
	***j* = 6**		6822	7880
Electron affinity/kJ mol^{-1}		101	190.2	195
Bond energy/kJ mol^{-1}	**E–E**	299	235	330
	E–H	257	~ 240	305
	E–O	314	268	343
	E–F	389	335	285
	E–Cl	313	251	245

Tellurium can occur native and was discovered surprisingly early (1782) for an element that is not very familiar to many chemists. We should expect tellurium, ([Kr] $4d^{10}$)$5s^2 5p^4$, to show less metallic character than antimony, ([Kr]$4d^{10}$)$5s^2 5p^3$, but more than selenium, ([Ar]$3d^{10}$)$4s^2 4p^4$. Even in nature it exhibits oxidation states of both (IV) and (VI). It occurs as the oxide, TeO_2, as basic oxides, or as the oxoanions TeO_3^{2-} or TeO_4^{2-}, together with many 3d and 5p metals, and is normally obtained from the anode slime produced in copper extraction. The oxide is dissolved in alkali to give TeO_3^{2-} which is reduced electrolytically to tellurium.

Unlike selenium, elemental tellurium exists in only one form, which is like the network of hexagonal helical chains found in 'grey' selenium, and indeed it forms solid solutions with it. It conducts electricity much better than selenium does, but very much worse than antimony. It is used in alloys to improve mechanical, and catalytic, properties.

Its chemical behaviour, too, is between that of a metal and a non-metal, although nearer to the latter. It dissolves in dilute acids, in the presence of air, to give tellurium(IV) compounds, which do not disproportionate; indeed there seems to be no tellurium(II) in aqueous solution. With oxygen, fluorine, and chlorine, tellurium forms a variety of compounds in (II), (IV), and (VI) states, although the range is smaller than for selenium (or for sulfur). Both the dioxide and the trioxide are amphoteric. The trioxide is a strong oxidizing agent but

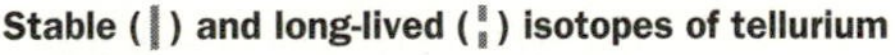

Stable (|) and long-lived (¦) isotopes of tellurium

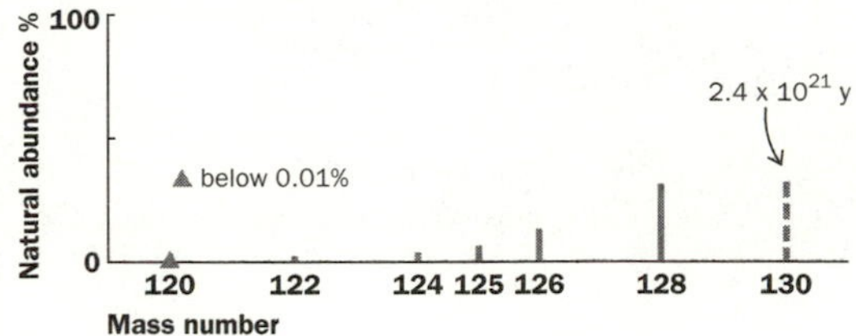

Te GE

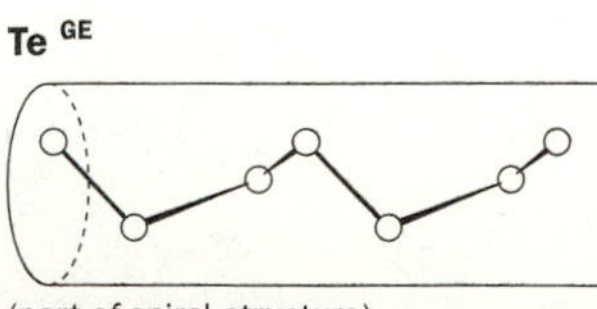

(part of spiral structure)

Tetrahalides of Te [GE]

$5s^2$

TeF_5 distorted octahedron with Te atom displaced from plane of the four F atoms towards the lone pair

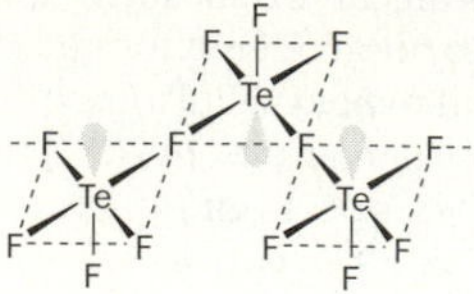

TeF_4 zigzag chain of corner-linked TeF_5 units, with lone pairs pointing alternately up and down

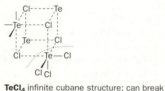

$TeCl_4$ infinite cubane structure; can break at dotted lines to give $TeCl_3^+$ and complex anions, e.g. $(TeCl_4)_4 - TeCl_3^+ \longrightarrow Te_3Cl_{13}^-$

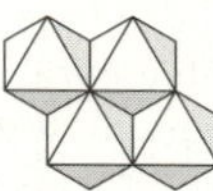

Te_4I_{16} formed by four edge-sharing TeI_6 octahedra

Subhalides of Te [GE]

formed by regular addition of halogen to Te spiral

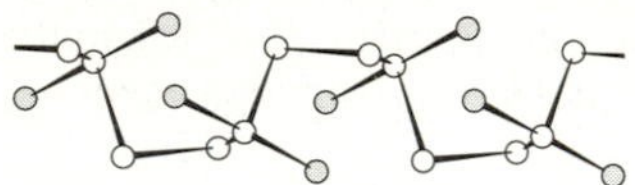

Te_3Cl_2

every third Te atom carries two Cl atoms

Te_2Br and **Te_2I**

alternate Te atoms are halogen bridged within the chain; the other Te atoms link two chains together

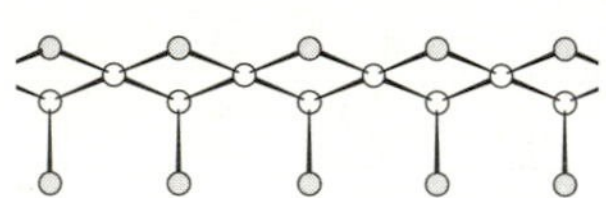

TeI, simpler (β) form

alternate Te atoms are iodine bridged as in Te_2I, the other Te atoms each carry an iodine

○ halogen ○ Te

Cation clusters of Te [GE]

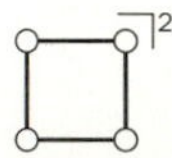

Te_4^{2+}, square

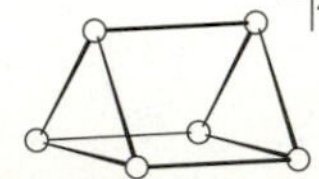

Te_6^{4+}, trigonal prism

less so than selenium(VI); it is reduced to the element by sulfur dioxide, and to H_2TeO_3 by hot hydrochloric acid. Tellurium(VI) acid is normally $Te(OH)_6$ [rather than H_2TeO_4 as would be expected by analogy with the selenium(VI) and sulfur(VI) acids]. When H_6TeO_6 is heated, water is lost, to give various condensed species, $H_2TeO_4.xH_2O$, of uncertain composition and structure.

All four tetrahalides, TeX_4, are known. The tetrafluoride is a chain of corner-linked, (roughly) square-pyramidal TeF_5 groups, which may also be envisaged as distorted octahedra as they have a stereochemically active inert pair, reminiscent of tin(II) and antimony(III) (see p. 387). The tetrachloride has an infinite cubane structure, which can lose successive $TeCl_3^-$ groups to give various complicated anions. The tetriodide is a tetramer of edge-shared TeI_6 octahedra. Tellurium also forms a family of subhalides in which halogen atoms are added regularly to the helical chains of the elemental form, for example, at every fourth or at every third tellurium atom. These subhalides have interesting electronic properties and have stimulated work on an element that has not been very widely studied.

Tellurium, like selenium, forms cluster cations such as the square Te_4^{2+}, and Te_6^{2+}, which is a trigonal prism. There are also mixed boat-shaped hexagonal rings, X_6^{2+}, containing cross-ring-bonded tellurium together with either sulfur or selenium. Although tellurium does not seem to form cluster anions on the scale shown by antimony or tin, the bent species Te_3^{2-} has been made (see p. 408).

Naturally, there is also a wide range of organotellurium compounds, although the element, unlike selenium, does not seem to have any essential biological role. Indeed, Sharpe has suggested that the relative lack of interest in tellurium chemistry may be owing to one aspect of its biochemical behaviour. If traces of it are accidentally ingested by humans, the body gets rid of it in exhalation and sweat, and in such a form that the smell of either is said to be even more revolting than the effects of selenium. Much intellectual or economic motivation would surely be necessary to overcome such a social disincentive.

For summary, see p. 403.

Summary

Electronic configuration

- ([Kr]$4d^{10}$)$5s^2 5p^4$

Element

- Metalloid; grey, chains; better electrical conductor than Se, worse than Sb

Occurrence

- Te^{IV} and Te^{VI} with oxygen, together with many 3d and 5p metals

Extraction

- From anode slime from copper manufacture; TeO_2 with alkali gives TeO_3^{2-}: reduced electrolytically

Chemical behaviour

- Oxidation states: (VI) (less oxidizing than Se^{VI}), (IV) and (II) (not in aqueous solution); less variety than for Se or S
- TeO_2 and TeO_3 amphoteric; oxoacids H_3TeO_3 and $Te(OH)_6$ (condensed forms made by heating)
- Halides: all TeX_4 (with stereochemically active inert pair); complex ions; subhalides by regular additions to chains in grey Te
- Cluster cations (like Se, unlike Sb and Sn)
- Organo derivatives

Uses

- In alloys, for mechanical properties, and catalysis
- Subhalides in electronic devices

Nuclear features

- Five main stable isotopes, and one very long-lived radioactive isotope

I Iodine

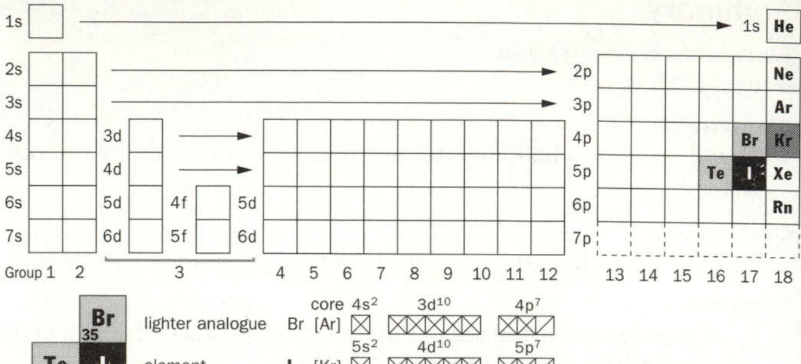

			core			
Br 35	lighter analogue	Br	[Ar]	$4s^2$	$3d^{10}$	$4p^7$
I 53	element	I	[Kr]	$5s^2$	$4d^{10}$	$5p^7$
Te 52	predecessor	Te	[Kr]	$5s^2$	$4d^{10}$	$5p^6$

	Predecessor	Element	Lighter analogue
Name	**Tellurium**	**Iodine**	**Bromine**
Symbol, Z	Te, 52	I, 53	Br, 35
RAM	127.60	126.90774	79.904
Radius/pm covalent	137	133.3	114.2
ionic, X^-		220	196
Electronegativity (Pauling)	2.1	2.7	2.96
Melting point/K	722.7	386.7	265.9
Boiling point/K	1263.0	457.50	331.93
$\Delta H^{\ominus}_{fus}$/kJ mol^{-1}	13.5	15.27	10.8
$\Delta H^{\ominus}_{vap}$/kJ mol^{-1}	50.63	41.67	30.0
Density (at 293 K)/kg m^{-3}	6240	4930	
Electrical conductivity/Ω^{-1}m^{-1}	2.293×10^2	7.299×10^{-8}	
Ionization energy for removal of jth electron/kJ mol^{-1} $j = 1$	869.2	1008.4	1139.9
$j = 2$	1795	1845.9	2104
$j = 3$	2698	3200	3500
Electron affinity/kJ mol^{-1}	190.2	259.2	324.7
Bond energy/kJ mol^{-1} E–E	235	151	193
E–H	~240	299	366
E–C		218	285
E–O	268	234	234
E–F	335	280	285
E–Cl	251	208	

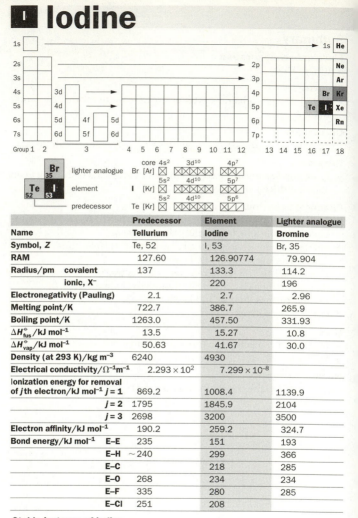

Stable isotopes of iodine

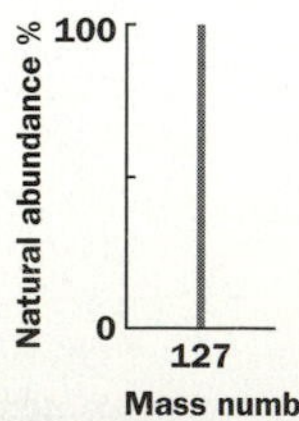

Long ago, when people put 'tincture of iodine' on cuts and scratches and some of them even learned Greek, it must have seemed strange that a word meaning 'violet' was used for a brown liquid. The name, of course, refers to the colour of the diatomic vapour, in which the gap between the π antibonding and σ antibonding orbitals is smaller than in the lighter halogens, and the I_2 molecule absorbs strongly in the yellow region of the visible spectrum. The violet colour persists when solid iodine is dissolved in a non-complexing solvent such as tetrachloromethane, but coordinating solvents such as the ethanol used in the 'tincture' change the energy gap between the orbitals and hence the colour.

Iodine, ([Kr]$4d^{10}$)$5s^25p^5$, with only one electron short of the inert gas configuration, would be expected to behave, like the rest of the halogens, as a typical non-metal; and we should certainly predict that it shows even less metallic character than its predecessor, tellurium, ([Kr]$4d^{10}$)$5s^25p^4$. Iodine is, indeed, non-metallic in most (but not all) of its chemical behaviour.

The iodide ion occurs in low concentrations in sea water, from which it is obtained by treatment either with the silver Ag^+ ion to give solid silver iodide, or with chlorine to give the stable brown triodide ion I_3^-. Silver iodide is converted by iron to iron(II) iodide, which is treated with chlorine to liberate iodine. Alternatively, the brine may be oxidized directly with chlorine and the triiodide ion is collected on an anion exchange resin. Iodine also occurs in Chilean saltpetre as the iodate ion IO_3^-, which is reduced by acidified sulfite solution to iodide. Treatment with further iodate gives iodine.

Iodine is extracted from the sea by certain seaweeds. It is also concentrated in our own thyroid gland to make the hormone thyroxin, and is used to treat thyroid disorders. Silver iodide is added to silver bromide to increase the sensitivity of photographic film (see p. 367), and iodine compounds have a number of laboratory uses as a reagent both for organic syntheses and for redox titrations (see p. 407).

At room temperature, iodine is a dark, shiny, volatile solid with a layer structure containing 'pairs' of iodine atoms (which are about 12% closer together than the other atoms within the layer). It dissolves not only in organic solvents (both solvating and non-solvating), but also in aqueous potassium iodide, to form I_3^-. In solid iodine, the layers are relatively closer together than in the similar crystals of bromine and chlorine, and there is some electrical conductivity within a layer. This slight tendency towards metallic character reminds us of antimony (see p. 393) and may be given a push by applying a pressure of 350 kbar, whereupon iodine,

Production of I_2

From 'brine' (concentrated sea water, Cl^- and some I^-)

$$I^-\left[+Cl^-\right] \xrightarrow{AgNO_3} AgI(s) \xrightarrow{Fe} Ag(s) + FeI_2(aq)$$

brine; just enough to react with the I^- ($AgNO_3$); clean scrap (Fe)

$FeI_2(aq) + Cl_2(g) \rightarrow I_2 + FeCl_2(aq)$

or brine + $Cl_2(g) \rightarrow I_3^-$ collect on anion exchange resin

From 'Chile saltpetre' ($NaNO_3$ containing $NaIO_3$)

$$2IO_3^- + 6HSO_3^- \rightarrow 2I^- + 6SO_4^{2-} + 6H^+$$

$$5I^- + IO_3^- + 6H^+ \rightarrow 3I_2\downarrow + 3H_2O$$

Solid I_2 [GE]

layer structure

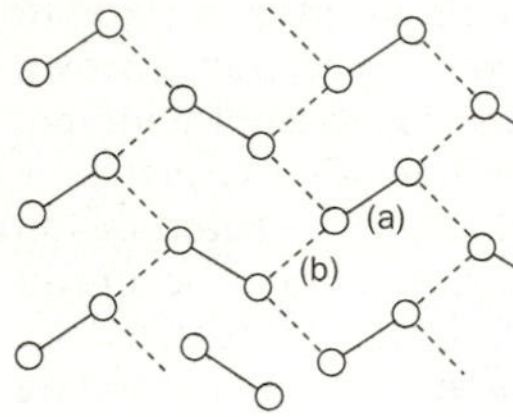

I–I distance /pm	
nearest in solid, (a)	271.5
next nearest in solid, (b)	350
in I_2 gas	266.6

Two interhalogen compounds of I [GE]

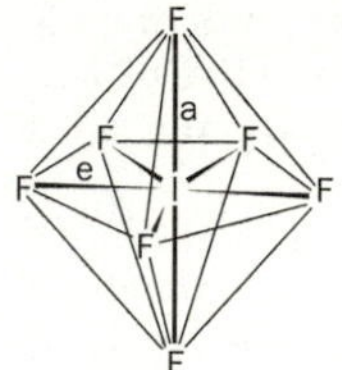

IF_7, pentagonal biprism (possibly slightly distorted).

The equatorial bonds (e) are about 8% longer than the axial ones (a)

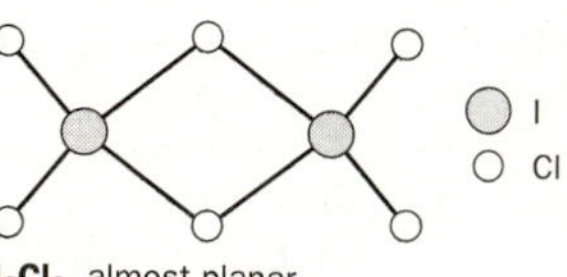

I_2Cl_6, almost planar

Iodate titrations

$IO_3^- + 5I^- + 6H^+ \rightarrow 3I_2 + 3H_2O$

$I_2 + 2S_2O_3^{2-} \rightarrow 2I^- + S_4O_6^{2-}$

apparently unlike tellurium or antimony, becomes a fully metallic conductor.

In many ways, iodine behaves just as we should expect of a large halogen, with lower ionization energies than its lighter analogues, weaker covalent bonds, and higher coordination numbers; iodine vapour and gaseous hydrogen iodide dissociate more readily, and iodine is the only halogen that forms a heptafluoride, IF_7. The larger iodide ion is more polarizable than the other halides, and so iodides tend to be more covalent or more molecular than other halides of the same element. Unlike the fluoride ion, which brings out high oxidation states in other elements, the iodide ion often combines only with lower oxidation states such as iron(II) and copper(I), since higher ones oxidize it to iodine. Strongly polarizing (class 'b') metal ions, such as silver(I) and mercury(II), form their strongest halide complexes with iodide. Partly because the iodide ion is less firmly hydrated than the other halide ions, iodides of heavy metals such as silver and lead are often insoluble in water; ease of oxidation of the iodide ion may also lead to low energy charge transfer transitions which are responsible for the colour of such compounds as PbI_2 (yellow) and HgI_2 (red).

Iodine, like chlorine and bromine, forms mono-interhalogen compounds, a trifluoride, and a pentafluoride, together with many triatomic anions such as IF_2^- and $IBrCl^-$, and various cations such as I_2Cl^+, IF_4^+, and IF_6^+. But, unlike the others, iodine also forms a dimeric species, the almost planar I_2Cl_6 with two I–Cl–I bridges; and, with its large radius, it can form, not only the pentagonal bipyramid IF_7, but also the anion IF_8^-.

Aqueous hydrogen iodide is, like the bromide and chloride, a strong acid: the weakness of the H–I bond more than makes up for the low free energy of hydration of the iodide ion. Again, like chlorine and bromine, iodine forms a range of halogen oxoacids and anions, although, as with bromine, the (III) state is not represented. The only common acids that have been isolated are two of iodine(VII), HIO_4 and H_5IO_6, and one ('iodic') of iodine(V), HIO_3. Dehydration of iodic acid gives I_2O_5, the most stable oxide of any halogen, with a complicated structure in which the iodine atoms are connected by both single and double oxygen bridges. (Iodine also forms various other, less stable, oxides, such as I_2O_4 and I_4O_9, of unknown structure.)

All the iodine oxoacids are less strongly oxidizing then their bromine and chlorine analogues, but, like the other HXO acids, hypoiodous acid, HIO, disproportionates. Soluble iodates are useful in volumetric analysis because the acidified solutions react with iodide ions to give iodine (or, more precisely, I_3^-), (see p. 405), which

The I_3^+ ion [GE]

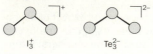

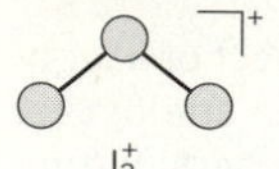

2−
Te_3^{2-}

Both ions have 20 valence electrons and are of similar shape, although I_3^+ has a tighter angle

Summary

Electronic configuration

- ([Kr]$4d^{10}$)$5s^2 5p^5$

Element

- Black shiny solid, readily volatilizes to purple diatomic gas
- Gives purple solution with non-complexing solvents, and brown with complexing solvents or aqueous KI (forms I_3^-)
- Becomes metallic at high pressure

Occurrence

- In sea water as I^- (low concentration); as deposits of IO_3^-

Extraction

- From sea water; with Ag^+ (AgI precipitated), converted to FeI_2 and treated with Cl_2; or with Cl_2 to give I_3^-, collected by ion exchange
- From IO_3^- by reduction to I^- with SO_3^{2-}, I_2 obtained by adding more IO_3^-

Chemical behaviour

- Less electronegative than lighter halogens, lower ionization energies, weaker bonds
- Iodide ion more polarizable; less soluble salts, stronger complexes with polarizing 'class b' cations
- High coordination number, e.g. in interhalogens, molecules, cations, and anions
- oxoacids of (VII) (various and complicated) and (V) (oxidizing) states; (I) disproportionates; (III) unknown
- Oxides: I_2O_5 fairly stable, others not
- Cations: I^+ if strongly complexed; various clusters
- Polyanions in great variety

Uses

- AgI in photography
- IO_3^- in volumetric analysis
- KI as a thyroid treatment
- Various compounds in organic syntheses

Nuclear features

- ^{127}I isotopically pure

can be titrated with a standard solution of thiosulfate. Starch is used as the indicator; iodine molecules enter the central cavity of a helical starch molecule, interact with the oxygen molecules on its inner surface, and form a long chain complex which is blue-black owing to charge transfer.

Periodic acid, HIO_4, is just one of a family of iodine(VII) oxoacids and anions. Most, unlike the tetrahedral ion IO_4^- or the square pyramidal IO_5^-, are based on an IO_6 octahedron which may dimerize (sharing either an edge or a face) or even trimerize, and at the same time be protonated on up to three of the oxygen atoms per iodine. Species such as $IO_3(OH)_3^{2-}$, $I_2O_8(OH)_2^{4-}$, or $H_7I_3O_{14}$ are formed, with complicated structures and confusing names; but they are better characterized than the analogous condensates of telluric acid (see p. 399).

Since iodine has lower ionization energies than the other halogens and can be pressed into being a metal, it seems to be a likely candidate for cation formation; and its polynuclear cations, such as I_2^+, I_3^+, I_5^+, and I_4^{2+}, are more numerous and stable than those reported for bromine or chlorine. (The ion I_3^+ has the same number of electrons as Te_3^{2-} and, like it, is bent rather than linear.) The monomeric cation I^+ seems to exist only if strongly complexed: in $I(py)_2^+$ and $I(crypt)^+$, or in the ionic solutions formed by $IClO_4$ (presumably I^+ and ClO_4^-) in diethyl ether.

Iodine also has a splendid variety of polyanions, and in this it differs from bromine and chlorine. Even the familiar triiodide ion can be linear or slightly bent, with I–I bonds of identical or slightly different lengths. Combinations of I_3^- ions and I_2 molecules lead to a menagerie of anions I_n^- (up to $n = 9$) and I_n^{2-} (up to $n = 8$), in various 'bent-rod' or 'bent-Y' shapes. The most complicated, so far, is I_{16}^{4-}, which is like an almost planar, more or less squared-off, meat hook. Iodine certainly puts its ready promotion of electrons to some interesting uses, even without its being squashed into a metal.

A complex polyiodide, I_{16}^{4-} ion GE

(schematic)

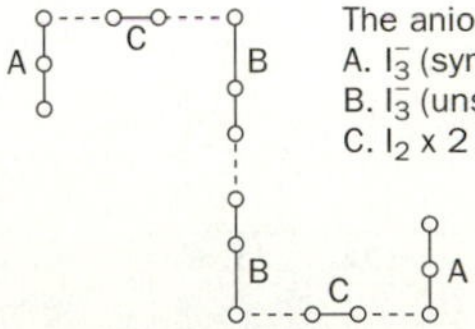

The anion contains:
A. I_3^- (symmetrical) x 2
B. I_3^- (unsymmetrical) x 2
C. I_2 x 2

Xe Xenon

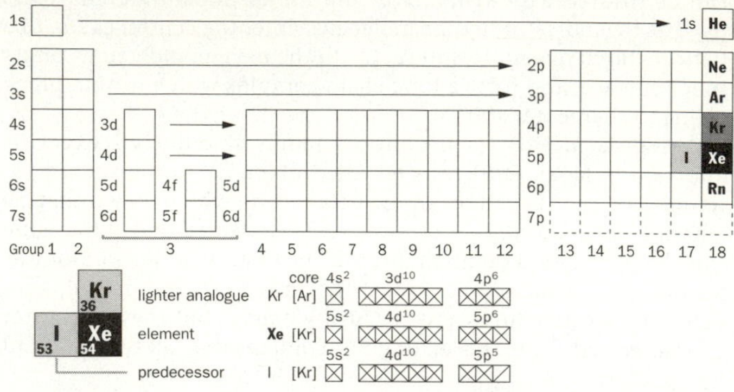

		Predecessor	Element	Lighter analogue
Name		Iodine	Xenon	Krypton
Symbol, Z		I, 53	Xe, 54	Kr, 36
RAM		126.90774	131.29	83.80
Radius/pm	atomic		218	
	covalent	133.3	209	189
	ionic, X^+		190	
Electronegativity (Pauling)		2.7	2.6	
Melting point/K		386.7	161.3	116.6
Boiling point/K		457.50	166.1	120.85
$\Delta H^\ominus_{fus}$/kJ mol^{-1}		15.27	3.10	1.64
$\Delta H^\ominus_{vap}$/kJ mol^{-1}		41.67	12.65	9.05
Ionization energy for removal of jth electron/kJ mol^{-1}	$j = 1$	1008.4	1170.4	1350.7
	$j = 2$	1845.9	2046	2350
	$j = 3$	3200	3097	3565
Bond energy/kJ mol^{-1}	E–O	280	84	
	E–F	208	133	50

Stable isotopes of xenon

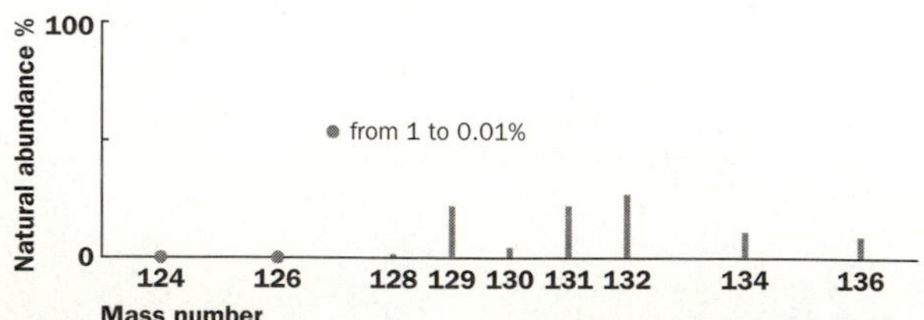

At xenon, ([Kr]$4d^{10}$)$5s^25p^6$, the outer shell is full; but by no means incapable of action. Krypton may not have persuaded us that 'inert' was an inappropriate label for the ns^2p^6 gases, but xenon is completely convincing.

Xenon (nine stable isotopes of it) is present in air, but as less than one part in ten million. It is separated by distillation and is used, together with other noble gases, to produce the different colours of 'neon' lights; and to provide fodder for those many experiments designed to make it form more and more compounds (in addition to the hydrates and clathrates that are also formed by argon and krypton, see p. 191). Naturally, it is denser than krypton, with higher melting and boiling points and a larger atom; and it was its size that led Pauling to suggest, in 1933, that it might be able to combine with fluorine. Its first ionization energy is not unreasonably high (about the same as that of bromine) but only weak bonds would be expected to be formed by so large an atom. The first xenon compound was not made until 1962, when $O_2^+PtF_6^-$ was prepared by treating dioxygen with platinum hexafluoride, which is a very fierce oxidizing agent. Since the first ionization energy of dioxygen is about the same as that of xenon, perhaps ...? Yes, it worked, and gave an orange solid, $Xe^+PtF_6^-$.

The next 30 years saw the creation of a wide range of xenon compounds, naturally almost all of them with fluorine or oxygen, or both. Xenon shows oxidation states of (II), (IV), (VI), and (VIII); it can form neutral species, cations, or anions; and, if only nearest neighbours are counted, it can have any coordination number from nought to eight. Examples are XeF_n (n = 2, 4, or 6), XeF_n^+ (n = 1, 3, 5, or 7), $FXeFXeF^+$, XeF_8^{2-}, XeO_n (n = 3 or 4), XeO_6^-, $XeOF_4$, XeO_2F_2, XeO_3F_2, $XeOF_5^-$, and XeO_3F^-. The difluoride, XeF_2, dissolves in water to give a strongly oxidizing solution, but decomposes in alkaline solution. The higher fluorides react with even traces of moisture to give the trioxide, XeO_3, which is highly explosive, and so work on xenon compounds needs stringent safety precautions.

Xenon shows great variety in its stereochemistry, much (but not all) of which is compatible with the VSEPR treatment (see p. 7). The tetrafluoride, XeF_4, for example, has its four bonding pairs in a square and its two lone pairs octahedrally above and below the plane. Gaseous XeF_6, however, with six bonding pairs and one lone pair, is a distorted octahedron, with the lone pair accommodated 'somewhere' in a constantly changing ('fluxional') arrangement. It is not surprising that solid XeF_6 is extremely complicated and, indeed, has four different forms. The bonding in XeF_2 and XeF_4 has been described in terms of linear 3-centre–2-electron bonds, but this treatment, too, runs into trouble with XeF_6; and discussion continues.

The varied stereochemistry of Xe GE

Shape	Example	Structure
Linear	XeF_2, $[FXeFXeF]^+$, $FXeOSO_2F$	—Xe—
Pyramidal	XeO_3	
T-shaped	$[XeF_3]^+$, $XeOF_2$	—Xe—
Tetrahedral	XeO_4	
Square	XeF_4	—Xe—
'See-saw'	XeO_2F_2	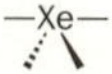 —Xe—
Trigonal bipyramid	XeO_2F_2	 —Xe—
Square pyramid	$XeOF_4$, $[XeF_5]^+$	
Octahedral	$[XeO_6]^{4-}$	
Distorted octahedral	$XeF_6(g)$, $[XeOF_5]^-$	
Square antiprismatic	$[XeF_8]^{2-}$	

Suggested MO for XeF_2 GE

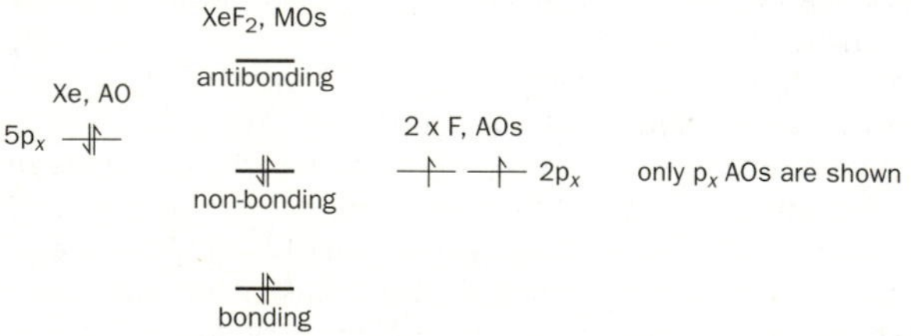

XeF$_2$, MOs

Xe, AO

antibonding

$5p_x$

2 x F, AOs

non-bonding

$2p_x$

only p_x AOs are shown

bonding

* It does not seem a healthy compound, as it has a half-life of only about two seconds at −98°C, but it does contain the first known metal–xenon bond.

The solid trioxide, XeO_3, although so violently explosive, can be dissolved in water to form a strongly oxidizing but kinetically sluggish solution (so inert that it takes days to oxidize Mn^{2+} ions to permanganate). Some salts of $HXeO_4^-$ or XeO_4^{2-} have been isolated, but in alkaline solution they disproportionate to xenon and xenon(VIII) or the perxenate ion XeO_6^{2-}. The tetroxide, XeO_4, is also highly explosive, but gaseous, with a tetrahedral molecule.

Understandably, it is a tempting challenge to try to coerce xenon into forming bonds with a wider variety of elements, and some modest success has been achieved. The two terminal fluorine atoms in the V-shaped ion $F-Xe-F-Xe-F^+$ have been replaced by $-N(SO_2F)_2$ groups. Compounds of xenon with other first row elements joined to very electronegative groups include the two unstable species $FXeBF_2$ and $Xe(CF_3)_2$, and so there is now evidence that xenon can bond to all the first row elements from boron to fluorine. The limits to which xenon chemistry is being pushed is illustrated by the compound* $XeCr(CO)_5$, which has been detected in solution.

Summary

Electronic configuration

- ([Kr]$4d^{10}$) $5s^25p^6$

Element

- Noble gas, ionization energy similar to that of bromine

Occurrence

- In air (< 10^{-5} %)

Extraction

- By fractional distillation

Chemical behaviour

- With very strong oxidizing agent can form compounds with F and/or O
- Oxidation states (O), (II), (IV), (VI), and (VIII)
- Coordination numbers: 0 to 8 inclusive, with varied geometry and enigmatic bonding
- XeO_3 highly oxidizing, explosive solid, but aqueous solution kinetically sluggish
- Can bond also with N, B, C, and Cr

Uses

- In 'neon' lights

Nuclear features

- Nine stable isotopes; commonest ^{132}Xe and ^{129}Xe ~ 26.5% each

Part VI

Filling the 6s, 5d, 4f, and 6p orbitals

The lanthanide period: caesium to radon

Cs Caesium

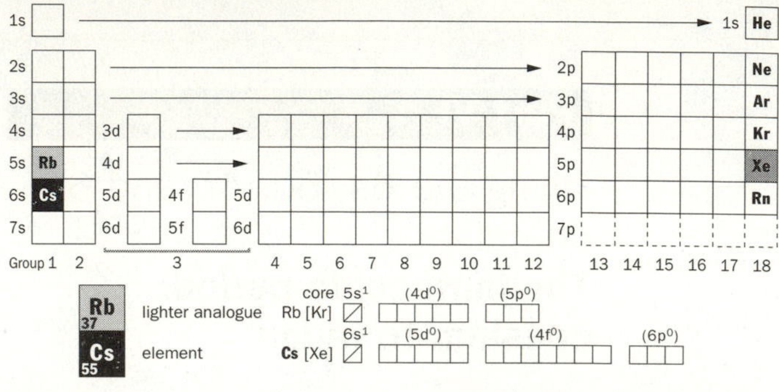

		Element	Lighter analogue
Name		**Caesium**	**Rubidium**
Symbol, *Z*		Cs, 55	Rb, 37
RAM		132.9054	85.8478
Radius/pm	**atomic**	265.4	247.5
	covalent	235	
	ionic, M^+	165	149
Electronegativity (Pauling)		0.8	0.82
Melting point/K		301.55	312.2
Boiling point/K		951.6	961
$\Delta H^{\ominus}_{fus}$/kJ mol^{-1}		2.09	2.20
$\Delta H^{\ominus}_{vap}$/kJ mol^{-1}		65.90	69.2
Density (at 293 K)/kg m^{-3}		1873	1532
Electrical conductivity/Ω^{-1}m^{-1}		5×10^6	8.0×10^6
Ionization energy for removal of *j*th electron/kJ mol^{-1}	***j* = 1**	375.5	403.0
	***j* = 2**	2420	2632
Electron affinity/kJ mol^{-1}		45.5	46.9
Dissociation energy of E_2(g)/kJ mol^{-1}		43.6	47.3
$E^{\ominus}$/V for $M^+(aq) \rightarrow M(s)$		–2.923	–2.924

Stable isotopes of caesium

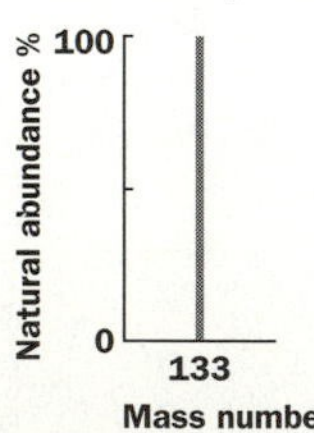

Caesium was discovered in spa waters in 1860, a little before rubidium and by the same spectroscopic method. It too was named after the colour of the dominant spectral line: its flame is sky blue. There is only one natural isotope, ^{133}Cs, and the frequency of one of its spectral transitions is used to define our unit of time, the second. The metal (which is gold coloured) is prepared by reducing the molten chloride with calcium, and is used on a small scale in radiation monitoring and in glass for specialist use.

Caesium, $[Xe]6s^1$, differs only a little from rubidium, $[Kr]5s^1$, and in ways that are mostly predictable from its greater size. As the heaviest ns^1 metal available in bulk (see francium, p. 525), caesium has the lowest first ionization energy of any accessible element. The ease with which the large Cs^+ ion is formed is of course accompanied by its less favourable hydration and the lower lattice energies of its salts, and trends in behaviour are not regular. Its (thermodynamic) power to reduce water is the same as that of potassium and rubidium; that is, between that of lithium and sodium but, unlike lithium, (see p. 37), it does so explosively, as its hydroxide is very soluble. It also reacts extremely rapidly with oxygen and, like rubidium, must be handled under argon.

Caesium, like its lighter analogues, forms blue solutions in liquid ammonia (see p. 129) and reacts with crypts large enough to accommodate it, to give caeside and electride complexes such as $Cs(crypt)^+Cs^-$ and $Cs(crypt)_2^+e^-$

The aquo caesium ion is more strongly adsorbed than most other M^+ ions on to many cation exchange resins because, as the largest naked M^+ cation, it has a small *hydrated* radius.

The size of the cation also has a complicated influence on the properties of solid compounds. Predictably, a high radius does not lead to strong ionic lattices. The hydroxide, CsOH, has a lower melting point than its lighter analogues, and the oxide, Cs_2O, has a layer structure. But Cs^+ can accommodate more neighbours than smaller M^+ cations; in the CsCl body-centred cubic structure, caesium is surrounded by eight chloride ions, whereas the Na^+ ion in the NaCl face-centred cubic structure has a coordination number of only six (see p. 184).

The large caesium ion is used to prepare crystals containing anions such as polyhalide ions like ICl_4^-, which are liable to decompose, since, from the Kaputskinskii equation (see p. 36), the increase in lattice energy accompanying the decrease in anion size is less with large cations than with smaller ones. Like potassium and rubidium, the chief oxide of caesium is CsO_2, containing the bulky but unstable superoxide ion, O_2^-.

Approximate hydrated radii of ([core]ns^0)$^+$ cations[cw]

M^+	Li^+	Na^+	K^+	Rb^+	Cs^+
~ r_{hyd}/pm	340	276	232	228	228

Caesium forms a large number of other oxides and non-stoichiometric oxide phases. One, Cs_2O_3, like the similar oxides of potassium and rubidium, has been formulated as $Cs_4^+O_2^{2-}(O_2^-)_2$, but with electron delocalization between three equivalent oxygen atoms. Like rubidium, caesium forms a number of bizarre sub-oxides, based on M_6O octahedra. The formulae and structures are complex and different for the two metals. One suboxide is $Cs_{11}O_3$ which consists of three Cs_6O octahedra, each sharing two adjacent faces, while another, 'Cs_7O', contains the same $Cs_{11}O_3$ group, together with 10 caesium atoms. Other suboxides are said to be yet more complicated.

Summary

Electronic configuration

- [Xe]$6s^1$

Element

- Soft, very reactive, yellow metal; blue flame

Occurrence

- In spa waters

Extraction

- Reduction of molten chloride with calcium

Chemical behaviour

- Similar to K and Rb, but lower ionization energy; Cs^+ has larger radius than Rb^+ and lower lattice energies, and stabilities of complex ions (and hydrated radius).
- Violent reaction with water (CsOH very soluble)
- Blue solutions with liquid ammonia
- Complexes with large crowns and crypts
- Coordination number 8 in CsCl
- Stabilizes large anions
- Main oxide CsO_2, but many others, including suboxides

Uses

- Definition of unit of time
- Radiation monitoring
- Glass manufacture

Nuclear features

- ^{133}Cs isotopically pure

Ba Barium

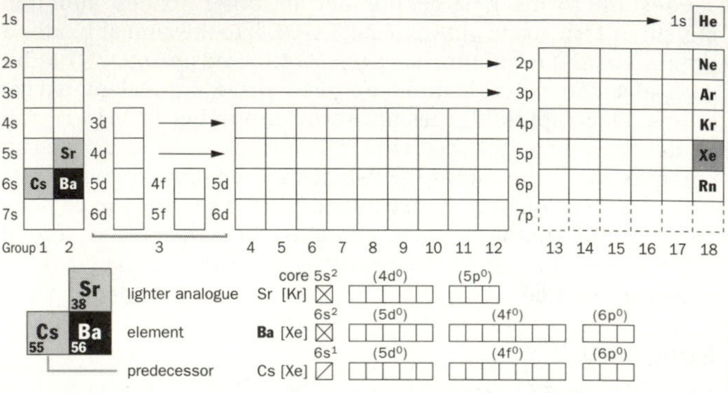

		Predecessor	Element	Lighter analogue
Name		Caesium	Barium	Strontium
Symbol, Z		Cs, 55	Ba, 56	Sr, 38
RAM		132.9054	137.237	87.62
Radius/pm	atomic	265.4	217.3	215.1
	covalent	235	198	192
	ionic, M^{2+}		143	127
Electronegativity (Pauling)		0.8	0.9	0.95
Melting point/K		301.55	1002	1042
Boiling point/K		951.6	1910	1657
$\Delta H^{\ominus}_{fus}$/kJ mol^{-1}		2.09	7.66	9.16
$\Delta H^{\ominus}_{vap}$/kJ mol^{-1}		65.90	150.9	138.91
Density (at 293 K)/kg m^{-3}		1873	3594	2540
Electrical conductivity/Ω^{-1}m^{-1}		5×10^6	2×10^6	4.348×10^6
Ionization energy for removal of jth electron/kJ mol^{-1}	$j = 1$	375.5	502.8	549.5
	$j = 2$	2420	965.1	1064.2
Electron affinity/kJ mol^{-1}		45.5	−46	−146
$E^{\ominus}$/V for $M^{2+}(aq) \rightarrow M(s)$			−2.92	−1.085

Stable isotopes of barium

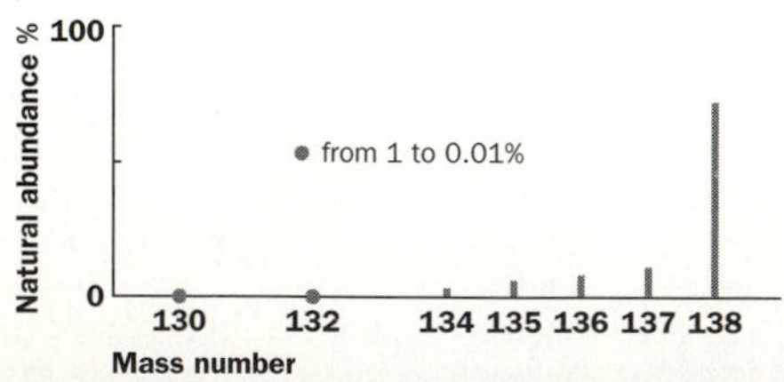

Barium, [Xe]$6s^2$, occurs mainly as its insoluble sulfate. It was first isolated in 1808 using, as for many other s^1 and s^2 metals, the (then) newly available technique of electrolysis. Today, the metal is usually obtained by converting the sulfate to the oxide, BaO, and reducing this with aluminium; but electrolytic reduction of the molten chloride is still used for smaller scale production. The main uses of barium depend on the combination of its high atomic mass with the very low solubility of the sulfate, which is used as a heavy slurry in oil drilling. As its large number of electrons make barium very opaque to X-rays, barium sulfate is also used for X-ray diagnosis of abnormalities in the digestive tract.

The relationship between barium and its predecessor caesium, [Xe]$6s^1$, is very similar to that between strontium, [Kr]$5s^2$, and rubidium, [Kr]$5s^1$. The ns^2 metal has a much higher first ionization energy, but readily loses its second electron to give the M^{2+} (ns^0) ion, which is smaller than the corresponding M^+ (ns^0) ion, as well as having twice its charge. So the M^{2+} ion forms much stronger lattices and more stable complexes with water and other ligands; and even though barium is toxic, its sulfate is so insoluble as to be safe. Barium is also used in ceramics and is a component of the superconductor '1–2–3' (see p. 279). Its salts give an apple-green flame and are used in fireworks.

As with caesium and rubidium, any differences between the behaviour of barium and strontium are a result of the fine balance between the lower ionization energy of the heavier atom, and the lower energy evolved when its ion interacts with other species in a lattice or complex. Barium, like strontium, is a very reactive metal, and as it reacts with both oxygen and nitrogen, it too is used as a 'getter'. With nickel, however, it forms an alloy that is stable enough to be used in sparking plugs. Like both caesium and strontium, it forms a blue solution in liquid ammonia (see p. 129). There are several barium oxides: certainly $Ba^{2+}O^{2-}$ and the peroxide $Ba^{2+}O_2^{2-}$, and probably also the superoxide $Ba^{2+}(O_2^-)_2$ and the ozonide $Ba(O_3^-)_2$, which, if substantiated, would be further illustration of the ability of large cations to discourage the decomposition of unstable anions (see p. 203). Certainly, the carbonate, $BaCO_3$, decomposes at a higher temperature than its lighter analogues such as $SrCO_3$ and $CaCO_3$.

Increase in cation size of the ns^2 metals decreases the heat (and free energy) of hydration more markedly than it decreases lattice energy, especially with large doubly charged anions; so the barium salts of sulfate, carbonate, and chromate are much less soluble in water than those of strontium or calcium. The larger Ba^{2+} is also less polarizing than Sr^{2+} or Ca^{2+}, and appears to be unhydrolysed in

Decomposition temperatures[SAL] of some metal carbonates, MCO_3

M	Mg	Ca	Sr	Ba
$T_{decomp.}$/°C	300	840	1100	1300

Stability constants, *K*, of some M^{2+} complexes[SC] with $EDTA^{4-}$

M^{2+}	Mg^{2+}	Ca^{2+}	Sr^{2+}	Ba^{2+}
log *K*	8.96	10.59	8.63	7.76
	too small to accommodate 6 donor atoms	6 donor atoms accepted		

Since the influence of M^{2+} on stability constant depends on:

(1) M^{2+}–L interaction

(2) ΔH°_{hyd} of M^{2+}

(3) ΔS°_{hyd} of M^{2+}

no simple electrostatic relationship would be expected. Factor (1), however, is often dominant.

solution. Barium hydroxide can be used in volumetric analysis as a strong base, and has one great advantage over, say, sodium hydroxide; it is never contaminated with carbonate ions, since any carbon dioxide absorbed from the air is converted to insoluble barium carbonate. We would rightly guess that the large barium ion forms weaker complexes than those of calcium or strontium. Even so, some are quite stable; and $EDTA^{4-}$ (see p. 202) combines with Ba^{2+} sufficiently strongly to dissolve even the notoriously insoluble barium sulfate.

Summary

Electronic configuration

- [Xe]$6s^2$

Element

- Reactive metal with apple-green flame

Occurrence

- Mainly as $BaSO_4$ (very insoluble)

Extraction

- Convert $BaSO_4$ to BaO and reduce with Al
- Electrolysis of molten chloride

Chemical behaviour

- Very similar to Sr but less polarizing
- Reacts with dioxygen and dinitrogen
- Oxides BaO and BaO_2 (and maybe higher ones)
- $BaCO_3$ and $BaSO_4$ very insoluble
- Ba^{2+} (aq) unhydrolysed and forms weaker complexes than Sr^{2+}, $Ba(OH)_2$ strong base
- Gives blue solution in liquid ammonia

Uses

- Ba as a getter
- Ba–Ni alloy in sparking plugs
- Ba salts in fireworks
- $BaSO_4$ as slurry for oil drilling and X-ray diagnosis
- $Ba(OH)_2$ as base in volumetric analysis
- In superconductors and other ceramics

La Lanthanum

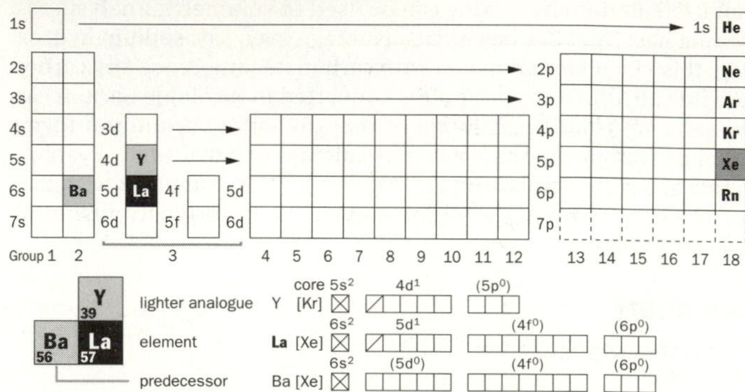

		Predecessor	Element	Lighter analogue
Name		**Barium**	**Lanthanum**	**Yttrium**
Symbol, *Z*		Ba, 56	La, 57	Y, 39
RAM		137.237	138.9055	88.90585
Radius/pm	**atomic**	217.3	187.7	181
	covalent	198	169	162
	ionic, M^{3+}		122	106
Electronegativity (Pauling)		0.9	1.1	1.22
Melting point/K		1002	1194	1795
Boiling point/K		1910	3730	3611
$\Delta H^{\ominus}_{fus}$/kJ mol^{-1}		7.66	10.04	17.2
$\Delta H^{\ominus}_{vap}$/kJ mol^{-1}		150.9	399.6	393.3
Density/kg m^{-3}		3594 (293 K)	6145 (298 K)	4469 (293 K)
Electrical conductivity/Ω^{-1}m^{-1}		2×10^6	1.754×10^6	1.754×10^6
Ionization energy for removal of *j*th electron/kJ mol^{-1}	***j* = 1**	502.8	538.1	616
	***j* = 2**	965.1	1067	1181
	***j* = 3**		1850	1980
	***j* = 4**		4819	5963
$E^{\ominus}$/V for $M^{3+}(aq) \rightarrow M(s)$			−2.38	−2.37

Stable | and long-lived (•) isotopes of lanthanum

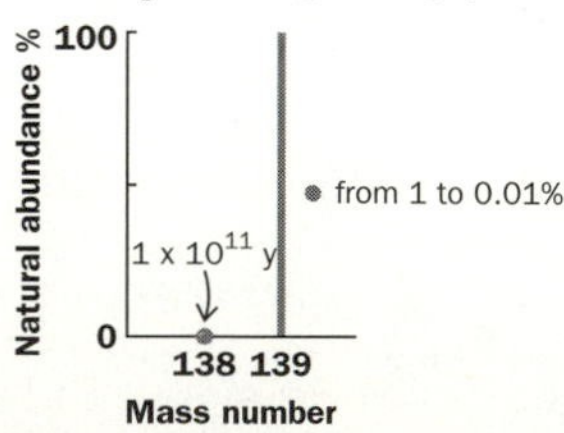

As expected, lanthanum, which follows barium, [Xe]$6s^2$, has the electronic structure [Xe]$5d^1 6s^2$ and so, formally, is the first member of the third (5d) transition series. But, like yttrium, [Kr]$4d^1 5s^2$, it is most familiar as the triply charged ion ([core]$(n-1)d^0 ns^0)^{3+}$ which forms predominantly ionic compounds, mainly with fluoride, oxide, and oxygen-containing anions such as phosphate, silicate, and carbonate.

We rightly expect lanthanum to be an electropositive metal, less so than barium but more so than yttrium. It tarnishes rapidly, burns easily in air, reacts not only with dilute acids but also (slowly) with water, and combines with most non-metals, often at room temperature. The ion La^{3+} (aq) is only very slightly hydrolysed [to a similar extent to Ca^{2+} (aq)] and the hydroxide $La(OH)_3$ is not amphoteric.

Many of the ways in which lanthanum differs from yttrium depend directly on size. The larger La^{3+} ion forms weaker complexes, but can accommodate more groups round it, as many as nine in the chloride and in the almost insoluble fluoride (and also in the hydroxide, but in this case it does resemble yttrium). In the hydrated sulfate, $La_2(SO_4)_3.9H_2O$, some metal ions even house twelve nearest neighbour oxygen atoms. Lanthanum has a much lower melting point than yttrium; maybe its 6s electrons are less available for bonding than the 5s electrons of yttrium.

Lanthanum forms some unexpected compounds that are formally divalent, such as LaH_2 and LaI_2. Unlike LaH_3 and LaI_3, which are insulators, the 'lanthanum(II)' compounds are good conductors and probably contain the La^{3+} ion and two X^- ions, together with one electron in the conduction band; so they are best written as $La^{3+} (X^-)_2 e^-$.

Although the (impure) oxide of lanthanum has been known for over 150 years, the (impure) metal was first isolated only in the 1920s, and was not obtained pure until after the Second World War. Its tardy isolation suggests a problem, as does its name, which means 'to lie hidden'. The difficulty in obtaining pure lanthanum is owing to the fact that the elements that follow it do not have their additional electrons in the 5d orbitals, as we might have supposed by analogy with the 3d and 4d transition metals which follow scandium and yttrium. Instead, the seven 4f orbitals are occupied to give a set of fourteen successors to lanthanum; only then is a second electron added to the 5d shell. So, the next element, cerium, is [Xe]$4f^1 5d^1 6s^2$. The 4f orbitals lie well inside the 6s and 5d ones, and they are progressivly stabilized by the increasing nuclear charge. In many of the fourteen post-lanthanum metals (but not all, see p. 452–3), the 4f orbitals are more stable than the 5d, so

Ionic radii of some M^{3+} ions with six coordination HKK

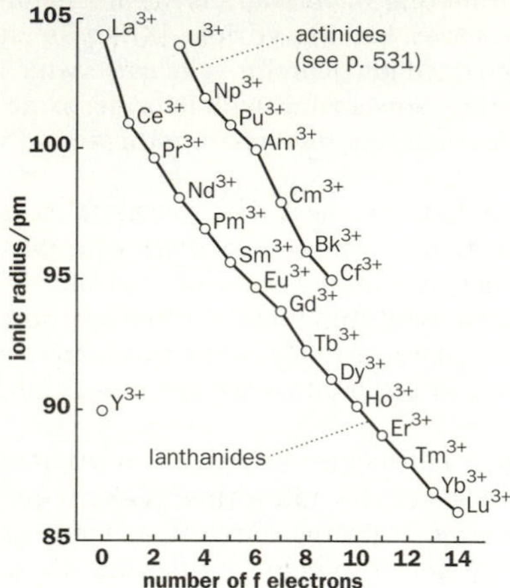

Effect of lanthanide contraction on heat of hydration of M^{3+} GE

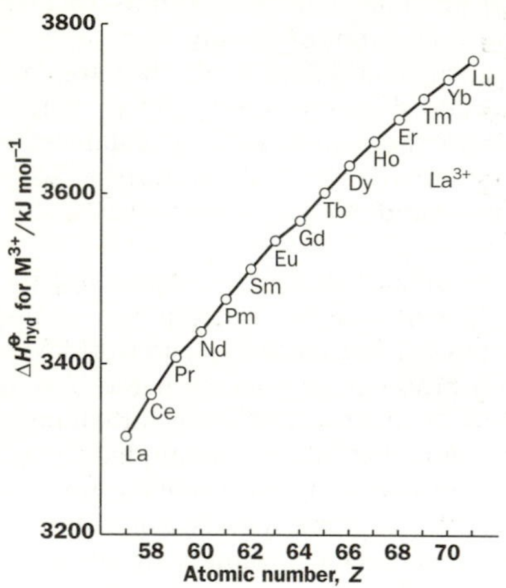

* Lanthanides with even atomic numbers are much more abundant than those with odd ones.

that the usual configuration is [Xe]$4f^{x+1}5d^0 6s^2$, rather than [Xe]$4f^x 5d^1 6s^2$ as it is for lanthanum ($x = 0$) and cerium ($x = 1$). For example, after cerium comes praseodymium ([Xe]$4f^3 5d^0 6s^2$).

It is not surprising that, as the additional electrons are added to deeply buried orbitals, they have much less effect on the behaviour of the element than if they were added to more outlying shells. When one or both of the 6s electrons are removed, the f electrons in the resulting ion are greatly stabilized. Energy for ionization of one more electron, from the 5d or more often the 4f orbital, is recouped by the combination of the M^{3+} ions with other species, but, unlike the d electrons in many transition metals, the remaining f electrons are usually held so firmly by the nucleus that they take no part in further reactions. So all the 4f elements show (III) as their main oxidation state, and for many of them it is the only oxidation state. As the fourteen elements from cerium, f^1, to lutetium, f^{14}, resemble both lanthanum, f^0, and each other more closely than does any previous sequence of elements, they are often given a communal name (which is not always either officially approved or consistently defined). Here we shall call the elements cerium to lutetium 'lanthanides', and use the general symbol Ln.

Although the lanthanides show many similarities to lanthanum, they do not behave identically to it. As the nuclear charge goes up without electrons being added to the 'outside' of the atom, the atomic radius gets progressively smaller, and this 'lanthanide contraction' causes a gradual deviation in size-dependent properties from those of lanthanum. On the other hand, each lanthanide, like any other element, has its individual electronic configuration which determines its chemical behaviour and, for some of them, causes gross individual differences from lanthanum, particularly for those elements near the beginning, the mid-point, and the end of the series.

Since most lanthanides behave very like lanthanum in most of their reactions, it is not surprising that they occur together in nature* and that it is very difficult to separate one from another. The situation is worsened by the fact that, while the lighter lanthanides are only marginally smaller than lanthanum, the heavier, smaller ones have radii similar to that of yttrium, and separation from that, too, is difficult. The almost regular shrinkage in radius of the fifteen elements $4f^{0-14}$ naturally causes gradual changes in some of their properties, and these trends can be exploited in order to separate the lanthanides from lanthanum, from yttrium, and from each other. So this is a convenient point to discuss the similarities between these elements and those varying properties that depend mainly on size. The more marked of the individual

Magnetic moments, μ_{eff}, of Ln^{3+} ions [CW]

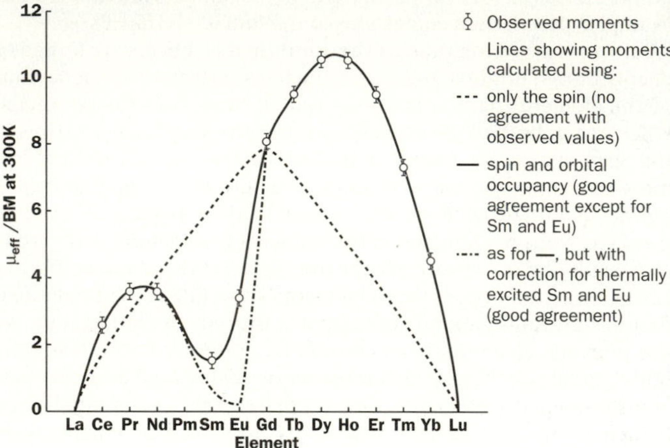

* Using extreme conditions, all lanthanides can be forced to form very unstable Ln^{2+} ions and some can be oxidized to Ln^{4+} in complex fluorides. Examples of more stable Ln^{II} and Ln^{IV} compounds are given later (see p. 452–5).

differences will be mentioned in the brief discussion of each element.

The lanthanide metals, like lanthanum, are electropositive and reactive, with their chemistry dominated by the Ln^{3+} ions, $([Xe]4f^n5p^06s^0)^{3+}$. The ions, like the atoms, shrink with increasing atomic number, the heavier ones can accommodate fewer nearest neighbours, often eight as opposed to nine for those with lower values of *n*. Bonding is mainly ionic, with a preference for very electronegative donor atoms such as fluorine and oxygen, and the strength of binding naturally increases slightly as the Ln^{3+} ion gets (heavier and) smaller. Optical and magnetic measurements show that electrons in the deeply buried f orbitals are not greatly affected by changes in the ionic environment, and there is no f orbital splitting large enough to persuade an ion to adopt a low spin configuration. As the f orbitals can hold up to seven unpaired electrons, high magnetic moments are commonplace; and they can often be calculated precisely from the electronic configuration of the ion, provided that the contribution of the orbital momentum is taken into account.

The fluorides and oxalates of the lanthanides are insoluble in water and so are useful for separating these elements from others. Some oxygen donor chelating anions react with Ln^{3+} ions to give complexes that are volatile and useful as intermediates in syntheses.

As lanthanide ions have no d electrons available for donation, they do not (except under extreme conditions) form carbonyls or π bonded organometallic complexes. Their complexes with cyclopentadienyl and similar ligands are ionic, and of various structures for the different elements.

A few lanthanides can form aqueous ions* Ln^{2+} or Ln^{4+} in addition to the usual Ln^{3+} and these elements can readily be separated from the others using conventional reduction or oxidation. In the majority, however, methods of separation depend on the very small differences in energy accompanying some two-phase process. For almost the first 160 years of lanthanide chemistry, only fractional crystallization was available, and often employed hydrated double salts of large oxoanions such as sulfate, nitrate, or bromate. Change in ionic radius changes the solubility by altering the energy balance between lattice energy, hydration enthalpy, and hydration entropy, but the net effect on the solubility is small and many (up to 20 thousand!) crystallizations are required for adequate separation.

Methods that are (somewhat) less tedious have been used since about 1950. Hydrated lanthanide ions, being triply charged, can be adsorbed at negatively charged sites on a column of a strong acid

Stability constants, K, of $LnEDTA^-$ complexes [SC]

in 0.1 M $K^+(NO_3^-$ or $Cl^-)$ at 25°C

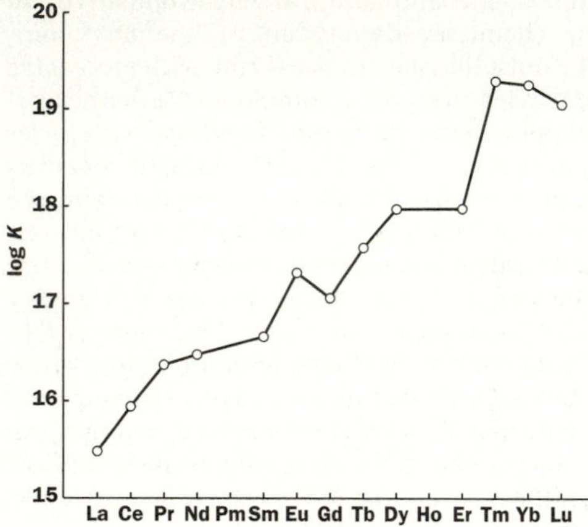

The broad increase of K with Z roughly follows the decrease in $r_{Ln^{3+}}$, but is less regular, possibly because of:

- variable, albeit low, complexing with background anion, NO_3^- or Cl^-
- different and opposing, effects of $r_{M^{3+}}$ on $\Delta H^\ominus$ and $\Delta S^\ominus$ of hydration
- experimental error (which was probably minimal in this meticulous Swiss work)

(e.g. sulfonate) ion exchange resin by displacing less highly charged cations from it. As the hydrated ions are of slightly different sizes, they are adsorbed with slightly differing strengths. They can be removed ('eluted') from the column by using a highly charged anionic complexing agent, such as $EDTA^{4-}$ or $citrate^{3-}$, which neutralizes their positive charge. The heavier lanthanides, with the smaller Ln^{3+} ions, form the stronger complexes, and so come off the resin first. By using several successive adsorptions and elutions, pure samples of lanthanide salts can be obtained. A similar method, which can be carried out continuously, involves extracting the lanthanides from aqueous solution into a liquid complexing agent such as an organic phosphate; again the heavier lanthanides are extracted more readily.

The pure (or fairly pure) metals may be obtained by reducing an anhydrous halide: either electrolytically using the chloride, or with calcium metal using the fluoride. Since the metals are so reactive, an argon atmosphere is used. For some purposes, the lanthanides need not be separated. A mixture of chlorides of the lighter lanthanides can be reduced electrolytically to the cerium-rich alloy misch metal, which, when mixed with 25% of iron, has long been used to make flints for lighters, to remove oxygen and sulfur from steels, and to improve the mechanical properties of various alloys. Lanthanum itself is used to make the alloy $LaNi_5$, which avidly absorbs hydrogen to give $LaNi_5H_6$, and also desorbs it rapidly, and so is used both to store hydrogen and to remove it from gas mixtures. The mixed oxides $La_{2-x}Ba_xCuO_4$ were the first known 'high temperature' (i.e. 35K) superconductors, although replacement of lanthanum by yttrium gives '1–2–3' which is superconducting at much higher temperatures (see p. 279). Many lanthanides, with f orbitals rich in unpaired electrons, are used in the manufacture of permanent magnets, which might seem odd because f electrons are too deeply buried to interact magnetically with the f electrons of a neighbouring atom. However, if a lanthanide is alloyed with a d transition metal, its f electrons can be aligned with the d electrons of the transition metal, which can interact with the f electron of another lanthanide atom and so on, to give a magnetically ordered domain as in $SmCo_5$ (see p. 439). The use of individual lanthanides in magnetic and optical devices is increasing rapidly.

For data and summary, see pp. 451–9.

Ce Cerium

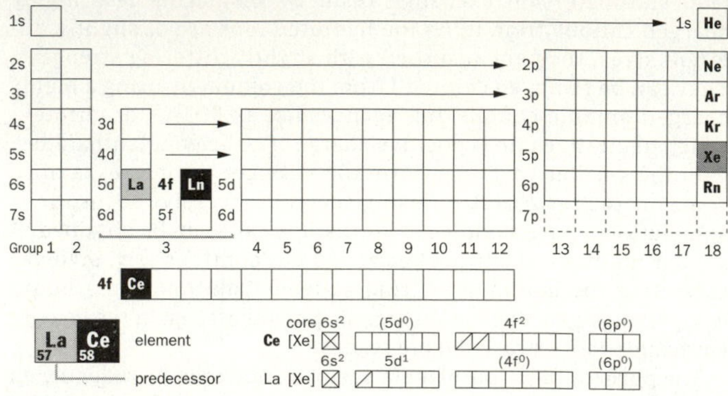

Stable isotopes of cerium

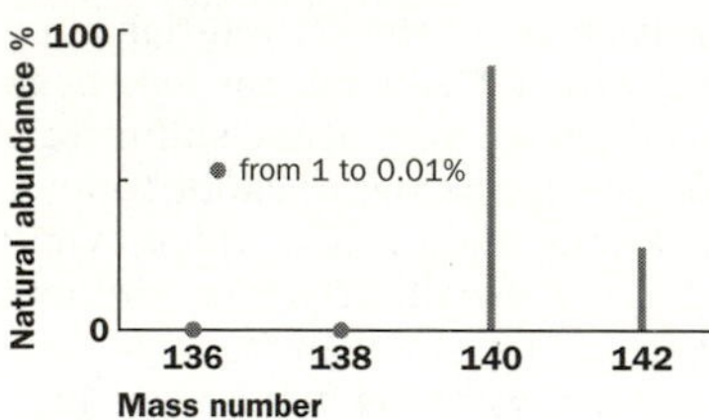

Cerium is the most familiar of the lanthanides and the most abundant (being five times as common as lead). There is still discussion about whether its electronic structure is $[Xe]4f^1 5d^1 6s^2$ (in which the nuclear charge is not yet great enough to make the 4f orbitals more stable than the 5d ones) or $[Xe]4f^2 5d^0 6s^2$; the 4f and 5d orbitals are certainly of very similar energy. As with all the 4f elements, the most stable oxidation state is (III), which for cerium gives the colourless ion Ce^{3+}, $4f^1$. Radii of both the atom and the ion are lower than those of lanthanum, but only slightly so. The 4f shell may also be emptied to give cerium(IV), which can exist in a number of fluoride or oxygen environments. The ion Ce^{4+}(aq) is readily hydrolysed and aqueous solutions of cerium (IV) probably contain anionic complexes with oxo counterions, such as $Ce(NO_3)_6^{2-}$. These are strongly oxidizing and are thermodynamically unstable; but, unless catalysed, they react too sluggishly to oxidize water and so are useful oxidizing agents for titrations. Cerium, alone of the 4f elements, forms the dioxide, rather than Ln_2O_3, when it burns in air (but see p. 444). Pure CeO_2 is white and rather inert, but the yellow hydrate, $CeO_2.nH_2O$, dissolves more readily. The oxide, CeO_2, is used as a glass polish and as a coating for 'self-cleaning' ovens. The iodate and phosphates of Ce^{IV} are insoluble in water and are used in the separation of cerium from Ln^{3+} ions (see p. 429). Cerium is the main component of 'misch metal' (see p. 431) and its other useful alloys include $CeNi_5$, which catalyses addition of hydrogen to carbon monoxide.

For data and summary, see pp. 451–9.

Pr Praseodymium

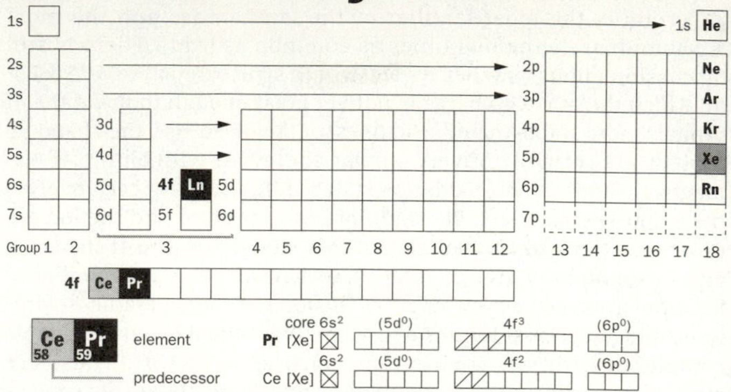

Stable isotopes of praseodymium

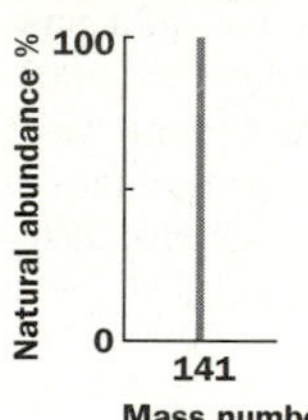

The nuclear charge on praseodymium, [Xe]$4f^36s^2$, is high enough to stabilize the 4f orbitals relative to the 5d. Since the f electrons are more embedded in the core than the 5d, they are less able to contribute to metallic bonding, so metallic praseodymium has a lower boiling point and heat of vaporization than lanthanum or cerium, in which the 5d electrons are somewhat freer. Even though more energy is needed to remove the fourth electron, praseodymium, like cerium, can form the oxide and fluoride in oxidation state (IV), and it burns in air to give Pr_6O_{11}. But these compounds are even more highly oxidizing than the cerium(IV) ones and, since they do, in practice, oxidize water, they cannot exist in aqueous solution. The dominant oxidation state of (III) gives the green ion Pr^{3+}(aq), f^2. Since the 4f electrons in both the atom and the cation Pr^{3+} have the same spin, these species have magnetic moments. Alloys of praseodymium are used for making permanent magnets. The ion Pr^{3+} is one of a number of lanthanide ions used as a 'shift reagent' because the changes that its electron spins cause in the NMR spectra of some of its organic ligands give us information about their structure.

For data and summary, see pp. 451–9.

Nd Neodymium

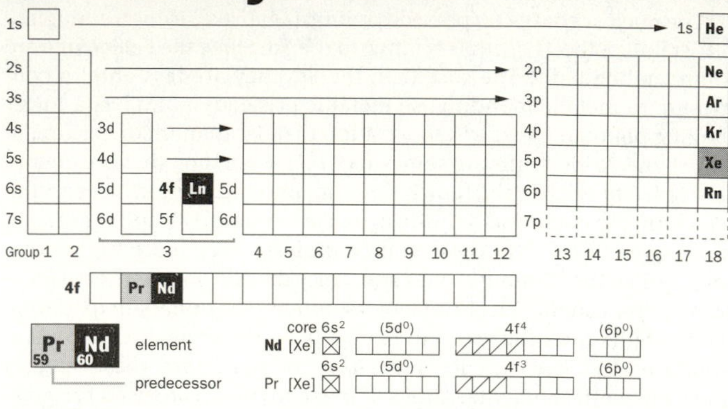

Stable (|) and long-lived (¦) isotopes of neodymium

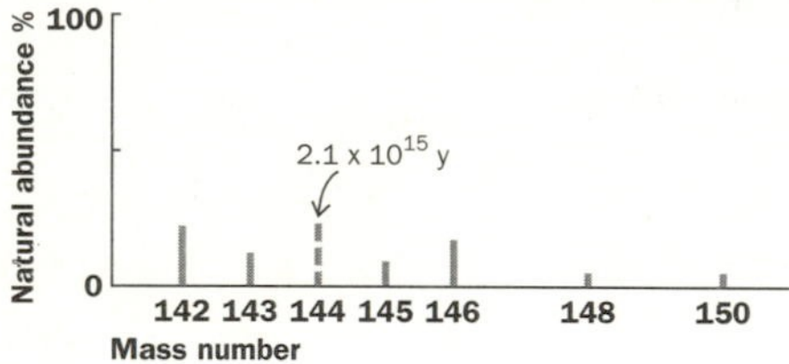

With the electronic configuration $[Xe]4f^4 6s^2$, neodymium behaves much as expected. The f electrons become more localized in the core and the energy of vaporization of neodymium is lower than that of praseodymium, and its (IV) state (in Cs_3NdF_7) is even more oxidizing. There are a few examples of neodymium(II), such as NdO, $NdCl_2$, and NdI_2, but these are strongly reducing. Neodymium is used in alloys for permanent magnets, in Nd_2Co_7 as a catalyst for dehydrogenation, and in glass manufacture. Its usual ion, Nd^{3+}, f^3, is violet-red and, when 1% of it is incorporated in a mixed oxide of yttrium and aluminium, it is used as a laser in the near-infrared.

For data and summary, see pp. 451–9.

Pm Promethium

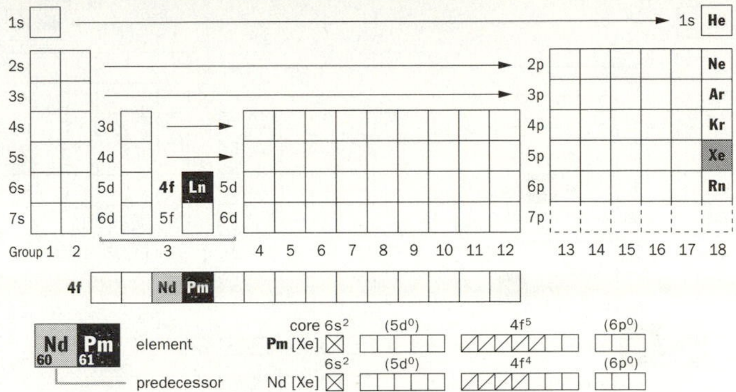

Key isotopes of promethium

Mass number	145	146	147	149	151
Half-life	17.7 y	5.53 y	2.6234 y	53.1 h	28.4 h

Promethium has no stable isotopes; the longest lived, ^{145}Pm, has a half-life of 17.7 years. Promethium-147, with a half-life of 2.6 years, is formed by spontaneous fission of uranium-238 and occurs in minute traces in uranium ores (less than one part in 10^{17} in African pitchblende). It can be extracted, on the milligram scale, from the fission products of nuclear reactors, and can be made artificially by neutron bombardment of neodymium, from which it is separated by cation exchange (see p. 431).

Surprisingly, quite a lot is known about promethium, which behaves predictably for a light lanthanide of configuration $[Xe]4f^5 6s^2$. Many promethium(III), f^4, compounds have been made and have colours in the pink–red–mauve range. The metal has been obtained by reduction of its (insoluble) fluoride with lithium. The other trihalides are soluble, but the oxalate and hydroxide are insoluble. Apparently there is even a use for it: in miniature batteries.

For data and summary, see pp. 451–9.

Sm Samarium

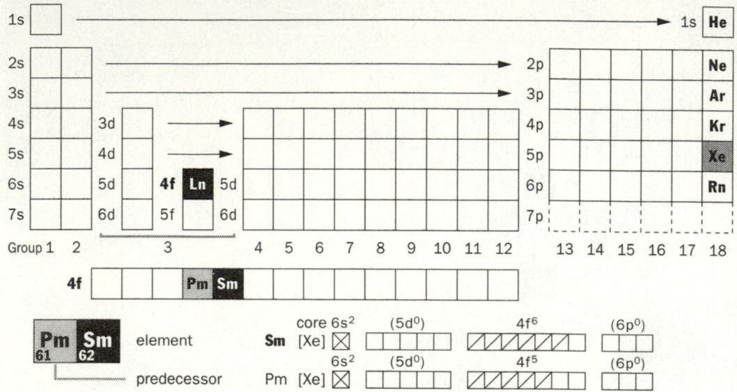

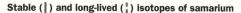

Stable (|) and long-lived (¦) isotopes of samarium

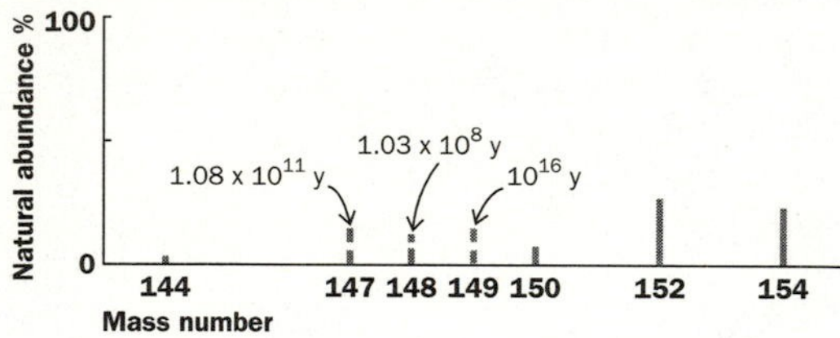

Like other (stable) lanthanides, samarium is used for permanent magnets, glasses, catalysis (for example as Sm_2Co_7, see also neodymium, p. 437), and electronic devices. The alloy $SmCo_5$ is a ferromagnet over ten thousand times more powerful than iron. As it has a very high tendency to absorb neutrons, samarium is also used in control rods in nuclear reactors.

Although we cannot compare samarium ([Xe]$4f^6 6s^2$) in detail with its scarce, unstable predecessor, we can see that it differs in a number of ways from neodymium ([Xe]$4f^4 6s^2$). Its 4f electrons are even more inaccessible and it has a higher third ionization energy and a very much lower heat of vaporization. The usual oxidation state is (III), which gives a pale yellow fluorescent ion Sm^{3+}, f^5. Samarium(IV) is unknown but there are several compounds of samarium(II), including the blood-red ion Sm^{2+}, and even the difluoride. There are three samarium oxides, SmO, Sm_2O_3, and the mixed valence $Sm^{II}Sm^{III}{}_2O_4$. Perhaps the greatest oddity of samarium is the sulfide, $Sm^{II}S$, which is normally a black semiconductor, but can be changed into a golden metallic conductor by applying high pressure or, for reasons as yet unknown, by scratching a single crystal of it. How excited the alchemists would have been.

For data and summary, see pp. 451–9.

Eu Europium

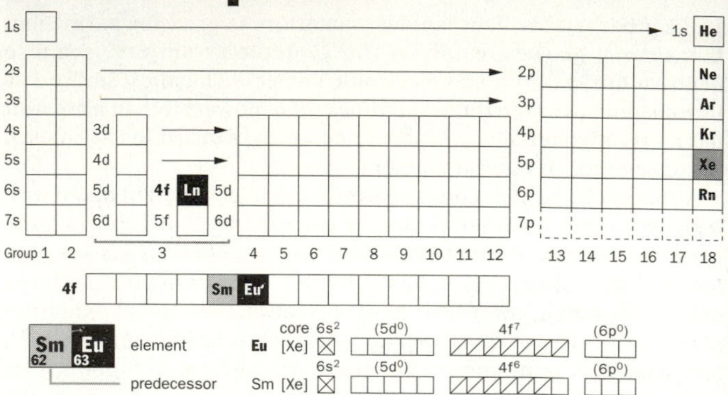

Stable isotopes of europium

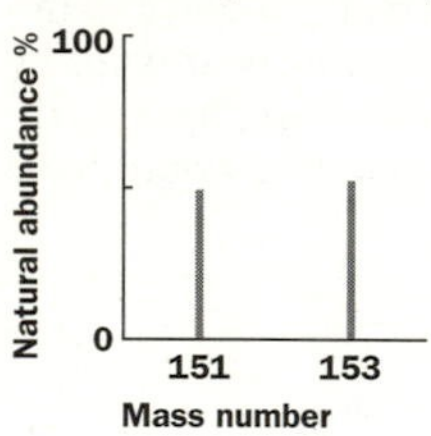

Metal radii of lanthanides [GE]

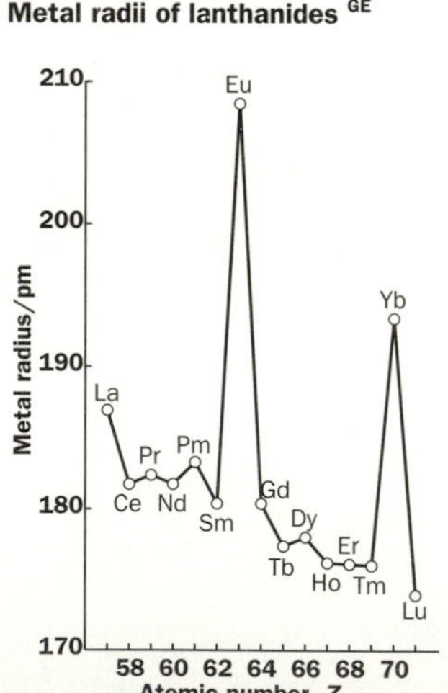

Third ionization energy, I_3 for $Ln^{2+} \rightarrow Ln^{3+} + e^-$ [GE]

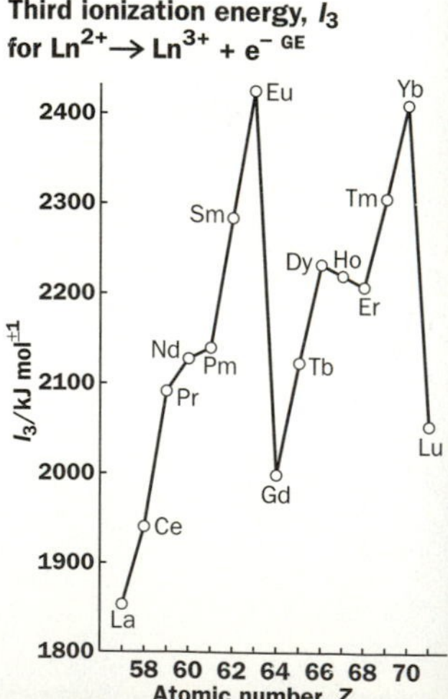

We can guess a lot about europium from its configuration, [Xe]$4f^7 6s^2$. The f orbitals are well stabilized within the electron core, and there is, as yet, no electron pairing between them. In many ways, europium metal behaves as if it is Eu^{2+}, $2e^-$, with a much larger metallic radius than any other lanthanide (see also ytterbium, p. 449), with a high third ionization energy, and with a very low heat of vaporization. The pale greenish yellow Eu^{2+} ion, unlike other Ln^{2+} ions, can exist in water, although oxidation by it is thermodynamically favoured. The metal dissolves in liquid ammonia to give a blue solution reminiscent of those formed by ([core]ns^2) metals; and europium sulfate, like barium sulfate, is very insoluble in water. With reducing ligands such as hydride and iodide, europium forms only the EuX_2 derivative. But, like the other lanthanides, europium forms a variety of stable compounds in oxidation state (III). The Eu^{3+}, $4f^6$, ion is colourless, because it absorbs ultraviolet rather than visible light. At temperatures near absolute zero it has a very low magnetic moment, which might seem odd for an ion with six unpaired electrons; but the contribution from the spins of the electrons is almost cancelled by that from their angular motion. At higher temperatures, some electrons become excited and the counteraction is less complete (see p. 428).

Most recent uses of europium are those that involve spectra. The Eu^{2+} ion is used as an X-ray intensifier in medicine, while Eu^{3+} is used as a 'shift reagent' (see p. 435). Both ions also fluoresce when excited electronically, and are used in phosphors in digital displays, and in green (Eu^{2+}) and red (Eu^{3+}) TV tubes. Like samarium, europium is used in nuclear reactor control rods, as it readily captures neutrons.

For data and summary, see pp. 451–9.

Gd Gadolinium

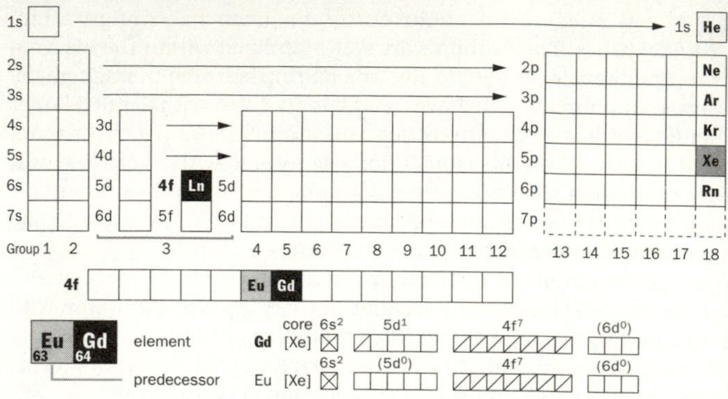

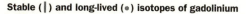

Stable (|) and long-lived (•) isotopes of gadolinium

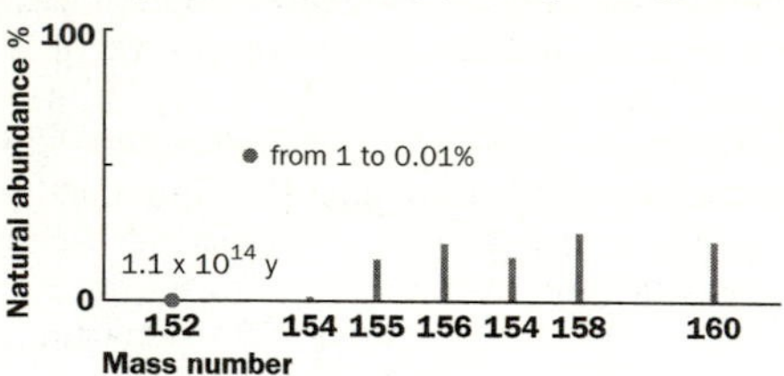

There is a hiccup in the electronic configuration of gadolinium, $[Xe]4f^7 5d^1 6s^2$, which is further evidence for the delicate balance between the energies of the 4f and 5d electrons: although the 4f orbitals are increasingly stabilized relative to the 5d as the nuclear charge becomes greater, it is favourable to put the eighth (f + d) electron into a 5d orbital rather than to subject it to the repulsion caused by pairing it in a 4f one. The 5d electron makes its presence felt by entering the conduction band, giving the metal dramatically higher melting and boiling points than europium, and similarly high heats of fusion and vaporation. Naturally, less energy is needed to remove this electron and so the third ionization energy falls (by 414 $kJ\ mol^{-1}$) from europium to gadolinium.

So it is no surprise that the chemistry of gadolinium is dominated by the Gd^{3+}, f^7, ion. This ion, like Eu^{3+}, is colourless but absorbs strongly in the ultraviolet. The iodide, GdI_2, is a metallic conductor of the type $Ln^{3+}(I^-)_2e^-$ (see p. 425). Although there is a lower chloride, $GdCl_{1.5}$, this contains Gd_6 clusters rather than gadolinium(II) or (I).

Many of the uses of gadolinium are by now predictable: for magnets, alloys, and electronic devices. As it captures neutrons extremely well (more than eight times as effectively as samarium), gadolinium is also used both in control rods and in neutron radiography. Its complex* with $DTPA^{5-}$ (which is very stable and virtually non-toxic) is used for imaging in NMR body scanners, particularly to detect brain tumours.

For data and summary, see pp. 451–9.

* Diethylene Triamine Penta-acetic Acid is an extended version of EDTA, with three donor nitrogen atoms and five donor oxygens.

Tb Terbium

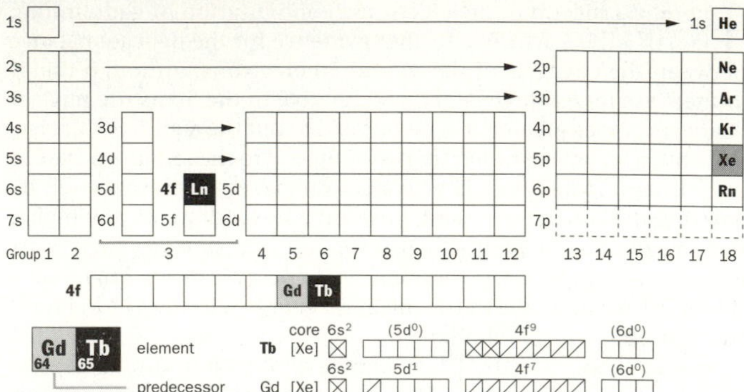

Stable isotopes of terbium

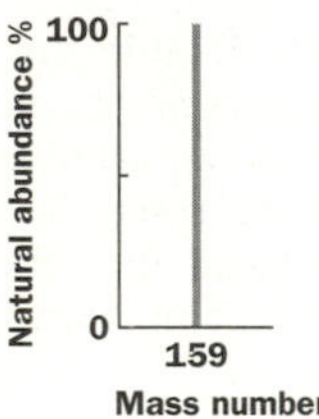

With nine (f + d) electrons, the hiccup is passed and the electronic structure reverts to $5d^0$, giving terbium the configuration $[Xe]4f^9 6s^2$. But the heat of vaporization of terbium is higher than that of gadolinium, and so the balance between increasing nuclear charge, decreasing size, and availability of outer electrons for the conduction band must again be very fine.

The chemistry of terbium is mainly that of the Tb^{3+}, f^8, ion which, in aqueous solution, is very pale pink. It fluoresces green and is used as a phosphor in TV tubes and digital displays, and also as an X-ray intensifier. Terbium also shows some (small) tendency to adopt the f^7configuration. It burns in air to the mixed oxide Tb_4O_7 (rather than to the usual Ln_2O_3) and can become pure terbium (IV) in TbO_2 and TbF_4. These exist only as solids, and are such strong oxidizing agents that they attack water.

For data and summary, see pp. 451–9.

Dy Dysprosium

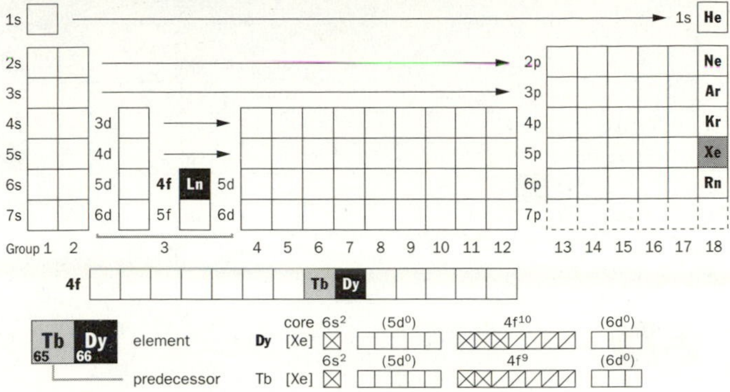

Stable isotopes of dysprosium

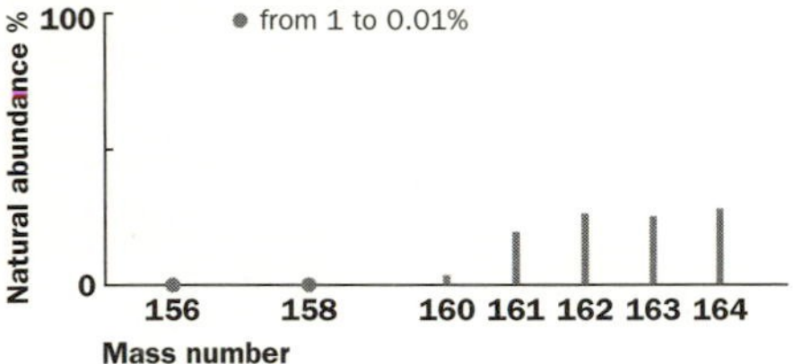

At dysprosium, $[Xe]4f^{10}6s^2$, the heat of vaporization has fallen a bit. The metal is, like its neighbours, used in alloys for magnets. Apart from a complex fluoride $Cs_3Dy^{IV}F_7$ and the formally divalent halides $Dy^{3+}(X^-)_2e^-$, the chemistry of dysprosium is mainly that of the Dy^{3+}, f^9, ion, which is pale yellow in aqueous solution.

For data and summary, see pp. 451–9.

Ho Holmium

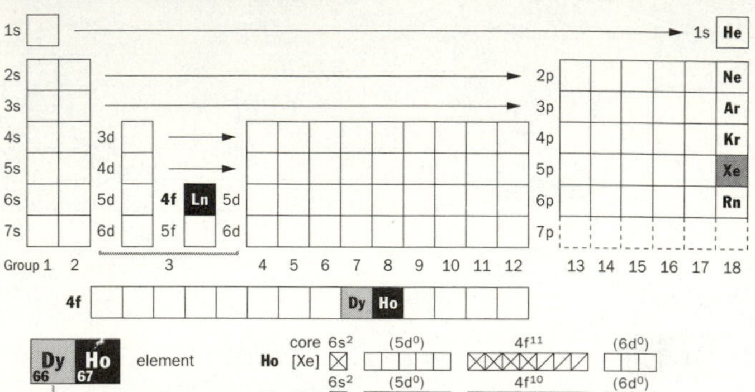

Stable isotopes of holmium

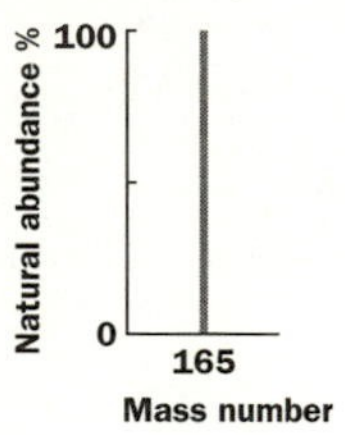

Holmium, $[Xe]4f^{11}6s^2$, continues the heavy lanthanide trend of decreased radius, lower coordination number, and lower heat of vaporization. It, too, is used in magnets. It seems to restrict its oxidation state to (III), giving a yellow Ho^{3+}, f^{10}, ion in aqueous solution.

For data and summary, see pp. 451–9.

Er Erbium

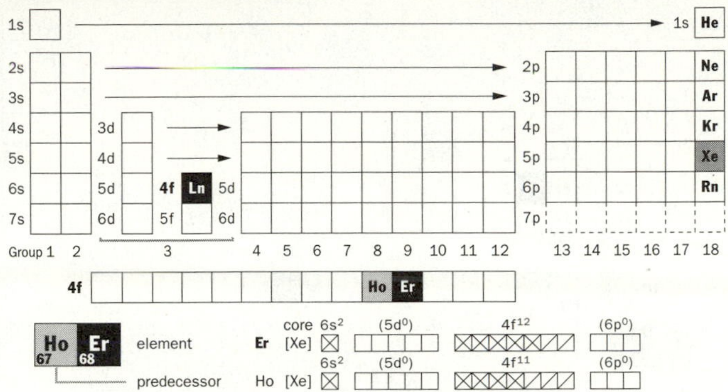

Stable isotopes of erbium

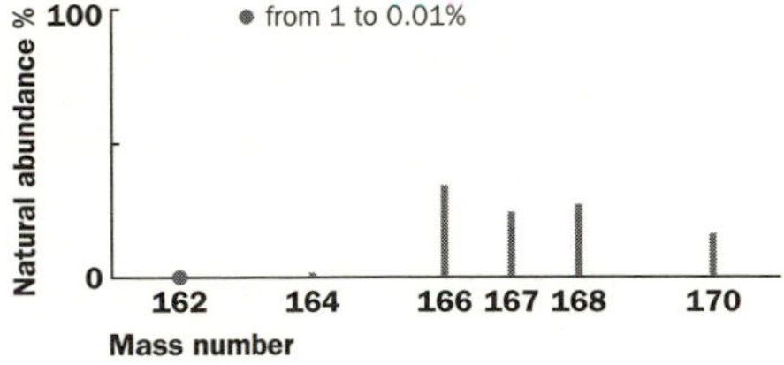

The configuration $[Xe]4f^{12}6s^2$ produces no unpredictable behaviour. The familiar trends continue and erbium, which gives a rose pink Er^{3+}, f^{11}, ion in aqueous solution, shows no sign of adopting any oxidation state other than (III). It is used in alloys, and also for making glass that absorbs infrared.

For data and summary, see pp. 451–9.

Thulium

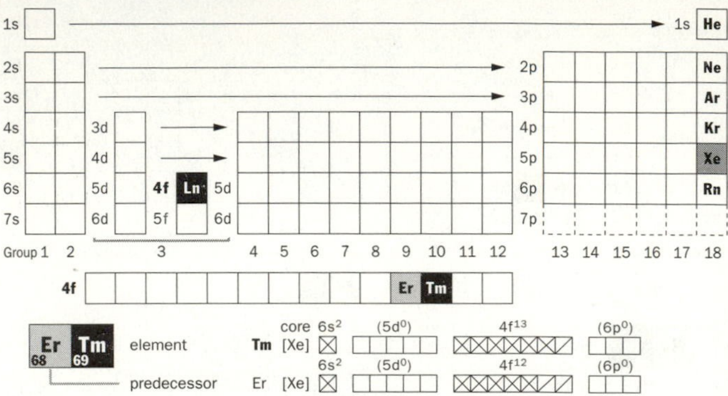

Stable isotopes of thulium

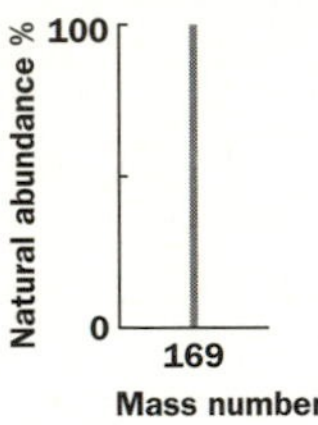

Thulium, $[Xe]4f^{13}6s^2$, is the second rarest of the lanthanides, after promethium. As expected, its heat of vaporization is markedly lower, and its third ionization energy higher, than that of erbium, showing that the (f + d) electrons are even less available for metallic bonding. Unlike holmium and erbium, it forms the familiar lanthanide dihalides $Tm^{3+}(X^-)_2e^-$ with metallic conductivity; but in its other compounds it is unambiguously Tm^{III}. The aqueous Tm^{3+}, f^{12}, ion is pale green. In zinc sulfide the Tm^{3+} ion fluoresces blue, but not well enough to be used as a phosphor for TV screens. Thulium is also used in X-ray equipment.

For data and summary, see pp. 451–9.

Yb Ytterbium

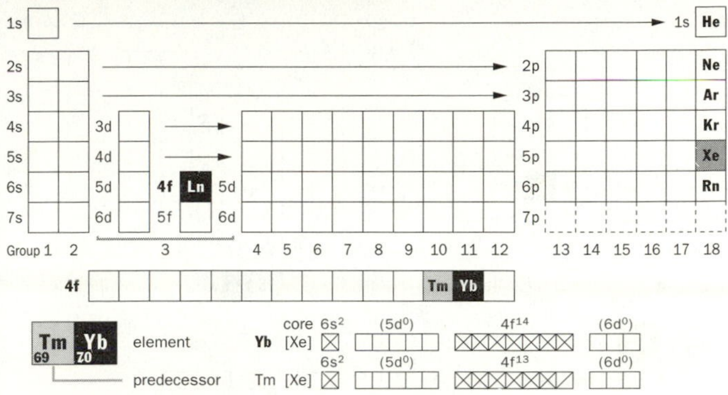

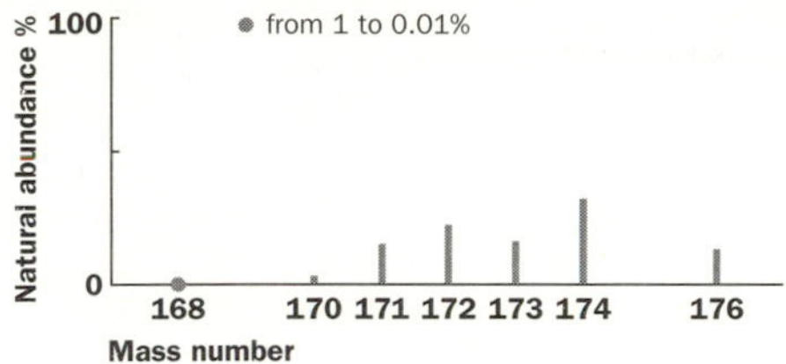

Stable isotopes of ytterbium

Natural abundance % 100 0
● from 1 to 0.01%
168 170 171 172 173 174 176
Mass number

If the 4f orbitals of the heavy lanthanides are indeed more stable than the 5d, they will be full at ytterbium, which, with a configuration of [Xe]$4f^{14}6s^2$, is similar in many ways to europium [Xe]$4f^7 6s^2$ (see p. 441). The radii of these two metals are much larger than those of neighbouring lanthanides, suggesting a leaning towards Ln^{2+} rather than Ln^{3+}. Ytterbium, in particular, has a very low heat of vaporization and both have relatively high energies for the third ionization step. Ytterbium, like europium, resembles the [core]ns^2 metals by forming blue solutions in liquid ammonia.

The dominant feature of ytterbium chemistry is the Yb^{3+}, f^{13}, ion, which is colourless in water. Because of the high nuclear charge, the hydroxide, $Yb(OH)_3$, unlike those of the other lanthanides, is slightly amphoteric and dissolves in hot concentrated sodium hydroxide. There are also a number of compounds of ytterbium(II), such as YbO and even YbF_2, but the yellow Yb^{2+}, f^{14}, ion, unlike Eu^{2+}, f^7, is oxidized by water. Ytterbium is used in gauges to measure (mechanical) stress.

For data and summary, see pp. 451–9.

Lu Lutetium

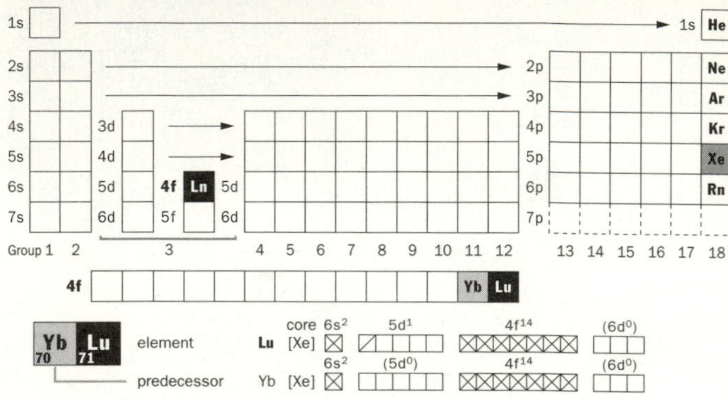

Stable ( ) and long-lived ( ) isotopes of lutetium

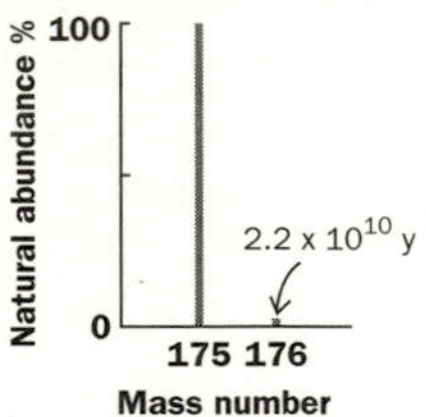

There need be no discussion about the electronic configuration of lutetium, $[Xe]4f^{14}5d^{1}6s^{2}$: the 4f shell is full, and the 5d singly occupied. As there are three electrons unambiguously available for the conduction band, the metal is hard and dense and its heat of vaporization is much (269 kJ mol^{-1}) higher than for ytterbium. As expected, it seems to form compounds only of lutetium(III). The aqueous Lu^{3+} ion, f^{14}, is colourless; and $Lu(OH)_3$, like $Yb(OH)_3$, dissolves in hot concentrated alkali.

Although for many years the main use of lutetium was the study of the behaviour of lutetium, the element is now proving a promising catalyst in organic synthesis.

For data and summary, see pp. 451–9.

The lanthanide (4f) series: trends and summaries La–Lu

The $[Xe]4f^n5d^16s^2$ metals are electropositive and reactive, forming mainly ionic M^{3+} compounds with a preference for hard ligands. As *n* increases, the f electrons become more withdrawn into the core and the ionic radius decreases, producing small differences in chemical behaviour. So, although the lanthanides often occur together, they can be separated by various frequently repeated two-phase techniques. A few lanthanides, near the beginning, centre, and end of the series, which also form M^{4+} or M^{2+} ions, can be separated more easily. The metals are often prepared by reducing the molten chloride. The f electrons are deeply embedded in the electron core: they are little affected by ligands and, since there may be up to seven unpaired electrons in an atom or ion, lanthanides have many magnetic (and optical) applications. Some lanthanides combine with reducing ligands such as iodide or hydride ions to give 'LnX_2' compounds which are good conductors, probably $Ln^{3+}.e^-.2X^-$. Summaries and data of individual elements are given in the table that follows.

Comparative summaries of elements

Element	La	Ce	Pr	Nd
Outer configuration	$4f^05d^16s^2$	$4f^25d^06s^2$ or $4f^15d^16s^2$	$4f^36s^2$	$4f^46s^2$
Atomic number	57	58	59	60
Relative atomic mass	138.9055	140.115	140.90765	144.24
Terrestrial abundance /ppm	32	68	9.5	38
Isotopes: mass no. [%]	139 [99.91] 138 [0.09]	140 [88.48] 142 [11.08] 138 [0.25] 136 [0.19]	140 [100]	142 [27.13] 144 [23.80] 146 [17.9] 143 [12.18] 145 [8.30] 148 [5.76] 150 [5.60]
Radius/pm: atomic	187.7	182.5	182.8	182.1
covalent	169	165	165	164
Ln^{3+} [other]	122	107 [Ce^{4+} 94]	106 [Pr^{4+} 92]	104
Electronegativity (Pauling)	1.1	1.1	1.1	1.1
*j*th ionization energy/$kJ mol^{-1}$				
j = 1	538.1	527.4	523.1	529.6
2	1067	1047	1018	1035
3	1850	1949	2086	2130
4	4819	3547	3761	3899
Melting point/K	1194	1072	1204	1294
Boiling point/K	3730	3699	3785	3341
$\Delta H^{\ominus}_{fus}$/kJ mol^{-1}	10.04	8.87	11.3	7.113
$\Delta H^{\ominus}_{vap}$/kJ mol^{-1}	399.6	313.8	332.6	283.7
Electrical resistivity /Ωm (298 K)	57×10^{-8}	73×10^{-8}	68×10^{-8}	64×10^{-8}
$E^{\ominus}$/V for:				
$Ln^{3+} \rightarrow Ln^0$	−2.38	−2.34	−2.35	−2.32
$Ln^{4+} \rightarrow Ln^{3+}$		1.74	3.2	4.9
$Ln^{3+} \rightarrow Ln^{2+}$				2.6
Ln^{3+} ion, f^x	f^0	f^1	f^2	f^3
Colour	Colourless	Colourless	Green	Red-violet

La to Lu: [Xe]$4f^x5d^y6s^2$

Pm	Sm	Eu	Gd	Tb
$4f^56s^2$	$4f^66s^2$	$4f^76s^2$	$4f^75d^16s^2$	$4f^96s^2$
61	62	63	64	65
~145	150.36	151.965	157.25	158.92534
trace	7.9	2.1	7.7	1.1
none stable	152 [26.6] 154 [22.6] 147 [15.0] 149 [13.8] 148 [11.3] 150 [7.4] 144 [3.1]	153 [52.2] 151 [47.8]	158 [24.84] 160 [21.86] 156 [20.47] 157 [15.65] 155 [14.80] 154 [2.18] 152 [0.20]	159 [100]
181	180.2	204.2	180.2	178.2
–	166	185	161	159
106	100 [Sm^{2+} 111]	98 [Eu^{2+} 112]	97	93 [Tb^{4+} 81]
	1.2	–	1.2	–
535.9 1052 2150 3970	543.3 1068 2260 3990	546.7 1085 2404 4110	592.5 1167 1990 4250	564.6 1112 2114 3839
1441	1350	1095	1586	1629
~3000	2064	1870	3539	3396
12.6	10.9	10.5	15.5	16.3
–	191.6	175.7	311.7	391
$\sim 50 \times 10^{-8}$ (273 K)	94×10^{-8}	90.0×10^{-8}	134×10^{-8}	114×10^{-8}
–2.29	–2.30 –1.55	–1.99 –0.35	–2.28	– 2.31 3.1
f^4 Pink-red	f^5 Pale yellow	f^6 Colourless	f^7 Colourless	f^8 V. pale pink

Continued on pp. 454–9

Element	Tb	Dy	Ho	Er
Outer configuration	$4f^96s^2$	$4f^{10}6s^2$	$4f^{11}6s^2$	$4f^{12}6s^2$
Atomic number	65	66	67	68
Relative atomic mass	158.92534	162.50	164.93032	167.26
Terrestrial abundance /ppm	1.1		1.4	3.8
Isotopes: mass no. [%]	159 [100]	164 [28.2] 162 [25.5] 163 [24.9] 161 [18.9] 158 [0.10] 156 [0.06]	165 [100]	166 [33.6] 168 [26.8] 167 [22.95] 170 [14.9] 164 [1.61] 162 [0.14]
Radius/pm: atomic	178.2	177.3	176.6	175.7
covalent	159	159	158	157
Ln^{3+} [other]	93 [Tb^{4+} 81]	91	89	89
Electronegativity (Pauling)	–	1.2	1.2	1.2
***j*th ionization energy/kJmol^{-1}**				
***j* = 1**	564.6	571.9	580.7	588.7
2	1112	1126	1139	1151
3	2114	2200	2204	2194
4	3839	4001	4100	4115
Melting point/K	1629	1685	1747	1802
Boiling point/K	3396	2835	2968	3136
$\Delta H^{\ominus}_{fus}$/kJ mol^{-1}	16.3	17.2	17.2	17.2
$\Delta H^{\ominus}_{vap}$/kJ mol^{-1}	391	293	251.0	292.9
Electrical resistivity /Ωm (298 K)	114×10^{-8}	57×10^{-8}	87.0×10^{-8}	87×10^{-8}
$E^{\ominus}$/V for:				
$Ln^{3+} \rightarrow Ln^0$	−2.31	−2.29	−2.33	−2.32
$Ln^{4+} \rightarrow Ln^{3+}$	3.1	5.7		
$Ln^{3+} \rightarrow Ln^{2+}$		−2.5		
Ln^{3+} ion, f^x	f^8	f^9	f^{10}	f^{11}
Colour	V. pale pink	Pale yellow	Yellow	Rose-pink,

Tm	Yb	Lu
$4f^{13}6s^2$	$4f^{14}6s^2$	$4f^{14}5d^16s^2$
69	70	71
168.93421	173.04	174.967
0.48	3.3	0.51
169 [100]	174 [31.8] 172 [21.9] 173 [16.12] 171 [14.3] 176 [12.7] 170 [3.05] 168 [0.13]	175 [97.41] 176 [2.59]
174.6	194	173.4
156	170	156
87 [Tm^{2+} 94]	86 [Yb^{2+} 113]	85
1.2	–	1.3
596.7 1163 2285 4119	603.4 1176 2415 4220	523.5 1340 2022 4360
1818	1097	1936
2220	1466	3668
18.4	9.20	19.2
274	159	428
79×10^{-8}	29×10^{-8}	79×10^{-8}
–2.32 –2.3	–2.22 –1.05	–2.30
f^{12} Pale green	f^{13} Colourless	f^{1} Colourless

Continued on pp. 456–9

Element	La	Ce	Pr	Nd
$Ln^{3+}e^-$ behaviour	LaI_2, LaH_2	–	–	NdI_2
Ln^{II}	–	–	–	NdO strongly reducing, $NdCl_2$
Ln^{IV}	–	CeO_2 (burn Ce in air), anionic Ce^{IV} complexes with O and F, oxidizing	With O and F but v. oxidizing	$CsNdF_7$ v. strongly oxidizing
Points of interest	High coordination numbers	Most abundant Ln	Burns in air to Pr_6O_{11}	
Uses	• Alloy $LaNi_5$ absorbs H_2 • $La_{2-x}Ba_xCuO_4$ was first 'high temp' super-conductor (now superceded)	• Ce – in lighter flints, – as S and O scavenger • $CeNi_5$ catalyst • CeO_2 – glass polish – oven coating	• Pr – in alloys – magnets • Pr^{3+} – NMR shift reagent	• Nd – magnets • Nd^{3+} – IR laser

Pm	Sm	Eu	Gd	Tb
			GdI_2	
	SmF_2, SmO, SmS, Sm^{II}, $Sm^{III}_2O_4$, Sm^{2+} (red)	Eu^{2+} (aq) greenish yellow		
				TbO_2 and TbF_4 v. strongly oxidizing
• No stable isotope • Extracted from reactors • Made artificially	$Sm^{II}S$ becomes metallic conductor at high pressure or on scratching	Resembles Ba: • $EuSO_4$ v. insoluble • Blue solution in liq. NH_3 • Eu^{3+} almost diamagnetic at 0 K	$GdCl_{1.5}$ contains Gd_6 clusters	Burns in air to Tb_4O_7
• Batteries	• Sm in nuclear control rods • $SmCo_5$ v. strong ferromagnet	• Eu in nuclear control rods • Eu^{2+} X-ray intensifier • Eu^{3+} – red phosphor – NMR shift reagent	• Gd – in nuclear control rods – in neutron radiography • Gd^{3+} – NMR – imaging	• Tb^{3+} – green phosphor i[illegible] TV – X-ray intensi[illegible]

Continued on pp. 45

Element	Tb	Dy	Ho	Er
$Ln^{3+}e^-$ behaviour		DyX_2		
Ln^{II}				
Ln^{IV}	TbO_2 and TbF_4 v. strongly oxidizing	Cs_3DyF_7		
Points of interest	Burns in air to Tb_4O_7	–	–	–
Uses	• Tb^{3+} – green phosphor in TV – X-ray intensifier	• Magnetic alloys	• Magnetic alloys	• Alloys • IR-absorbent glass

	Tm	Yb	Lu
	TmX_2		–
		Yb^{2+} (yellow) reduces water	–
	High I_3	• $Yb(OH)_3$ slightly amphoteric • High I_3 • Blue solution with liq. NH_3	$Lu(OH)_3$ slightly amphoteric
	• In X-ray equipment • Tm^{3+} blue phosphor	• In stress gauges	• Catalysis

Hf Hafnium

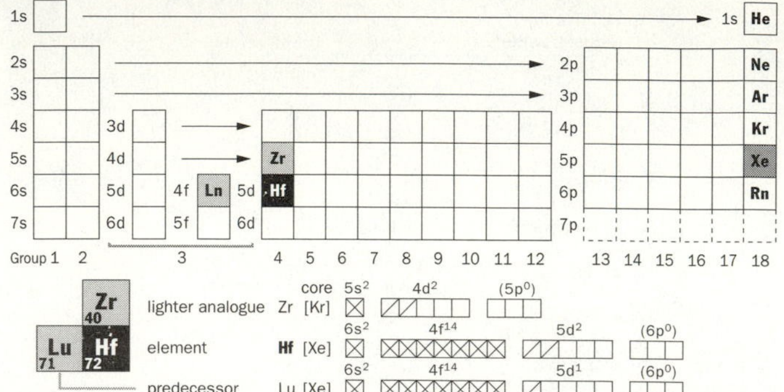

		Predecessor	Element	Lighter analogue
Name		**Lutetium**	**Hafnium**	**Zirconium**
Symbol, *Z*		Lu, 71	Hf, 72	Zr, 40
RAM		174.967	178.49	91.224
Radius/pm	**atomic**	173.04	156.4	160
	covalent	156	144	145
	ionic, M^{4+}	85	84	87
Electronegativity (Pauling)			1.3	1.33
Melting point/K		1936	2503	2125
Boiling point/K		3668	5470	4650
$\Delta H^{\ominus}_{fus}$/kJ mol^{-1}		19.2	25.5	23.0
$\Delta H^{\ominus}_{vap}$/kJ mol^{-1}		428	661.1	581.6
Density (at 293 K)/kg m^{-3}		9840	13 310	6506
Electrical conductivity /Ω^{-1}m^{-1}		1.266×10^6	3.175×10^6	2.375×10^6
Ionization energy for removal of *j*th electron/kJ mol^{-1}	***j* = 1**	523.5	642	660
	***j* = 2**	1340	1440	1267
	***j* = 3**	2022	2250	2218
	***j* = 4**	4360	3216	3313
$E^{\ominus}$/V for $M^{4+}(aq) \rightarrow M(s)$			−1.70	

Stable isotopes of hafnium

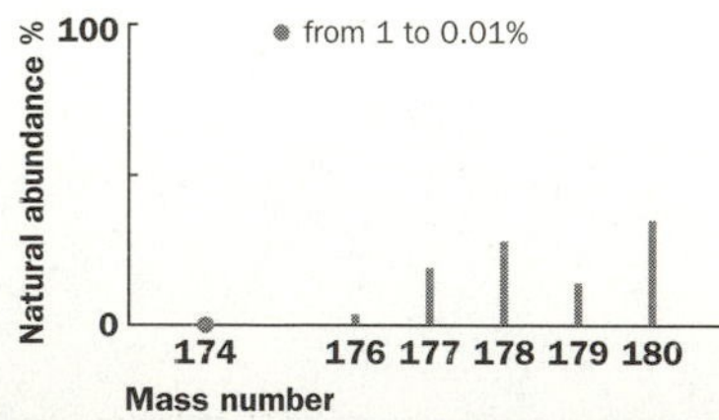

Hafnium is even more similar to zirconium (each with a d^2 outer shell) than lanthanum ([Xe]$6s^25d^1$) is to yttrium ([Kr]$5s^24d^1$); and the reason is not hard to find. Between lanthanum and hafnium come the fourteen 4f elements which, as we have seen, have slightly smaller atoms with each increase in atomic number. This lanthanide contraction has repercussions beyond the 4f series and one of its effects is that the elements of the third transition series ([Xe]$4f^{14}$)$5dn6s^2$ have (except for lanthanum) radii that are very similar to those with the same value of *n* in the second transition series, [Kr]$4d^n5s^2$. The effect is particularly marked for hafnium, ([Xe]$4f^{14}$)$5d^26s^2$, and zirconium, [Kr]$4d^25s^2$, which have almost identical radii. We should expect two elements with the same outer configuration and atoms of the same size to behave in almost exactly the same way. These two elements always occur together in nature, with hafnium, on average, about one-fiftieth as abundant; and since their chemical behaviour is almost indistinguishable, total purification of either element is extremely difficult. Solvent extraction of the nitrates or thiocynates is used for successive separations. As expected, the heavier hafnium has higher melting and boiling points than zirconium, and also than lutetium, because hafnium has two d electrons rather than one.

The many hafnium isotopes do, of course, have totally different *nuclear* properties from those of zirconium; and hafnium metal, unlike zirconium, absorbs neutrons particularly well. This makes hafnium useful as a neutron moderator, but it also means that the zirconium used for uranium dioxide cladding (see p. 337) should contain as little hafnium as possible.

There seems to be one other, not very fundamental, way in which hafnium differs from zirconium. In the hafnium tetracyclopentadienyl compound, two rings are σ bonded, and two joined by five-point π bonding (as in the analogous titanium compound). In all other ways, the chemistry of hafnium appears to mirror that of zirconium. Although we shall find many similarities between corresponding pairs of elements in the 4d and 5d series, the extreme resemblance of hafnium to zirconium is remarkable for two elements differing by as much as 32 in atomic number.

For summary, see p. 492–3.

Ta Tantalum

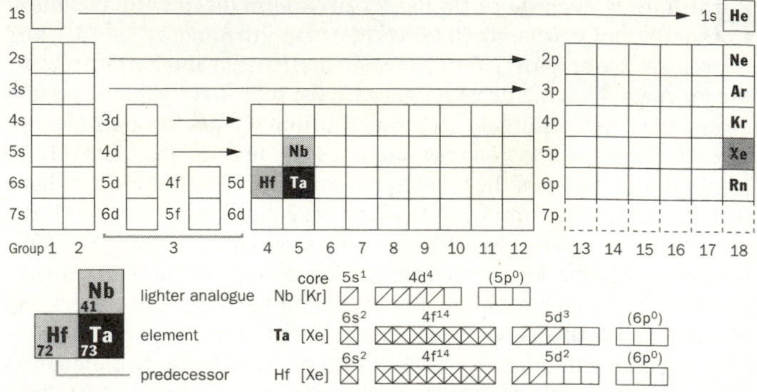

		Predecessor	Element	Lighter analogue
Name		**Hafnium**	**Tantalum**	**Niobium**
Symbol, *Z*		Hf, 72	Ta, 73	Nb, 41
RAM		178.49	180.9479	92.90638
Radius/pm	**atomic**	156.4	143	142.9
	covalent	144	134	134
	ionic, M^{3+}		72	
	M^{4+}	84	68	74
Electronegativity (Pauling)		1.3	1.5	1.64
Melting point/K		2503	3269	2741
Boiling point/K		5470	5698 ± 100	5015
$\Delta H^{\ominus}_{fus}$/kJ mol^{-1}		25.5	31.4	27.2
$\Delta H_{vap}^{\ominus}$/kJ mol^{-1}		661.1	753.1	696.6
Density (at 293 K)/kg m^{-3}		13 310	16 654	8570
Electrical conductivity/Ω^{-1}m^{-1}		3.175×10^6	8.032×10^6	8×10^6
Ionization energy for removal of *j*th electron/kJ mol^{-1}	***j* = 1**	642	761	664
Electron affinity/kJ mol^{-1}			14	86.2

Stable isotopes of tantalum

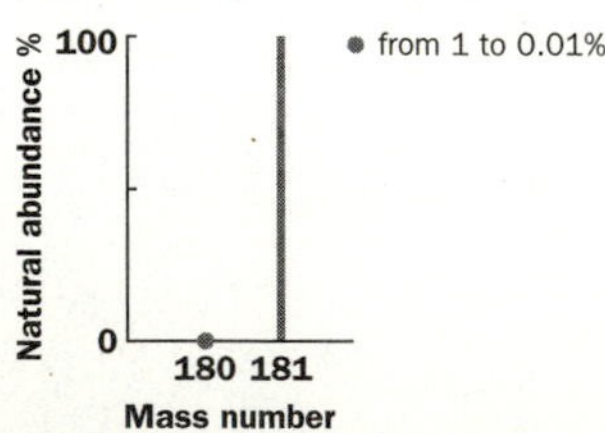

Thanks to the lanthanide contraction, we have another 5d element with radii almost exactly the same as those of its 4d counterpart, but of course more dense and with higher melting and boiling points. As expected from its extra d electron, the metallic bonding is stronger in tantalum than in hafnium. The behaviour of tantalum, ([Xe]$4f^{14}$)$5d^3 6s^2$, is predictably very similar to that of niobium [Kr]$4d^4 5s^1$ and the two elements always occur together (see p. 341). However, tantalum is not quite such a 'carbon copy' of niobium as hafnium is of zirconium. Although it makes no difference to chemical behaviour (see p. 341), we should note that, in all the 5d transition metals with less than eight 5d electrons, the 6s orbital is full, giving a configuration of ([Xe]$4f^{14}$)$5d^n 6s^2$, unlike the most common [Kr]$4d^{n+1} 5s^1$ configuration in the 4d series.

Tantalum and niobium can be separated via their fluorides, whether by fractional crystallization (of $K_3NbOF_5.2H_2O$ and K_2TaF_7), solvent extraction, or ion exchange of the complex fluoroanions. Tantalum is even more resistant to corrosion than is niobium and so is used in chemical plants and for surgical implants. As its tightly packed oxide film is also a good insulator, tantalum is used for making capacitors and filaments. The oxide Ta_2O_5 is even more resistant to chemical attack than is Nb_2O_5 and dissolves only in hydrofluoric acid, oleum, and fused alkalis. Its structure is described as 'very complex' and its various forms, although they too contain MO_6 octahedra, are not necessarily the same as those of Nb_2O_5. Many other tantalum species are, however, very like their niobium analogues (see p. 341), including halides based on TaX_6 octahedra; but there seems to be no tetrafluoride of tantalum. There are also $Ta_6O_{19}^{8-}$ ions, made up of six TaO_6 octahedra, and octohedral Ta_6 clusters forming the centre of $Ta_6X_{12}{}^{n+}$ ions. The sulfide, TaS_2, is, like NbS_2, graphitic and can fit other molecules between its layers.

Tantalum, like niobium, forms a hexacarbonyl anion, $Ta(CO)_6^-$, and can combine with four cyclopentadienyl groups. Organic derivatives of tantalum are more stable than their niobium analogues and they have been made in greater variety. Many have useful catalytic properties, and some have bizarre structural features, like multiple (even triple) Ta–Ta bonds, and dihydrogen as a bridging group. But such chemical oddities are, unfortunately, outside the scope of this book.

For summary, see pp. 492–3.

W Tungsten (or Wolfram

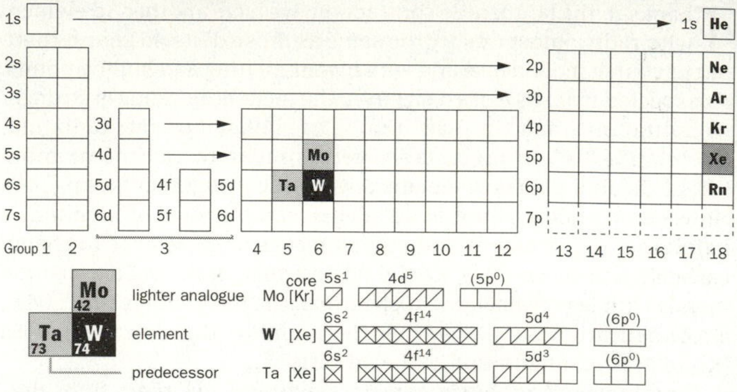

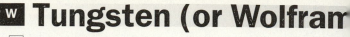

		Predecessor	Element	Lighter analogue
Name		**Tantalum**	**Tungsten**	**Molybdenum**
Symbol, *Z*		Ta, 73	W, 74	Mo, 42
RAM		180.9479	183.85	95.94
Radius/pm	**atomic**	143	137	136.2
	covalent	134	130	129
	ionic, M^{4+}	68	68	
Electronegativity (Pauling)		1.5	2.4	2.2
$\Delta H^{\ominus}_{fus}$/kJ mol^{-1}		31.4	35.2	27.6
$\Delta H^{\ominus}_{vap}$/kJ mol^{-1}		753.1	799.1	594.1
Density (at 293 K)/kg m^{-3}		16 654	19 300	10 220
Electrical conductivity/Ω^{-1}m^{-1}		8.032×10^6	1.770×10^7	
Ionization energy for removal of *j*th electron/kJ mol^{-1}	***j* = 1**	761	770	685
Electron affinity/kJ mol^{-1}		14	78.6	72.0

Stable isotopes of tungsten

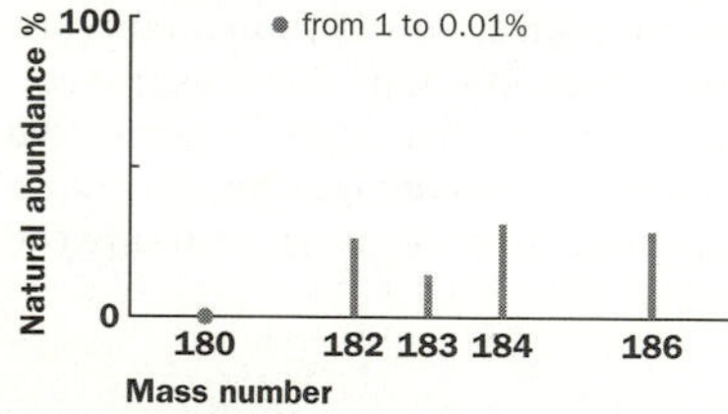

Tungsten, ([Xe]$4f^{14}$)$5d^4 6s^2$, like its predecessors, hafnium and tantalum, has atoms of very similar size to its 4d analogue, molybdenum, [Kr]$4d^5 5s^1$. It is of course denser, with higher transition points; indeed it is the highest melting of all metals. It seems that all its 5d electrons are available for bonding. Although it behaves like molybdenum in many ways, there are more differences between these two elements than between tantalum and niobium (themselves less like each other than hafnium and zirconium).

Tungsten is obtained from the ore $CaWO_4$ (which often contains molybdenum as well) via the trioxide, which can be reduced with hydrogen. The metal, which must be worked by powder metallurgy, is widely used for light bulb filaments. The carbide is extremely hard and is used for drill tips and abrasives.

Tungsten shows the same range of oxidation states as molybdenum, but with the higher ones even more favoured. There are seven different forms of the oxide WO_3, all based on corner-shared WO_6 octahedra. Reduction again gives blue oxides containing varying amounts of the (V) state; and in the dioxide there are bonded pairs of tungsten atoms.

Condensation reactions of the tungstates are slower than those of the molybdates, and result in different types of species, such as $W_{12}O_{42}^{10-}$ and $W_{12}O_{40}^{6-}$, which consist of blocks of three edge-sharing octahedra, assembled in various ways. Heteropolytungstates are formed with the same guest atoms as in the heteropolymolybdates (see p. 347).

The 'tungsten bronzes', $M^{I}_{x}WO_3$, are more stable than their molybdenum analogues, and better known. They too have metallic conductivity and lustre. Their (intervalence charge transfer) colours vary from yellow, $Na_{0.9}WO_3$, through to black, $Na_{0.3}WO_3$, demonstrating dramatically the increasing proportion of tungsten in lower oxidation states. Sulfides of tungsten are like those of molybdenum, but less important.

Many of the halides of the two elements are also similar; but the tungsten compounds WCl_6, WBr_6, and WBr_5 exist, although the analogous molybdenum ones do not, which indicates that tungsten(VI) is less likely to be reduced than molybdenum(VI). The greater preference of tungsten for higher oxidation states is also demonstrated by tungsten(III) which is more strongly reducing than molybdenum(III). The trihalides of tungsten MX_3 unlike those of molybdenum, contain octahedral M_6 clusters with a halogen atom above each of the twelve edges, and one more outside each of the six corners; but with molybdenum only the (II) halides contain clusters. The anion $W_2Cl_9^{3-}$, unlike $Mo_2Cl_9^{3-}$, is diamagnetic and probably contains a W–W bond.

One of the simpler isopolytungstate ions, $W_4O_{16}^{8-}$ GE

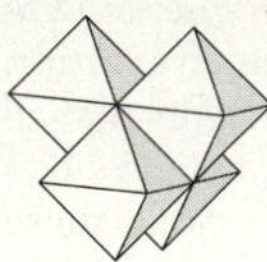

Tungsten, unlike molybdenum, seems to form no aquo cations, even hydrolysed polynuclear ones. Like chromium and molybdenum, it has six (d + s) electrons and forms an 18-electron dibenzene sandwich and a hexacarbonyl compound, which is more stable than the corresponding molybdenum carbonyl. Although it will be intriguingly difficult to rationalize the differences shown by two such similar elements in the way that they pack octahedra together, tungsten lacks the human appeal of those elements on which our own being depends: unlike molybdenum it appears to play no part in living processes.

For summary, see pp. 492–3.

Re Rhenium

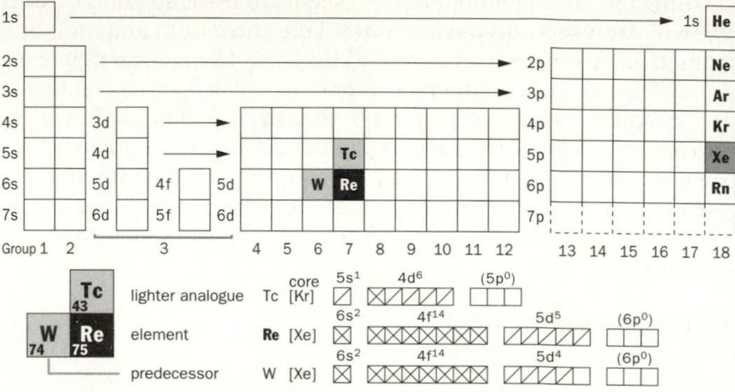

		Predecessor	Element	Lighter analogue
Name		**Tungsten**	**Rhenium**	**Technetium**
Symbol, *Z*		W, 74	Re, 75	Tc, 43
RAM		183.85	186.207	98.9062
Radius/pm	**atomic**	137	137	135.8
	covalent	130	128	
	ionic, M^{4+}	68	72	72
Electronegativity (Pauling)		2.4	1.9	1.9
Melting point/K			3453	2445
Boiling point/K			5900	5150
$\Delta H^{\ominus}_{fus}$/kJ mol^{-1}		35.2	33.1	23.81
$\Delta H^{\ominus}_{vap}$/kJ mol^{-1}		799.1	707.1	585.22
Density (at 293 K)/kg m^{-3}		19 300	21 020	
Electrical conductivity/Ω^{-1}m^{-1}		1.770×10^7	7.692×10^7	4.4×10^6
Ionization energy for removal of *j*th electron/kJ mol^{-1}	***j* = 1**	770	760	702
	***j* = 2**		1260	1472
	***j* = 3**		2510	2850
	***j* = 4**		3640	
Electron affinity/kJ mol^{-1}		78.6	14	96

Stable (|) and long-lived (¦) isotopes of rhenium

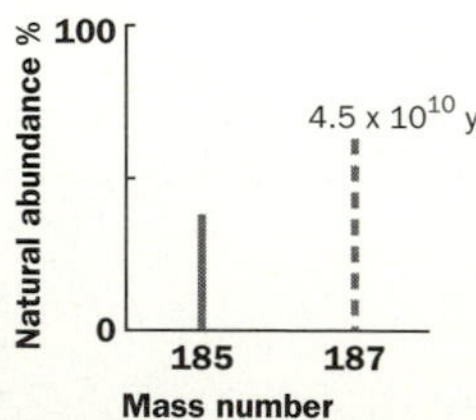

Rhenium, ([Xe]$4f^{14}$)$5d^5 6s^2$, was the last naturally occurring element to be discovered, in 1925. It is even less abundant than its 'artificial' analogue, technetium, $4d^5 5s^2$; but, despite its rarity and price, it has been more widely studied. Like technetium, rhenium forms a layer sulfide, ReS_2, traces of which occur in molybdenum sulfide ores. It is extracted by roasting the flue dust of molybdenum plants to volatilize rhenium as the oxide Re_2O_7; conversion to ammonium perrhenate, NH_4ReO_4, followed by treatment with hydrogen at high temperature, gives the metal.

Many of the properties of rhenium are predictable from those of tungsten, ([Xe]$4f^{14}$)$5d^4 6s^2$, and of technetium, [Kr]$4d^6 5s^1$. Rhenium has (slightly) lower values of melting and boiling points, and heat of atomization, than tungsten; the 5d electrons are starting to 'enter the core'. The metal is hard and resistant to corrosion, and must be worked by powder metallurgy; but, like technetium, it dissolves in bromine water, to form the perrhenate ion $Re^{VII}O_4^-$. Rhenium is used, together with platinum, as a catalyst in the production of low-lead petrol.

Rhenium, like technetium, has almost no cation chemistry and favours higher oxidation states. The oxide ReO_2 and the sulfides Re_2S_7 and ReS_2 are similar to their technetium counterparts, and the ReO_4^- ion is only weakly oxidizing. The perrhenate ion, like TcO_4^- but unlike MnO_4^-, is colourless because it is more difficult to transfer an electron to rhenium(VII) or technetium(VII) than to manganese(VII) and so the charge transfer absorption for the heavier metals involves ultraviolet radiation rather than visible light, which is of lower energy. The volatile oxide Re_2O_7 resembles Tc_2O_7 in the vapour phase, but has a more complex structure in the solid. Rhenium also forms a nine-coordinate hydride anion, ReH_9^{2-}.

There are, however, several ways in which the behaviour of rhenium does not mirror that of technetium, continuing the trend in which, for a pair of early 4d and 5d transition metals with the same number of $[(n-1)d + ns]$ electrons, the differences between them

The anion ReH_9^{2-} GE

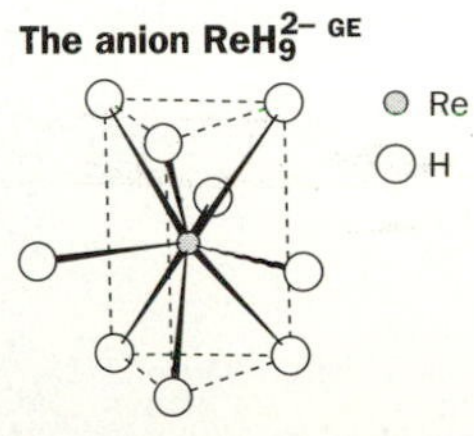

ReO_3 GE

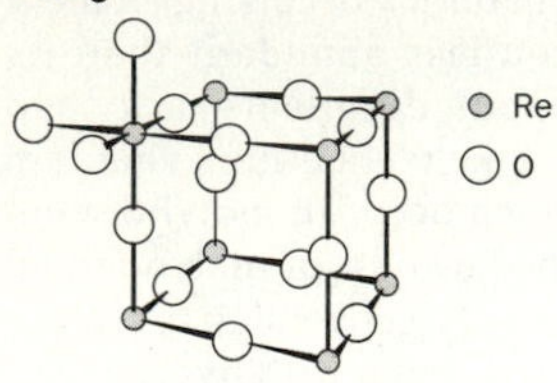

$(ReCl_3)_3$ and $Re_3Cl_{12}^{3-}$ MM

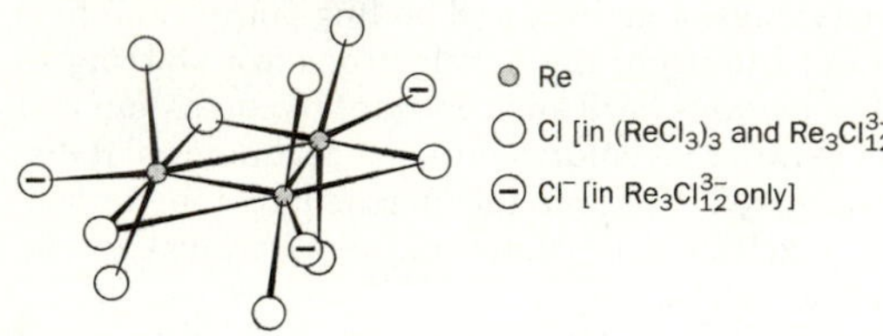

$Re(C_5H_5)_2H$ GE

(an 18-electron compound)

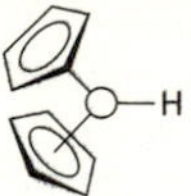

seem to increase across the series. The lowest oxidation state of uncomplexed rhenium is (III) but, like technetium(II), it disproportionates to the (IV) state and the metal. Rhenium resembles tungsten in forming a trioxide, ReO_3, of corner-linked ReO_6 octahedra, which can be reduced to form mixed oxide 'bronzes'.

Many halides of rhenium differ from those of technetium for both high and low oxidation states. Rhenium is the only metal to form a thermally stable compound with as many as seven fluoride atoms; the structure of ReF_7 is the same as that of IF_7. There is a dimeric chloride, Re_2Cl_{10}, with two edge-sharing $ReCl_6$ octahedra (but no Re–Re bond). The 'trichloride' $Re^{III}Cl_3$ is, in fact, a trimer, containing a triangular cluster of doubly bonded rhenium atoms, with three halogen bridges in the same plane. The other halogens lie almost above and almost below the rhenium atoms, which can each add an extra chloride, to give the anion $Re_3Cl_{12}^{3-}$. Rhenium forms similar species with bromine and iodine. The dirhenium chloro ion $Re_2Cl_8^{2-}$ contains a quadruple Re–Re bond and is much more stable than the technetium analogue, which is readily reduced to $Tc_2Cl_8^{3-}$.

On the other hand, the 18-electron carbonyl, $Re_2(CO)_{10}$, is much less stable than the technetium one, and ignites spontaneously in air. While rhenium, like technetium, seems not to form a cyclopentadienyl sandwich, it does add two (tilted) rings together with a hydrogen atom. Indeed, bonds from metal to hydrogen are a feature of rhenium chemistry, as illustrated by its unique ReH_9^{2-} ion and by some of its many complexes, which are outside the range of this book. It is good that its scarcity and expense has not discouraged work on rhenium, which, with its stable (VII) state, its clusters and quadruple bonds, its hydride complexes, and its high, odd coordination numbers, has so much of interest to show us. And who would have imagined that a metal almost as refractory and corrosion resistant as tungsten would have dissolved in *bromine water*?

For summary, see pp. 492–3.

Os Osmium

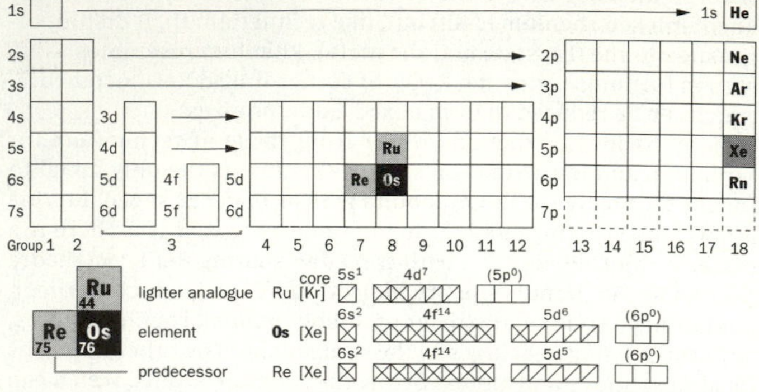

		Predecessor	Element	Lighter analogue
Name		**Rhenium**	**Osmium**	**Ruthenium**
Symbol, *Z*		Re, 75	Os, 76	Ru, 44
RAM		186.207	190.2	101.07
Radius/pm	**atomic**	137	135	134
	covalent	128	126	124
	ionic, M^{2+}		89	
	M^{3+}		81	177
	M^{4+}	72	67	65
Electronegativity (Pauling)		1.9	2.2	2.2
Melting point/K		3453	3327	2583
Boiling point/K		5900	5300	4173
$\Delta H^{\ominus}_{fus}$/kJ mol^{-1}		33.1	29.3	23.7
$\Delta H^{\ominus}_{vap}$/kJ mol^{-1}		707.1	627.6	567.8
Density (at 293 K)/kg m^{-3}		21 020	22 590	12 370
Electrical conductivity/Ω^{-1}m^{-1}		7.692×10^7	1.231×10^7	1.316×10^7
Ionization energy for removal of *j*th electron/kJ mol^{-1}	***j* = 1**	760	840	711
Electron affinity/kJ mol^{-1}		14	106	101

Stable (|) and long-lived (|) isotopes of osmium

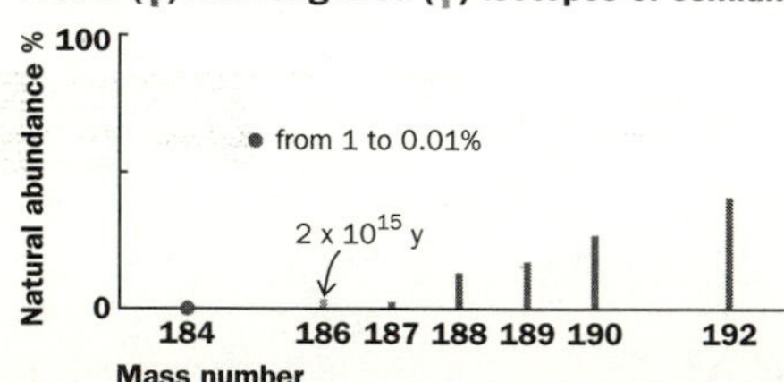

Osmium smells, or so its name would suggest to those who know ancient Greek. The smell, said to be like ozone, is that of the volatile tetroxide (tetrahedra of OsO_4) formed when finely divided osmium is exposed to air. It is a substance best avoided: very toxic, it oxidizes damp organic matter, such as the human eye if so allowed, and is itself reduced to black particles of OsO_2. As expected, osmium, ([Xe]$4f^{14}$)$5d^6 6s^2$, has a somewhat lower transition points and heat of atomization than rhenium, ([Xe]$4f^{14}$)$5d^5 6s^2$. Although all the outer electrons are free enough to be involved in forming the tetroxide, the higher oxidation states of osmium are more easily reduced than those of rhenium. Osmium(IV) is the normal product of the reduction of the tetroxide, which, with concentrated hydrochloric acid, gives chlorine and $H_2Os^{IV}O_6$. The highest halide is OsF_7 (not OsF_8 as was once thought), but, unlike ReF_7, it is unstable. Osmium, like rhenium, seems to have no simple cation chemistry.

We should expect some similarities between osmium and ruthenium ([Kr]$4d^7 5s^1$) since their outer configurations would be the same for any compound in an oxidation state above (I). Osmium does indeed occur with ruthenium and, like it, is obtained as a powder from the waste products of platinum manufacture. It, too, is used as a hardener of precious alloys and as a catalyst.

Osmium also behaves like ruthenium in many ways. Both elements are unusual in forming tetroxides, and neither has an oxide below $M^{IV}O_2$. Both have pyrites-type disulfides $M^{II}(S_2^{-II})$. The two elements form a range of halides including the tetrameric $(MX_5)_4$ rings and MX_6^{2-} anions (but see also below). The d^4 and d^5 complexes of both osmium and ruthenium are all low spin since the same factors are at work in the 5d series as in the 4d elements. The carbonyls, $Os(CO)_5$ and $Os_3(CO)_{12}$, resemble those of ruthenium, and there are also higher carbonyls containing from five to eight metal atoms; and osmium too has a rich chemistry with cyclopentadienyl.

However, the difference between a 5d metal and its 4d analogue increases across the series, and osmium is in no way an 'identical twin' to ruthenium. Osmium tetroxide is reduced only by concentrated hydrochloric acid, whereas even the dilute acid reduces ruthenium tetroxide. While both tetroxides dissolve in alkali, they give different species: $Os^{VIII}O_4(OH)_2^{2-}$, but $Ru^{VI}O_4^{2-}$. Osmium's highest monomeric halides are OsF_7, $OsCl_5$, $OsBr_4$, and OsI_3, while its lowest are OsF_4, $OsCl_3$, $OsBr_3$, and OsI. For ruthenium, on the other hand, the highest ones are RuX_3 (except for the fluoride, RuF_6) and the lowest ones are RuX_2 (except for the fluoride, RuF_3). For osmium, the most usual oxidation state is (IV), as opposed to (III) for ruthenium.

Two Os carbonyls [GE]

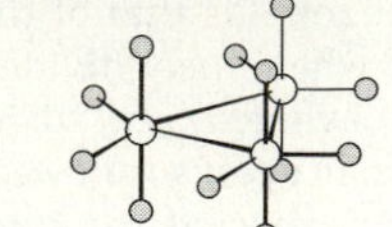

$Os_3(CO)_{12}$
has no bridging CO
[like $Ru_3(CO)_{12}$]

○ Os
● C of CO
(O omitted)

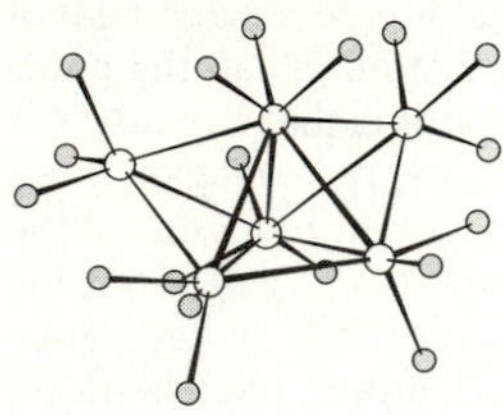

$Os_6(CO)_{18}$
also has no bridging CO

The central Os_6 cluster is formed from an Os_4 tetrahedron (base thickened in diagram) with additional Os atoms above two of its four faces

Osmium forms a number of complexes containing the linear double bonded oxocation $Os^{VI}O_2^{2+}$, which appear to be better established than their few ruthenium analogues. But, as yet, osmium lacks any chemistry of lower oxidation state complexes which could even begin to rival the interest of the ruthenium ones.

For summary, see pp. 492–3.

Ir Iridium

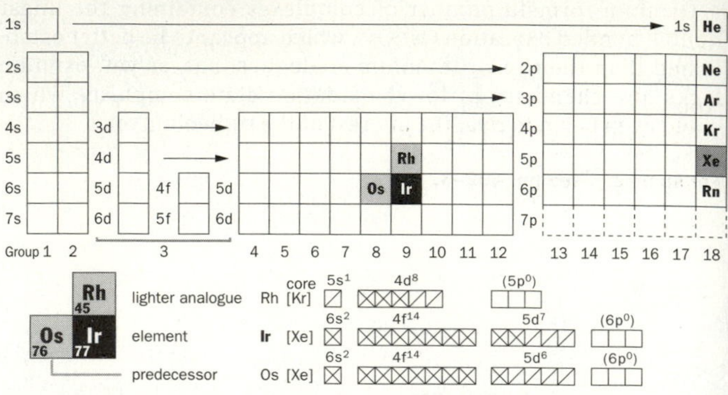

		Predecessor	Element	Lighter analogue
Name		**Osmium**	**Iridium**	**Rhodium**
Symbol, *Z*		Os, 76	Ir, 77	Rh, 45
RAM		190.2	192.22	102.90550
Radius/pm	**atomic**	135	135.7	134.5
	covalent	126	126	125
	ionic, M^{2+}	89	89	86
	M^{3+}	81	75	75
	M^{4+}	67	66	67
Electronegativity (Pauling)		2.2	2.2	2.3
Melting point/K		3327	2683	2239
Boiling point/K		5300	4403	4000
$\Delta H^{\ominus}_{fus}$/kJ mol^{-1}		29.3	26.4	21.55
$\Delta H^{\ominus}_{vap}$/kJ mol^{-1}		627.6	563.6	495.4
Density (at 293 K)/kg m^{-3}		22 590 (293 K)	22 560 (290 K)	12 410 (293 K)
Electrical conductivity/Ω^{-1}m^{-1}		1.231×10^7	1.887×10^7	2.217×10^7
Ionization energy for removal of *j*th electron/kJ mol^{-1} *j* = 1		840	880	720
Electron affinity/kJ mol^{-1}		106	151	109.7
$E^{\ominus}$/V for $M^{3+}(aq) \rightarrow M(s)$			1.156	0.76

Stable isotopes of iridium

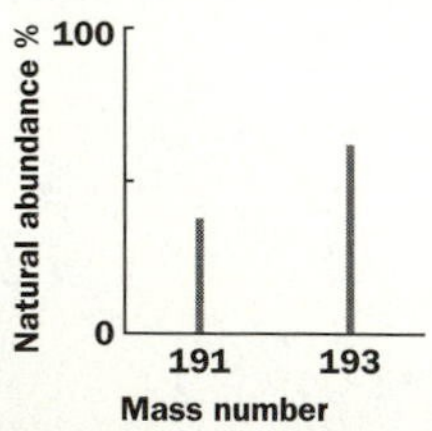

Iridium, ([Xe]$4f^{14}$)$5d^7 6s^2$, is in many ways very predictable. Following osmium, it is a rare, corrosion-resistant metal, obtained (like rhodium, with difficulty) from platinum residues. Although very expensive, it finds some use in hard alloys. Its name, from Iris, the goddess of rainbows, suggests that its compounds have a variety of colours. Its increasingly withdrawn 5d electrons make its metallic bonding weaker than in osmium, and its higher oxidation states less stable. It forms no tetroxide or heptafluoride. With an electronic structure of $(5d + 6s)^9$, it shows many similarities to rhodium $(4d + 5s)^9$, but it continues to illustrate the trend that the differences between a 5d metal and its 4d counterpart increase across the series.

Iridium, like rhodium, has high oxidation state fluorides, IrF_6, $(IrF_5)_4$, and IrF_4, as well as all the trihalides, and three hexahalo anions $Ir^{III}X_6^{3-}$ (with no fluoride). But iridium, unlike rhodium, also forms three $Ir^{IV}X_6^{2-}$ ions (with no iodide), and for the chloride and bromide the redox behaviour is much the same in the two oxidation states. Oxygen, too, brings out a higher oxidation state in iridium than in rhodium, its most important oxide being IrO_2, in contrast to Rh_2O_3. In water, iridium is probably present as a polymeric iridium(IV) hydroxo cation, while rhodium can exist as the simple aquo cation, $Rh(H_2O)_6^{3+}$. Iridium gives 'softer' metal ions than rhodium, and also forms a variety of hydride complexes, indeed, more than any other transition metal.

Like rhodium, iridium(I) can complex with triphenylphosphine (PPh_3) groups and one of these, too, is a useful catalyst. 'Vaska's compound' $Ir^{I}Cl.CO.(PPh_3)_2$ is square planar, with the two PPh_3 groups '*trans*' (that is 'opposite') to each other, and it can add two more groups, becoming iridium(III) if need be, to give various octahedral complexes. Since one of its many adducts is dioxygen, Vaska's compound has been studied as a possible oxygen carrier.

The most stable, simple compound of iridium with carbon monoxide is $Ir_4(CO)_{12}$, which contains the tetrahedral Ir_4 cluster but

Vaska's compound

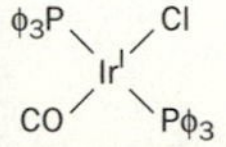

$Ir_4(CO)_{12}$ GE

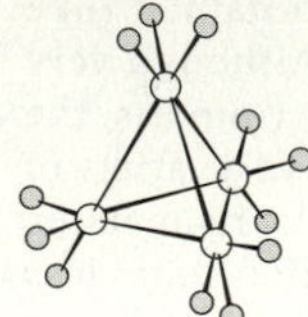

○ Ir
● C of CO
(O omitted)

Bridging carbonyls in $M_4(CO)_{12}$

M	Ru	Rh	Os	Ir
bridging CO groups	0	3	0	0
r_M/pm	134	134.5	135	135.7

no bridging carbonyl groups. In this respect, it resembles the carbonyls of osmium and ruthenium, but differs from the carbonyl-bridged $Rh_4(CO)_{12}$. It is sometimes suggested that the heavier elements are too large for a metal–metal bond to be spanned by a carbonyl bridge; but this explanation seems unlikely since all four of these elements are of extremely similar size.

Iridium ions, like those of many other transition metals, form complexes with bipyridyl. Although this ligand, being a π acceptor, stabilizes lower oxidation states (see p. 253), it can also complex with higher ones, such as iridium(IV). The octahedral complex $Ir(bipy)_3^{4+}$ has been studied by cyclic voltammetry, an electrochemical technique that enables us to see the way in which a species is reduced in solution. This complex was found to be reduced, in seven separate steps, to $Ir^{III}(bipy)_3^{3-}$, showing that all eight complexes, from $Ir(bipy)_3^{4+}$ to $Ir(bipy)_3^{3-}$ can exist in solution, although some are present only fleetingly and have not yet been captured in a crystal lattice.

For summary, see pp. 492–3.

Pt Platinum

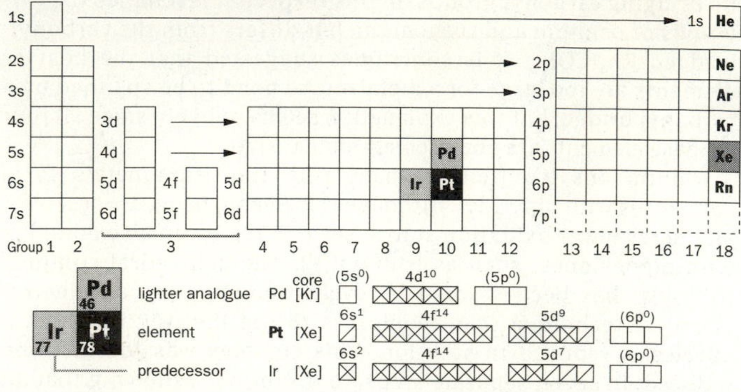

		Predecessor	Element	Lighter analogue
Name		**Iridium**	**Platinum**	**Palladium**
Symbol, *Z*		Ir, 77	Pt, 78	Pd, 46
RAM		192.22	195.08	106.42
Radius/pm	**atomic**	135.7	138	137.6
	covalent	126	129	128
	ionic, M^{2+}	89	85	86
	M^{4+}	66	70	64
Electronegativity (Pauling)		2.2	2.3	2.2
Melting point/K		2683	2045	1825
Boiling point/K		4403	4100 ± 100	3413
$\Delta H^{\ominus}_{fus}$/kJ mol^{-1}		26.4	19.7	17.2
$\Delta H^{\ominus}_{vap}$/kJ mol^{-1}		563.6	510.5	393.3
Density/kg m^{-3}		22 560 (290 K)	21 450 (293 K)	12 020 (293 K)
Electrical conductivity/$\Omega^{-1}m^{-1}$		1.887×10^7	9.434×10^6	9.259×10^6
Ionization energy for removal of *j*th electron/kJ mol^{-1}	***j* = 1**	880	870	805
	***j* = 2**		1791	1875
Electron affinity/kJ mol^{-1}		151	205.3	53.7

Stable (|) and long-lived (•) isotopes of platinum

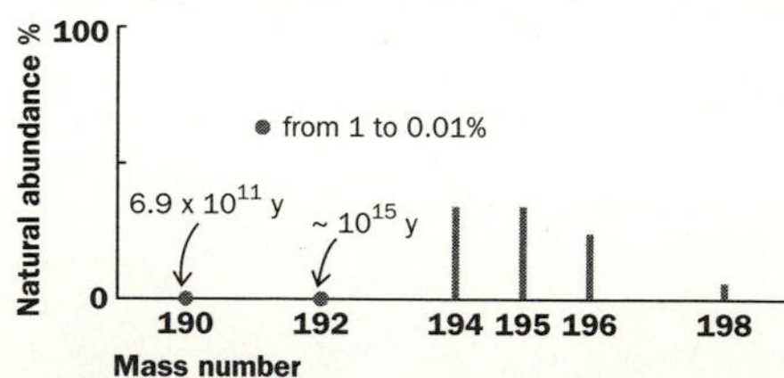

Platinum, like its three predecessors, occurs native but it is much less rare. It was used to make jewellery (by primitive powder technology) in Ecuador before the Spanish conquest, and probably also in ancient Egypt. As we have seen, it occurs with the other so-called 'platinum metals' (ruthenium, rhodium, palladium, osmium, and iridium), which must be removed during purification. Like other members of the set, it is used as a catalyst, for example for oxidation of ammonia to nitric acid, and for lessening the pollution from car exhaust. It, too, is very resistant to corrosion and has been used to make crucibles, resistance wires, and electrodes, including the catalytic 'platinum black' needed for the hydrogen [H_2(g), H^+(aq)] electrode. As the metal has the same coefficient of thermal expansion as glass, it is also used to make permanent glass to metal seals.

The metallic bonding in platinum is, predictably, weaker than that in iridium and stronger than that in palladium. The electronic configuration of the isolated atom is ([Xe]$4f^{14}$)$5d^9 6s^1$; comparison with the outer configurations of its predecessor iridium ($5d^7 6s^2$) and of its lighter analogues, palladium ($4d^{10} 5s^0$) and nickel ($3d^8 4s^2$), shows how finely the energy levels are balanced. Although platinum is attacked by fused alkali and aqua regia (see p. 489), it is even less reactive than palladium and is resistant to many reagents. However, it does form low-melting eutectics with many elements (such as boron, lead, phosphorus, arsenic, and bismuth), and compounds of these should not be heated under reducing conditions in a platinum crucible in case it collapses.

In many ways platinum behaves predictably for a metal nearly at the end of the third transition series. The differences between platinum, $5d^9 6s^1$, and palladium, $4d^{10} 5s^0$, are more marked than between previous elements and platinum has fewer oxidation states than earlier 5d elements. As usual, fluorine brings out the highest state, to give PtF_6 and PtF_6^-, and platinum, like iridium, forms a tetrameric $(Pt^{V}F_5)_4$, but this disproportionates. The instability of the (VI) state is shown by the fierce oxidizing power of PtF_6, which converts dioxygen to $O_2^+PtF_6^-$ and xenon to $XePtF_6$ (see p. 411).

Platinum forms a wide variety of compounds in the (II) state, in which it shows 'class b' behaviour (see p. 315). These are even more kinetically inert and thermodynamically stable than their palladium counterparts, and all are diamagnetic. [But unlike palladium, platinum forms no difluoride, doubtless because platinum(II) is more readily oxidized than palladium(II).] Since the complexes are so reluctant to swap one ligand for

Isomers of $[Pt(NH_3)_2Cl_2]_x$

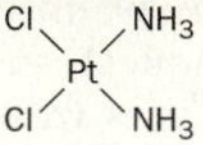

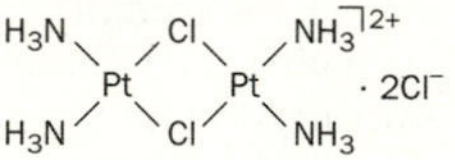

A stack of $Pt(CN)_4^{x-}$ ions SM

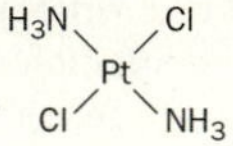

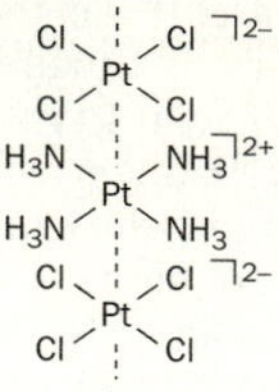

in $K_2Pt(CN)_4\,X_{0.3} \cdot 3H_2O$

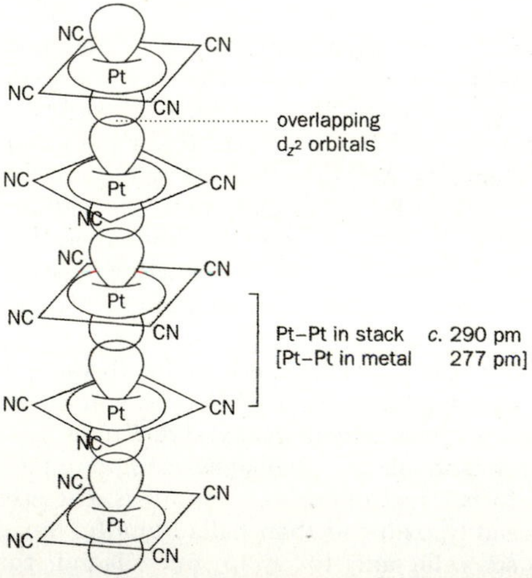

another,* different isomers are not in rapid equilibrium with each other, and the individual solids can be prepared. For example, there are four with the empirical formula $Pt(NH_3)_2Cl_2$. Two are monomeric isomers, one *cis* and one *trans*. The *cis*-isomer ('cisplatin') is used to treat some cancers and it is known that, on diplacement of its chlorine atoms, it combines with DNA so as to inhibit the division of malignant cells. There are also two ionic species, the dimer $[(NH_3)_2Pt(2Cl).Pt(NH_3)_2]^{2+}2Cl^-$ which contains two bridging chlorine atoms, and $[Pt(NH_3)_4^{2+}]$ $[PtCl_4^{2-}]$ in which the square ions are stacked, like a pile of tiles, one above the other, with interaction between the d_{z^2} orbitals. These platinum(II) complexes often have palladium(II) counterparts, which have been less thoroughly studied.

Despite the diversity of platinum(II) compounds, platinum, unlike palladium, favours the (IV) state. Its main oxide is PtO_2, and it forms all four tetrahalides. Although we might have expected the favoured oxidation state to decrease from iridium to platinum, the $5d^6$ configuration of platinum(IV) has the advantage of the very high ligand field stabilization of the octahedral low spin state, which is much more marked than in iridium(IV), $5d^5$. The platinum(IV) complexes are, like those of platinum(II), both thermodynamically stable and kinetically inert; but platinum(IV) shows greater preference for harder donors like oxygen. As with chromium(III) and cobalt(III), the slowness of ligand exchange has made it possible to make detailed studies of isomerism in platinum complexes. Some species contain platinum in both oxidation states, for example $[Pt^{II}(NH_3)_4]^{2+}[Pt^{IV}Cl_6]^{2-}$ and its isomer $[Pt^{IV}(NH_3)_4Cl_2]^{2+}[Pt^{II}Cl_4]^{2-}$.

An interesting compound of square-planar platinum in a non-integral oxidation state is $K_2Pt(CN)_4.X_{0.3}.3H_2O$ (where X is chloride or bromide). The $Pt(CN)_4^{x-}$ squares are again stacked in a pile (but now alternately staggered by 45°), and there is such strong interaction between the platinum atoms that they are almost as close together as in the metal itself. Moreover, the complex is an electronic conductor in the direction of the linked metal atoms and is known as a 'one-dimensional metal'. It even has a copper coloured metallic lustre.

---

* The actual rate of ligand exchange in square complexes depends neither on the group that is coming nor on the one that is leaving, but on the group in the position trans to the site where the swap occurs. This so-called *trans effect* allows us to choose a synthetic route likely to give the isomer we need, and is fully discussed in many more advanced textbooks.

The Pt ethanoate tetramer [CW]

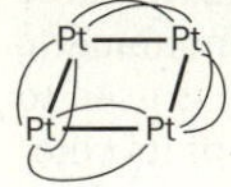

the Pt tetramer has M–M bonds

unlike the Pd ethanoate trimer (see p. 364)

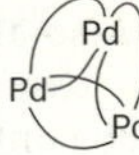

= CH_3COO^-

Some Pt carbonyl anions [GE]

The $Pt_6(CO)_{12}^{2-}$ ion

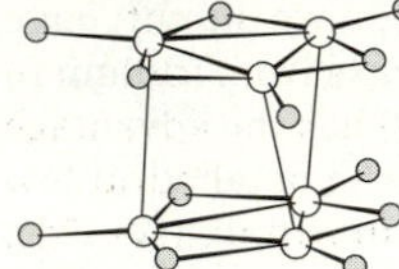

The Pt_{19}^{4-} cluster in $Pt_{19}(CO)_{22}^{4-}$

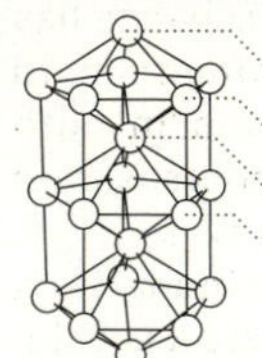

Site	No. of Pt of this type	No. of CO carried by each Pt terminal	bridging
Apex	2	1	0
Base	10	1	1
Centre	2	0	0
Waist	5	0	2

$Pt_{19}(CO)_{22}^{4-}$ attachment of CO to different Pt sites (schematic)

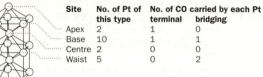

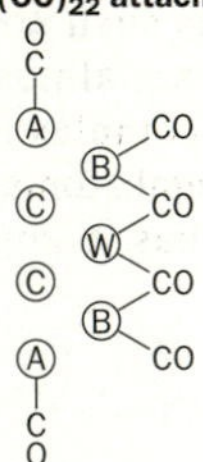

Platinum also differs from palladium in many of its reactions with organic ligands. With ethanoate, for example, it forms a tetramer with metal–metal bonds (rather than a trimer without metal–metal bonds). Platinum, in both (II) and (IV) states, forms more stable σ bonds with carbon than do other transition metals, and these are often stabilized by a π acceptor ligand such as a phosphine. Platinum(II) also forms π bonded complexes with such ligands as ethene (which, unlike palladium, it favours over allyl ligands). The compound $K[PtCl(C_2H_4)].H_2O$, known as Zeiss's salt, was first made in 1831; it is useful in syntheses since the attached ethene is readily oxidized and then set free.

Platinum, like palladium, forms no simple stable carbonyl; but it does form a number of larger anionic ones, which all contain bridging carbonyls and triangular Pt_3 clusters. Examples are $Pt_6(CO)_{12}^{2-}$, $Pt_9(CO)_{18}^{2-}$, and, to end with something really outlandish, $Pt_{19}(CO)_{22}^{4-}$, in which two of the metal atoms are entirely encapsulated by others.

For summary, see pp. 492–3.

Au Gold

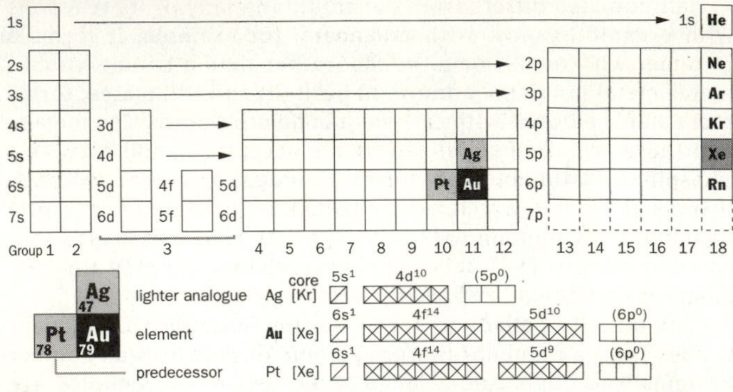

		Predecessor	Element	Lighter analogue
Name		**Platinum**	**Gold**	**Silver**
Symbol, *Z*		Pt, 78	Au, 79	Ag, 47
RAM		195.08	196.96654	107.8682
Radius/pm	**atomic**	138	144.2	144.4
	covalent	129	134	134
	ionic, M^+		137	113
	M^{3+}		91	
Electronegativity (Pauling)		2.3	2.5	1.9
Melting point/K		2045	1337.58	1235.08
Boiling point/K		4100 ± 100	3080	2485
$\Delta H^\ominus_{fus}$/kJ mol^{-1}		19.7	12.7	11.3
$\Delta H^\ominus_{vap}$/kJ mol^{-1}		510.5	324.4	255.1
Density/kg m^{-3}		21 450	19 320	10 500
Electrical conductivity/Ω^{-1}m^{-1}		9.434×10^6	4.255×10^7	6.289×10^7
Ionization energy for removal of *j*th electron/kJ mol^{-1}	***j* = 1**	870	890.1	731
	***j* = 2**	1791	1980	2073
Electron affinity/kJ mol^{-1}		205.3	222.8	125.7
$E^\ominus$/V for	**$M^{3+}(aq) \rightarrow M(aq)$**		1.36	
	$M^+(aq) \rightarrow M(s)$		1.83	0.80

Stable isotopes of gold

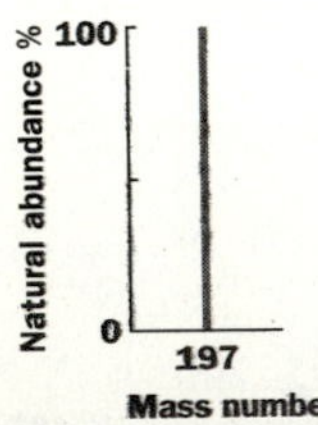

GOLD. What other chemical element conjures up such aspiration, be it for immense wealth, a royal crown, or athletic victory? What other element goaded generations of alchemists to search for the 'Philosopher's Stone' with which they could transform 'baser' metals into it? How can an electronic configuration of ([Xe]$4f^{14}$)$5d^{10}6s^1$ lead to the idea of virtually unattainable El Dorado, a mythical 'Place of Gold'? Platinum, ([Xe]$4f^{14}$)$5d^9 6s^1$, resembles gold in both abundance and untarnishability, but not in romantic appeal. Some of the prehistory of gold is a result of chance: it lay around in nuggets that could easily be seen and collected. Thanks to the marked withdrawal of the d electrons into the core, the metallic bonding is due mainly to the 6s electron and so gold is soft and could easily be cold-worked in prehistoric times. But the energy of the filled d orbital is only a little below that of the half-filled s orbital, and an electron can be promoted by fairly low energy radiation: in fact by violet light, giving gold its characteristic 'warm' colour. The high nuclear charge of gold keeps its outer electrons well under control, and so the metal is attacked only by oxidizing reagents that also contain strong complexing agents. A gold ring can be dug from an acid peat bog after several thousand years and still look as new. Many of gold's past and present uses are based on its resistance to corrosion, for example, for storage and transfer of wealth, jewellery, ceremonial objects, and dentistry. Fine films of it are also used for heat-reflecting screens, and colloidal gold (of colour dependent on particle size, but often red) was used to colour Venetian glass.

Since it is now rare to find gold nuggets, gold is extracted from powdered rock, either with mercury (forming gold amalgam) or with potassium cyanide, together with air or hydrogen peroxide. The gold is oxidized to gold(I), which is stabilized by forming a strong complex with cyanide [linear $Au^{I}(CN)_2^-$]. The metal is then obtained by adding zinc and, if necessary, is purified by electrolysis.

Gold has the same radius and outer electronic structure, of $(n\text{–}1)d^{10}\,ns^1$, as silver and predictably resembles it in having weak metallic bonding, high thermal and electrical conductivity, and resistance to corrosion by damp air. However, it continues the trend of increasing difference between the 5d transition metals and their 4d analogues; and gold differs much more from silver than, say, osmium does from ruthenium. Gold is the most electronegative of metals and can be persuaded to form a simple anion if combined with a very electropositive metal such as caesium. Solid Cs^+Au^- is an insulator, while the melt is an ionic conductor. However, gold has almost no cationic chemistry. The only acid that dissolves it without forming complex anions is H_2SeO_4 (see p. 311).

Extraction of gold

$$\underset{\text{crude}}{Au} \xrightarrow[CN^-]{air} Au(CN)_2^-$$

$$2Au(CN)_2^- + Zn \rightarrow Zn(CN)_4^{2-} + 2Au \downarrow$$

Chloroanions of Au in '$CsAuCl_3$' GE

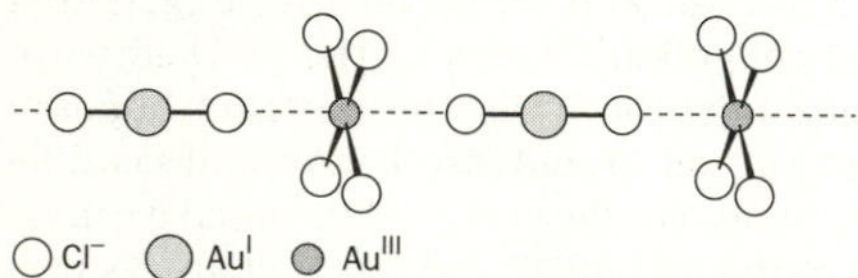

Trihalides of Au GE

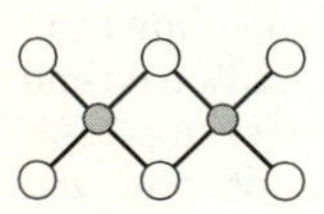

planar dimeric Au_2Cl_6 and Au_2Br_6

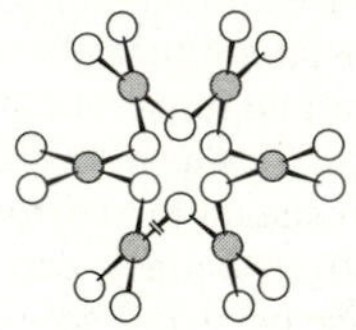

spiral $(AuF_3)_\infty$

* So called, because the alchemists thought that only a 'royal liquid' could dissolve a metal as 'noble' as gold or platinum; they did not know about H_2SeO_4. A mixture of concentrated nitric and hydrochloric acids, the nitric acid acts as a strong oxidizing agent, while the chloride ion complexes with, and so stabilizes, the gold(III) formed.

Gold also dissolves in aqua regia,* which gives $Au^{III}Cl_4^-$, a square-planar d^8 ion, very strongly stabilized by the ligand field. The hydroxo ion $Au(OH)_4^-$ may exist in alkaline solutions. Unlike silver, which usually adopts the (I) state and forms (II) and (III) compounds only rarely, gold favours the (III) state, although it also forms a range of (I) complexes such as the cyanide and some mixed valence (I) and (III) compounds analogous to $Ag^IAg^{III}O_2$ (see p. 369). So '$CsAuCl_3$' does not contain gold(II), nor indeed any gold cation, but should be written as $(Cs^+)_2[Au^ICl_2^-]$ $[Au^{III}Cl_4^-]$.

The favoured oxidation states are well illustrated, as is often the case, by the halides. Gold has three trihalides: dimeric, planar Au_2X_6 with chloride and bromide, and AuF_3, which forms a spiral chain. With iodine it gives only AuI, but there are no other monohalides. Fluorine can force gold up to the (V) state in AuF_5 and AuF_6^-; silver, on the other hand, forms AgX for all halides and also AgF_4^{2-} (see p. 369).

Gold(I), although less common than silver(I), is also a 'class b' metal, which favours sulfur over oxygen. Organic sulfur compounds of gold(I) have been used to treat rheumatoid arthritis, and it is thought that the gold may interact with the –SH groups of proteins.

Gold is unusual amongst transition metals in forming some moderately stable alkyl compounds, such as R_2AuBr, which do not contain a stabilizing π acceptor ligand such as a phosphine derivative. But it does combine with such ligands to give compounds containing gold clusters, which are often based on an Au_6 'chair', perhaps with other gold atoms around it. It also forms an Au_{13} cluster, in which a central gold atom is surrounded by an icosahedron of twelve others.

In the cautionary fable, King Midas turned all he touched into gold, so creating great wealth and untold misery. Gold, alone of all the chemical elements, attracts such myth making. But, although its resistance to corrosion is essential for its role in world finance, it is not always inactive, 'as good as gold'; and when stimulated to react it can give some gold medal surprises. Would we have predicted that one metallic element of legendary resistance to corrosion could form a monatomic *an*ion, a pentafluoride, and a 13 membered cluster?

For summary, see pp. 492–3.

The third (5d) transition series: trends and summaries Hf–Au

Between the 4d and 5d series, the atomic number increases by 32 (rather than by 18 as between the 3d and 4d series). As this large increase in nuclear charge is accompanied by the addition of fourteen 4f electrons, which screen very ineffectively, the outer orbitals are sucked towards the core; and as a result of this 'lanthanide contraction' the radius of any 5d element is extremely similar to that of its 4d analogue. The 5d metals are, of course, denser and have higher melting and boiling points, but we should expect the third transition series to resemble the second very closely, and certainly much more closely than 4d metals resemble those of the 3d series. Many 5d metals, like their 4d analogues, are used as catalysts. At the beginning of the second and third series, the similarities are very marked, the pairs of metal (such as zirconium and hafnium, or niobium and tantalum) occur together and since their chemical behaviour is almost identical they may be very difficult to separate. As with the 4d elements, hardness, resistance to corrosion, and maximum oxidation number increase from the beginning of the 5d series to the middle, but then decline. But the difference between a 5d metal and its 4d analogue increases markedly across the series, with the 5d element favouring higher oxidation states. For example, the main oxidation state of platinum is (IV) and that of gold is (III), as opposed to (II) for palladium and (I) for silver; and there are no 4d analogues of such compounds as WCl_6, WBr_5, ReH_9^{2-}, OsO_4, or PtF_6. In low formal oxidation states, the 5d elements show an even greater tendency to form metal–metal bonds and clusters than do their 4d counterparts, but a slightly lower tendency to form carbonyl derivatives with −CO− bridges. Since the brief summaries that follow will focus on the ways in which a 5d metal *differs* from its 4d analogue, rather than stress the many similarities between them, they should be read in conjunction with the section on trends and summaries of the 4d elements (see p. 451–5).

Comparative summaries of elements

Element	Hf	Ta	W	Re
Configuration	([Xe]$4f^{14}$)$5d^2 6s^2$	([Xe]$4f^{14}$)$5d^3 6s^2$	([Xe]$4f^{14}$)$5d^4 6s^2$	([Xe]$4f^{14}$)$5d^5 6s^2$
(Analogous 4d element)	(Zr, [Kr]$4d^2 5s^2$)	(Nb, [Kr]$4d^4 5s^1$)	([Mo, [Kr]$4d^5 5s^1$)	(Tc, [Kr]$4d^6 5s^1$)
Chemical similarity to 4d analogue	Very close indeed (separated by solvent extraction with NO_3^- or SCN^-)	Very close (separated via F^- complexes, by solubility, solvent extraction, or ion exchange)	Fairly close • found together same oxidation states • iso- and hetero-polyanions • 'bronzes' • layer disulfide, • dibenzene sandwich and hexacarbonyl	Fairly close • dissolves in Br_2 (aq) to give Re^{VII} • very little cation chemistry • M_2O_7(g) and MO_3(s) similar
Main differences from 4d analogue	• Hf strong neutron absorber	• Ta even more corrosion resistant • different M_2O_5 structures • no TaF_4 known • more organic derivatives known	• Higher states less oxidizing for W; lower ones more reducing • (VI) state condensations slower, with different products • no aquo cations • WCl_6, WBr_6, WBr_5 known • WX_3 contain W_6 clusters • $W_2Cl_9^{3-}$ has W–W bond • no known biological role	• Extracted from Mo flue dust • higher oxidation states favoured • Re^{VII} weakly oxidizing; lowest state is Re^{III} (disporportionate • Re_2O_7(s) more complex than Tc_2O_7(s) • 'bronzes' (like W) • high CN e.g. ReF_7 ReH_9^{2-} • polymeric halides Re_2Cl_{10} (no Re–R bond), Re_3Cl_9 (Re cluster), and (stable); $Re_2Cl_8^{2-}$ (quadruple Re–Re bond) • $Re_2(CO)_{10}$ more stable than Tc_2(C • many organic deri with Re–H
Uses	Neutron moderator	Chemical plants, surgical implants Capacitors Filaments (Ta_2O_5 surface insulator)	Light bulb filaments Carbide for drill tips and abrasives	Catalyst
Other points of interest	• $Hf(C_5H_5)_4$ has two σ and two π ligands (like Tr, but unlike Zr)		• Highest m.p. of all metals (all d electrons available)	• Less hard than W (d electrons entering core)

Hf to Au: $[Xe]4f^{14}5d^{y}6s^{2}$ or $[Xe]4f^{14}5d^{y+1}6s^{1}$

	Os	Ir	Pt	Au
	$([Xe]4f^{14})5d^{6}6s^{2}$ (Ru, $[Kr]4d^{7}5s^{1}$)	$([Xe]4f^{14})5d^{7}6s^{2}$ (Rh, $[Kr]4d^{8}5s^{1}$)	$([Xe]4f^{14})5d^{9}6s^{1}$ (Pd, $[Kr]4d^{10}5s^{0}$)	$([Xe]4f^{14})5d^{10}6s^{1}$ (Ag, $[Kr]4d^{10}4s^{1}$)
	Moderate • occurs with Ru • highest oxide OsO_4, lowest OsO_2, OsS_2 has pyrites structure • similar halides, carbonyls, and derivs • oxocation OsO_2^+	Moderate • (VI), (V), (IV), states with F^- • IrX_3 with all X^- • Ir^I complexes, e.g. with $P\Phi_3$	Some • Pt^{II} complexes stable and inert; many isomers • no stable simple carbonyl	Very little • soft metal, high conductivity • Au(I), CN = 2, linear • Class 'b'
	• Most usual state (IV) [cf. for (III) Ru] • OsO_4 sol. alkali without reduction (RuO_4 reduced to RuO_4^{2-}) • highest halides OsF_7 (unstable), $OsCl_5$, $OsBr_5$, and OsI_3; lowest OsF_4, $OsCl_3$, $OsBr_3$, and OsI	• IrX_4 (exc. I), stability similar to IrX_3 • main oxide IrO_2 (cf. Rh_2O_3) • in water: polynuclear hydroxo Ir^{IV} [cf. Rh^{3+}(aq)] • $Ir_4(CO)_{12}$ has no bridging CO	• Much commoner • higher oxidation states favoured PtF_6 (strongly oxidizing); $(PtF_5)_4$ (disproportionates); Pt^{IV} most usual [cf. Pd^{II}) • (IV)/(II) mixed valence species • no PtF_2 • ethanoate tetrameric, with Pt–Pt bonds • favours ethene over allyl groups	• Almost no cation chemistry • v. resistant to acids (exc. aqua regia → $AuCl_4^-$) • (III) state dominant [some (I) and mixed (III)/(I)] • (V) state with F^- • fairly stable alkyls (without π acceptors) • Au_6 and Au_{13} clusters with π ligands • Au^- in Cs^+Au^-
	Catalyst Alloy hardener	Alloy hardener	Crucibles, wires, electrodes, jewellery, seals to glass. Catalyst.	Jewellery Dentistry Heat-reflecting screens.
	• OsO_4 volatile; v. strongly oxidizing	• Coloured compounds • metal bonds weaker than in Os (electron withdrawal) • Vaska's compound, complexes with O_2	• $K_2Pt(CN)_4X_{0.3}\cdot 3H_2O$ 1D metal	• Extracted as $Au(CN)_2^-$ • most electronegative metal • AuF_3 spiral chain; Au_2X_6 (Cl + Br) planar; no AuI_3

Hg Mercury

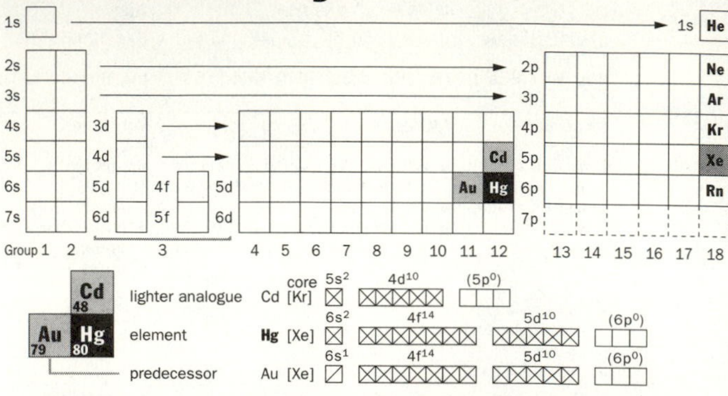

		Predecessor	Element	Lighter analogue
Name		**Gold**	**Mercury**	**Cadmium**
Symbol, *Z*		Au, 79	Hg, 80	Cd, 48
RAM		196.96654	200.59	112.411
Radius/pm	**atomic**	144.2	160	148.9
	covalent	134	144	141
	ionic, M^{2+}		112	103
Electronegativity (Pauling)		2.5	2.0	1.7
Melting point/K		1337.58	234.28	594.1
Boiling point/K		3080	629.73	1038
$\Delta H^{\ominus}_{fus}$/kJ mol^{-1}		12.7	2.331	6.11
$\Delta H^{\ominus}_{vap}$/kJ mol^{-1}		324.4	59.15	99.87
Density (at 293 K)/kg m^{-3}		19 320	13 546	8650
Electrical conductivity/Ω^{-1}m^{-1}		4.255×10^6	1.063×10^6	1.464×10^7
Ionization energy for removal of *j*th electron/kJ mol^{-1}	***j* = 1**	890.1	1007.0	867.6
	***j* = 2**	1980	1809.7	1631
	***j* = 3**		3300	3616
Electron affinity/kJ mol^{-1}		222.8	−18	−26
$E^{\ominus}$/V for	**$M^{2+}(aq) \rightarrow M(l)$**		0.8535	−0.402
	$M^{2+}_2(aq) \rightarrow M(l)$		0.7960	

Stable isotopes of mercury

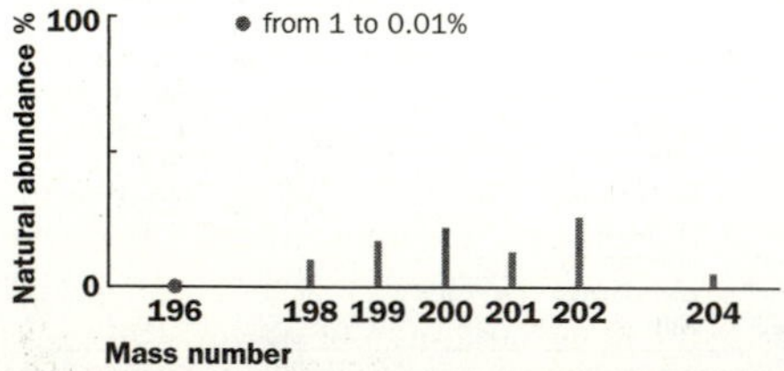

Mercury, even though 'messenger to the gods', can hardly rival the romantic appeal of gold. But an element commonly called quick (i.e. living) silver can hardly be dull; and even its symbol Hg recalls its poetic Latin name, *hydragyrum*, or liquid silver. How can a metallic element of greater mass than gold still be liquid at room temperature?

Following gold, mercury, ($[Xe]4f^{14}5d^{10})6s^2$, has full outer d and s shells. As its large nuclear charge is only poorly shielded by the 4f shell, the d electrons are (like those in cadmium, $[Kr]4d^{10}5s^2$) unavailable for use. Only the 6s electrons contribute to the metallic bonding, which is very weak. Not only does mercury have a very low melting point (−39°C) but the liquid is very volatile, giving a monomeric vapour. By metallic standards, mercury has a low electrical conductivity, but is, none the less, used as a liquid contact; and its density and high coefficient of thermal expansion make it suitable for barometers and thermometers.

We have seen that the similarity between a pair of 4d and 5d analogues decreases across the transition series, and that gold differs in many ways from silver, partly on account of having much more tightly bound d electrons. There are also marked differences between mercury and cadmium, but often in degree rather than in kind.

The main oxidation state of mercury is (II) and it shows strong 'class b' behaviour, with a preference for heavier donor atoms. Mercury occurs as the sulfide HgS, red cinnabar, from which it is extracted by roasting the ore with air, iron, or lime. Its main use is to form an amalgam with sodium for use in chemical industry, and as vapour in the greenish white mercury street lamps. It also forms amalgams with many other metals and was previously used for the extraction of native silver and gold. But the lighter transition metals do not form amalgams, and so mercury can be stored in iron flasks.

The affinity of mercury for sulfur enables it to react with −SH groups in living organisms. It is used as a pesticide, but is also very toxic to humans, attacking the nervous system. Industrial waste that contains mercury is a severe environmental hazard.

Mercury does not tarnish in air, nor combine appreciably with oxygen below 350°C. At slightly higher temperatures, it forms the oxide HgO, which owes its red colour to the fact that the charge transfer from the oxygen to the reducible mercury needs so little energy that it occurs in the visible region. (Compounds of cadmium, which is less easily reduced, are less strongly coloured.) Mercury does not, however, bind oxygen very strongly and the oxide decomposes at 400°C, a reaction that allowed Priestley and

Stability constants of some Hg^{II} halides[SC]

$\log K_n$ and $\log \beta_4$ for $HgX_n^{(2-n)+}$ in 0.5 M $NaClO_4$ at 25°C

log K_n	n =	1	2	3	4	log β_4
X^- =	Cl^-	6.74	6.48	0.95	1.05	15.22
	Br^-	8.94	7.94	2.27	1.75	20.90
	I^-	12.87	10.95	3.67	2.37	29.86

Data show:

the increasing preference of Hg^{2+} for heavier halide ligands

the very high stability of HgI_4^{2-}

the high preference of Hg^{2+} for twofold (linear) coordination

Disproportionation of 'Hg^I'

In acid solution, for disproportionation

$$Hg_2^{2+}(aq) = Hg^{2+}(aq) + Hg(l) \qquad K_{disp} \sim 0.01$$

If we add a ligand, L, which reacts at least 100-fold more favourably with Hg^{II} than with Hg^I, the equilibrium constant for the reaction

$$Hg_2^{2+}(L)\,(aq) \rightarrow Hg^{2+}(L)\,(aq) + Hg(l)$$

will be greater than 1, and the Hg^I will disproportionate e.g. when L = S or 4 × I.

The Hg_2^{2+} ion

In solids (linear XHg–HgX) Hg–Hg = 260 ± 10 pm

[In metallic Hg (12 coordinate) Hg–Hg = 300 pm]

Hg_2^{2+} ion persists in gas phase

* Lavoisier was later guillotined after the French Revolution on a trumped-up charge, but in reality for the sin of being born into wrong social class.

Lavoisier to prepare oxygen from the air in the late eighteenth century.*

Mercury, unlike cadmium, does not reduce dilute acids to hydrogen; but it will dissolve in concentrated oxidizing acids. In dilute aqueous solution it exists as Hg^{2+}(aq), which can be hydrolysed, eventually giving zigzag chains with linear O–Hg–O links. Mercury(II), like silver(I) and gold(I), has a d^{10} configuration and favours a two-coordinate linear geometry. Mercury difluoride has an ionic lattice and is totally hydrolysed by water; but the other dihalides exist in solution mainly as linear HgX_2 molecules. They do, however, form very stable tetrahedral ions with excess of halide: the tetraiodo anion, HgI_4^{2-}, is the most stable purely inorganic complex known. Mercury(II) can also be surrounded by six groups in an octahedron that is squashed so that two bonds are much shorter than the other four, and so the metal ion is almost in its favoured linear state.

We have seen that metal clusters are commonly formed by members of the third transition series, although metal–metal bonds are not restricted to these elements. But the dimeric mercury(I) 'cluster' is unusual in being so stable in aqueous solution. There is a delicate balance between Hg_2^{2+}(aq) and a mixture of Hg^{II} and the metal; in the absence of ligands other than water, disproportionation is unfavourable, but only just. If we add a reagent that stabilizes the (II) state relative to the (I) state, disproportionation occurs and mercury is deposited; the mercury(II) may be stabilized as a very insoluble solid such as the sulfide, HgS, or a very stable complex ion such as HgI_4^{2-}. But mercury clusters are not limited to the familiar Hg_2^{2+} ion. The analogous linear trimer and tetramer, Hg_3^{2+} and Hg_4^{2+}, can also be made by oxidizing mercury with arsenic pentafluoride in liquid sulfur dioxide. A lower proportion of mercury gives a gold solid with metallic lustre (named 'alchemists' gold') and formula $Hg_x^{2+}(AsF_6)_2$ where x may be non-integral and varies between 3 and 8; it contains long strands of Hg_x^{2+} ions perpendicular to each other, with the anions in between. There is a similar silvery substance that contains mercury in hexagonal sheets.

In many ways mercury behaves like a heavier but softer version of cadmium, with stronger class b behaviour and even more withdrawn d electrons; but it is in its 'cluster' chemistry that it shows both its individuality and its relationship with its predecessor, gold.

For summary, see p. 499.

Summary

Electronic configuration

- $([Xe]4f^{14}5d^{10})6s^2$

Element

- Metal, liquid at room temperature; fairly volatile; stable to air, water, and dilute acid

Occurrence

- As sulfide HgS

Extraction

- Roasting sulfide with air, iron, or lime

Chemical behaviour

- Does not tarnish in air; forms HgO between 350°C and 400°C
- Soluble in concentrated oxidizing acids
- Forms amalgams with many s and p and heavy d block elements
- Main oxidation state (II), Hg^{2+} polarizing, class b ion; linear and tetrahedral coordination; strong complex with I^-; highly toxic
- 'Hg(I)' is Hg_2^{2+}, just stable to disproportionation in water
- Linear 'clusters' Hg_x^{2+} ($x = 3$ or 4 and non-integral up to 8)

Uses

- As amalgam: in manufacture of Cl_2 and NaOH, and for extraction of gold
- As vapour in street lamps
- As liquid in switches, barometers, and thermometers

Nuclear features

- Seven stable isotopes

Tl Thallium

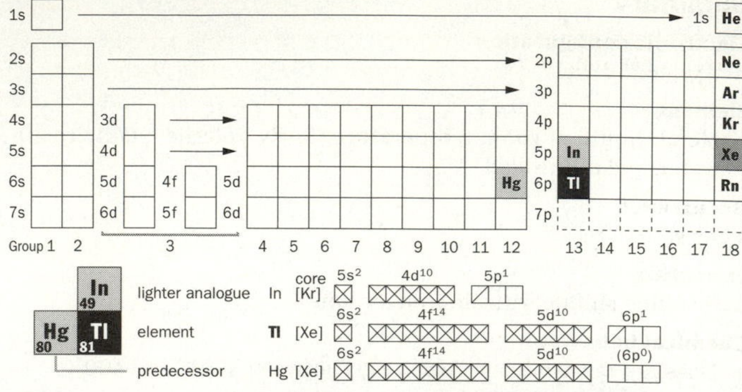

		Predecessor	Element	Lighter analogue
Name		**Mercury**	**Thallium**	**Indium**
Symbol, *Z*		Hg, 80	Tl, 81	In, 49
RAM		200.59	204.3833	114.82
Radius/pm	**atomic**	160	170.4 (α)	162.6
	covalent	144	155	150
	ionic, M^+		149	132
	M^{3+}		105	92
Electronegativity (Pauling)		2.0	1.6 [Tl^I]	1.8
			2.0 [Tl^{III}]	
Melting point/K		234.28	576.7	429.32
Boiling point/K		629.73	1730	2353
$\Delta H^{\ominus}_{fus}$/kJ mol^{-1}		2.331	4.31	3.27
$\Delta H^{\ominus}_{vap}$/kJ mol^{-1}		59.15	162.1	226.4
Density/kg m^{-3}		13 546 (293 K)	11 850 (293 K)	7310 (298 K)
Electrical conductivity/Ω^{-1}m^{-1}		1.063×10^6	5.555×10^6	1.195×10^7
Ionization energy for removal of *j*th electron/kJ mol^{-1}	***j* = 1**	1007.0	589.3	558.3
	***j* = 2**	1809.7	1971.0	1820.6
	***j* = 3**	3300	2878	2704
$E^{\ominus}$/V for	**$M^{3+}(aq) \rightarrow M^+(aq)$**		1.25	−0.444
	$M^+(aq) \rightarrow M(s)$		−0.3363	−0.338

Stable isotopes of thallium

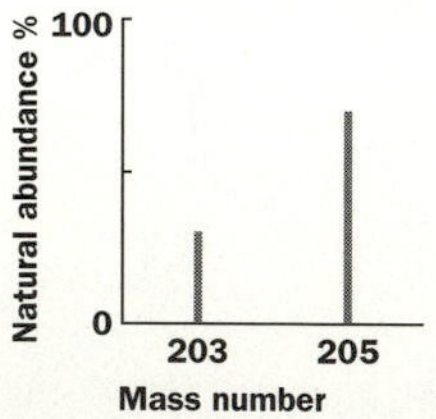

Thallium, ($[Xe]4f^{14}5d^{10})6s^26p^1$, has a strong chemical individuality. Like indium, ($[Kr]4d^{10})5s^25p^1$, it was discovered spectroscopically and named (from the Greek for green shoot) after its dominant spectral line. Both elements can form compounds in the (III) and (I) states (see p. 381); but, for thallium the (I) state is by far the more stable. The Tl^+ ion is about the same size as Rb^+, which it resembles in some ways; but with its much higher nuclear charge, it also shows some similarity to Ag^+. Like its predecessor, mercury, (and its successor, lead), thallium is very toxic and was formerly used to kill ants and rats; but since its compounds neither taste nor smell, it is very dangerous, and so is little used nowadays (except, like lead, for optical materials).

Thallium metal is much less reactive than rubidium, but much more so than silver or mercury. The metal is tarnished by damp air to TlOH. It dissolves in dilute acids to give the simple cation Tl^+(aq) which appears to remain unhydrolysed if the pH is raised. Indeed, TlOH is as strong a base as the hydroxides of potassium or rubidium. Thallium also forms a water-soluble carbonate and many other salts that resemble those of potassium and rubidium, although they are often less soluble in water. Thallium is found in nature replacing potassium in feldspars and micas; but, like indium, it is also found in sulfide ores, and is extracted from lead flue dust. The process involves precipitation of TlCl, followed by conversion to the soluble Tl_2SO_4, which is electrolysed to give the metal; the solubilities of these thallium(I) salts resemble those of AgCl and Rb_2SO_4.

As with indium, thallium can form a variety of compounds with sulfur, selenium, and tellurium. Most contain thallium(I), with or without some thallium(III), although Tl_2Se_3 appears to contain only thallium(III). The diversity is increased by the presence of polysulfide ions S_n^{2-}. Thallium(I) forms both an oxide, Tl_2O (which is hygroscopic and takes up water to become TlOH), and a peroxide, Tl^IO_2. There is also a thallium (III) oxide, Tl_2O_3, and a mixed oxide, $(Tl^I)_3\,Tl^{III}O_3$.

Like indium, thallium exhibits both oxidation states also in its halides, but again the (I) state is the more stable. The thallium(I) halides resemble the silver halides in their solubility in water, their colour, and their sensitivity to light; but unlike the silver halides, they do not dissolve in ammonia. (Thallium also resembles silver in forming a coloured, insoluble chromate.) The trihalides differ somewhat from those of indium and are less stable. The fluoride TlF_3 is hydrolysed by water, and the chloride and bromide are nonvolatile. Thallium(III) is too strongly oxidizing to exist together with the reducing iodide ion, and the triiodide actually contains thallium(I) as $Tl^+I_3^-$. Thallium(III) chloride, like $InCl_3$, can accept chloride ions to give $TlCl_4^-$, $TlCl_5^2$, and $TlCl_6^{3-}$, which can share one

A mixed valence Tl sulfide, Tl_4S_3 or $(Tl^+)_3[Tl^{III}S_3]^{3-}$ GE

(compare In_4Se_4 on p.000; for each formula unit, $Tl^{III}:Tl^{I}$ = 1:3, while $In^{III}:In^{I}$ = 3:1)

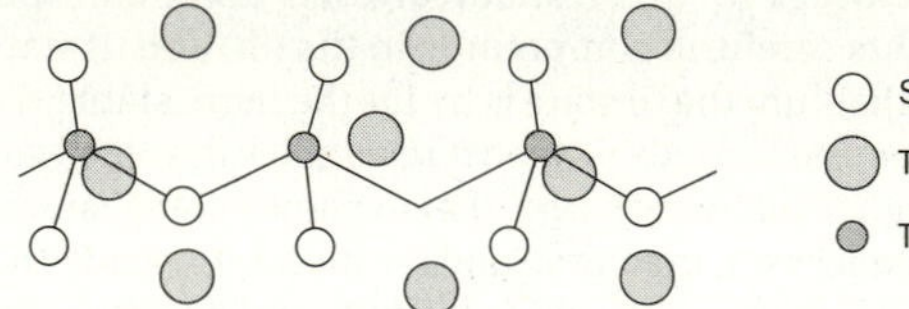

face with a second $TlCl_6^{3-}$ octohedron to give $Tl_2Cl_9^{3-}$. The dibromide and dichloride, like their indium analogues, are mixed valence compounds $Tl^I[Tl^{III}X_4]$.

Thallium forms a number of organometallic compounds. The linear ion $(H_3C-Tl-CH_3)^+$ is stable to air and water, and resembles $H_3C-Hg-CH_3$, which has the same number of electrons. Thallium also combines with a single cyclopentadienyl ring, interacting with all five carbon atoms.

Despite its formal similarity to indium, the striking feature of thallium is its preference for the (I) state; Tl^{3+} is a strong oxidizing agent which must be made electrolytically. The non-involvement of the $6s^2$ electrons (leading to a preferred oxidation state equal to the number of 6*p* electrons) is not restricted to thallium. Known as the 'inert pair effect', this is also a feature of the chemistry of lead ($6p^2$) and bismuth ($6p^3$). Some would claim that mercury ($6p^0$) also shows inert pair behaviour by its lack of reactivity and its monomeric vapour, in which it exhibits an oxidation state of zero. Lighter ns^2np^x metals may also form a few species in oxidation state *x*, but $(x + 2)$ is usually the favoured state (but see pp. 285–9).

We cannot attribute this inert pair effect to any single chemical cause. If we speculate as to why the decomposition of, say, a metal trichloride into the monochloride and chlorine should be more favourable for thallium than for indium, we have to consider two main differences between them: the energy that the ns^2 electrons must be given if they are to be able to react, and the energy given out when one bond is formed from chlorine to M^I, as opposed to three bonds being formed from chlorine to M^{III}. Neither of these factors can readily be either measured or estimated. The energy of electron promotion depends on nuclear charge, atomic radius, and shielding; and all three terms are intimately related. For atoms such as thallium, with high nuclear charge, the increased attraction of the nucleus on the inner s and p electrons stabilizes and accelerates them to such an extent that, by relativity theory, their mass increases. The s and p orbitals contract, which increases their screening power; this makes the d and f orbitals change in shape also, and the energy levels not only change, but also split. The strength of covalent bonds also depends on the size, shape, and energy of the orbitals, and so shows an equally complicated dependence on the electronic configuration. But the behaviour of the metals in the series $6s^26p^{0-3}$ demonstrates that the stabilization to be gained by forming three M–Cl bonds in MCl_3 (as opposed to one M–Cl bond in MCl, plus the Cl–Cl link in Cl_2) is not enough to justify disruption of the $6s^2$ pair. However, for the lighter elements, which form stronger bonds in both MCl_3 and MCl, there seems to be

a greater energy advantage in the higher oxidation state, and this can often be enough to overcome the promotion energies of the $5s^2$ or $4s^2$ electrons. Any further attempt to explain the inert pair effect is beyond the scope of this book.

Summary

Electronic configuration

- ([Xe]$4f^{14}5d^{10}$)$6s^26p^1$

Element

- Soft, dense metal, tarnishes in damp air; soluble in dilute acids; very toxic

Occurrence

- As sulfides

Extraction

- From lead flue dust, via TlCl (insol) and Tl_2SO_4(aq)., gives metal on electrolysis

Chemical behaviour

- Dominant state is (I); Tl^+ unhydrolysed, TlOH strong base; halides resemble AgX; salts with oxoanions are like those of K^+ and Rb^+, but less soluble
- Range of oxides: Tl^I_2O, Tl^IO_2, $Tl^{III}_2O_3$, and $(Tl^I)_3Tl^{III}O_3$; and of compounds with S, Se, and Te
- Tl^{III} strongly oxidizing; TlF_3 hydrolysed; TlX_3 in Cl^- and Br^- give anionic complexes: $Tl^ITl^{III}X_4$
- Organometallics: $Tl(CH_3)_2^+$ and π bonded $Tl(C_6H_5)$

Uses

- Formerly as pesticide
- In optical materials

Nuclear features

- $^{205}Tl:^{203}Tl \sim 7:3$

Pb Lead

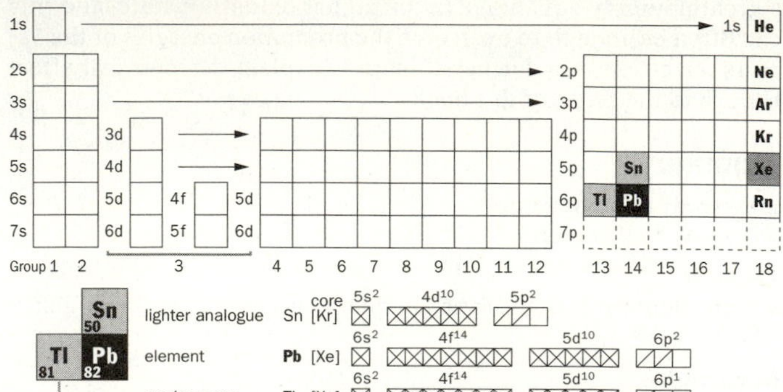

		Predecessor	Element	Lighter Analogue
Name		**Thallium**	**Lead**	**Tin**
Symbol, *Z*		Tl, 81	Pb,82	Sn, 50
RAM		204.3833	207.2	118.710
Radius/pm	**atomic**	170.4 (α)	175	140.5
	covalent	155	154	140
	ionic, M^{2+}		132	93
	M^{4+}		84	74
Electronegativity (Pauling)		1.6 [Tl^{I}]	2.3	2.0
		2.0 [Tl^{III}]		
Melting point/K		576.7	600.65	505.118
Boiling point/K		1730	2013	2543
$\Delta H^{\ominus}_{fus}$/kJ mol^{-1}		4.31	5.121	7.20
$\Delta H^{\ominus}_{vap}$/kJ mol^{-1}		162.1	179.4	290.4
Density (at 293 K)/kg m^{-3}		11 850	11 350	5750 (α)
				7310 (β)
Electrical conductivity/Ω^{-1}m^{-1}		5.555×10^6	4.843×10^6	9.091×10^6
Ionization energy for removal of *j*th electron/kJ mol^{-1}	***j* = 1**	589.3	715.5	708.6
	***j* = 2**	1971.0	1450.4	1411.8
	***j* = 3**	2878	3081.5	2943.0
	***j* = 4**		4083	3930.2
	***j* = 5**		6640	
Electron affinity/kJ mol^{-1}			35.1	116
Bond energy/kJ mol^{-1}	**E–E**		100	195
	E–H		180	< 314
	E–C		130	225
	E–O		398	557 Sn^{II}
	E–F		314	322
	E–Cl		244	315
$E^{\ominus}$/V for	**$M^{4+}(aq) \rightarrow M^{2+}(aq)$**		1.69	0.15
	$M^{2+}(aq) \rightarrow M(s)$		–0.1251	–0.137

It is no surprise that lead, ([Xe]$4f^{14}5d^{10}$)$6s^2 6p^2$, following mercury and thallium, is a soft, dense metal with toxic compounds in which it often exhibits the 'inert pair effect' by retaining its $6s^2$ electrons. But lead has a much longer history even than mercury. For nearly nine thousand years it has been used in glazes and as a flexible waterproof material. The Babylonians used it for water retention in their 'hanging gardens', the Romans for plumbing, and their descendants world-wide for roof-lining, window glazing, and protection of electric cables. Many of these uses are declining as cheaper, lighter, less toxic materials become available, and the main use of metallic lead today is in car batteries and 'non-lead-free' solders. Organolead 'anti-knock' agents in petrol are, happily, on the decline, for lead exerts its poisonous action cumulatively, attacking, in particular, thiol sites on enzymes.

Lead is fairly abundant and is generated as the stable end-product of two natural radioactive decay series. Since these produce different isotopes, the relative atomic mass of lead can be stated only very imprecisely. Lead occurs as the carbonate, the sulfate, and, most importantly, the sulfide, from which it is obtained. The ore is roasted in air to give the oxide, which can be reduced to lead either with coke or with more lead sulfide.

The resistance of lead to corrosion is mainly a result of the inert surface layer of oxide, chloride, or carbonate that forms if it is exposed to air; and, similarly, it does not react appreciably with those acids such as hydrochloric and sulfuric that form insoluble salts with it. However, since lead nitrate and ethanoate salts are soluble in water, it does dissolve in these acids, and predictably gives lead(II) compounds; lead(IV) ethanoate can also be made but it is strongly oxidizing. Indeed, lead(IV) is a slightly stronger oxidizing agent even than thallium(III).

The greater 'inertness' of the $6s^2$ pair, relative to the $5s^2$ pair is emphasized by comparing lead with tin, ([Kr]$4d^{10}$)$5s^2 5p^2$. Direct reaction with all the halogens gives lead dihalides PbX_2 (as opposed to tin tetrahalides SnX_4), and with sulfur and selenium we find only PbS and PbSe (whereas, with a higher proportion of reagent, tin gives higher compounds). The lead dihalides have less complicated structures than the tin ones, doubtless because the $6s^2$ pair is less involved in the bonding. They are photosensitive, like the halides of silver and thallium, with much ionic character. The iodide is bright yellow, through charge transfer from the iodide to the (fairly) easily reducible lead ion.

The only stable lead tetrahalide that can be made is, predictably, the fluoride, PbF_4. Although lead, like mercury and thallium, does not form a stable hydride, alkyl derivatives, R_2PbH_2 and R_3PbH, have been prepared.

Stable isotopes of lead

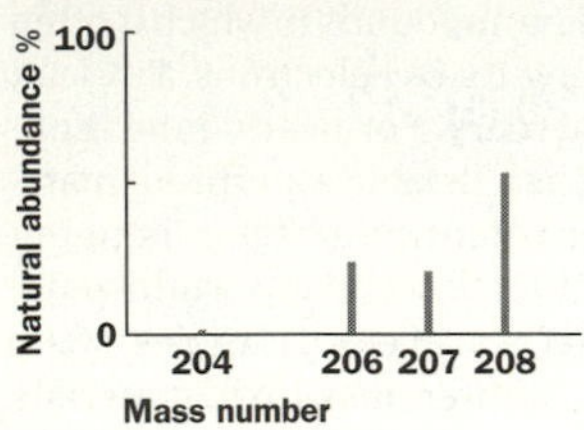

The Pb_6O cluster in $Pb_6O(OH)_6^{4+}$ GE

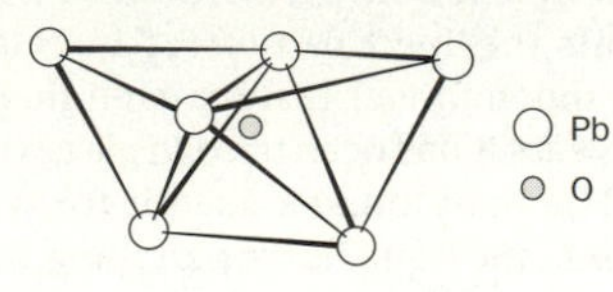

The $Pb(C_5H_5)_2$ chain GE

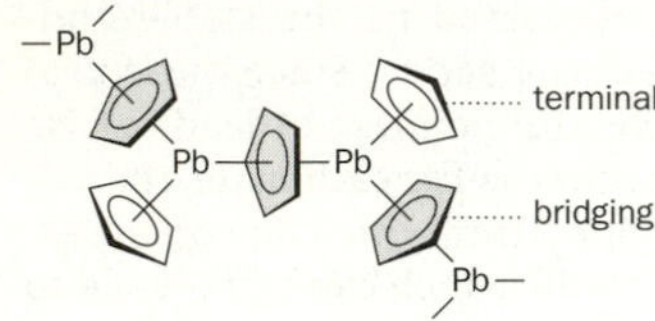

Summary

Electronic configuration

- ($[Xe]4f^{14}5d^{10}$)$6s^2 6p^2$

Element

- Soft, dense metal in air; surface tarnish prevents further corrosion, and in HCl and H_2SO_4 protected by insoluble salts; dissolves in nitric and ethanoic acids; very toxic

Occurrence

- As carbonate, sulfate, and sulfide

Extraction

- From sulfide, roasted to oxide, and reduced with coke or more PbS

Chemical behaviour

- Favours (II) state; with X_2 gives PbX_2, with simpler structures than SnX_2; Pb^{2+} solutions hydrolysed to polynuclear cations

Both lead(II) and lead(IV) form simple oxides, $Pb^{II}O$ and $Pb^{IV}O_2$ (which is formed in car batteries and is used as an oxidant). The other important lead oxide is the mixed valence 'red lead', $Pb^{II}_2Pb^{IV}O_4$, which is used as a metal primer, as a pigment, and in various capacities in the glass, rubber, and plastics industries; and lead also forms a number of less important oxides. Hydrolysis of lead(II) solutions gives polymeric cations such as $[Pb_6O(OH)_6]^{4+}$, the Pb_6 cluster being a Pb_4 tetrahedron that shares each of two of its faces with one other similar tetrahedron. The oxygen atom sits inside the central tetrahedron, and one hydroxy group is poised over each face of the two end tetrahedra. There is, apparently, no simple hydroxide.

Lead seems to resemble tin more in its exotic compounds than in its simple ones. It forms metal–metal bonds, both to itself (such as in the anion Pb_5^{2-} in liquid ammonia) and to metal atoms bound to carbonyl groups (see p. 389). Like tin, lead rather surprisingly adopts the (IV) state in its alkyl and aryl derivatives, even when lead(II) is the starting material. These compounds (such as tetraethyl 'anti-knock', notorious for its polluting presence in car exhaust) are less stable than the tin analogues, and fewer have been made. Also like tin, lead(II) can be incorporated into organometallic compounds if the ligands are large enough. For example, lead(II) forms a dicyclopentadienyl compound. Like that of tin? Not at all: the lead version is no simple monomer, but a chain with half of the rings acting as bridges between two lead atoms (with five-point attachments each way) and the other half attached, also by all five carbon atoms, but to only one lead. For all its venerable history, lead can still surprise us with some exciting new behaviour.

- Oxides $Pb^{II}O$, $Pb^{IV}O_2$, and $Pb^{II}{}_2Pb^{IV}O_4$
- (IV) state usually highly oxidizing: PbF_4 stable; no hydrides but alkyl derivatives, e.g. R_2PbH_2, known
- Cluster anions in liquid ammonia and bonds to metal atoms in carbonyls

Uses

- Metal in car batteries and solder
- Previously cable covering and for roofing and glazing (declining)
- Organolead 'anti-knock' agents

Nuclear features

- At end of two radioactive series; four stable isotopes

Bi Bismuth

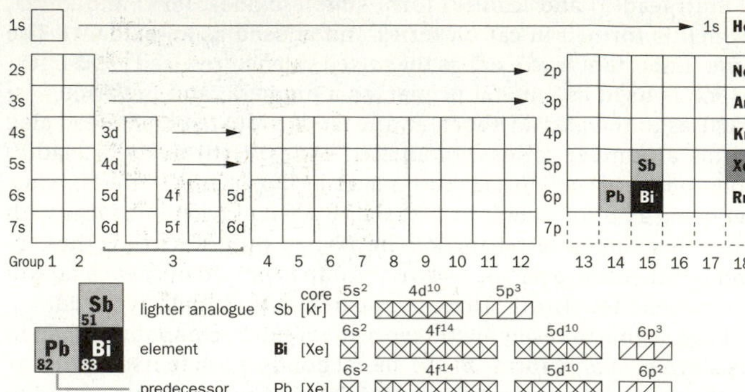

		Predecessor	Element	Lighter analogue
Name		**Lead**	**Bismuth**	**Antimony**
Symbol, *Z*		Pb, 82	Bi, 83	Sb, 51
RAM		207.2	208.9804	121.75
Radius/pm	**atomic**	175	155	182
	covalent	154	152	141
	ionic, M^{3+}		96	89
Electronegativity (Pauling)		2.3	2.0	2.05
Melting point/K		600.65	544.5	903.89
Boiling point/K		2013	1833 ± 5	1908
$\Delta H^{\ominus}_{fus}$/kJ mol^{-1}		5.121	10.48	20.9
$\Delta H^{\ominus}_{vap}$/kJ mol^{-1}		179.4	179.1	67.91
Density (at 293 K)/kg m^{-3}		11 350	9747	6691
Electrical conductivity/Ω^{-1}m^{-1}		4.843×10^6	9.363×10^5	2.564×10^6
Ionization energy for removal of *j*th electron/kJ mol^{-1}	***j* = 1**	715.5	703.2	833.7
	***j* = 2**	1450.4	1610	1794
	***j* = 3**	3081.5	2466	2443
	***j* = 4**	4083	4372	4260
	***j* = 5**	6640	5400	5400
Electron affinity/kJ mol^{-1}		35.1	91.3	101
Bond energy/kJ mol^{-1}	**E–E**	100	200	299
	E–H	180	194	257
	E–C	130	143	215
	E–O	398	339	314
	E–F	314	314	389
	E–Cl	244	285	313
$E^{\ominus}$/V for $M^{3+}(aq) \rightarrow M(s)$			0.317	

We should expect that bismuth, $([Xe]4f^{14}5d^{10})6s^26p^3$, which has one more outer electron than valence orbitals, would be slightly less metallic than lead, $([Xe]4f^{14}5d^{10})6s^26p^2$; but that, despite the much higher nuclear charge and the poor 4f shielding, it would be a little more metallic than antimony, $([Kr]4d^{10})5s5p^3$. We should also expect that the $6s^2$ electrons are often aloof from compound formation, so we should guess that bismuth tends to use only its three 5p electrons and favours the (III) state, which would be even more resistant to oxidation than its inert pair predecessors, lead(II) and thallium(I) (see pp. 507 and 501).

Natural bismuth is isotopically pure, ^{209}Bi, and is used as a target in the nuclear synthesis of polonium and astatine (see pp. 515 and 519). Bismuth, like lead, occurs combined with oxygen or as the sulfide but it is occasionally found native. The ores are roasted to the oxide Bi_2O_3, which is reduced by coke or iron to the metal. Yes, bismuth is indeed a metal, although not a very good one. Its usual (α) form, like that of the other ns^2p^3 elements antimony and arsenic, is a layer structure of puckered hexagons; but, in bismuth, the distance between the layers is little more than the interatomic distance within them. Bismuth is brittle, with a low metallic conductivity, and (unlike other elements except gallium and germanium) it expands on freezing. It is said to look pinkish. Unlike its three $n = 6$ predecessors, it does not seem to be toxic to humans, and is used in pharmaceuticals that kill bacteria that cause peptic ulcers.

Bismuth, like several other 5p and 4p metals, is used in low melting alloys. But not all of its vast range of alloys have low melting points; some, such as Na_3Bi (m.p. 840°C), obviously involve chemical interaction since they are much less fusible than their component metals.

Bismuth forms weak covalent linkages to other elements. Like lead, it forms no stable hydride; BiH_3 was one of the first compounds to be discovered (in 1918), using radioactive tracers, but it decomposes at –45°C. Unlike lead, bismuth forms only unstable organoderivatives. Only few compounds of bismuth(V) are known, predictably with fluorine and oxygen. The pentafluoride, BiF_5, is explosively reactive; and the salt $NaBi^VO_3$ is so strong an oxidizing agent that it converts the Mn^{2+} to purple MnO_4^- and so is used as a spot test for manganese.

The most accessible compounds of bismuth are those in the 'inert pair' (III) state. All four trihalides exist, with BiF_3 having a much higher melting point than the others. They form a range of complex anions such as $BiX^{(n-3)-}$ (where n = 4, 5, or 6), together with others that are more complicated, such as $Bi_2Cl_8^{2-}$ (two square

Stable isotopes of bismuth

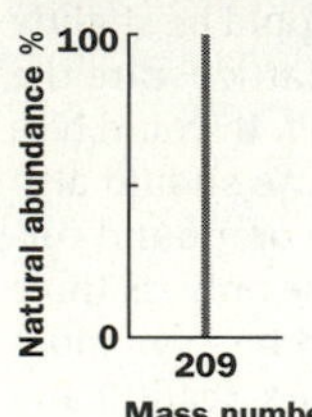

The cation $Bi_6(OH)_{12}^{6+}$ GE

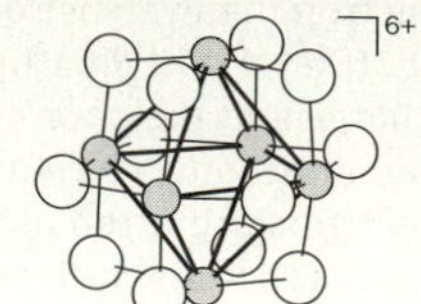

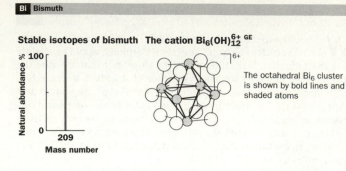

The octahedral Bi_6 cluster is shown by bold lines and shaded atoms

Some cationic Bi clusters GE

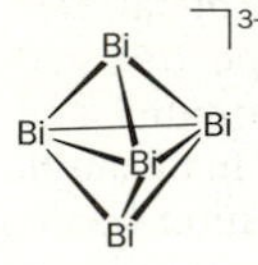

Bi_5^{3+}

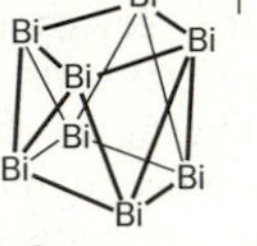

Bi_8^{2+}

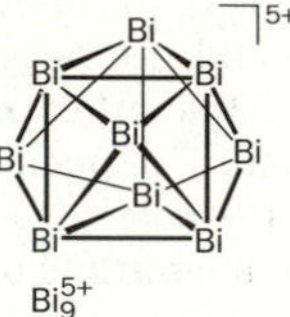

Bi_9^{5+}

Summary

Electronic configuration

- ([Xe]$4f^{14}5d^{10}$)$6s^26p^3$

Element

- Brittle, pinkish metal, but with puckered hexagon layers and low electronic conductivity; fairly low melting, expands on freezing; stable to air and water; dissolves in concentrated nitric acid; non-toxic to humans

Occurrence

- With oxygen or as sulfide; occasionally native

Extraction

- By reduction of Bi_2O_3 with coke or iron

pyramids *trans*-fused at a basal edge) and $Bi_2Cl_9^{2-}$ (two face-shared octahedra). In many ways, bismuth(III) compounds are like those of a fairly electropositive metal. The oxide Bi_2O_3 (a complicated polymer) is insoluble in alkali but dissolves in acid to give the ion Bi^{3+} (aq), which forms salts resembling those of lanthanum, since the two M^{3+} cations are of similar size. Mild hydrolysis gives $Bi_6(OH)_{12}^{6+}$ (a Bi_6 octahedron with hydroxy bridges above each edge), followed by the solid hydroxide, $Bi(OH)_3$, which does not dissolve in alkali. The sole sulfide, Bi_2S_3, is (unlike Sb_2S_3) insoluble in excess sulfide solution. The trihalides are not as readily hydrolysed as those of less metallic elements like antimony.

Bismuth does not form cluster anions on the scale of lead or tin, or of its earlier *n*p^5 analogues, (though Bi_4^{2-} and $Bi_2Sn_4^{2-}$ have been reported). It does, however, form a range of bizarre cluster cations with frameworks (and probably also multicentre two-electron bonds) reminiscent of the boranes (see p. 000). In molten $BiCl_3$, bismuth produces such species as Bi^+, Bi_3^+, Bi_5^{3+}, Bi_8^{2+}, and Bi_9^{5+}. These give exotic compounds such as $Bi_{24}Cl_{28}$ (which contains two Bi_9^{5+} clusters, four $BiCl_5^-$ ions, and one $Bi_2Cl_9^{2-}$ ion). The unexpected cation Bi^+ (perhaps with an 'inert' 6p^2 pair, as well as the 6s^2?) can be trapped by adding hafnium tetrachloride to give $Bi^+Bi_9^{5+}[HfCl_6^{2-}]_3$: an improbable looking compound, which would have amazed chemists of earlier generations.

Chemical features

- (III) state favoured; Bi^{3+}(aq) hydrolysed to $Bi_6(OH)_{12}^{6+}$ and then to $Bi(OH)_3$ (not amphoteric); all BiX_3 halides and various complex anions (not easily hydrolysed)
- (V) state very oxidizing; BiF_5 explosive, $NaBiO_3$ oxidizes manganese (II) to (VII)
- BiH_3 and organoderivatives very unstable
- Clusters, mainly cations (but some anions)

Uses

- In low melting alloys
- In pharmaceuticals
- For nuclear syntheses

Nuclear features

- Isotopically pure ^{209}Bi

Po Polonium

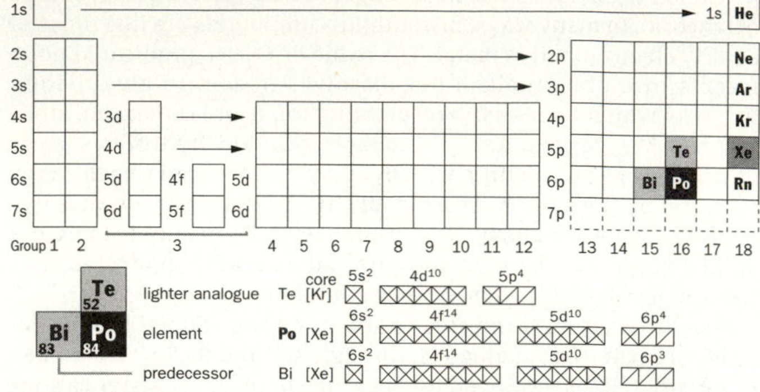

		Predecessor	Element	Lighter analogue
Name		**Bismuth**	**Polonium**	**Tellurium**
Symbol, *Z*		Bi, 83	Po, 84	Te, 52
RAM		208.9804	209	127.60
Radius/pm	**atomic**	155	167	143.2
	covalent	152	153	137
	ionic, M^{4+}		65	97
	X^{2-}		230	211
Electronegativity (Pauling)		2.0	2.0	2.1
Melting point/K		544.5	527	722.7
Boiling point/K		1883 ± 5	1235	1263.0
$\Delta H^{\ominus}_{fus}$/kJ mol^{-1}		10.48	10	13.5
$\Delta H^{\ominus}_{vap}$/kJ mol^{-1}		179.1	100.8	50.63
Density (293 K)/kg m^{-3}		9747	9320 (α)	6240
Electrical conductivity/Ω^{-1}m^{-1}		9.363×10^5	7.143×10^5	2.293×10^2
Ionization energy for removal of *j*th electron/kJ mol^{-1}	***j* = 1**	703.2	812	869.2
Electron affinity/kJ mol^{-1}		91.3	183	190.2
$E^{\ominus}$/V for M^{2+}(aq) → M(s)			0.37	

Key isotopes of polonium

Mass number	209	210	211	216	218
Half-life	105 y	138.4 d	0.52 s	0.15 s	3.11 min

* 520 kJ h^{-1} g^{-1}

Polonium has no stable isotope and, since its single naturally occurring one (^{210}Po, generated by radium decay) has a half-life of only 138.38 days, it is far from abundant. It occurs as one part in *ten million* in uranium ore, and it is from this that Polish-born Marie Curie painstakingly extracted and characterized it in 1898 by its radioactive properties. The same isotope can now be produced artificially by bombarding very pure bismuth ^{209}Bi with neutrons and distilling off the remaining bismuth. The polonium decays by emitting α-particles (helium nuclei) of such high energy that they heat up the sample, often destroying the compound or the solvent, and pose a severe radiation hazard. So, polonium can only be studied using stringent safety precautions and time-consuming techniques.

Despite these many hurdles, we know quite a lot about polonium, which, as we should expect from its electronic configuration of ([Xe]$4f^{14}5d^{10}$)$6s^26p^4$, resembles both bismuth ([Xe]$4f^{14}5d^{10}$)$6s^26p^3$ and tellurium ([Kr]$4d^{10}$)$5s^25p^4$. Polonium is, rather surprisingly, more metallic than bismuth; both its allotropes have fully 3D structures with low metallic conductivity. But, like tellurium, it combines with electropositive metals to form salts such as Na_2Po, in which it is in oxidation state (–II), forms an extremely unstable hydride H_2Po, and combines directly with oxygen, fluorine, and chloride. Polonium can also form compounds in the (II), (IV), and (VI) states, but the (VI) state is strongly oxidizing (and in this polonium resembles selenium more than tellurium).

Unlike tellurium, but like bismuth, polonium dissolves (readily) in acids. Unlike bismuth, however, it does not lose all its 6p electrons but forms pink solutions containing Po^{2+}(aq). [Is this another sign of an 'inert p pair', as in Bi^+ (see p. 513)?] However, the colour rapidly changes to yellow as Po^{2+} is oxidized to Po^{4+}, apparently by oxidizing species produced by the action of its α-particles on water.

The (IV) state appears to be the most favoured. Although polonium forms a dichloride and a dibromide, it forms all four tetrahalides. Its main oxide is PoO_2 which, although amphoteric, is more basic than tellurium dioxide; and it also forms a number of simple polonium(IV) salts, such as $Po(SO_4)_2$.

One might think that so transitory an element as polonium would be useless but this is far from true. Its radioactive energy can be profitably harnessed. As polonium decays it generates much heat* which can provide thermoelectric power for spacecraft; and without any moving parts. With a target of beryllium, ^9Be, the electrons emitted by polonium can also be used to generate neutrons.

For summary, see p. 517.

Summary

Electronic configuration

- ([Xe]$4f^{14}5d^{10}$)$6s^2 6p^4$

Element

- Radioactive; 3D metal, available only in very small quantities; high energy α-emissions impede study

Occurrence

- Traces in uranium ore

Nuclear synthesis

- By neutron bombardment of ^{209}Bi

Chemical behaviour

- Combines with O_2, F_2, and Cl_2
- Po^{2-} in Na_2Po very unstable
- Dissolves in acids to Po^{2+}, oxidized to Po^{4+} by radiolysis products
- (IV) state favoured, PoO_2 and $Po(SO_4)_2$
- (VI) state strongly oxidizing (like Se)

Uses

- α-emissions provide thermoelectric power for spacecraft, and neutrons with a ^9Be target

At Astatine

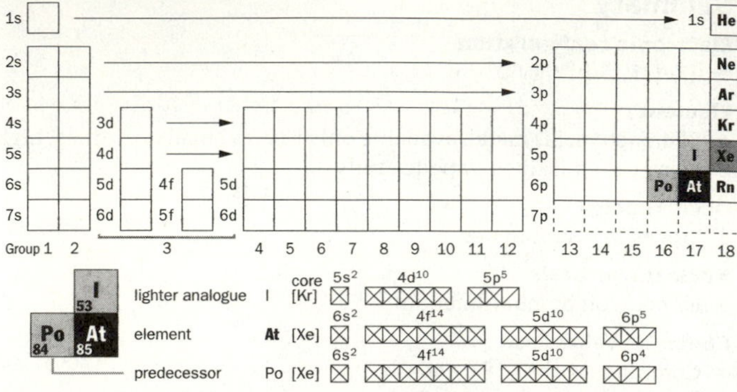

	Predecessor	Element	Lighter analogue
Name	**Polonium**	**Astatine**	**Iodine**
Symbol, Z	Po, 84	At, 85	I, 53
RAM	~ 209	~ 210	126.90774
Radius/pm ionic, X^-		227	220
Electronegativity (Pauling)	2.0	2.2	2.7
ΔH°_{fus}/kJ mol^{-1}	10	23.8	15.27
Ionization energy for removal of jth electron/kJ mol^{-1} $j = 1$	812	930	1008.4
$j = 2$		1600	1845.9
Electron affinity/kJ mol^{-1}	183	270	259.2
Bond energy/kJ mol^{-1} E–E		110	151
E°/V for $M_2(s) \rightarrow 2M^-(aq)$		0.2	0.535

Key isotopes of astatine

Mass number	210	211
Half-life	8.1 h	7.2 h

Summary

Electronic configuration

- ([Xe]$4f^{14}5d^{10}$)$6s^26p^5$

Element

- Radioactive and very short lived; volatile and soluble in CCl_4 (like I_2), but bulk properties not known (available in only 10^{-8} g samples)
- Studied by tracer techniques; high energy α-emission impede study

The heaviest halogen was well named 'unstable' as its two longest lived isotopes, ^{210}At and ^{211}At (both man-made), have half-lives of only 8.1 and 7.2 hours. Six other isotopes, with half-lives ranging from 54 s to about 10^{-4} s, are very minor components of radioactive decay series, but their natural abundance is negligible.

The creation of ^{211}At is similar to that of polonium ^{210}Po: the bombardment of bismuth ^{209}Bi, this time with α-particles, as the value of Z must now be increased by two (see p. 27). Since the largest sample yet made is about 5×10^{-8} g, the bulk properties of astatine are not known; but an amazing amount of information has been squeezed out of tracer studies, using solutions of 10^{-11}–10^{-15} mol dm^{-3}, and shows that astatine behaves as we might predict for a heavier version of iodine. The element is volatile, but less so than iodine and it, too, dissolves in tetrachloromethane. Astatine reacts with fairly strong reducing agents to give a soluble species (presumably At^-), which is coprecipitated with silver iodide or palladium iodide, and which can be changed back into the element using only a very weak oxidizing agent [following the decreasing stability, from F^- (aq) to I^- (aq) of the aqueous halide ion in relation to the element]. Astatine also reacts with moderate oxidizing agents to give a species (presumably AtO_3^-) that is coprecipitated with silver iodate. And there is an intermediate species (AtO^- or AtO_2^-) which, unlike its lighter analogues, does not disproportionate.

Astatine also forms interhalogen compounds, AtX, which can add a halide ion to give anions such as AtI_2^-, $AtBrI_2^-$, and $AtBrI^-$. Values of the equilibrium constants have, surprisingly, been measured: the complex AtI_2^- is much more stable even than the I_3^- ion.

Astatine, like iodine, is absorbed by the thyroid gland; but the study of astatine in a biochemical, or even an organic environment, has been little developed. This is owing to the astatine rather than to the researcher: its powerful α-emissions play havoc with chemical bonding, which is more vulnerable to energy changes in organic compounds than in many inorganic ones.

Nuclear synthesis

- By α-particle bombardment of ^{209}Bi

Chemical behaviour

- Seems typical heavy halogen
- Reduction gives At^-, coppt. with AgI, readily reoxidize to At(0)
- Moderate oxidation gives AtO^-, or AtO_2^- (both stable to disproportionation), and AtO_3^-, coppt. with $AgIO_3$
- Forms interhalogens AtX and anions AtX_2^- etc.

Rn Radon

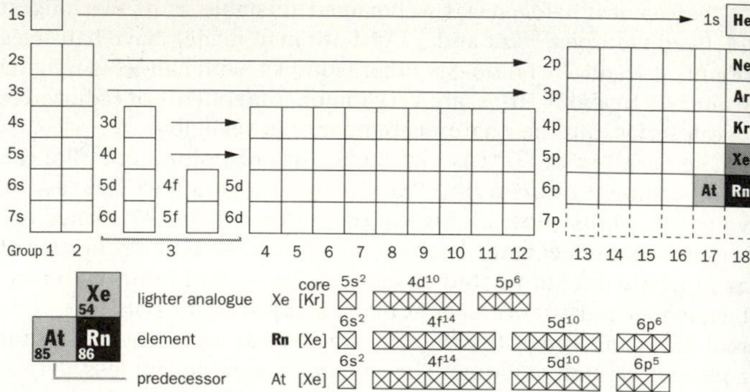

	Predecessor	Element	Lighter analogue
Name	**Astatine**	**Radon**	**Xenon**
Symbol, *Z*	At, 85	Rn, 86	Xe, 54
RAM	~ 210	~ 222	131.29
Melting point/K		202	161.3
Boiling point/K		211.4	166.1
$\Delta H^{\ominus}_{vap}$/kJ mol^{-1}		18.1	12.65
Ionization energy for removal of *j*th electron/kJ mol^{-1} *j* = 1	930	1037	1170.4

Key isotopes of radon

Mass number	219	220	222
Half-life	3.96 s	55.6 s	3.82 d

At radon, the $6s^2p^6$ shell is complete, giving us the heaviest noble gas, which presumably is even less an 'inert' gas than xenon. It certainly has a lower ionization energy. Unfortunately, we do not know much of its chemical behaviour as it has no stable isotope, and its least unstable one ^{222}Rn (obtained by decay of radium ^{226}Ra) has a half-life of only 3.82 days. Because of its radioactive decay there has been concern about even the trace level of radon that has seeped into buildings in regions of high natural radioactivity. Sealed tubes of it were formerly used in cancer treatment, but newer methods are more convenient.

Despite its short life-span, radon too has been forced into compound formation and tracer experiments have indicated the existence of RnF_2 and RnF^+. It also seems that the simple cation Rn^+ can replace Na^+ or K^+ on a fluorinated cation exchange resin.

Summary

Electronic configuration

- $([Xe]4f^{14}5d^{10})6s^26p^6$

Element

- Short lived, highly radioactive, heavy, noble gas; high energy α-emissions impede study

Occurrence

- In radioactive rocks (from decay of ^{226}Ra)

Extraction

- Fractionation of gases from pockets in rocks

Chemical behaviour

- Little studied but seems to form RnF_2 and RnF^+; and (unlike Xe) Rn^+

Part VII

Filling the 7s, 6d, 5f, and … orbitals

The actinide period: Francium to …?

Fr Francium

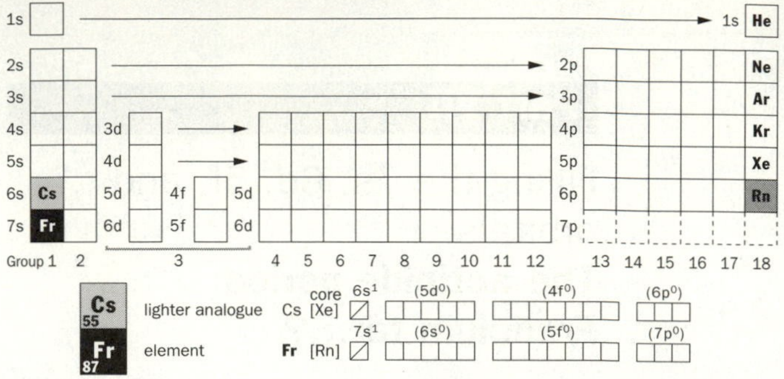

	Element	Lighter analogue
Name	**Francium**	**Caesium**
Symbol, *Z*	Fr, 87	Cs, 55
RAM	~ 222	132.9054
Melting point/K	300	301.55
Boiling point/K	950	951.6
Ionization energy for removal of *j*th electron/kJ mol^{-1} *j* = 1	400	375.5

Key isotopes of francium

Mass number	212	223
Half-life	20.0 min	21.8 min

Francium has been little studied, which is not surprising as its only naturally occurring (and its least unstable) isotope has a terrestrial abundance of only 2×10^{-18} parts per million, and a half-life of 21.8 minutes. It is formed as a very minor decay product of actinium, ^{227}Ac, itself a member of the uranium, ^{235}U, decay series, and was first identified (radiochemically) in 1939 by Marguerite Perey, who, in the tradition of Marie Curie, named it after her country. Tracer studies show that francium, $[Rn]7s^1$, predictably resembles caesium, $[Xe]6s^1$, and rubidium, $[Kr]5s^1$, in its ion exchange and solubility behaviour.

Summary

Electronic configuration

- $[Rn]7s^1$

Element

- Radioactive and very short lived, probably resembling caesium

Ra Radium

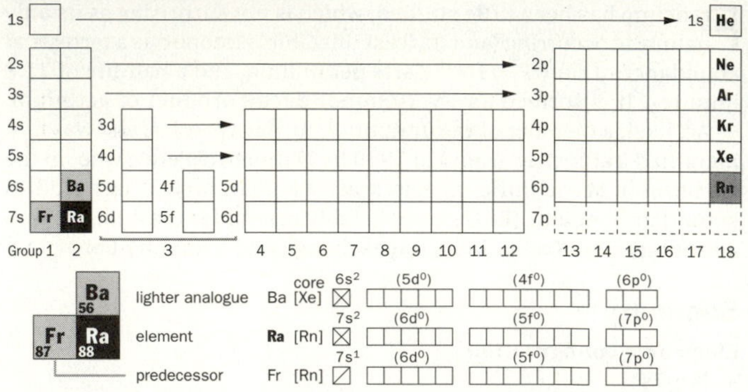

		Predecessor	Element	Lighter analogue
Name		**Francium**	**Radium**	**Barium**
Symbol, *Z*		Fr, 87	Ra, 88	Ba, 56
RAM		~ 222	226.0254	137.237
Radius/pm	**atomic**		223	217.3
	ionic, M^{2+}		152	143
Electronegativity (Pauling)			0.9	0.9
Melting point/K		300	973	1002
Boiling point/K		950	1413	1910
$\Delta H^{\ominus}_{fus}$**/kJ mol^{-1}**			7.15	7.66
$\Delta H^{\ominus}_{vap}$**/kJ mol^{-1}**			136.8	150.9
Density (at 293 K)/ kg m^{-3}			~ 5000	3594
Electrical conductivity/Ω^{-1}m^{-1}			1×10^6	2×10^6
Ionization energy for removal of *j*th electron/kJ mol^{-1}	***j* = 1**	400	509.3	502.8
	***j* = 2**		979.0	965.1
$E^{\ominus}$**/V for $M^{2+}(aq) \rightarrow M(s)$**			−2.916	−2.92

Key isotopes of radium

Mass number	223	224	226	228
Half-life	11.43 d	3.66 d	1600 y	5.75 y

Compared with its $7s^1$ predecessor, francium, radium, $[Rn]7s^2$, might seem relatively common; its terrestrial abundance is one part in one million million. Its epic extraction by Pierre and Marie Curie in 1898 grabbed the scientific imagination: from ten tonnes of the uranium ore, pitchblende, only one thousandth of a gram of radium could be gleaned. But at least it didn't disappear under their very eyes because, although all radium isotopes are radioactive, the longest lived one, ^{226}Ra, has a half-life of 1600 years.

We should, of course, expect radium to resemble barium, $[Xe]6s^2$, and there has been enough radium produced for us to know something of its bulk properties and its behaviour. Its first and second ionization energies are a little higher than those of barium, which might seem odd, until we remember that the nuclear charge has increased by 32, and that 4f electrons screen very ineffectively. But the Ra^{2+} ion is, none the less, larger than Ba^{2+}, and the metal itself is softer. Like barium, radium is a very electropositive metal, with exactly the same tendency to react with water. An oxide, RaO, a hydroxide, $Ra(OH)_2$, and various salts of Ra^{2+} have been made. Unlike barium, but like strontium and calcium, its flame colour is in the red range; but unlike the strontium flame, which is traditionally described as 'crimson', that of radium is said to be 'carmine'. Radium has been used even outside research labs: in luminous paint and for cancer therapy, but it has now been superceded in both these spheres.

Summary

Electronic configuration

- $[Rn]7s^2$

Element

- Radioactive metal; ^{226}Ra long lived; resembles barium

Occurrence

- In minute amounts with uranium

Uses

- Formerly in luminous paint and for cancer therapy

Ac Actinium

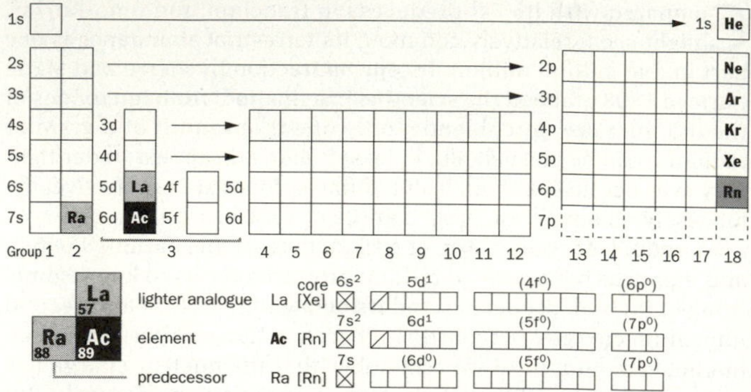

	Predecessor	Element	Lighter analogue
Name	**Radium**	**Actinium**	**Lanthanum**
Symbol, *Z*	Ra, 88	Ac, 89	La, 57
RAM	226.0254	~ 227	138.9055
Radius/pm **atomic**	223	187.8	187.7
ionic, M^{3+}	152	118	122
Electronegativity (Pauling)	0.9	1.1	1.1
Melting point/K	973	1320 ± 50	1194
Boiling point/K	1413	3470 ± 300	3730
ΔH°_{fus}**/kJ mol^{-1}**	7.15	14.2	10.04
ΔH°_{vap}**/kJ mol^{-1}**	136.8	293	399.6
Density/kg m^{-3}	~ 5000 (293 K)	10 060 (293 K)	6145 (298 K)
Ionization energy for removal of *j*th electron/kJ mol^{-1} $j = 1$	509.3	499	538.1
$j = 2$	979.0	1170	1067
$j = 3$		1900	1850
E°**/V for** $M^{3+}(aq) \rightarrow M(s)$		−2.13	−2.38

Key isotopes of actinium

Mass number	225	227	228
Half-life	10.0 d	21.77 y	6.13 h

Actinium, with a configuration of [Rn]$6d^1 7s^2$, is the first member of a fourth (6d) transition series. As with all known elements heavier than lead, it has no stable isotopes; but traces are formed during the radioactive decay of naturally occurring elements. Since only 0.2 mg of actinium is present in one tonne of uranium ore, the element is best obtained by nuclear synthesis: radium is bombarded with neutrons and the actinium separated from other radiation products by cation exchange.

We know quite a lot about actinium even though it is available in only milligram quantities and its radioactivity poses severe difficulties. Radiation damage to the researcher must be avoided by stringent safety precautions, such as working by remote control. But radiation may also affect the sample itself, either by heating it or, if the decay produces high energy γ-rays, even by breaking chemical bonds ('radiolysis'). Although the most useful isotope of actinium, ^{227}Ac, decays fairly sedately ($t_{\frac{1}{2}} = 2.77$ y), problems are caused by one of its decay products, ^{227}Th, which emits powerful γ-rays.

The known behaviour of actinium is much as we should expect. The element is a soft, reactive metal that glows in the dark, and often reacts like lanthanum, [Xe]$5d^1 6s^2$. Actinium needs slightly less energy than lanthanum for removal of the electron from the more extended d orbital, but slightly greater energy for ionization of electrons from the more penetrating s orbital. Its compounds seem restricted to oxidation state (III), with the dihydride being a metallic conductor, doubtless $Ac^{3+}(H^-)_2e^-$.

Actinium might also be expected to resemble lanthanum as the launch pad element for a series of successors with increasingly filled 5f orbitals. The fourteen 'actinide' (An) elements that follow it are indeed like the lanthanides in many ways, but by no means in all. Here we shall compare the two series in outline, so that we have a framework for brief discussion of individual elements.

The experimental problems of actinide chemistry are horrendous, and there are few research centres with the very expensive facilities needed for such work. Only thorium and uranium occur naturally, while a few of the others can be obtained from radioactive waste. The rest have to be obtained by nuclear synthesis; as the atomic number increases, the difficulty and costs rise sharply, and the yield plummets. Since many different elements are produced in nuclear reactions, the one needed must be obtained by very fine-tuned separation, usually by cation exchange. Often, the isotope that is easiest to synthesize is not the longest lived, which would probably be the one most convenient to work with; and obviously any isotope that decays by electron emission cannot be

Radioactive decay schemes, involving elements Np to Pb [P]

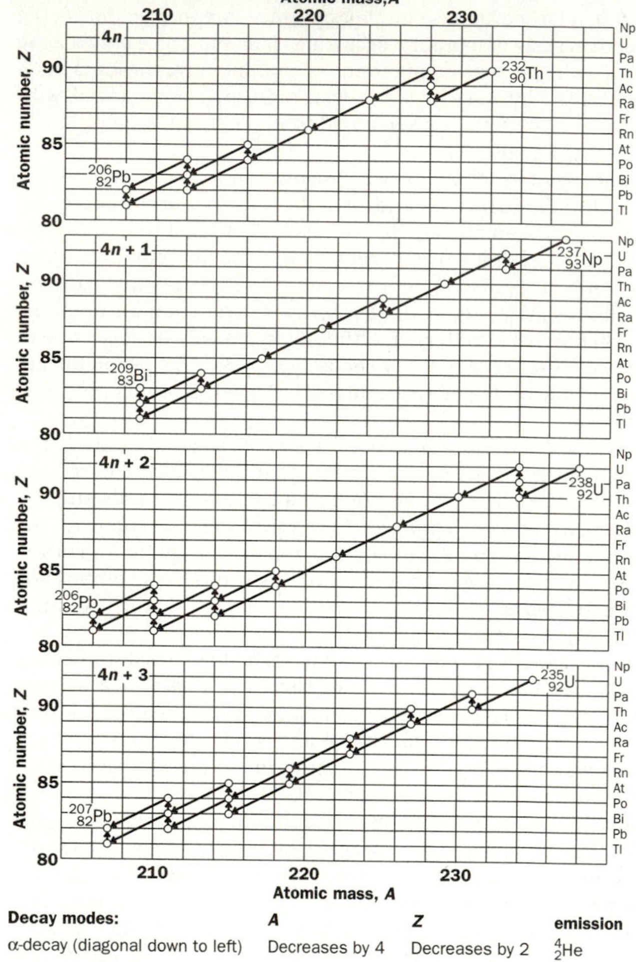

Decay modes:	*A*	*Z*	emission
α-decay (diagonal down to left)	Decreases by 4	Decreases by 2	$^{4}_{2}He$
β-decay (vertically up)	Unchanged	Increases by 1	$^{0}_{-1}e$
[γ-decay (same point)	Unchanged	Unchanged	Energy]

Series names

In any one series, mass numbers of all members have the same remainder, *C* when divided by 4, since this quantity can change only by α-decay. So, the four series are labelled ($4n + C$), where $C = 0$, 1, 2 or 3

used for studying high oxidation states! Once the element has been made, the experiments must be designed so that information can be obtained safely, and often very quickly, from extremely small quantities: well below the range of any non-tracer chemical technique (see 'availability' in summary pp. 566–9). Trace quantities are frequently adsorbed on to solid surfaces (either of the vessel or of a precipitate) and self-heating may complicate the results. The intense radioactivity of most actinides is, however, essential to their study because, by observing the products, speed, and energy of their decay, we can identify the isotope formed and track its role in chemical reactions; but the knowledge so gleaned is much narrower, and less confident, than that for more conventional elements.

Those actinides that have been made in moderate bulk are, like the early lanthanides and yttrium, reactive electropositive metals. Actinide compounds are often synthesized via hydrides, which may be ionic $An^{3+}(H^-)_3$, metallic $An^{3+}(H^-)_2e^-$, or non-stoichiometric. The atomic radius of the actinides decreases with increasing atomic number (see p. 426) and, since the ions are large, they can accommodate a large number of surrounding groups in a variety of arrangements. They form a range of complexes, particularly with fluoride ions and oxygen donors, and also with three or even four groups (such as cyclopentadienyl) that bind through all five carbon atoms; but no stable complexes with carbon monoxide or cyanide ions are formed under normal conditions. Thus far, the outer electrons of the actinides seem to behave much like those of the lanthanides, and there are also some specific similarities between analogous 4f and 5f elements.

The main (non-nuclear) differences between the actinides and the lanthanides seem to arise from the greater availability of their f orbitals. The electrons in the 5f shell, unlike the 4f electrons of the lanthanides, are not entirely buried in the atom core, but approach the periphery of the atom. The variation in metallic radii is less easy to rationalize. The magnetic and spectroscopic properties of the actinides are much more complex than those of the lanthanides, and extremely difficult to interpret. This is partly because the energies of the 5f orbitals (like the 3d, 4d, and 5d orbitals, but unlike the 4f orbitals) are markedly influenced by the presence of ligands. In the early actinides, the energies of the 5f, 6d, 7s, and 7p orbitals are all close together. It is not always easy to know the electronic configuration of an actinide atom, and any combination of these outer orbitals may be involved in bonding. This allows a much greater variety of oxidation states than is possible for the lanthanides: from uranium (6d + 5f = 4) to americium (6d + 5f = 7)

Radioactive decay schemes, involving elements Np to Pb (cont.) [P]

Higher actinides

Most isotopes (not all) decay by α- or β- emission and eventually become a member of one of the series, that can be identified from the remainder, *C*, of *A*/4.

For $^{242}_{94}Pu$, $C = 2$ and it decays to $^{238}_{92}U$ at the start of the $(4n + 2)$ series.

$^{242}_{94}Pu \xrightarrow{\alpha\text{-decay}} {}^{4}_{2}He + {}^{238}_{92}U$ $(4n + 2)$

For $^{241}_{94}Pu$, $C = 1$ and so it joins the $(4n + 1)$ series

$^{241}_{94}Pu \xrightarrow{\beta\text{-decay}} {}^{0}_{-1}e + {}^{241}_{95}Am \xrightarrow{\alpha\text{-decay}} {}^{4}_{2}He + {}^{237}_{93}Np$ $(4n + 1)$

Some isotopes go through many decay steps before joining a traditional series

$^{252}_{98}Cf \xrightarrow{\alpha} {}^{248}_{96}Cf \xrightarrow{\alpha} {}^{244}_{94}Pu \xrightarrow{\alpha} {}^{240}_{92}U \xrightarrow{\beta} {}^{240}_{93}Np \xrightarrow{\beta} {}^{240}_{94}Pu \xrightarrow{\alpha} {}^{236}_{92}Np \xrightarrow{\alpha} {}^{232}_{90}Th$ $(4n)$

Metal radii of actinides [GE]

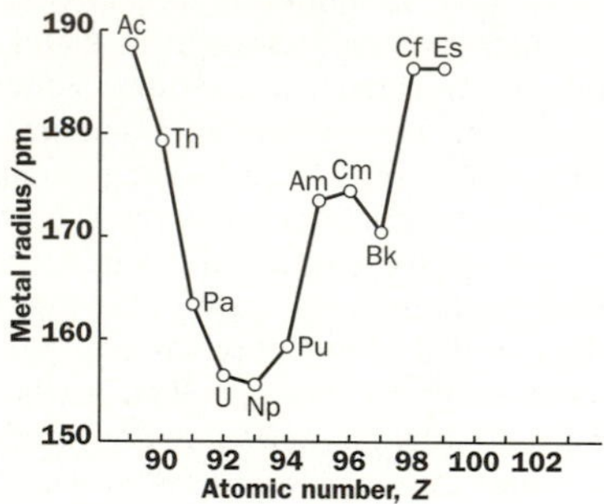

Oxidation states [HKK] of Ln and Ac

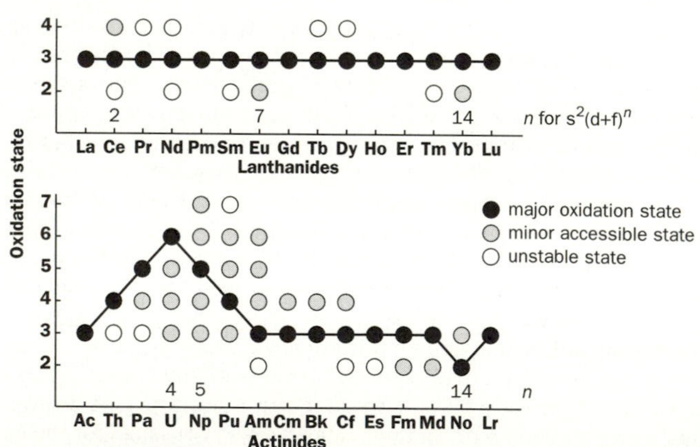

states from (III) to (VI) are known, with an occasional appearance of both (II) and (VII). Such variety is reminiscent of early 3d elements, such as vanadium, chromium, or manganese, rather than of the 4f series. In the later actinides, however, the 5f orbitals become increasingly contained within the core. The predominant oxidation state of (III), in which the $7s^2 6d^1$ electrons are ionized, is supplemented mainly by those states that involve the $5f^7$ or $5f^{14}$ configuration. So the properties of these short-lived, heavy actinides, in so far as they are known, seem to be very similar to those of the lanthanides.

In the brief outlines of the chemistry of individual actinide elements, we shall assume this general framework and highlight only points of special interest.

Cation exchange separation of some Ln^{3+} and Ac^{3+} ions

The eluant was α-hydroxy butyrate

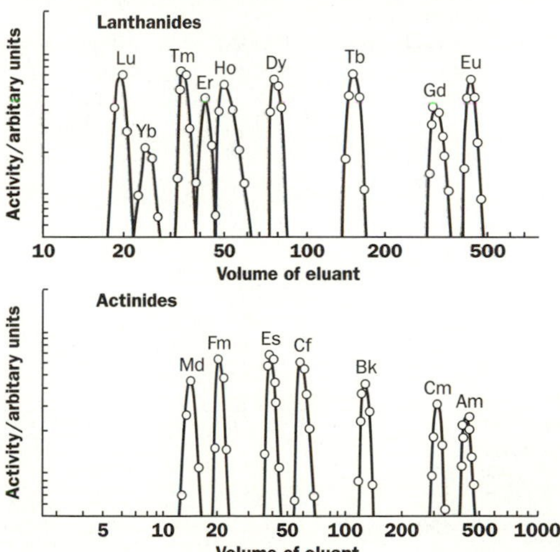

[From J. J. Katz, L. R. Morss, G. T. Seaborg, in *The chemistry of actinide elements* Vol. 2, pp. 1131–1133. J. J. Katz, L. R. Morss and G. T. Seaborg (eds.). Chapman & Hall, NewYork, 1986, Reproduced with permission.]

For summary, see pp. 565–9.

Th Thorium

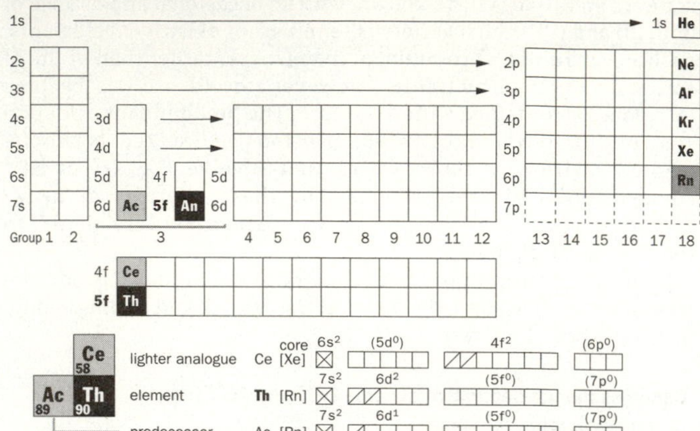

		Predecessor	Element	Lighter analogue
Name		**Actinium**	**Thorium**	**Cerium**
Symbol, *Z*		Ac, 89	Th, 90	Ce, 58
RAM		~ 227	232.0381	140.115
Radius/pm	**atomic**	187.8	179.8	182.5
	ionic, M^{3+}	118	101	107
	M^{4+}		99	94
Electronegativity (Pauling)		1.1	1.3	1.1
Melting point/K		1320 ± 50	2023	1072
Boiling point/K		3470 ± 300	~5060	3699
$\Delta H^{\ominus}_{fus}$/kJ mol^{-1}		14.2	< 19.2	8.87
$\Delta H^{\ominus}_{vap}$/kJ mol^{-1}		293	543.9	313.8
Density (293 K)/kg m^{-3}		10 060	11 720	(various forms)
Electrical conductivity/Ω^{-1}m^{-1}			7.692×10^6	1.370×10^6
Ionization energy for removal of *j*th electron/kJ mol^{-1}	***j* = 1**	499	587	527.4
	***j* = 2**	1170	1110	1047
	***j* = 3**	1900	1978	1949
	***j* = 4**		2780	3547
$E^{\ominus}$/V for $M^{4+}(aq) \rightarrow M(s)$			−1.83	1.72

Long-lived (¦) isotopes of thorium

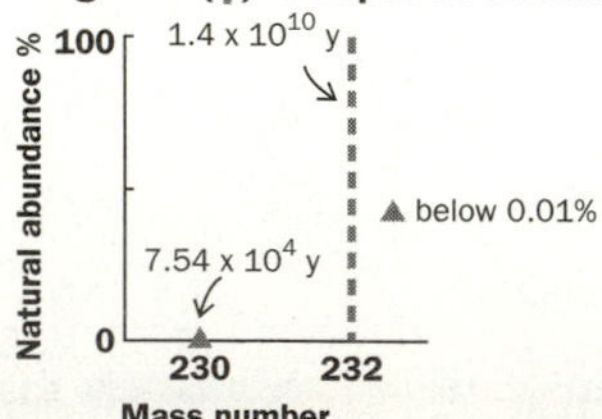

Thorium, [Rn]$6d^2 7s^2$, is no typical actinide. It occurs naturally and is more common than, say, tin, iodine, or silver. Although radioactive (series $4n + 2$), its common isotope, ^{323}Th, is extremely long lived ($t_{\frac{1}{2}} = 1.4 \times 10^{10}$y). The 6d orbitals of thorium, like those of actinium, are more stable than the 5f. Ionization of all the d and s valence electrons gives (IV) as the main oxidation state.

Thorium occurs with lanthanum, from which it is separated by extraction into TBP (a liquid organic phosphate) from acid solution. The metal is obtained by reducing the dioxide with calcium or the tetrafluoride with magnesium, but, since it is very reactive, an argon atmosphere must be used. Thorium, like lanthanide metals, is used in alloys. Its refractive oxide ThO_2 is used (still) for gas mantles. The ion Th^{4+}(aq) has the highest known charge of any ion; although it is large enough to avoid excessive charge density, it is hydrolysed if the pH is increased and eventually gives $Th(OH)_4$. The tetrafluoride is not soluble in aqueous fluoride solutions, but an unexpected variety of structurally diverse complex fluorides can be made by dry methods. In $(NH_4)_3ThF_7$, for example, there are chains of ThF_9 groups, while $(NH_4)_5ThF_9$ contains ThF_8^{4-} anions and fluoride ions. However, thorium shows more typical actinide behaviour in its lower iodides, which are metallic conductors, and probably $Th^{4+}(I^-)_2(e^-)_2$ and $Th^{4+}(I^-)_3e^-$.

For summary, see pp. 565—9.

Pa Protactinium

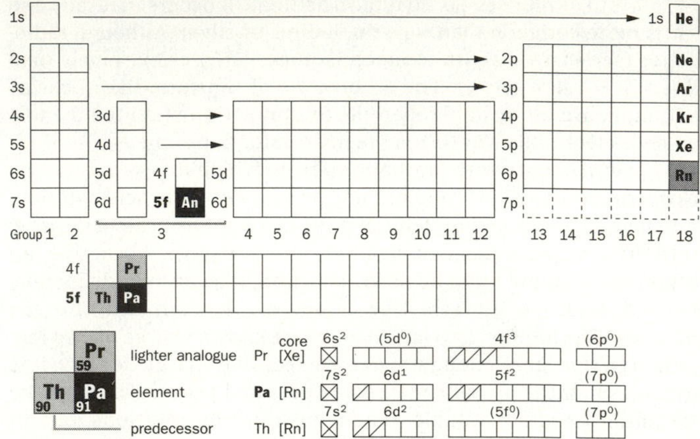

		Predecessor	Element	Lighter analogue
Name		**Thorium**	**Protactinium**	**Praseodymium**
Symbol, *Z*		Th, 90	Pa, 91	Pr, 59
RAM		232.0381	231.03588	140.90765
Radius/pm	**atomic**	179.8	160.6	182.8
	ionic, M^{3+}	101	113	106
	M^{4+}	99	98	92
Electronegativity (Pauling)		1.3	1.5	1.1
Melting point/K		2023	2113	1204
Boiling point/K		~ 5060	~ 4300	3785
$\Delta H^{\ominus}_{fus}$/kJ mol^{-1}		< 19.2	16.7	11.3
$\Delta H^{\ominus}_{vap}$/kJ mol^{-1}		543.9	481	332.6
Electrical conductivity/Ω^{-1}m^{-1}		7.692×10^6	5.650×10^6	1.471×10^6
Ionization energy for removal of *j*th electron/kJ mol^{-1}	***j* = 1**	587	568	523.1
$E^{\ominus}$/V for $M^{4+}(aq) \rightarrow M(s)$		−1.83	−1.46	−0.96

Key isotopes of protactinium

Mass number	231	232	233	234
Half-life	3.27×10^4 y	1.31 d	27.0 d	6.7 h

It might seem odd that *prot*actinium comes *after* actinium; but, being an actinide, it is radioactive and, when its most stable isotope, ^{231}Pa, decays, it emits helium nuclei and gives actinium (series $4n+3$). The process is not too rapid for survival ($t_{\frac{1}{2}} = 3.27 \times 10^4$ y) and traces of protactinium occur naturally in uranium ores. About 130 g of the element has been obtained,* almost all from 60 tonnes of sludge from uranium extraction. It can also be separated in small quantities from spent nuclear fuel.

The nuclear charge of protactinium is large enough to start stabilizing the 5f orbitals relative to the 6d, and so the atom has the configuration $[Rn]5f^1 6d^2 7s^2$, or even $[Rn]5f^2 6d^1 7s^2$. Although it forms some f^1 compounds, such as Pa^{4+}(aq.), PaF_4, $PaCl_4$, and PaO_2, its preferred oxidation state is (V), with no f or d electrons, as in Pa_2O_5, PaF_5, $PaCl_5$, and various compounds containing the linear PaO_2^+ ion, typical of the lighter actinides. In aqueous solution, protactinium (V) hydrolyses to a colloidal mess, except in the presence of fluoride, when complexes up to PaF_8^{3-} can be formed. Although, as yet, we know rather little about protactinium, its meagre chemistry shows how greatly it differs from the corresponding lanthanide (praseodynium $[Xe]5d^0 4f^3 6s^2$) in which the (III) state is dominant, but which can be oxidized up to (IV).

For summary, see pp. 565–9.

* By UKAEA (United Kingdom Atomic Energy Authority), who distributed it for the research that has produced almost all our knowledge of the element

U Uranium

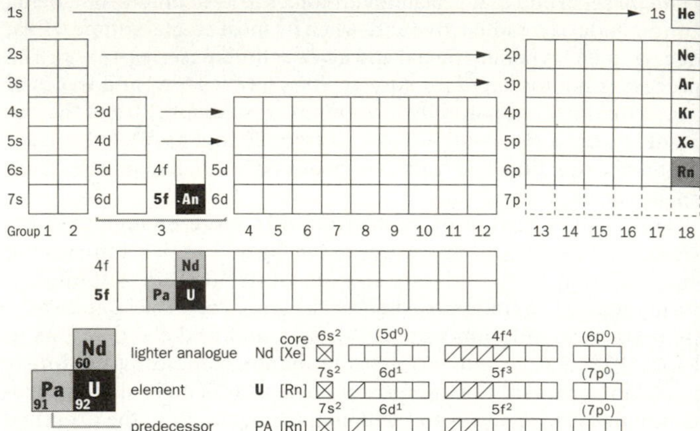

	Predecessor	Element	Lighter analogue
Name	**Protactinium**	**Uranium**	**Neodymium**
Symbol, *Z*	Pa, 91	U, 92	Nd, 60
RAM	231.03588	238.0289	144.24
Radius/pm atomic	160.6	138.5	182.1
ionic, M^{3+}	113	103	104
M^{4+}	98	97	
Electronegativity (Pauling)	1.5	1.4	1.1
Melting point/K	2113	1405.5	1294
Boiling point/K	~4300	4018	3341
ΔH°_{fus}/kJ mol^{-1}	16.7	15.5	7.113
ΔH°_{vap}/kJ mol^{-1}	481	422.6	283.7
Density (at 293 K)/kg m^{-3}		18 950	7007
Electrical conductivity/Ω^{-1}m^{-1}	5.650×10^6	3.247×10^6	1.562×10^6
Ionization energy for removal of *j*th electron/kJ mol^{-1} *j* = 1	568	584	529.6
***j* = 2**		1420	1035

Long-lived isotopes of uranium

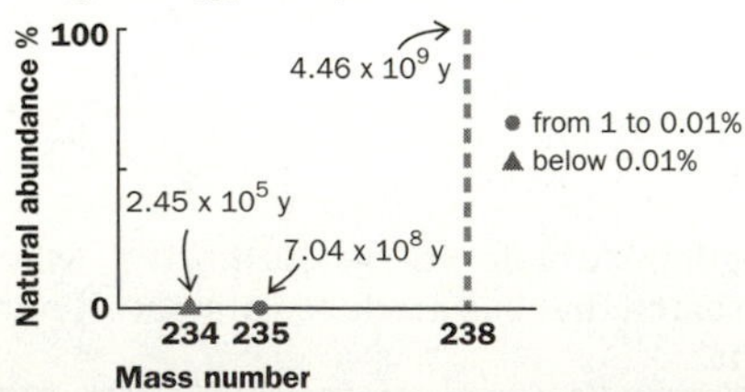

Uranium, which has two long-lived isotopes, occurs naturally. Although it is about as common as tin, it is very thinly spread. It is found combined with oxygen as U_3O_8, and often sparsely in phosphate ores. It is extracted, via the UO_2^{2+} ion, by reducing the tetrafluoride with magnesium.

Uranium is by far the most widely studied actinide and owes much of its high profile to the isotope ^{235}U, which is the only naturally occurring nucleus that can be split when bombarded by neutrons. This fission of a heavy nucleus into two lighter ones releases a large amount of energy, together with other neutrons which, if conditions are right, may themselves produce fission. The fission may cascade to give an atomic explosion (of sombre implications for humankind); or, with adequate control of the number and energy of the neutrons and the nuclear fuel, it may be channelled to produce a steady supply of nuclear power; or, if the supply of either neutrons or fuel is inadequate, the fission may fizzle out. Although the many technical, economic, political, and ethical problems raised by either nuclear weapons or nuclear power are outside the scope of this book, every chemist will be aware of the environmental problems that would be posed by sudden, or insidiously gradual, release of radioactive material, or planned disposal of radioactive waste.

The fissionable uranium-235 nucleus constitutes only 0.72% of natural uranium and must be separated from the bulk uranium-238. Some of the various techniques that are available depend simply on the small difference in mass. Uranium, with its large size and wide range of oxidation states, forms a hexafluoride UF_6 which, despite its high RMM, is volatile. As the lighter $^{235}UF_6$ molecules diffuse more rapidly than those of $^{238}UF_6$, they are also concentrated nearer the centre of a gas centrifuge, so the two isotopes can, in principle, be separated. But more refined separation has become possible. Laser beams can now be so precisely tuned that a sample of UF_6 can be irradiated at exactly the energy needed to break a $^{235}U{-}F$ bond, while leaving a $^{238}U{-}F$ bond intact. The result is a mixture of solid $^{235}UF_5$ and gaseous $^{238}UF_6$. Lasers can also be used to ionize ^{235}U atoms in the metal vapour, to give $^{235}U^+$ ions, which can be collected at a negatively charged electrode. The uranium-238 is again unaffected.

The high oxidation states shown in UF_6 and UF_5 are typical of the lighter actinides. Uranium, $[Rn]5f^16d^37s^2$ (or $[Rn]5f^36d^17s^2$), can exist in all oxidation states from (III) to (VI). Although the most common halides are UX_4, the dominant state of uranium is (VI), usually in the form of the linear 'uranyl' ion UO_2^{2+}, which has long been used to impart a yellow-green fluorescence to glass. Although it is only a

The uranyl group UO_2^{2+} in $UO_2Cl_4^{2-}$ GE

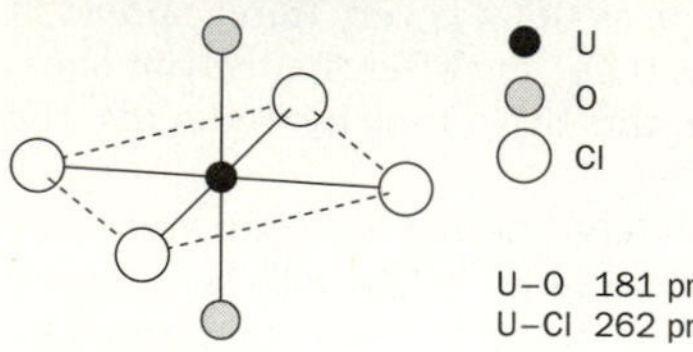

'Uranocene'

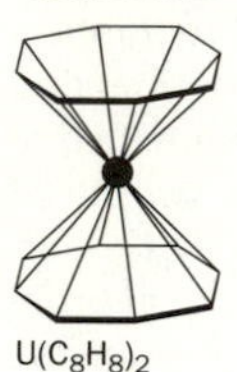

$U(C_8H_8)_2$

very weak oxidizing agent, the uranyl ion can be reversibly reduced to the (V) state, probably as UO_2^+. In solution, there is rapid electron exchange also between U^{3+} and U^{4+}, as again only a single electron is involved. However, the redox equilibrium between uranium(IV) and (V) is very slow, as would be expected, since if uranium(V) is indeed UO_2^+, two U–O bonds must be made or broken.

Many complexes of uranium have been made and their structures and spectra serve as a database for comparison with those of its less accessible neighbours. They illustrate how higher coordination can be achieved with larger metal ions. Although the actinides form $An(C_5H_5)_3$ and $An(C_5H_5)_4$ compounds with cyclopentadienyl, rather than $An(C_5H_5)_2$ sandwiches, they do form 'metallocene' sandwiches, very similar to ferrocene (see p. 74), with the larger cyclo-octatetrene, C_8H_8, which binds with all eight of its carbon atoms. Uranium also forms a 'superphthalocyanine' in which the UO_2^{2+} ion is coordinated (equatorially) by five planar nitrogen atoms, instead of by the usual four.

For summary, see pp. 565–9.

Np Neptunium

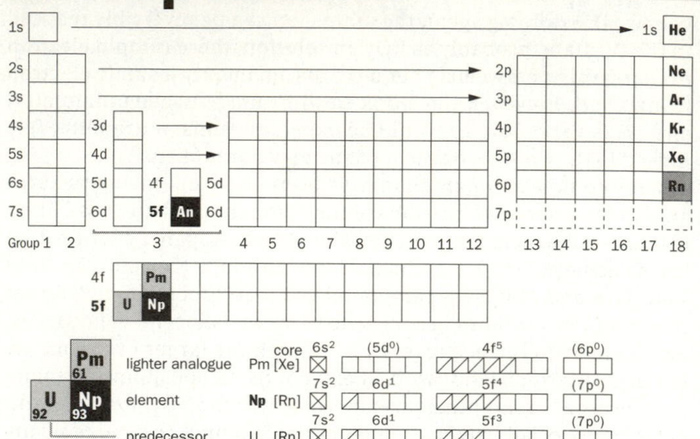

		Predecessor	Element	Lighter analogue
Name		**Uranium**	**Neptunium**	**Promethium**
Symbol, *Z*		U, 92	Np, 93	Pm, 61
RAM		238.0289	237.0482	~ 145
Radius/pm	**atomic**	138.5	131	181
	ionic, M^{3+}	103	110	106
	M^{4+}	97	95	
Electronegativity (Pauling)		1.4	1.4	
Melting point/K		1405.5	913	1441
Boiling point/K		4018	4175	
$\Delta H^{\ominus}_{fus}$/kJ mol^{-1}		15.5	9.46	12.6
$\Delta H^{\ominus}_{vap}$/kJ mol^{-1}		422.6	336.6	
Density/kg m^{-3}		18 950 (293 K)	20 250 (293 K)	7220 (298 K)
Electrical conductivity/Ω^{-1}m^{-1}		3.247×10^6	8.197×10^5	
Ionization energy for removal of *j*th electron/kJ mol^{-1}	***j* = 1**	584	597	535.9

Key isotope of neptunium

Mass number	237
Half-life	2.14×10^6 y

No element after uranium occurs in appreciable quantities in nature. Neptunium, which is obtained mainly as a by-product of the nuclear fuel industry, has been studied much less than its economically important neighbours, uranium and plutonium. But it illustrates a number of interesting trends in actinide chemistry.

Neptunium has an even wider range of oxidation states than uranium, and with xenon trioxide (a *very* strong oxidant, see p. 413) gives the octahedral ion $Np^{VII}O_6^{5-}$, which can exist as the solid lithium salt or in strongly alkaline solution. The smaller neptunium(VII) anion, NpO_5^{3-}, can be prepared in lattices with large cations such as Ba^{2+}. However, in neptunium, with configuration of $[Rn]5f^5 6d^0 7s^2$ or $[Rn]5f^4 6d^1 7s^2$, the 5f electrons are becoming increasingly stable relative to the 6d. In contrast to uranium, the most stable state of neptunium is (V) rather than (VI). The hexafluoride decomposes to the pentafluoride on heating, and the ion NpO_2^+ is much better established than UO_2^{2+}. A strong oxidizing agent such as cerium(IV) is needed to convert it to NpO_2^{2+}. A number of strongly reducing compounds of neptunium(III), such as NpH_3, have been made, together with various complexes, mainly in the (IV) and (VI) states, many of which predictably resemble those of uranium (see p. 539). There is also an oxide, NpO.

For summary, see pp. 565–9.

Pu Plutonium

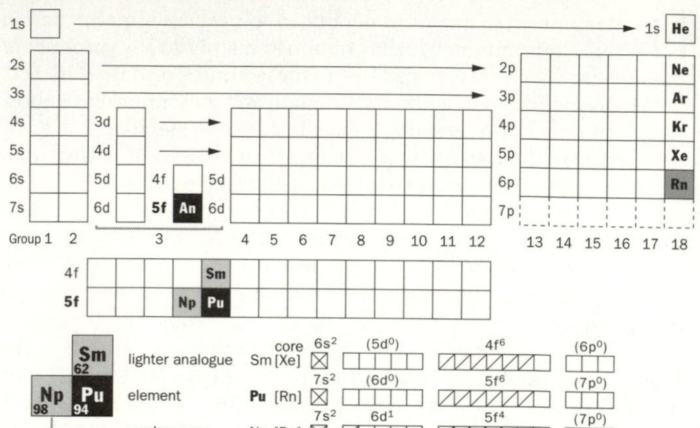

	Predecessor	Element	Lighter analogue
Name	**Neptunium**	**Plutonium**	**Samarium**
Symbol, *Z*	Np, 93	Pu, 94	Sm, 62
RAM	237.0482	~ 244	150.36
Radius/pm ionic, M^{3+}	110	108	111
M^{4+}	95	93	100
Electronegativity (Pauling)	1.4	1.3	1.2
Melting point/K	913	914	1350
Boiling point/K	4175	3505	2064
$\Delta H^{\ominus}_{fus}$/kJ mol^{-1}	9.46	2.8	10.9
$\Delta H^{\ominus}_{vap}$/kJ mol^{-1}	336.6	343.5	191.6
Density/kg m^{-3}	20 250 (293 K)	19 840 (α) (298 K)	7529 (293 K)
Electrical conductivity/Ω^{-1}m^{-1}	8.197×10^5	6.757×10^5	1.064×10^6
Ionization energy for removal of *j*th electron/kJ mol^{-1} $j = 1$	597	585	543.3

Key isotopes of plutonium

Mass number	239	242	244
Half-life	24.11 y	3.76×10^5 y	8.2×10^7 y

With regard to high profile, plutonium has pride of place amongst the man-made actinides, for the same reason that uranium stands out from those few others that occur naturally. It has a fissionable isotope. Plutonium-239 is formed by neutron bombardment of uranium-238, which is always present in nuclear reactors; and the fissionable plutonium can be made in a quantity greater than the nuclear fuel consumed in order to produce the bombarding neutrons. So, plutonium-239 enables us to 'breed' nuclear fuel. Of course, conditions have to be strictly controlled, as do safety precautions. Any plutonium we might ingest becomes lodged in our bones and liver, where the helium nuclei emitted cause grave damage. The antidote is a strongly complexing ligand (of a type related to EDTA), administered at once so that it can bind most of the plutonium before this becomes localized in any organ.

Another isotope, ^{238}Pu, is actually implanted into the body (though it is carefully shielded first). The self-heating effect (see p. 531) is used, via a thermocouple, to generate electricity for pacemakers; and also to provide power for satellites.

Preparation of ^{239}Pu and other nucleides[P]

$^{238}_{92}U(n, \gamma)\,^{239}_{92}U \xrightarrow{\beta} {}^{239}_{93}Np \xrightarrow{\beta} {}^{239}_{94}Pu$

$^{235}_{92}U(n, \gamma)\,^{236}_{92}U(n, \gamma)\,^{237}_{92}U \xrightarrow{\beta} {}^{237}_{93}Np(n, \gamma)\,^{238}_{93}Np \xrightarrow{\beta} {}^{238}_{94}Pu$

Nuclear bombardment reactions are often written in the shorthand,

target nucleus (bombarding particle , particle or radiation emitted) resultant nucleus

So, the first bombardment is

$^{238}_{92}U + {}^{1}_{0}n \rightarrow {}^{239}_{92}U$ which decays

$^{239}_{92}U \xrightarrow[\text{decay}]{\beta} {}^{0}_{-1}e + {}^{239}_{93}Np$ which also decays

$^{239}_{93}Np \xrightarrow[\text{decay}]{\beta} {}^{0}_{-1}e + {}^{239}_{94}Pu$

Purification

Add HNO_3 to give isotopes of

Pu^{VI}		Pu^{IV}	separate by
Np^{VI}	Add $ONO^- \rightarrow$	Np^{V}	solvent extraction
U^{VI}		U^{VI}	with TPB

Pu^{IV} in TPB + mild aqueous reducing agent $\rightarrow Pu^{3+}(aq)$

This process, based on large differences in redox behaviour, can be repeated for further purification.

The outer electrons of plutonium, [Rn]$5f^6 6d^0 7s^2$, behave predictably, with yet further stabilization of the 5f orbitals. Plutonium shows all oxidation states from (II) to (VII), and all except the two extremes can exist in the same solution, as Pu^{3+}, Pu^{4+}, PuO_2^+, and PuO_2^{2+}. The (VII) state is similar to that of uranium, and the (II) state is represented by the oxide PuO and the hydride PuH_2. But the most stable state for plutonium is (IV), in contrast to neptunium(V) and uranium(VI), and these differences are exploited in separating plutonium from uranium and other actinides. Oxidation of plutonium residues with O_2F_2 gives the volatile hexafluoride PuF_6, but this (unlike UF_6) can readily be decomposed to give solid PuF_4. Similarly, if a mild oxidizing agent is added to a solution containing a mixture of Pu^{4+} and U^{4+} ions, the plutonium is unaffected, although the uranium is oxidized to UO_2^{2+}. In many other ways plutonium behaves very similarly to uranium; but the differences between the two, as well as being of practical use, illustrate how much more variety is shown by the actinides than by the analogous lanthanides.

For summary, see pp. 565–9.

Am Americium

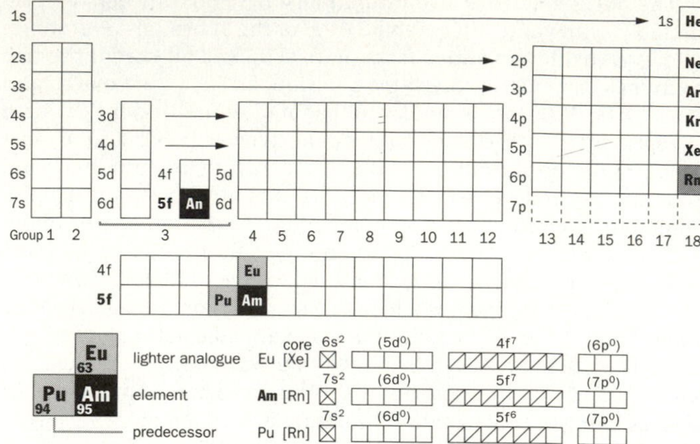

	Predecessor	Element	Lighter analogue
Name	**Plutonium**	**Americium**	**Europium**
Symbol, *Z*	Pu, 94	Am, 95	Eu, 63
RAM	~ 244	~ 243	151.965
Radius/pm atomic		184	204.2
ionic, M^{3+}	108	107	112
M^{4+}	93	92	98
Electronegativity (Pauling)	1.3	1.3	
Melting point/K	914	1267	1095
Boiling point/K	3505	2880	1870
$\Delta H^{\ominus}_{fus}$/kJ mol^{-1}	2.8	14.4	10.5
$\Delta H^{\ominus}_{vap}$/kJ mol^{-1}	343.5	238.5	175.7
Density (at 293 K)/kg m^{-3}	19 840 (α)	13 670	5243
Electrical conductivity/Ω^{-1}m^{-1}	6.757×10^5	1.470×10^6	1.111×10^6
Ionization energy for removal of *j*th electron/kJ mol^{-1} *j* = 1	585	578.2	546.7

Key isotopes of americium

Mass number	241	243
Half-life	432.2 y	7.37×10^3 y

In many ways americium, [Rd]$5f^7 6d^0 7s^2$, resembles its predecessor, plutonium. Both are obtained by intentional neutron bombardment and are also found in nuclear fuel residues. Americium, too, has one fairly long-lived isotope, ^{243}Am ($t_{\frac{1}{2}} = 7.37 \times 10^3$y) and is used as a small, portable source of power. Like the earlier actinides, it shows a wide range of oxidation states; but the balance between them is different. States (V) and (IV) disproportionate in aqueous solution, but since the reactions involve oxygen atom transfer, the changes are slow (see p. 541). The higher states, (V) and (VI), are reduced by the products of radioactive decay (see p. 529); the (VII) state is much less accessible than for plutonium or neptunium. Unlike plutonium, americium does not form even an unstable hexafluoride; its highest halide is AmF_4. Like the earlier actinides, it forms a dioxide, but this is less stable than Am_2O_3, and for americium the (III) state is the dominant one. As with europium ([Xe]$4f^7 5d^0 6s^2$), americium can also exist in oxidation state (II), present in chloride melts, and in the oxide AmO, which is similar to that of plutonium.

For summary, see pp. 565–9.

Cm Curium

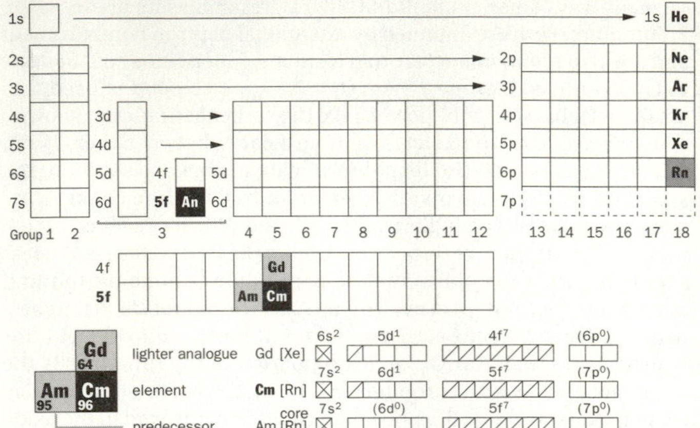

	Predecessor	Element	Lighter analogue
Name	**Americium**	**Curium**	**Gadolinium**
Symbol, *Z*	Am, 95	Cm, 96	Gd, 64
RAM	~ 243	~ 247	157.25
Radius/pm ionic, M^{2+}		119	
M^{3+}	107	99	97
M^{4+}	92	88	
Electronegativity (Pauling)	1.3	1.3	1.2
Melting point/K	1267	1610 ± 40	1586
Density/kg m^{-3}	13 670 (293 K)	13 300 (293 K)	7900.4 (at 298 K)
Ionization energy for removal of *j*th electron/kJ mol^{-1} $j = 1$	578.2	581	592.5
$E^\ominus$/V for $M^{4+}(aq) \rightarrow M^{3+}(aq)$		3.2	
$M^{3+}(aq) \rightarrow M(s)$		−2.06	−2.2

Key isotopes of curium

Mass number	242	244	245	246	247	248
Half-life	162.9 d	18.11 y	8.5×10^3 y	4.78×10^3 y	1.56×10^7 y	3.4×10^5 y

Research on the heavier actinides is even more challenging than the study of the earlier members of the series, not least because decreasing amounts are available. Not all the later actinides can be extracted from spent nuclear fuels and some must be made by particle bombardment of a target, usually of a slightly earlier actinide which is itself available in only very small quantities. For curium, the particle may again be a neutron, or it may be an α-particle (helium nucleus) of four times the mass. Although curium has been prepared in gram quantities, it is difficult to study its solutions, since they are evaporated to dryness by self-heating. Like plutonium and americium, it is used as a power source. Curium has the configuration $[Rn]5f^7 6d^1 7s^2$ and, like gadolinium, $[Xe]4f^7 5d^1 6s^2$ (see p. 443), shows a preference for oxidation state (III). Although curium also forms CmO, CmO_2, and CmF_4, it shows no trace of oxidation states above (IV).

For summary, see pp. 565–9.

Bk Berkelium

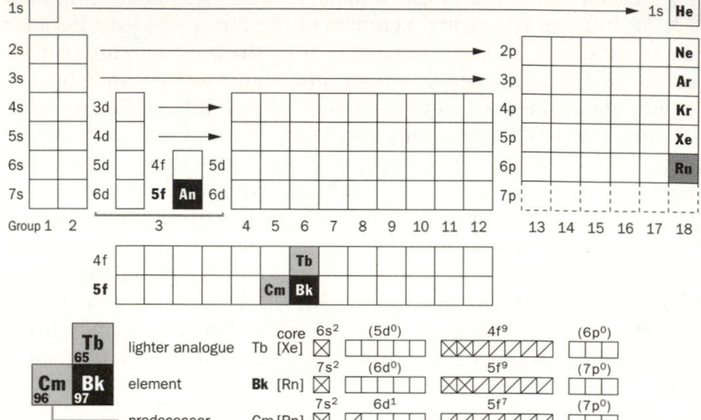

	Predecessor	Element	Lighter analogue
Name	**Curium**	**Berkelium**	**Terbium**
Symbol, *Z*	Cm, 96	Bk, 97	Tb, 65
RAM	~ 247	~ 247	158.92534
Radius/pm ionic, M^{2+}	119	118	
M^{3+}	99	98	93
M^{4+}	88	87	81
Electronegativity (Pauling)	1.3	1.3	
Density (at 293 K)/kg m^{-3}	13 300	14 790	8229
Ionization energy for removal of *j*th electron/kJ mol^{-1} $j = 1$	581	601	564.6
$E^{\ominus}$/V for $M^{4+}(aq) \rightarrow M^{3+}(aq)$	3.2	1.67	
$M^{3+}(aq) \rightarrow M(s)$	–2.06	–2.01	

Key isotopes of berkelium

Mass number	247	249
Half-life	1.4×10^3 y	320 d

Production of ^{247}Bk by ion bombardment[P]

$$^{244}_{96}\text{Cm} + {}^{4}_{2}\text{He}^{2+} \rightarrow {}^{247}_{97}\text{Bk} + {}^{1}_{1}\text{H}^{+}$$

α-particle proton

or, in shorthand, $^{244}_{96}\text{Cm}\ (\alpha,\ p)\ ^{247}_{97}\text{Bk}$

Berkelium can either be made by neutron or ion bombardment, or be extracted from nuclear fuels. It is available in only tens of milligram quantities; and its longest lived isotope, berkelium-247, is produced in only trace amounts. With a configuration of [Rn]$5f^86d^17s^2$ or [Rn]$5f^96d^07s^2$, its dominant oxidation state is (III). It resembles curium in forming also BkO, BkO_2, and BkF_4. Although berkelium (IV) is strongly oxidizing, it is more stable than the (IV) state of curium or americium; this is reminiscent of terbium(IV) (see p. 444), although berkelium(IV), unlike terbium(IV), can exist in solution. Moreover, berkelium has an unusually low radius in the metal, compatible with partial delocalization of four electrons rather than of the expected three.

For summary, see pp. 565–9.

Cf Californium

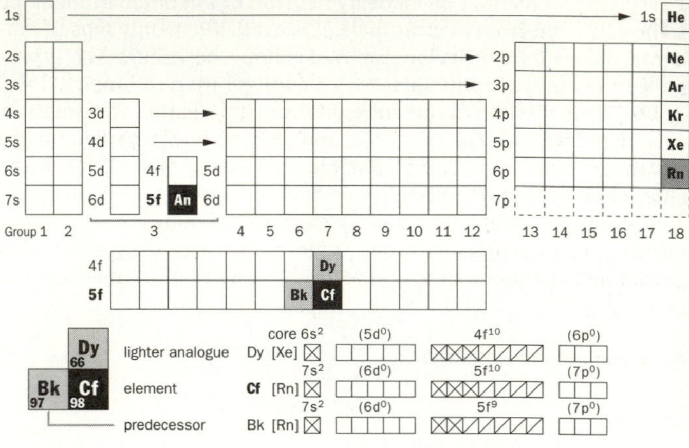

	Predecessor	Element	Lighter analogue
Name	**Berkelium**	**Californium**	**Dysprosium**
Symbol, *Z*	Bk, 97	Cf, 98	Dy, 66
RAM	~ 247	~ 251	162.5
Radius/pm ionic, M^{2+}	118	117	
M^{3+}	98	98	91
M^{4+}	87	86	
Electronegativity (Pauling)	1.3	1.3	1.2
Ionization energy for removal of *j*th electron/kJ mol^{-1} *j* = 1	601	608	571.9
$E^{\ominus}$/V for $M^{3+}(aq) \rightarrow M^{2+}(aq)$		−1.6	
$M^{2+}(aq) \rightarrow M(s)$		−2.1	
$M^{3+}(aq) \rightarrow M(s)$	−2.01	−1.9	−2.29

Key isotopes of californium

Mass number	249	251	252
Half-life	351 y	890 y	2.64 y

Californium, like berkelium, is made by neutron or α-particle bombardment and is available in similarly modest amounts. With a configuration of [Rn]$5f^{10}6d^{0}7s^{2}$, it exists predominantly as californium(III), and its few known compounds in the (II) and (IV) states are less stable than the analogous ones of berkelium.

For summary, see pp. 565–9.

Es Einsteinium

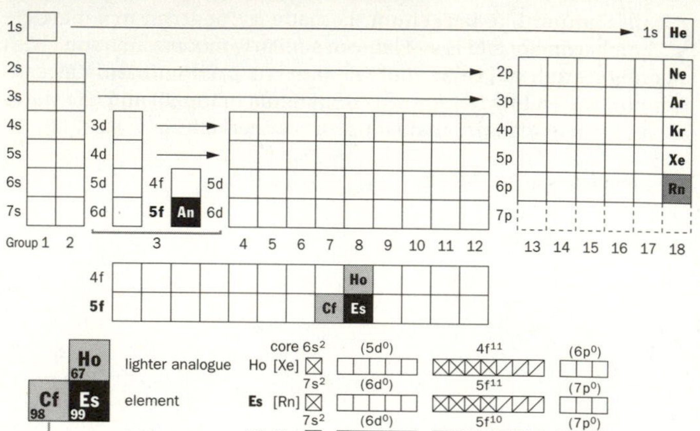

	Predecessor	Element	Lighter analogue
Name	**Californium**	**Einsteinium**	**Holmium**
Symbol, *Z*	Cf, 98	Es, 99	Ho, 67
RAM	~ 251	~ 254	164.93032
Radius/pm ionic, M^{2+}	117	116	
M^{3+}	98	98	89
M^{4+}	86	85	
Electronegativity (Pauling)	1.3	1.3	1.2
Ionization energy for removal of *j*th electron/kJ mol^{-1} *j* = 1	608	619	580.7
$E^⊖$/V for $M^{3+}(aq) \rightarrow M^{2+}(aq)$	−1.6	−1.5	
$M^{2+}(aq) \rightarrow M(s)$	−2.1	−2.2	
$M^{3+}(aq) \rightarrow M(s)$	−1.9	−1.9	−2.33

Key isotopes of einsteinium

Mass number	252	253	254	255
Half life	1.29 y	20.47 d	275 d	39.8 d

Decay [P] of Fm to Es

$$^{251}_{100}\text{Fm} \xrightarrow{\text{EC}} {}^{251}_{99}\text{Es}$$

$$^{253}_{100}\text{Fm} \xrightarrow{\text{EC}} {}^{253}_{99}\text{Es}$$

EC is electron capture

Einsteinium was discovered in the fall-out from the 1952 thermonuclear explosion in the Pacific; two of its isotopes, ^{251}Es and ^{253}Es are generated when the fermium isotopes of the same mass number decay by electron capture (see p. 192). Einsteinium is now made on the tenth of a milligram scale, by neutron and ion bombardment. It is a typical later actinide, of configuration $[Rn]5f^{11}6d^{0}7s^{2}$, with almost all of its few known compounds in the (III) state.

For summary, see pp. 565–9.

Fm Fermium

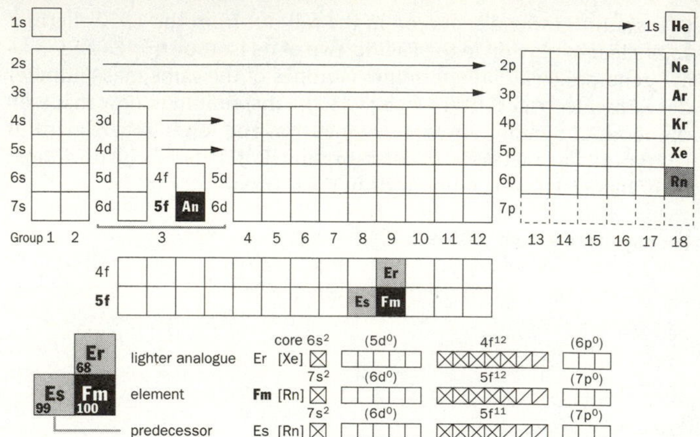

	Predecessor	Element	Lighter analogue
Name	**Einsteinium**	**Fermium**	**Erbium**
Symbol, *Z*	Es, 99	Fm, 100	Er, 68
RAM	~ 254	~ 257	167.26
Radius/pm ionic, M^{2+}	116	115	
M^{3+}	98	97	89
M^{4+}	85	84	
Electronegativity (Pauling)	1.3	1.3	
Ionization energy for removal of *j*th electron/kJ mol^{-1} *j* = 1	619	627	588.7
$E^{\ominus}$/V for $M^{3+}(aq) \rightarrow M^{2+}(aq)$	−1.5	−1.15	
$M^{2+}(aq) \rightarrow M(s)$	−2.2	−2.37	
$M^{3+}(aq) \rightarrow M(s)$	−1.9	−1.96	−2.32

Key isotopes of fermium

Mass number	253	254	255	257
Half-life	3 d	3.24 h	20.1 h	100.5 d

Fermium was also discovered in the fall-out from the 1952 explosion in the Pacific and, like einsteinium, fermium-257 can also be made (in milligram quantities) by neutron bombardment. But this isotope immediately absorbs a further neutron to give fermium-258, which decays within seconds by spontaneous nuclear fission. In order to avoid this problem, the synthesis of fermium and of the still heavier actinides must be carried out without using neutrons. Instead, the target element is bombarded by α-particles, or more often by heavier ions, so as to jump the fermium-258 hurdle. Yields are discussed in terms not of milligrams, but of the number of atoms, and the elements can be studied only in solution. Fermium is available in batches of about 10^{11} atoms, which might not sound too inadequate until we realise that it is less than 10^{-12} of a mole. Even so, it has been established that, as we should expect, fermium, $[Rn]5f^{12}6d^{0}7s^{2}$, like einsteinium, exists predominantly in oxidation state (III).

For summary, see pp. 565–9.

Md Mendelevium

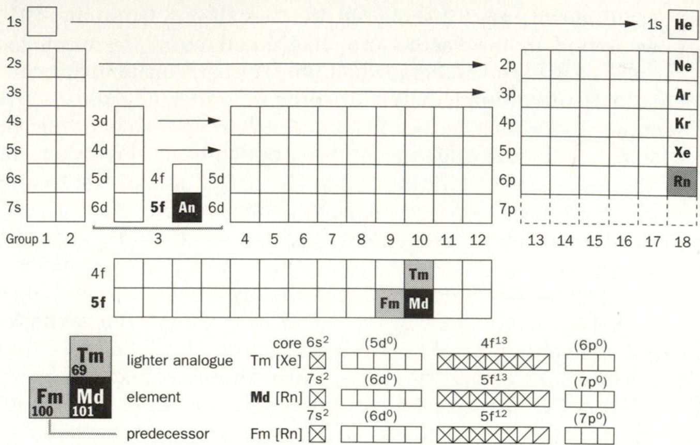

	Predecessor	Element	Lighter analogue
Name	**Fermium**	**Mendelevium**	**Thulium**
Symbol, *Z*	Fm, 100	Md, 101	Tm, 69
RAM	~ 257	~ 258	168.93421
Radius/pm ionic, M^{2+}	115	114	
M^{3+}	97	96	87
M^{4+}	84	84	
Electronegativity (Pauling)	1.3	1.3	
Ionization energy for removal of *j*th electron/kJ mol^{-1} *j* = 1	627	635	596.7
$E^{\ominus}$/V for $M^{3+}(aq) \rightarrow M^{2+}(aq)$	−1.15	−0.15	
$M^{2+}(aq) \rightarrow M(s)$	−2.37	−2.4	
$M^{3+}(aq) \rightarrow M(s)$	−1.96	−1.65	−2.32

Key isotope of mendelevium

Mass number 258

Half-life 56 d

Production of Md by α-bombardment [P]

$$^{253}_{99}\text{Es} + ^{4}_{2}\text{He} \rightarrow ^{256}_{101}\text{Md} + ^{1}_{0}\text{n}$$

$$\downarrow \text{EC}$$

$$^{256}_{100}\text{Fm}$$

Mendelevium is obtained by bombarding einsteinium with α-particles in an accelerator; and as einsteinium is somewhat hard to come by, we should expect the availability of mendelevium to be very low. Typically, only a million atoms are produced per experiment. It seems that, although mendelevium, [Rd]$5f^{13}6d^07s^2$, favours the (III) state, mendelevium(II) may also be stable.

For summary, see pp. 565–9.

No Nobelium

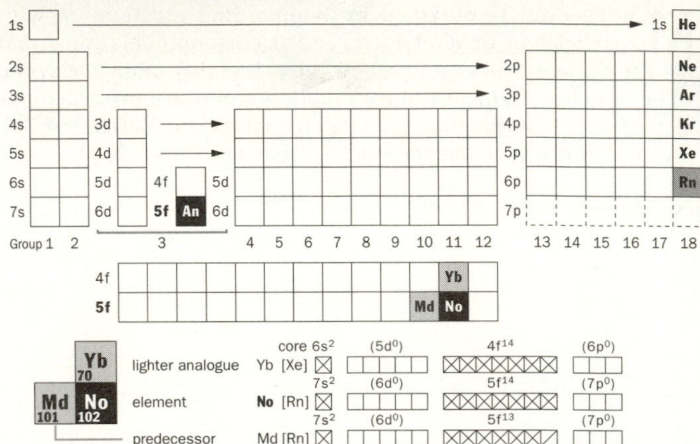

	Predecessor	Element	Lighter analogue
Name	**Mendelevium**	**Nobelium**	**Ytterbium**
Symbol, *Z*	Md, 101	No, 102	Yb, 70
RAM	~ 258	~ 259	173.04
Radius/pm ionic, M^{2+}	114	113	113
M^{3+}	96	95	86
M^{4+}	84	83	
Electronegativity (Pauling)	1.3	1.3	
Ionization energy for removal of *j*th electron/kJ mol^{-1} *j* = 1	635	642	603.4
$E^{\ominus}$/V for $M^{3+}(aq) \rightarrow M^{2+}(aq)$	−0.15	1.4	−1.05
$M^{2+}(aq) \rightarrow M(s)$	−2.4	−2.5	

Key isotope of nobelium

Mass number 259

Half-life 58 min

Production of No by heavy ion bombardment [P]

$$^{241}_{94}Pu + ^{16}_{8}O \rightarrow ^{253}_{102}No + 4^{1}_{0}n$$

$$^{246}_{96}Cm + ^{12}_{6}C \rightarrow ^{254}_{102}No + 4^{1}_{0}n$$

$$^{238}_{92}U + ^{22}_{10}Ne \rightarrow ^{255}_{102}No + 5^{1}_{0}n$$

To make nobelium we must use heavy ion bombardment, such as firing oxygen nuclei on to a plutonium target, or carbon nuclei on to curium. The main product has a half-life of three minutes. Yields of about a thousand atoms are obtained but these are enough to tell us that nobelium, $[Rn]5f^{14}6d^{0}7s^{2}$, exists predominantly as nobelium(II), which is more stable to oxidation than ytterbium(II) ($4f^{14}$) and can be converted to nobelium(III) only by using a reagent as powerful as permanganate.

For summary, see pp. 565–9.

Lr Lawrencium

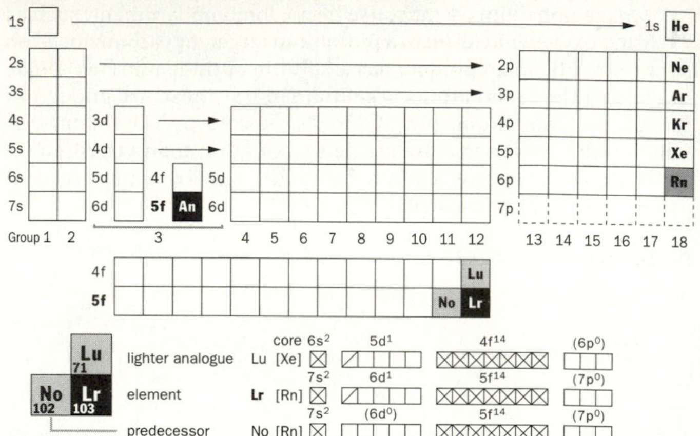

	Predecessor	Element	Lighter analogue
Name	**Nobelium**	**Lawrencium**	**Lutetium**
Symbol, *Z*	No, 102	Lr, 103	Lu, 71
RAM	~ 259	~ 260	174.967
Radius/pm ionic, M^{2+}	113	112	
M^{3+}	95	94	85
M^{4+}	83	83	
Electronegativity (Pauling)	1.3	1.3	
$E^\ominus$/V for $M^{3+}(aq) \rightarrow M(s)$	–1.2	–2.06	–2.30

Key isotope of lawrencium

Mass number	260
Half-life	3 min

Production[P] of Lr

$$^{252}_{98}Cf + ^{10}_{5}B \rightarrow ^{257}_{103}Lr + 5\,^{1}_{0}n$$

$$^{252}_{98}Cf + ^{11}_{5}B \rightarrow ^{257}_{103}Lr + 6\,^{1}_{0}n$$

This final actinide can be synthesized (about ten atoms at a time) by bombarding californium with boron nuclei. The isotope formed has a half-life of three minutes. As lawrencium has the configuration ([Rn]$5f^{14}$)$6d^1 7s^2$, it is not surprising that, like lutetium ($4f^{14}$, see p. 450), it seems to exist only in the (III) state.

For summary, see pp. 565–9.

The actinide (5f) series: trends and summaries Th–Lr

The $[Rn](5f + 6d)^{n+1}7s^2$ elements resemble the lanthanides, in that they are electropositive and reactive. Most have a (III) state which shows an 'actinide contraction' (see p. 426); and some form $An^{3+}\,2X^-e^-$ hydrides and iodides. But the actinides differ from the lanthanides in that their f valence electrons are less withdrawn into the core (and so are more similar in energy to their d valence electrons). As a consequence, the early actinides show a much wider range of oxidation state than do the lanthanides, often favouring a state higher than (III); and there is more variation across the series, with the heavier actinides being more like the lanthanides than the lighter ones.

The nuclear instability of the actinides causes many problems, broadly increasing with atomic number: short supply, short life time, self-heating, radiolysis, β-decay, reduction, health and safety, and EXPENSE. But their radioactivity allows them to be studied by tracer methods, and to be identified by their decay characteristics.

Summaries of individual elements are given in the table that follows, together with data, where available.

Comparative summaries of elements

Element	Th	Pa	U	Np
	Thorium	Protactinium	Uranium	Neptunium
Outer configuration $5f^x6d^y7s^2$	$6d^27s^2$	$5f^16d^27s^2$	$5f^36d^17s^2$	$5f^46d^17s^2$
Atomic number	90	91	92	93
Relative atomic mass	232.0381	231.03588	283.0289	237.0482
Source	From La ores by TBP extraction and reduction of ThO_2 or ThF_4	Uranium ores and nuclear waste	Widely and thinly distributed; extracted from oxygen compounds via U^{VI} and UF_4 by reduction with Mg	From decay of Pu; in nuclear fuel industry
Availability	Slightly rarer than Pb	130 g in all	Extracted on large scale	
Longest lived isotope; mass no. half-life	^{232}Th almost isotopically pure 1.4×10^{10}y	231 3.27×10^4y	238 4.46×10^8y	237 2.14×10^6y
Ionic radius/pm	Th^{4+}, 99	Pa^{4+}, 98	U^{4+}, 97 U^{3+}, 103	Np^{3+}, 110
Electronegativity (Pauling)	1.3	1.5	1.38	1.36
Ionization energy for *j*th electron				
***j* = 1**	587	568	584	597
2	1110		1420	
3	1978			
4	2780			
Dominant oxidation state	**IV** with O and F	**V** PaO_2^+, Pa_2O_5, PaX_5, complexes	**VI** UO_2^{2+} UF_6	**V** NpO_2^+(aq) Np_2O_5 fluorocomplexes
Other oxidation states	• Lower iodides	• IV PaO_2 PaX_4	• V UO_2^+ • IV • III	• VI NpO_2^{2+}, NpO_3, NpF_6 • VII NpO_6^{5-}, NpO_5^{3-} strongly oxidizing • IV NpO_2 halides • III Np^{3+}(aq) halides • II NpO
Points of interest	Complex fluorides $Th^{4+}(I^-)_3e^-$ $Th^{4+}(I^-)_22e^-$ metallic conductors		UO_2^{2+} linear UF_6 volatile Organometallics with large ligands	
Uses	• Th in alloys • ThO_2 in gas mantles		• Nuclear fuel • Nuclear weapons • Glass colourant	

Th-Lr: $[Rn]5f^x6d^y7s^2$

Pu	Am	Cn	Bk	Cf
Plutonium	Amricum	Curium	Berkelium	Californium
$5f^67s^2$	$5f^77s^2$	$5f^76p^17s^2$	$5f^97s^2$	$5f^{10}7s^2$
94	95	96	97	98
244	243	247	247	251
Neutron bombardment of ^{238}U	Neutron bombardment; nuclear fuel residues	^1_0n or ^4_2He bombardment	Nuclear fuel waste; neutron or ion bombardment	^1_0n or ^4_2He bombardment
Generated in nuclear reactors		Gram quantities	0.01 g scale	0.01 g scale
244 8.2×10^7 y	243 7.37×10^3 y	247 1.56×10^7 y	247 1.4×10^3 y	251 890 y
Pu^{4+}, 93 Pu^{3+}, 108	Am^{3+}, 107	Cm^{3+}, 99	Bk^{3+}, 98	Cf^{3+}, 98
1.28	1.3	1.3	1.3	1.3
585	578.2	581	601	608
IV $Pu^{4+}(aq)$, PuO_2 halides	**III** $Am^{3+}(aq)$ Am_2O_3 halides	**III** $Am^{3+}(aq)$, Am_2O_3, halides	**III** halides	**III** $Cf^{3+}(aq)$,
• VII Li_5PuO_6, $PuO_5^{3-}(aq)$ • VI PuO_2^{2+}, PuF_6 • V PuO_2^+ • III Pu^{3+}, Pu_2O_3 halides • II PuO, PuH_2	• VI reduced by decay products • V disproportionates; reduced by decay products • IV disproportionates slowly in water • II AmO, $AmCl_2$	• IV CmO_2, CmF_4, $Cm^{4+}(aq)$ • II CmO	• IV BkO_2, BkF_4 strongly oxidizing • II BkO	• IV unstable • II unstable
PuF_6 readily decomposed		Self-heating evaporates solutions	Low metallic radius Bk^{IV} more stable than Am^{IV} or Cm^{IV}	
• 'Breeder' of nuclear fuel • Nuclear weapons • Power source for pacemakers and satellites	• Portable power source	• Power source		

Element	Cf	Es	Fm	Md
	Californium	Einsteinium	Fermium	Mendelevium
Outer configuration $5f^x6d^y7s^2$	$5f^{10}7s^2$	$4f^{11}7s^2$	$5f^{12}7s^2$	$5f^{13}7s^2$
Atomic number	98	99	100	101
Relative atomic mass	251	252	257	(258)
Source	${}^1_0n$ or ${}^4_2He$ bombardment	Nuclear fall-out; ${}^1_0h$ and ion bombardment	Nuclear fall-out; ion bombardment	${}^4_2He$ bombardment of Es
Availability	0.01 g scale	10^{-4} g scale	10^{-12} mole batches	10^6 atom batches
Longest living isotope; mass no.	251	252	257	258
half-life	890 y	1.29 y	100.5 d	56 d
Ionic radius/pm	Cf^{3+}, 98	Es^{3+}, 98	Fm^{3+}, 97	Md^{3+}, 96
Electronegativity (Pauling)	1.3	1.3	1.3	1.3
Ionization energy for *j*th electron				
***j* = 1**	608	619	627	635
2				
3				
4				
Dominant oxidation state	**III** $Cf^{3+}(aq)$, halides	**III**	**III** $Fm^{3+}(aq)$	**III** $Md^{3+}(aq.)$
Other oxidation states	• IV unstable • II unstable	• II	?	• II?
Points of interest			${}^{257}Fm$ absorbs neutrons to give ${}^{258}Fm$ (decays v. rapidly)	
Uses				

No	Lr
Nobelium	Lawrencium
$5f^{14}7s^2$	$5f^{14}6d^17s^2$
102	103
(259)	(260)
Heavy ion bombardment by O on Pu, or C on Cm	Bombarding Cf with B nuclei
10^3 atom batches	10 atom batches
259 58 min	260 3 min
No^{3+}, 95	Lr^{3+}, 94
1.3	1.3
642	
III No^{3+}(aq.)	**III**
• II No^{2+}(aq.)	
No^{2+}(aq.) more difficult to oxidize than Yb^{2+}	

Rf Trans-actinides

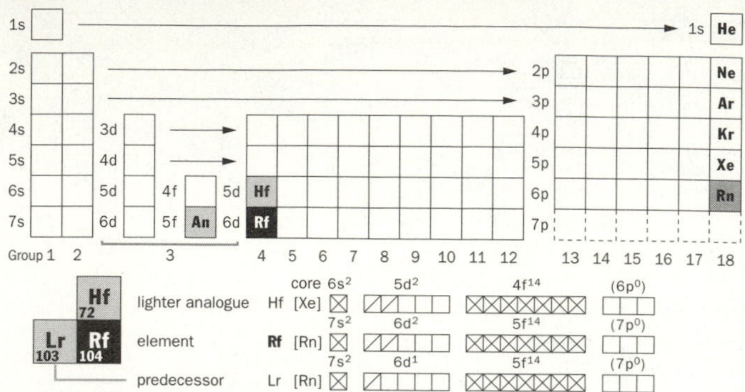

Key isotopes of rutherfordium

Mass number	257	259	261
Half-life	4.8 s	~ 3.1 s	~ 65 s

Recommended names and symbols of elements 104–109

Element	Name	Symbol
104	Rutherfordium	Rf
105	Dubnium	Db
106	Seaborgium	Sg
107	Bohrium	Bh
108	Hassium	Hs
109	Meiterium	Mt

System for naming newly synthesized elements

Digit	0	1	2	3	4	5	6	7	8	9
Classical root*	nil	un	bi	tri	quad	pent	hex	sept	oct	enn

* the initial letter of the root is used to form the symbol.

Systematic names of elements 110–112 [with possible successors, 113–120]

Atomic number	Name	*Symbol
110	Ununnilium	Uun
111	Unununium	Uuu
112	Ununbium	Uub
113	Ununtrium	Uut
114	Ununquadium	Uuq
115	Ununpentium	Uup
116	Ununhexium	Uuh
117	Ununseptium	Uus
118	Ununoctium	Uuo
119	Ununennium	Uue
120	Unbinilium	Ubn

The search for higher elements in the 6d transition series and beyond involves bombarding stable heavy nuclei with increasingly heavy ions. The speed of these missiles must be carefully chosen: they must have enough momentum to provide the activation energy for a major shuffle of nucleons in order to accommodate the incoming particle, but not have so much energy that they destroy the target nuclei. New synthetic nuclei are identified by their decay products, and this is more reliable if they decay by α-particle emission than if they undergo spontaneous fission. It is now established that elements up to $Z = 112$ have been synthesized and we even know something about their behaviour. Element-104 for example, forms a volatile tetrachloride, similar to that of hafnium (see p. 461) and quite unlike the salt-like trichlorides of the lanthanides and actinides. Element-106 has also been reported to form compounds similar to those found in the 5d series. However, we should not expect the behaviour of the fourth (6d) transition series to mirror that of the third. At these high atomic numbers, the changes outlined for thallium ($Z = 81$, see p. 501) are more pronounced and the relative energies of the orbitals are modified by the influence of the nuclear mass on the velocity of the electrons and by the different effect of the nuclear charge on the penetration, and hence on the screening, of s, p, d, and f shells. The technical difficulties of trans-actinide work are almost unimaginable. Element-109 is synthesized by bombarding bismuth-209 with nuclei of iron-58. One in 10^{14} collisions is successful, which gives a weekly yield of one atom and this has a half-life of only five milliseconds.

Element-112 is even more elusive. On 9 February 1996, at 10.37 p.m., a single atom was produced after three weeks of bombarding lead with 5×10^{18} zinc nuclei at 30 000 km s^{-1}. It existed for one-third of a millisecond and then decayed, first to element-110, which itself decayed rapidly to other trans-actinides, and eventually to fermium. (This of course also decays, but at this point its creators had lost the trail.)

There are also problems about even naming these elements, since research groups in different countries may synthesize different nuclei of the same element with, of course, quite different decay properties; and it is by these that the isotopes, and hence the elements, are identified. Those laboratories that report synthesis of a new element also suggest its name, and rival claims, such as rutherfordium and kurtchatovium for element 104, are assessed by an international committee. In February 1997 names and symbols were recommended for the first six post-actinide elements: numbers 104 to 109. Elements 110, 111, and 112, together with any later arrivals, are given preliminary names in Latin mixed with

Key isotopes of dubnium

Mass number	258	262
Half-life	4 s	34 s

Greek, based on the digits of their atomic numbers. On this system element 110 is called ununnilium and given the symbol Uun, while element 111 is unununium, of symbol Uuu; in the spoken language, however, the names one-one-o and one-one-one seem easier on the tongue and mind.

Theoretical studies of nuclear structure suggest that elements of certain atomic numbers, such as 114 [presumably ([Rn]$5f^{14}6d^{10}$)$7s^27p^2$)] and 126*, might have nuclei that are much more stable than their high mass would suggest. It is at these isolated regions of atomic number, which are optimistically termed 'islands of stability', that current research is directed in order to provide yet more atomic diversity, by creating some so-called super-heavy elements.

* Of who knows what configuration, maybe with 5g electrons.

Copyright acknowledgements

The author and publisher thank the following copyright holders for permission to reproduce copyright material:

Academic Press: for material marked **PIC** from W. W. Porterfield, *Inorganic Chemistry*, 2nd edn. (Academic Press, 1993).

Addison Wesley Longman: for material marked

HKK from J. E. Huheey, E. A. Keiter and R. L. Keiter, *Inorganic Chemistry*, 4th edn. (Harper Collins, New York, 1993) from INORGANIC CHEMISTRY: PRINCIPLES OF STRUCTURE AND REACTIVITY, 4th Edition by James H. Huheey, Ellen A. Keiter and Richard L. Keiter. Copyright © 1993 by Harper Collins College Publishers. Reprinted by permission of Addison-Wesley Educational Publishers Inc.

PM from R. V. Parish, *The Metallic Elements* (Longman, London, 1977). Reprinted by permission of Addison Wesley Longman Ltd.

R from H. Rossotti, *The Study of Ionic Equilibria* (Longman, London, 1978). Reprinted by permission of Addison Wesley Longman Ltd.

S from A. G. Sharpe, *Inorganic Chemistry*, 3rd edn., (Longman, Harlow, UK, 1992). Reprinted by permission of Addison Wesley Longman Ltd.

American Chemical Society: for material on p. 22, from P. X. Armandarez and K. Nakomoto (1966), *Inorganic Chemistry*, **5** 796, Figure 1.

Butterworth-Heinemann Ltd: for material marked **GE** from N. N. Greenwood and A. Earnshaw: *Chemistry of the Elements* (Butterworth-Heinemann Ltd., Oxford, 1984).

Cambridge University Press: for material marked:

D from W. E. Dasent, *Inorganic Energetics*, 2nd edn. (Cambridge University Press, 1982).

J from D. A. Johnson, *Some Thermodynamic Aspects of Inorganic Chemistry*, 2nd edn. (Cambridge University Press, 1982).

Chapman & Hall, Ltd.: for material marked:

MM from K. M. Mackay, *Modern Chemistry*, 4th edn., (Blackie, Glasgow, 1989), pp. 74, 258, 287, and 305 for material on p. 553 from Katz, J. J., Seaborg, G. T., and Morss, L. R., eds., *The*

Chemistry of the Actinide Elements (Chapman & Hall, New York, 1986), vol. 2, pp. 1131–1133; Fig. 14.3, p. 1132.

SM from L. Smart and E. Moore, *Solid State Chemistry* (Chapman & Hall, London, 1992), (Fig. 1.32, p. 32; Figs 2.9 and 2.10, p. 87; Fig. 4.10, p. 168; Fig. 1.26, p. 199; Fig. 7.7, p. 245; Fig. 7.8, p. 249; Figs 8.6 and 8.7, p. 265).

Chemistry in Britain: For material on p. 325 from, *Chem. Br*, (1971), p. 205, Figure 6.

Elsevier Science SA: for material on p. 40. Reprinted from *Journal of Organometallic Chemistry*, **2**, (E. Weiss and E. A. C. Lucker), Figure 2, p. 197, 1964, with kind permission from Elsevier Science S.A. P.O. Box 564, 1001 Lausanne, Switzerland.

W H Freeman & Co: for material marked:

AEC from ATOMS, ELECTRONS AND CHANGE by Atkins. Copyright © 1991 by P. W. Atkins. Used with permission of W. H. Freeman and Company

AB from GENERAL CHEMISTRY 2/E by Atkins and Beran. Copyright © 1992 by P. W. Atkins and J. A. Beran. Used with permission of W. H. Freeman and Company.

Dr A. G. Massey: for material marked:

M from A. G. Massey, *Main Group Chemistry* (Ellis Horwood, Chichester, UK, 1990).

Oxford University Press for material marked:

AP from P. W. Atkins, *Physical Chemistry*, 5th edn., (Oxford University Press, Oxford, 1994).

E from J. Emsley, *The Elements*, 2nd edn. (Clarendon Press, Oxford, 1991).

C from P. A. Cox, *The Elements* (Oxford University Press, Oxford, 1989).

SAL from D. F. Shriver, P. W. Atkins, and C. H. Langford, *Inorganic Chemistry*, 2nd edn., (Oxford University Press, Oxford, 1994).

Royal Society of Chemistry, London: for material on p. 140, from J. W. Akitt and J. M. Elders (adapted as in A. Massey, *Main Group Chemistry*, Ellis Harwood (1990), p. 178).

John Wiley & Sons Inc.: for material marked **CW** from F. A. Cotton and G. Wilkinson, *Advanced Inorganic Chemistry*, 5th edn. Copyright © John Wiley & Sons, New York, 1988. Reprinted by permission of John Wiley and Sons incorporated.

Every effort has been made to trace copyright holders. (Apologies are offered to those copyright holders whom we have been unable to trace). Any omissions notified to Oxford University Press, will be corrected in future printings.

Index